Turning Center Programming, Setup, and Operation

A Guide To Mastering The Use Of CNC Turning Centers

FOURTH EDITION

MIKE LYNCH

44 Little Cahill Road • Cary, IL 60013
Phone 847.639.8847 • Fax 847.639.8857
Internet: www.cncci.com
Email: lynch@cncci.com

Notice

The information in this book is meant to supplement, not replace, proper CNC machine usage training. In all cases, information contained in the machine tool builder's manual will supersede material presented in this book. Like any industrial equipment that requires skill to master, CNC machine tools pose some risk. The publisher advises readers to take full responsibility for their safety and know their limits, as well as the limits of the CNC machine tool being used. Before practicing the skills and techniques described in this book, read all safety advice presented in the machine tool builder's manuals, be sure that your equipment is well maintained, and do not take risks beyond your level of experience, aptitude, training, and comfort level.

Library of Congress Cataloging-in-Publication Data

Lynch, Mike

CNC Turning Center Programming, Setup, and Operation - 4th edition p. cm.

Includes index.

ISBN-13: 978-1492731337

ISBN-10: 1492731331

Distributed to the book trade by CNC Concepts, Inc.

Table of contents

Preface

CNC turning centers are among the most popular forms of CNC metal-cutting machine tools. Just about every company that uses metal-cutting CNC machines has at least one turning center.

It is the intention of this text to introduce beginners and experienced CNC people alike to programming, setup, and operation techniques used to utilize CNC turning centers. We will begin in a basic manner, ensuring that even newcomers to CNC will be able to follow the presentations. And we use a building blocks approach – so as you get deeper into the material – we'll be adding to what you already know. When you're finished, you will have a thorough understanding of what it takes to program, setup, and run a CNC turning center.

We use a *Key Concepts* approach to presenting CNC. The Key Concepts allow us to minimize the number of major topics you must master in order to become proficient with CNC turning center usage. With this approach, there are ten Key Concepts related to CNC turning center programming, setup, and operation.

Key Concepts one through four provide the most important building blocks, giving you a way to organize your thoughts for this sophisticated type of CNC equipment. Starting with Key Concept Number Five (program formatting), we'll begin putting it all together.

Beginners should concentrate most on understanding the points made early in each Key Concept. It is as important to know *why* you are doing things as it is to know how to do them. Concentrate first on the whys. It will be impossible for beginners to totally memorize and comprehend every technique used with turning centers the very first time it is presented or read. Rest assured that if you can understand the basic reasoning behind why each CNC feature is required, it will be much easier to master the use of the feature. Once this basic reasoning is understood, it will be easy to review material to extract specific details of how each CNC feature is used – so you can start putting your CNC turning center/s to good use.

For experienced programmers, this text will provide you with alternatives. You'll be able to compare your current methods to our recommendations. If you have previous CNC turning center experience for other types of CNC controls, you will find it relatively easy to adapt what you already know to the most popular CNC control in the industry, the Fanuc control. If you have programming experience for other types of machine tools, such as machining centers or wire EDM machines, this text will help you adapt what you know to CNC turning centers.

As experienced programmers know, there are many ways to utilize CNC equipment. This text will show you one or two safe ways to accomplish your tasks. You can use your own common sense and past experience to develop your own style.

Prerequisites

Rest assured that this text will cover CNC turning centers from the ground up. We will assume that you have absolutely no previous experience with CNC. However, there are certain things we do assume about students reading this text.

Basic machining practice experience

We will assume that you have at least some experience with basic machining practice as it applies to turning center operations. We will assume you possess knowledge of machining operations like rough and finish turning, rough and finish facing, rough and finish boring, threading, necking, drilling, and tapping. We will also assume that you understand cutting tools and related cutting conditions required to perform turning center operations safely and efficiently.

If you have experience with any form of conventional lathe, like an engine lathe or turret lathe, we think you will find it remarkably easy to learn how to program, setup, and operate a CNC turning center. Think of it this way: You already know what you want the machine to do. It will be a relatively easy task to learn how to tell the CNC turning center to perform your desired machining operations. This is why machinists make the best CNC programmers.

On the other hand, if you have no previous basic machining practice experience, or worse, no shop experience, your task will be much greater. You not only need to learn CNC programming, setup, and operation, you also need to learn basic machining practice. At the very least, you will need the help of an experienced machinist to explain the tasks you will be performing on your CNC turning center. More likely, you'll need more training in this area. If you have no previous machining practice experience, we strongly recommend that you enroll in a training program related to basic machining practice in conjunction with reading this text. Basic machining practice courses can be found at your local technical/vocational schools, colleges, and universities.

Math

The word *numerical* in computer numerical control implies that numbers are highly involved with CNC. Indeed, every CNC command includes numbers, and almost every CNC command requires an arithmetic calculation to be made. However, most calculations are quite simple to make. The types of arithmetic calculations required for the typical CNC turning center program include addition, subtraction, multiplication, and division. For more complex workpieces, some right-angle trigonometry may also be required.

This text will require very little in the way of math (though if you are using this text in conjunction with a live course, your instructor may wish to include more math). For the most part, we will simply assume that you can add, subtract, multiply, and divide. We will be teaching CNC turning center usage, not mathematics. For this reason, our examples and practice exercises will be quite simple regarding the math you need to know. However, we do not wish to understate the importance of the math required to prepare real-world programs for CNC turning centers. In real life, the CNC programmer must be prepared to perform rather complex calculations involving trigonometry in order to come up with coordinates needed in the CNC program. Or, some form of computer aided manufacturing (CAM) system can be used to reduce the amount of math required for programming.

Motivation

This should go without saying. We assume that you are motivated to learn. If you are highly motivated to learn about CNC turning centers, it will make your task much easier. Your motivation will help you overcome any problems you may have with learning the material in this text. With motivation, you'll stick to it until you understand.

Controls covered

Since the Fanuc control is the most popular CNC control available, specific examples in this text are given in Fanuc format. Keep in mind, however, that the Key Concepts approach we use throughout this text will make it possible for you to learn techniques that can be applied to just about any CNC control on the market. Also keep in mind that several control manufacturers claim that their controls are Fanuc compatible. These manufacturers include Yasnac, Mitsubishi Meldas, Tasnac, Mazatrol (EIA option), Haas, Fadal, and Flashcut (among others). Even if you do not have one of these controls, we truly feel that if you understand the basic concepts, and if you understand how specific techniques are applied to one particular control type, it will be relatively easy to adapt what you know to just about any CNC turning center control being used today.

There are several specific Fanuc control models covered by this text. In addition, two American companies have also marketed (or still do) Fanuc products in the U.S. (G.E. Fanuc and General Numerics). Here is a general list of Fanuc controls covered by this text in order of newest to oldest. Every current model Fanuc control is covered.

21T, 20T, 18T, 16T, 15T, 11T, 10T, 0T, 3T, 6T

Without getting into too much history of how Fanuc controls have evolved, you should also know that Fanuc has always enhanced each control model within its lifetime. Generally speaking, Fanuc adds a letter after the control model name to determine the actual control version (such as A, B, or C). The first release of a new control is the A version, and as new versions are introduced, the letter is changed (to B and C). For example, the 0T control began as a 0TA, and evolved into the 0TB, and 0TC as Fanuc continued to improve its capabilities.

An exception to this naming convention rule has to do with Fanuc's conversational turning center controls. For these controls, Fanuc designates that the control is conversational with an "F" designation (presumably, the letter F stands for FAPT). The 3TF, 10TF, and 15TF are examples of conversational turning center controls. While this text does not address conversational programming, each of these controls can also be programmed using the techniques we show in this text.

When it comes to programming as we teach it in this text, Fanuc controls can be broken into two categories. We designate them as programming style one and programming style two. Here is a more specific list of the controls within each programming style.

Programming style one:

0TA (and some 0TB controls)
3T and 3TC

Programming style two:

21T
20T
18TA
16TA
15TA, 15TF
11TA
10TA
0TB, 0TC
6TA, 6tB

While programming style two is slightly easier to work with, rest assured that the programming differences between style one and style two are extremely minor and that they will be discussed when appropriate during this text.

When it comes to machine tool setup and operation, the specific techniques used to operate the various control models mentioned above vary dramatically. For this reason, we include a set of operation procedures for most of the controls listed above. These procedures will show how to run a CNC turning center in a step-by-step manner. If you are using this text in conjunction with a formal CNC course, your instructor may have developed special operation handbooks for the machines owned by your school/company.

What about conversational controls?

As mentioned, Fanuc (and other control manufacturers) also makes a series of conversational CNC turning center controls, (the 3TF, 10TF, and 15TF). Though these controls can be programmed conversationally, keep in mind that all Fanuc conversational controls can also be programmed by conventional methods. This text will nicely cover the CNC side of these conversational controls.

Controls other than Fanuc

Though the techniques given during this text are specific to Fanuc controlled CNC turning centers, keep in mind that most CNC turning center controls are programmed with very similar techniques. In fact, since Fanuc is so very popular and holds the largest market share of CNC controls in the world, many CNC turning center control manufacturers are making sure that their controls are Fanuc compatible. This means that programs written for a Fanuc control will run in a control that is Fanuc compatible. For this reason, this text will even help you learn to write programs for controls other than Fanuc.

Limitations

Please take note that our first goal will be to acquaint you with the programming of two axis turning centers (X and Z). For the bulk of this text, only these two linear axes will be discussed. But more and more turning centers are being manufactured with more than two axes.

These machines include three axis turning centers that have *live tooling*. The third axis is a rotary axis (called the C axis) that is internal to the machine's spindle. Live tools (tools that rotate) like drills, taps, and end mills can be used to perform secondary machining operations that are more commonly thought of as milling machine (or machining center) operations.

There are also turning centers that have two spindles. These machines can perform machining operations on two workpieces (or each end of a workpiece) simultaneously. Some of these machines have *opposing* spindles to make it easy (and automatic) to transfer the workpiece from one spindle to the other. These machines will have at least four axes (X and Z for two spindles). Many additionally incorporate a rotary (C) axis in each spindle to accommodate live tooling, meaning these machines commonly have six axes.

While the bulk of this text will address the use of two axis turning centers, we do include a special appendix (after Lesson Twenty-Three) that addresses special machine types and special accessories used with turning centers. We will discuss the use of live tooling and sub-spindle turning centers in this lesson – along with certain important accessories – like bar feeders.

Instruction method

This text is commonly used with a formal CNC course. Possibly you are enrolled in a technical school's CNC course. Or maybe you are attending a company's in-plant CNC training course. In either case, you have an instructor making presentations and available to help you understand the material. Or you may be using this text in conjunction with a video or CD-rom course, meaning your instructor is recorded. Either way, audio presentations should help you understand the concepts more easily. And with video and CD-roms, you can review the material as often as necessary.

If you have purchased this text separately and do not have the benefit of an instructor, your task will be a little harder. While all of the key points made during live presentations *are included* in this text, you will be left on your own to study hard enough to grasp the material presented.

Scope

As the name of this text implies, we address all three tasks a person must master in order to utilize a CNC turning center, including programming, setup, and operation.

Programming is the act of preparing a series of commands that tell the CNC turning center how to machine a workpiece. It involves coming up with a machining process, selecting cutting tools, designing a workable workholding setup, and actually creating the CNC program. The method of programming we'll be using in this text is called *manual programming* (also called G code level programming). While there are other ways to create programs (using a computer aided manufacturing, or CAM, system, for example), it is with G code level manual programming techniques that you can be the most intimate with a CNC turning center – commanding everything the machine will do. Every CNC person must understand this form of programming in order to make modifications to the CNC program at the machine when a job is run.

Setup is the act of preparing the CNC turning center to run a series of workpieces (called a *job* or *production run*). Tasks involve (among other things) making the workholding setup, assembling and loading cutting tools, determining and entering certain offsets, loading the program, and verifying that the program is correct. We'll be discussing the related tasks in the approximate order that setups are made.

Operation actually involves two things. First, you must be comfortable with the general manipulation of a CNC turning center. This involves knowing the various components on the machine, the buttons and switches, and how to perform several important procedures. Second, you must be able to complete a production run once the setup is made. Tasks needed to complete a production run involve (among others) workpiece load and unload, cycle activation, measuring completed workpieces and making sizing adjustments if/when necessary, and dull tool replacement.

Key Concepts approach

This effective presentation method will allow you to organize your thoughts as you read this text. This text includes ten Key Concepts (six for programming and four for setup and operation). Here are several benefits to this presentation method.

1) Any good training program should *put a light at the end of the tunnel.* All students want to know where they stand throughout any training course. With our Key Concepts approach, you will always have a clear understanding of your progress throughout the text.

2) During each Key Concept, we will first present the main idea behind the concept. As stated earlier, we say it is at least as important to understand why you are doing things as it is to understand how to do them. Think of these early presentations for each Key Concept as the why. From there, we will present the specific techniques that are related to each concept.

3) The Key Concepts allow us to use a *building blocks approach* and present information in a very tutorial manner. We will be constantly building on previously presented information.

4) The Key Concepts approach allows us to limit the number of new ideas you must understand in order to grasp information presented within the text. Think of it this way: If you can understand but ten basic ideas, you will be well on your way to becoming proficient with CNC turning centers!

Lesson structure

We further divide the ten Key Concepts into twenty-eight lessons. This makes it possible to organize the most important topics related to turning center usage.

Practice, practice, practice!

In this edition, we've included lots of practice within the text to help you confirm your understanding of the presented material. Answers are provided right in the text, close to the exercise so you can check your own work.

We also make available a separate workbook/answer-book combination to accompany this text. After each lesson, there is a special note that asks you to do an exercise in the workbook. Many of the exercises include programming activities. Many technical school instructors who are using this text in their CNC curriculums use the workbook to allow you to do homework assignments during the class – and to confirm your understanding of the material included in this text.

Key Concepts and lessons

Here is a list of the ten Key Concepts and the twenty-eight lessons that comprise this text. Again, Key Concepts one through six are related to programming – Key Concepts seven through ten are related to setup and operation.

1: Know your machine from a programmer's viewpoint

1: Machine configurations
2: Understanding turning center speeds and feeds
3: General flow of the CNC process
4: Visualizing the execution of a CNC program
5: Program zero and the rectangular coordinate system
6: Determining program zero assignment values
7: Four ways to assign program zero
8: Introduction to programming words

2: You must prepare to write programs
9: Preparation steps for programming

3: Understand the motion types
10: Programming the three most basic motion types

4: Know the compensation types
11: Introduction to compensation
12: Geometry offsets
13: Wear offsets
14: Tool nose radius compensation

5: You must provide structure to your CNC programs
15: Introduction to program structure
16: Four types of program format

6: Special features that help with programming
17: One-pass canned cycles
18: G71 and G70 – rough turning and boring multiple repetitive cycles followed by finishing
19: G72-G75 – other multiple repetitive cycles
20: G76 – Threading multiple repetitive cycle
21: Working with subprograms
22: Special considerations for Fanuc 0T and 3T controls
23: Other special features of programming
---: Appendix – special machine types and accessories

7: Know your machine from a setup person or operator's viewpoint
24: Tasks related to setup and running production
25: Buttons and switches on the operation panels

8: Know the three basic modes of operation
26: The three modes of operation

9: Understand the importance of procedures
27: The key operation procedures

10: You must know how to safely verify programs
28: Program verification

Note once again that we present the programming discussions first. While you may be more interested in setup and operation, there are many programming-related topics that setup people and operators must understand. And it is during a presentation about programming that many CNC features are best introduced. When you complete the programming Key Concepts, you may be surprised at how many setup- and operation-relate topics you already understand.

Some readers may not be at all interested in programming. You can skip the programming-related lessons and jump right to the setup- and operation-related lessons. At the beginning of Key Concept Number Seven, we provide a list of important setup- and operation-related topics that are presented during programming. You can read just the listed presentations in the programming lessons.

Enjoy!

We at CNC Concepts, Inc. wish you the best of luck with this text. We hope you find it easy to understand our written presentations and the presentations of your instructor (live or recorded). Once completed, we hope this text makes your introduction to Fanuc controlled CNC turning centers as easy and enjoyable as possible.

Know Your Machine From A Programmer's Viewpoint

It is from two distinctly different perspectives that you must come to know your CNC turning center/s. Here in Key Concept One, we'll look at the machine from a programmer's viewpoint. Much later, during Key Concept Seven, we'll look at the machine from a setup person's or operator's viewpoint.

Key Concept Number One is the longest of the Key Concepts. It contains eight lessons:

1: Machine configurations
2: Understanding turning center speeds and feeds
3: General flow of programming
4: Visualizing program execution
5: Understanding program zero
6: Determining program zero assignment values
7: How to assign program zero
8: Introduction to programming words

A CNC programmer need not be nearly as intimate with a CNC turning center as a setup person or operator – but they must, of course, understand enough about the machine to create programs – to instruct setup people and operators – and to provide the related setup and production run documentation.

First and foremost, a CNC programmer must understand what the CNC turning center is designed to do. That is, they must understand the machining operations a turning center can perform. They must be able to develop a workable process (sequence of machining operations), select appropriate cutting tools for each machining operation, determine cutting conditions for each cutting tool, and design a workholding setup. All of these skills, of course, are related to basic machining practice – which as we state in the Preface – are beyond the scope of this text. For the most part, we'll be assuming you possess these important skills.

Though this is the case, we do include some important information about machining operations that can be performed on CNC turning centers throughout this text. For example, we discuss how to develop tool paths for machining operations in Lessons Nine and Ten. We provide a description of rough and finish turning and boring in Lesson Eighteen. Threading is discussed in Lesson Twenty. And in general, we provide suggestions about how machining operations can be programmed when it is appropriate. This information should be adequate to help you understand enough about machining operations to begin working with CNC turning centers.

If you've had experience with conventional (non-CNC) machine tools...

A CNC turning center can be compared to an engine lathe (or any kind of conventional lathe). Many of the same operations performed on a conventional lathe are performed on a CNC turning center. If you have experience with manually operated lathes, you already have a good foundation on which to build your knowledge of CNC turning centers.

This is why machinists make the best CNC programmers. With a good understanding of basic machining practice, you can easily learn to program CNC equipment. You already know *what* you want the CNC machine to do. It is a relatively simple matter of learning *how to tell the CNC machine* to do it.

If you have experience with machining operations like rough and finish turning, rough and finish facing, drilling, rough and finish boring, necking and threading – and if you understand the processing of machined workpieces – believe it or not, you are well on your way to understanding how to program a CNC turning center. Your previous experience has prepared you for learning to program a CNC turning center.

We can also compare the importance of knowing basic machining practice in order to write CNC programs to how important it is for a speaker to be well versed with the topic they will be presenting. If not well versed with their topic, the speaker will not make much sense during the presentation. In the same way, a CNC turning center programmer who is not well versed in basic machining practice will not be able to prepare a program that makes any sense to experienced machinists.

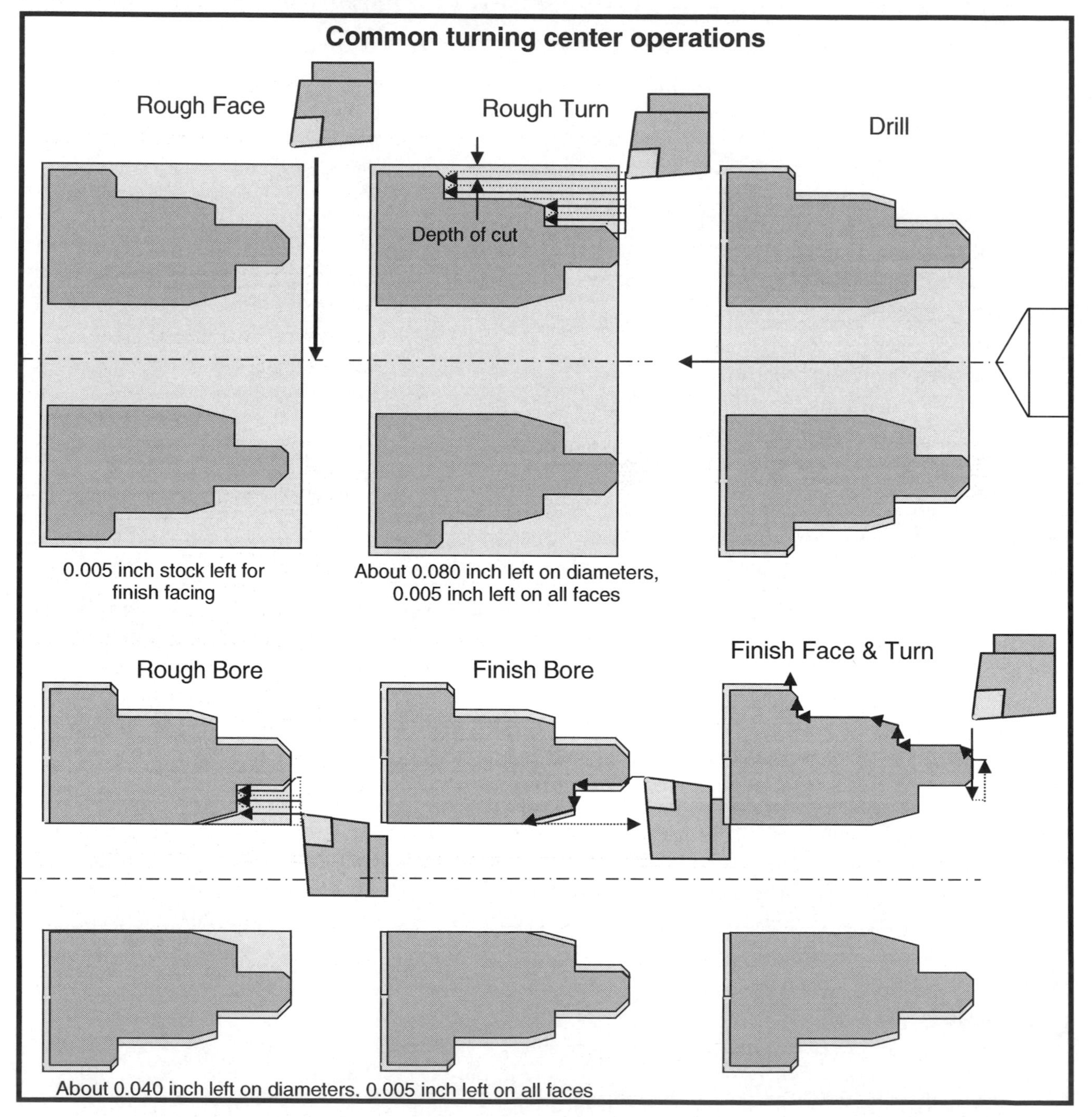

Lesson 1:

Machine Configurations

As a programmer, you must understand what makes up a CNC turning center. You must be able to identify its basic components – you must understand the moving components of the machine (called axes) – and you must know the various functions of your machine that are programmable.

Most beginners tend to be a little intimidated when they see a turning center in operation for the first time. Admittedly, there will be a number of new functions to learn. The first point to make is that you must not let the machine intimidate you. As you go along in this text, you will find that a turning center is very logical and is almost easy to understand with proper instruction.

You can think of any CNC machine as being little more than the standard type of equipment it is replacing with very sophisticated and automatic motion control added. Instead of activating things manually by hand-wheels and manual labor, you will be preparing a *program* that tells the machine what to do. Virtually anything that needs to be done on a true CNC turning center can be activated through a program – meaning anything you need the machine to do can be commanded in a program.

Types of CNC turning centers

There are several types of CNC turning centers. While at first glance there may appear to be substantial differences among the various types, all turning centers share several commonalities. We'll begin by describing the most popular type of CNC turning center – the *universal style slant bed turning center*. Because it is so popular, this is the machine type we will use for all examples in this text. We will then introduce several other types of turning centers, comparing them to the universal style slant bed turning center.

Universal style slant bed turning center

This style of turning center is called a *universal style* turning center because it can perform all three forms of turning applications – chucking work, shaft work, and bar work. This explains why it is the most popular type of turning center – it provides the most flexibility to CNC turning center users.

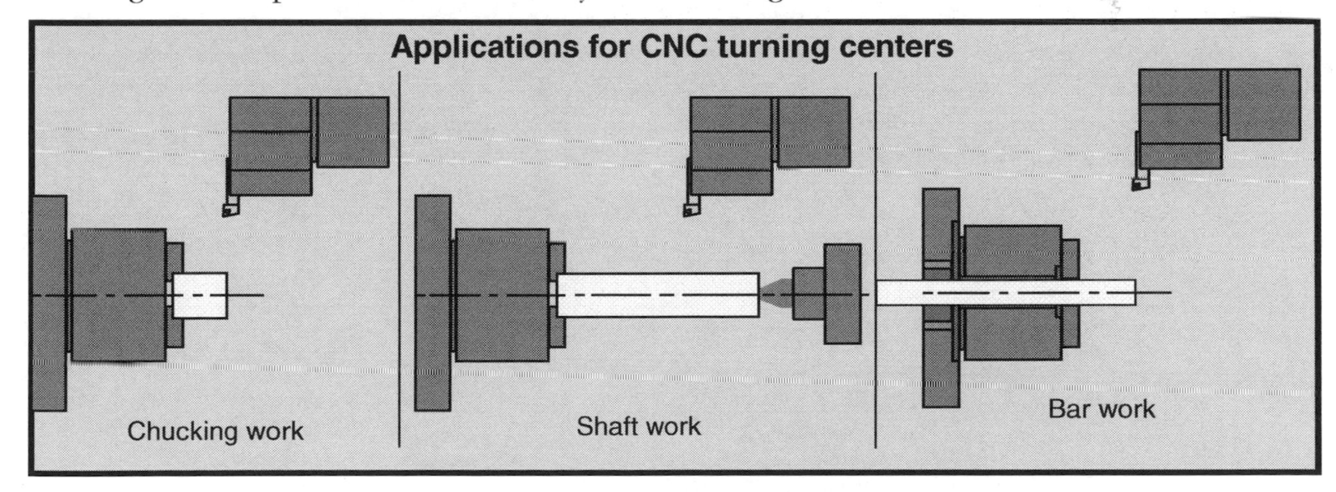

When raw material comes to the machine in the form of short slugs (like round bars cut to length), the application is called chucking (or chucker) work. The raw material is secured solely by the workholding device (commonly a three-jaw chuck).

With longer slugs (longer than about three to four times the raw material diameter), the workholding device by itself will not be sufficient to secure the workpiece for machining. For these applications, some form of work support device/s must be used (commonly a tailstock and/or steady-rest). This application is called shaft work.

With bar work, the raw material comes to the turning center in the form of a long bar (from four to fifteen feet long [1.2-5 meters], depending upon the type of bar feeder being used). Bar work requires a special bar support and feeding device (called a bar feeder). The bar is fed through the headstock and spindle into the working area. A workpiece is machined and cut off from the bar. The bar is then fed again for another workpiece to be machined.

Figure 1.1 shows a universal style slant bed turning center. The headstock houses a spindle to which the workholding device is mounted. Our illustration shows a three-jaw chuck, but other types of workholding devices can be used (collet chuck, expanding mandrel, etc.). To the right of the workholding device is the tailstock, which is used to support the right end of long workpieces – again, for shaft work. The turret of the turning center is used to hold cutting tools and it can be quickly rotated from one tool station to another. Current turning centers have turrets that hold from six to twelve cutting tools.

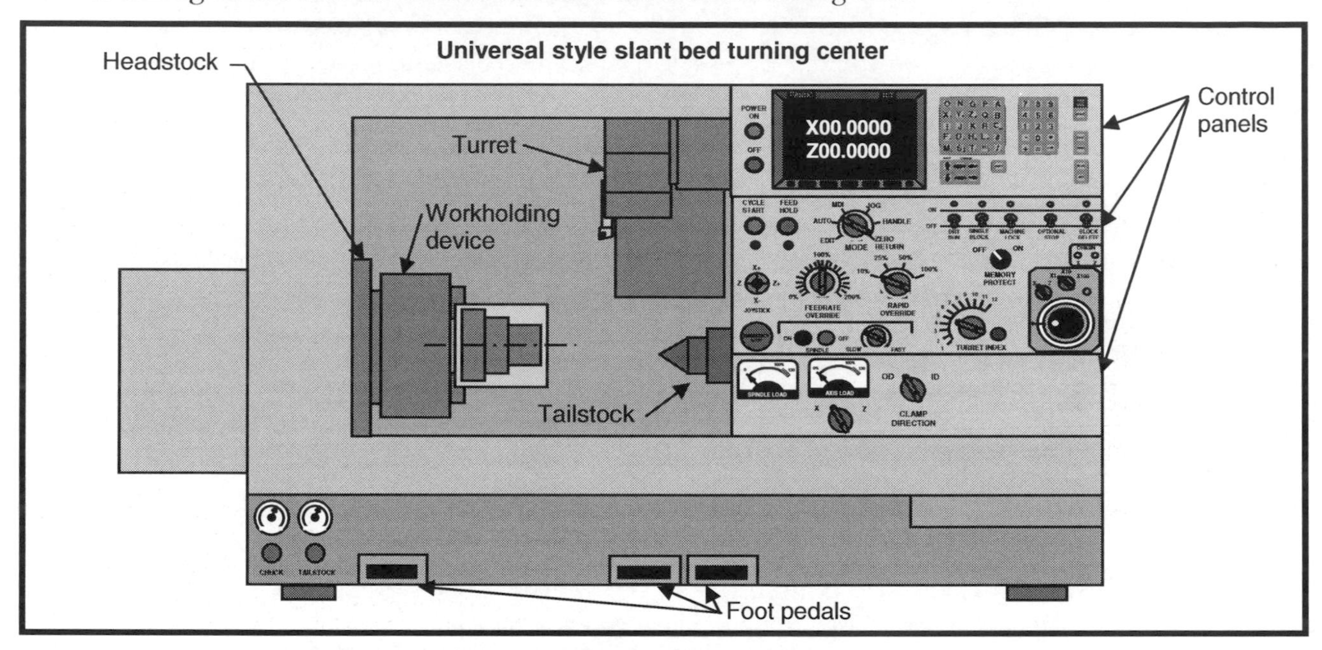

Figure 1.1 – A universal style slant bed turning center with its door removed

Directions of motion (axes) for a universal style slant bed turning center

All turning centers have at least two linear *axes* of motion. The turret (and cutting tool) will move along with these two axes. By *linear*, we mean the axis moves along a straight line.

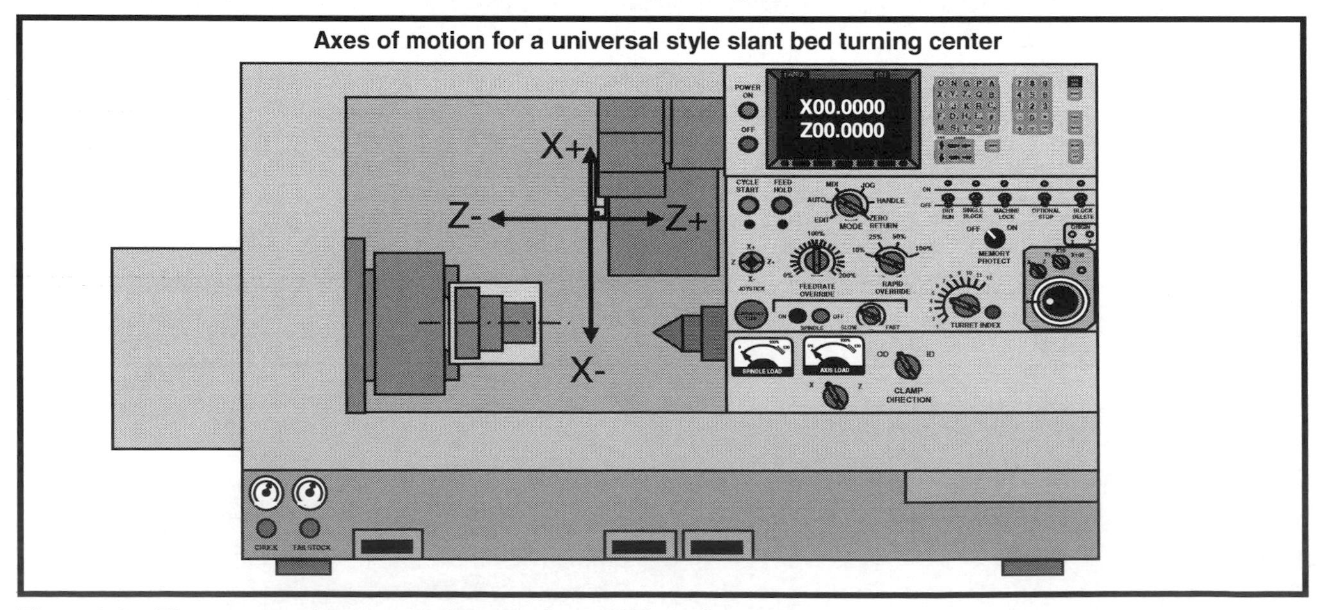

Figure 1.2 – The two most basic axes of motion for a CNC turning center

The *diameter-controlling* axis (up/down motion of the turret as shown in Figure 1.2) is the *X axis*. The *length-controlling* axis (right/left motion of the turret as shown in Figure 1.2) is the *Z axis*. Figure 1.2 shows these directions of motion along with the polarity (+/-) for each.

These two most basic directions of motion will remain exactly the same for almost all types of turning centers (only a handful of turning center manufacturers stray from what we show in Figure 1.2.) *The X axis will always be the diameter-controlling axis* – and X minus is always the direction that causes the cutting tool to move to a smaller diameter (toward the spindle centerline). *The Z axis will always be the length controlling axis* – and Z minus will always be the direction that causes the cutting tool to move toward the workholding device.

X is specified in diameter

Though we may be a little ahead of ourselves, the X axis is designated in *diameter* for almost all turning centers. That is, if a diameter of 3.0 inches must be machined, the designation for the X axis will be **X3.0**. There are some (especially older) turning centers that require the X axis to be specified with radial values. For these machines, the word **X1.5** will cause the tool to be positioned to a 3.0 inch diameter. Note that it is much easier to work with a turning center when the X axis if it is designated in diameter – which is why most current model turning centers do so.

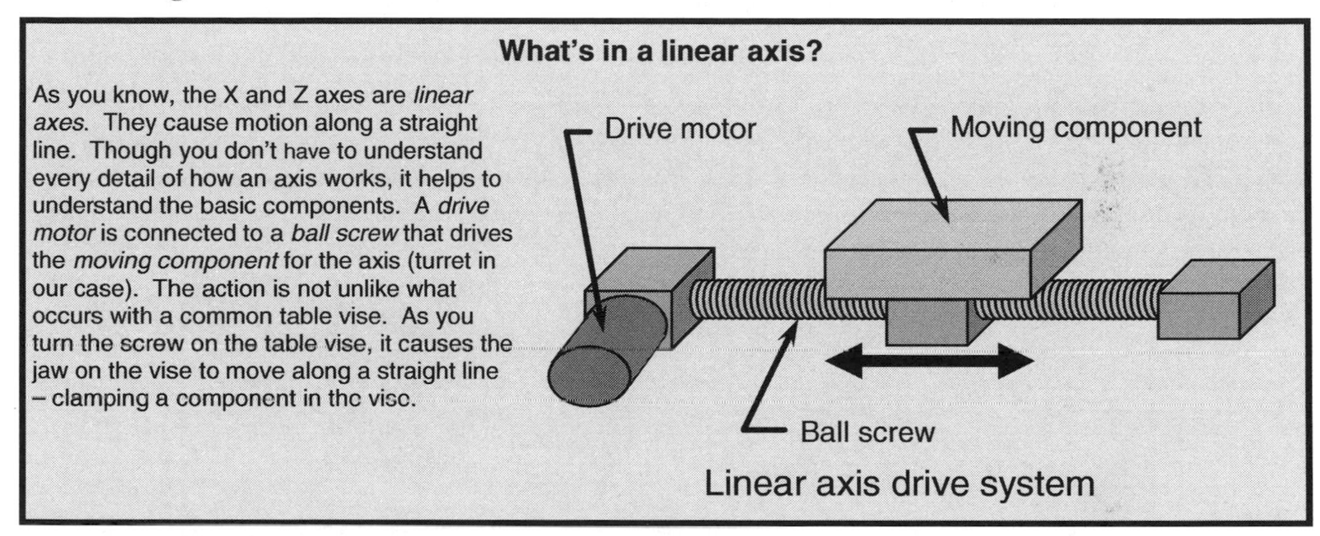

Live tooling for a universal style slant bed turning center

We have just described the most basic form of a universal style slant bed turning center. Again, this machine has two axes (X and Z) – and it can perform all three kinds of turning work (chucking work, shaft work, and bar work). The majority of universal style slant bed turning centers that are in use today are of this configuration.

There is, however, a special accessory called *live tooling* that can be equipped on all types of CNC turning centers (including the universal style slant bed turning center). This accessory, which is becoming quite popular, makes it possible for a turning center to perform machining operations that are more commonly associated with CNC *machining centers* (or milling machines).

These operations include drilling, tapping, reaming, and milling (among others). In essence, turning centers equipped with live tooling can perform both turning center operations and machining center operations – giving it the ability to more completely machine a workpiece. For many applications, this eliminates the need to perform secondary operations on another machine tool. Figure 1.3 shows a workpiece that requires live tooling if it is to be completely machined on a turning center.

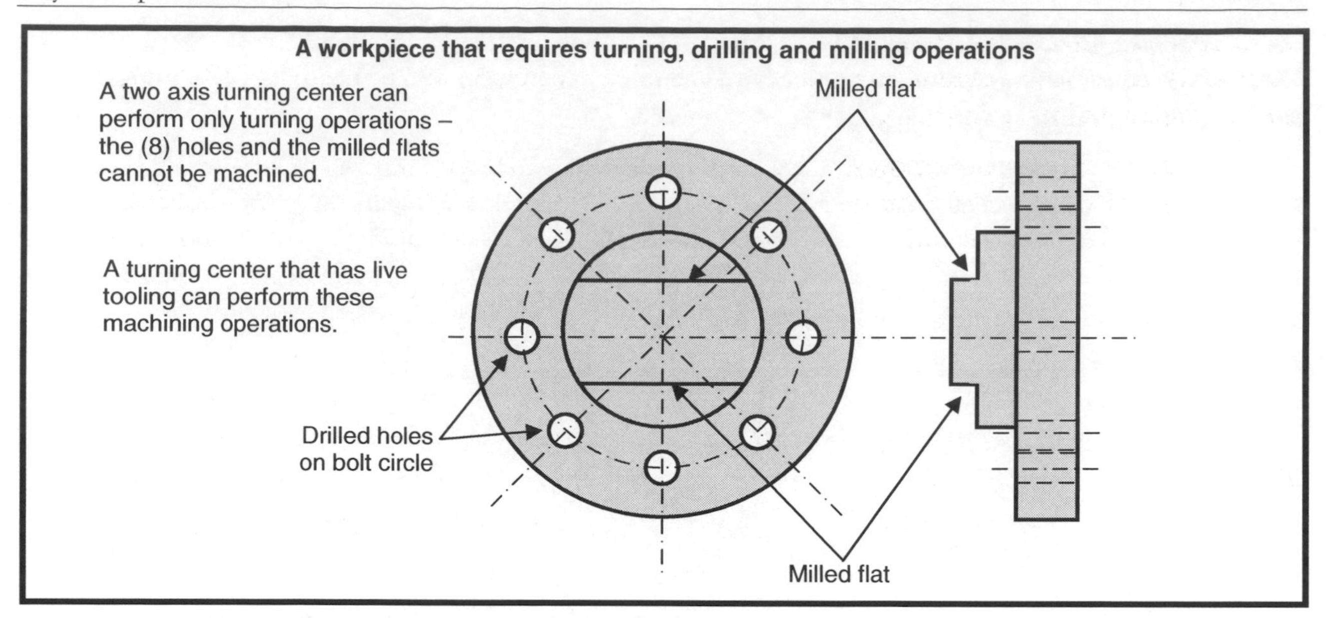

Figure 1.3 – A workpiece that requires live tooling

Turning centers that are equipped with live tooling have two additional features. First, as the name implies, they have a special device mounted within the turret that makes it possible to rotate cutting tools (again, like drills and end mills). Second, they have a special rotary axis (called the C axis) built into the spindle drive system. Figure 1.4 shows a universal style slant bed turning center that has live tooling. These turning centers are sometimes referred to as mill/turn machines.

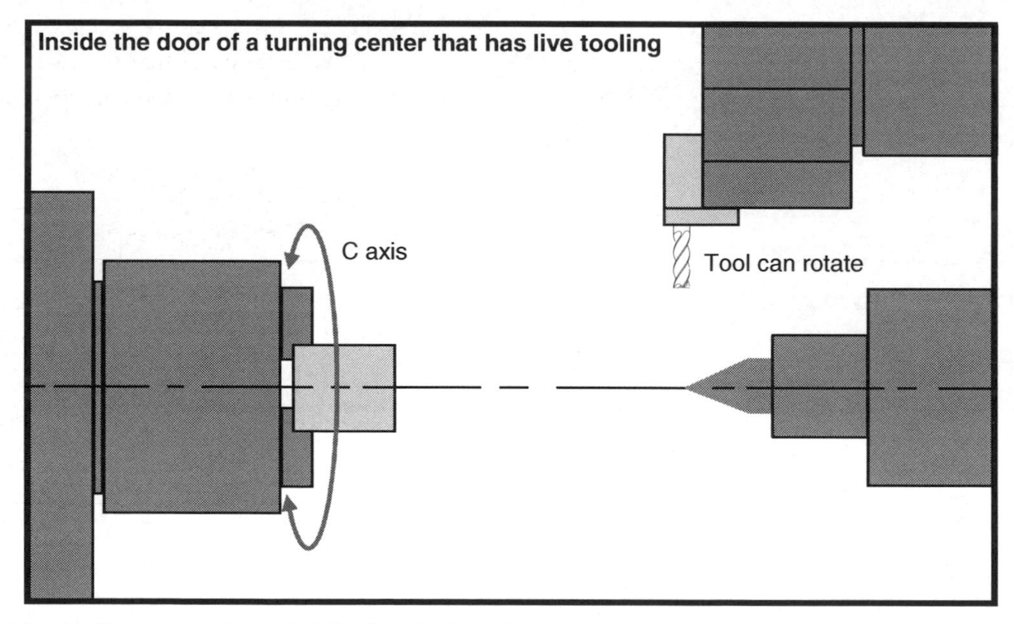

Figure 1.4 – Live tooling on a universal style slant bed turning center

The cutting tool depicted in Figure 1.4 points along the X axis. Another style of tool holder is usually available that allows the cutting tool to point along the Z axis (toward the chuck).

When the programmer selects the *live tooling* mode, the rotary axis within the spindle is engaged and the spindle within the turret will be used to rotate cutting tools. When the programmer selects the *normal turning mode*, these devices are disengaged.

The bulk of this text describes the programming of the normal turning mode (not using live tooling). We describe live tooling in detail in the Appendix after Lesson Twenty-Three.

Other types of CNC turning centers

The machine types we show from this point have a great deal in common with universal style slant bed turning centers. Commonalities include:

- **Axis direction** – X is always the diameter-controlling axis (toward and away from the spindle centerline). Z is always the length-controlling axis (toward and away from the workholding device).
- **X axis polarity** – X minus is always the direction toward the spindle centerline (getting smaller in diameter. Note: there are two machine builders that we know of that reverse X axis polarity for some of the machines they have produced (X minus is away from the spindle centerline).
- **Z axis polarity** – Z minus is always the direction toward the workholding device.
- **X is almost always specified in diameter** – If you need to machine a 4.0 inch diameter, the word X4.0 will be used. There are a few (especially older) machines that require X positions to be specified as a radial value (to turn a 4.0 inch diameter the word X2.0 will be used).
- **Turret/tool moves in each axis** – With one exception (the Swiss-type turning center), the turret (and cutting tool) will move in each axis. That is, the workpiece will remain stationary in X and Z while cutting tools will move to perform machining operations.
- **Live tooling** – All forms of CNC turning centers can be equipped with live tooling (as an optional feature).

Chucking style slant bed turning center

Figure 1.5 shows another common form of CNC turning center. Notice that the only difference between this machine and the universal style is the absence of the tailstock. This, of course, means that the machine cannot perform shaft work – it is limited to chucker and bar work. These machines usually have a limited Z axis stroke length.

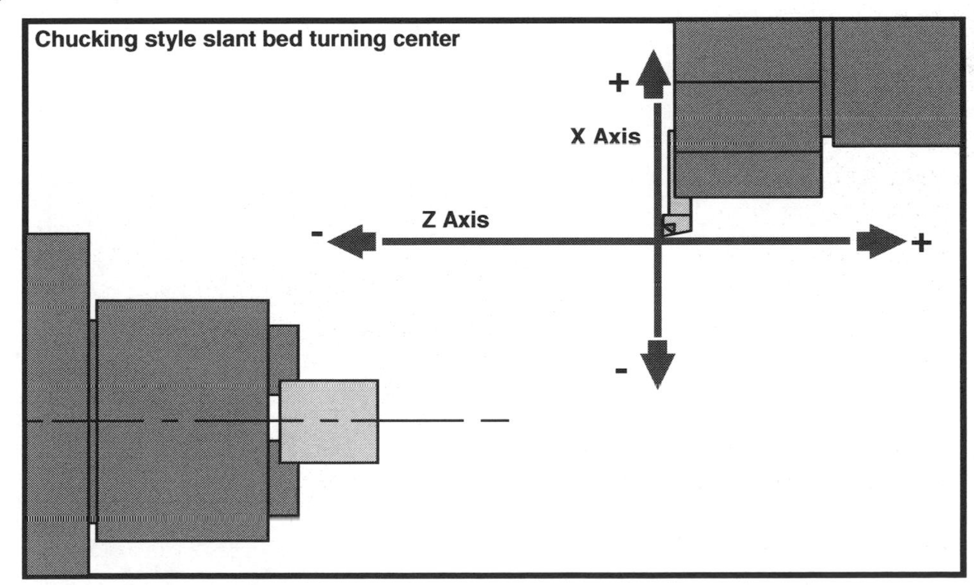

Figure 1.5 – Chucking style slant bed turning center (axis polarity reflects turret motion)

Twin spindle horizontal bed turning centers

This is a form of turning center that is becoming quite popular. Its popularity stems from the fact that two cycles can be running at the same time. This effectively doubles productivity – in essence, it is like having two machines in one – but it takes less floor space and it is more convenient for one operator to run.

Though there are exceptions, these machines are not equipped with tailstocks, meaning they are limited to chucker and bar applications. Figure 1.6 shows the configuration.

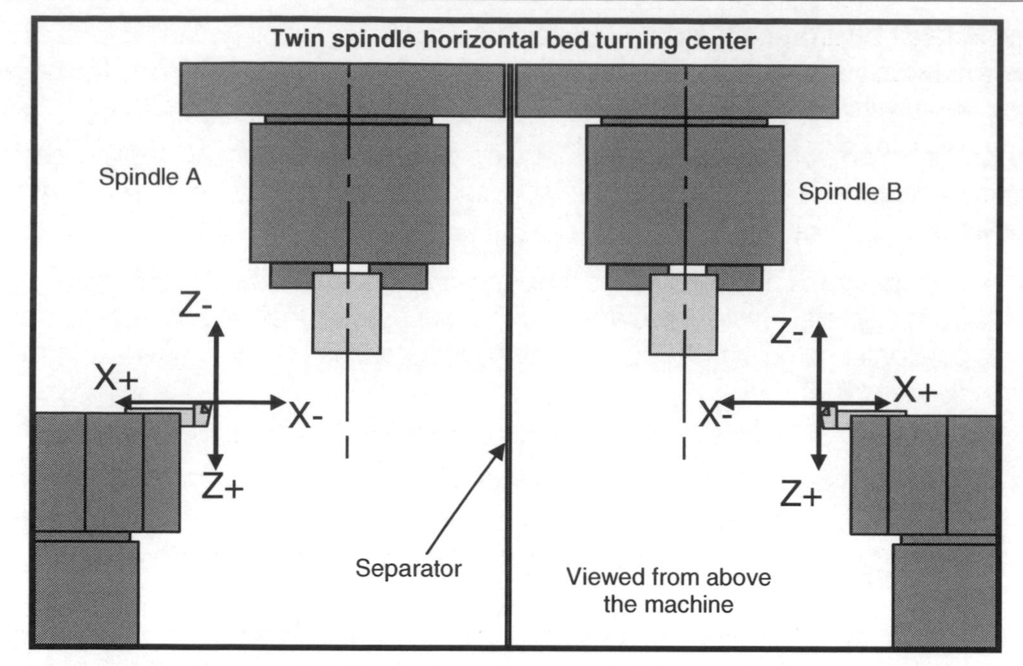

Figure 1.6 – Twin spindle horizontal bed turning center (axis polarity reflects turret motion)

Again, this kind of turning center is like having two machines in one. Each spindle is totally independent, meaning two different programs can be run at the same time. Examples of applications for this machine include running two identical operations, running first and second operation for a workpiece, and running two different workpieces. One or both spindles can even be used for bar work.

While this machine may look radically different from slant bed turning centers, notice that our given commonalities still apply. X is still the diameter controlling axis and X minus is still the direction going toward the spindle centerline (for both spindles). Z is still the length controlling axis and Z minus is the direction going toward the workholding device. And X is still specified in diameter. A program written for a chucking style or slant bed turning center could be run on either spindle of this machine.

Sub-spindle style turning centers

This form of two-spindle turning center has the spindles positioned in such a way that they face one another. Figure 1.7 shows one.

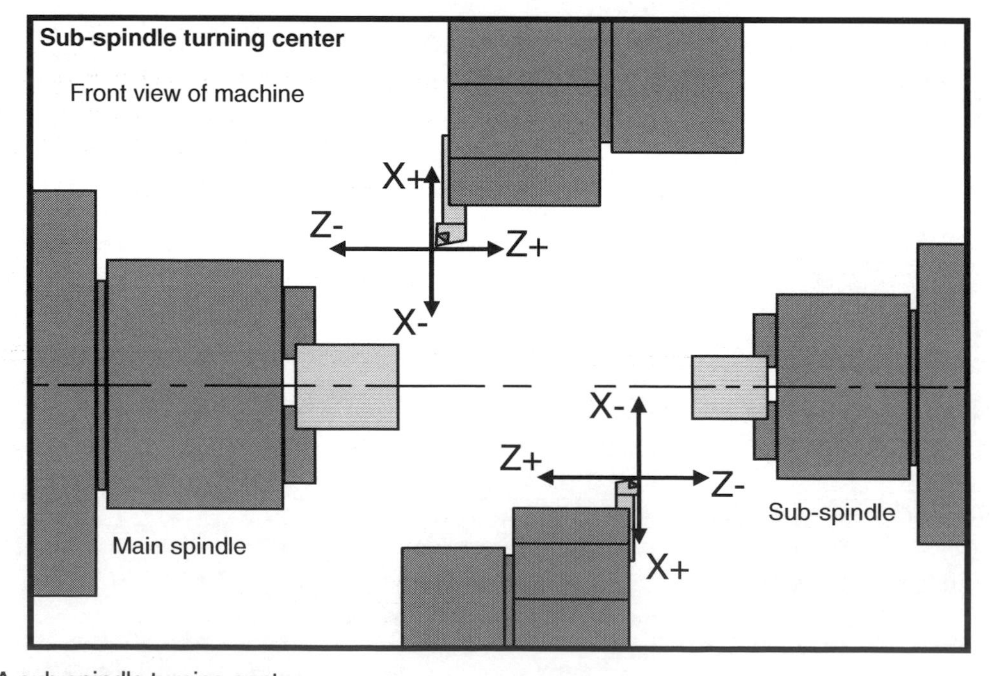

Figure 1.7 – A sub-spindle turning center

This design provides an advantage over the twin-spindle horizontal bed turning center just shown. The advantage has to do with transferring a workpiece from one spindle to the other. With many sub-spindle turning centers, the sub-spindle can move forward to (automatically) take a workpiece from the main spindle. This makes it possible to easily perform machining operations on the second end of the workpiece in the sub-spindle. Figure 1.8 shows how the transfer can take place.

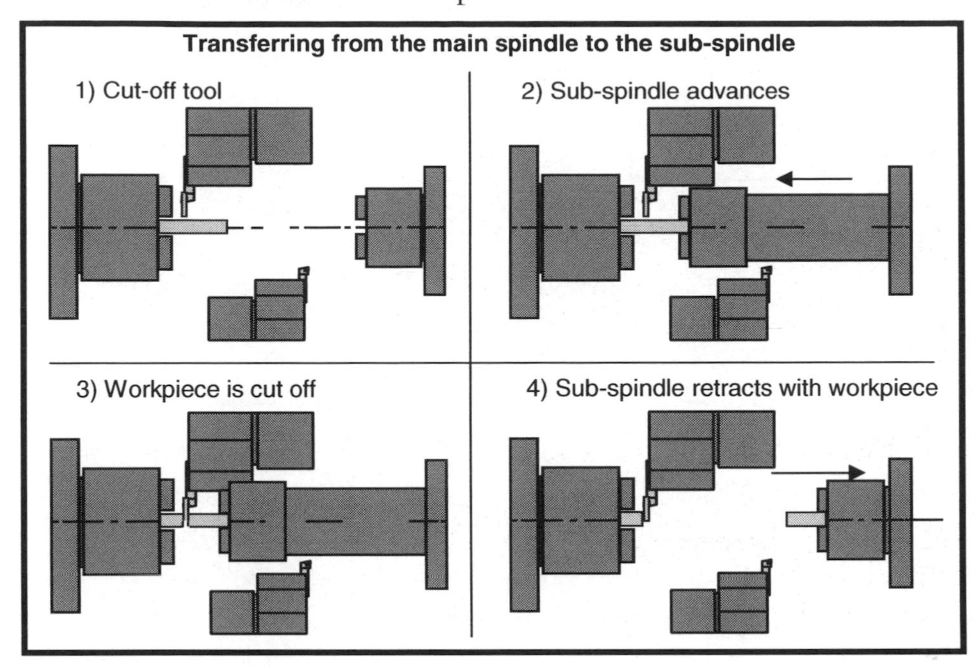

Figure 1.8 – Transferring a workpiece from the main spindle to the sub-spindle

In this application, the main spindle is usually being fed by a bar feeder (a bar application). After the bar feed, the main spindle machines the right side of the workpiece. When finished, the sub-spindle will advance to hold and support the workpiece during the cutoff operation (the sub-spindle runs at the same speed as the main spindle during the cutoff operation). After the cutoff, the sub-spindle retracts. The main spindle will feed another bar for the right side operation while the sub-spindle machines the left side of the previous workpiece. In this way, one completed workpiece is produced per cycle.

Vertical single spindle turning centers

This kind of turning center is commonly used to machine large diameter (but relatively short), heavy workpieces. When a workpiece is held in a horizontal orientation (as is the case with all turning centers discussed to this point), the weight of the workpiece will work against the work holding device (chuck). That is, workpiece weight will have a tendency to cause the workpiece to fall out of the chuck – making for unstable workholding. The heavier the workpiece, the more difficult it is to adequately hold it in a horizontal orientation.

While horizontal orientation may be fine for light workpieces, as the weight of the workpiece increases, it may become difficult (if not impossible) to properly hold and support the workpiece held in this manner. With a *vertical* turning center, the weight of the workpiece actually *helps* to stabilize it in the work holding device.

Figure 1.9 shows an example of a single spindle vertical turning center.

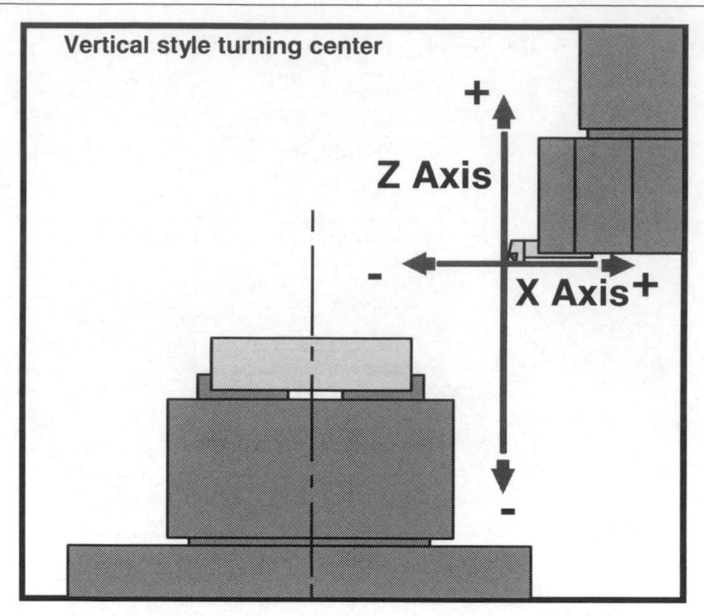

Figure 1.9 – Single spindle vertical turning center (axis polarity reflects turret motion)

Since workpieces to be run on this kind of machine are usually quite short, tailstocks are seldom equipped with this kind of turning center. Also, since the spindle is mounted vertically, this type of turning center cannot be used as a bar feed machine. Its only feasible application is chucking work.

Twin spindle vertical turning centers

The twin spindle vertical turning center has the same advantages as the twin spindle horizontal turning center discussed earlier. Figure 1.10 shows a drawing of one.

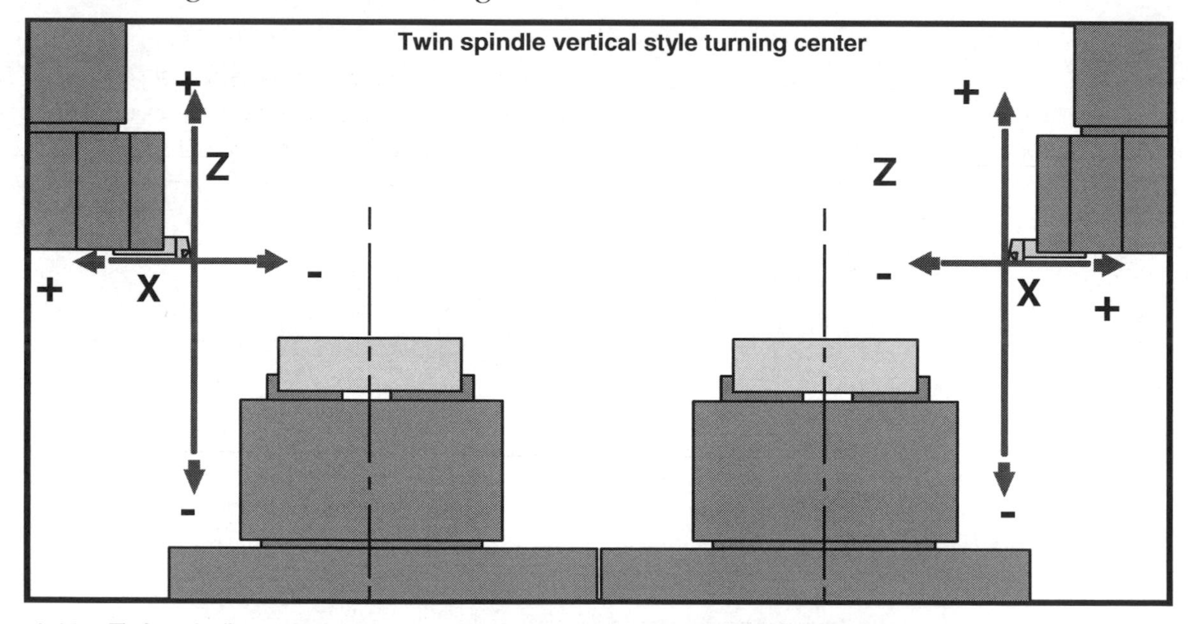

Figure 1.10 – Twin spindle vertical turning center (axis polarity reflects turret motion)

Again, think of this machine as like having two machines in one. A program written for one spindle can be run in the other.

Gang style turning centers

This form of turning center does not have a turret to hold cutting tools. Instead, the cutting tools are mounted on a table or *sub-plate*. While tool interference can sometimes be a problem, the advantage with this style of turning center is that there will be virtually no tool changing time. Figure 1.11 shows an example of this type of turning center.

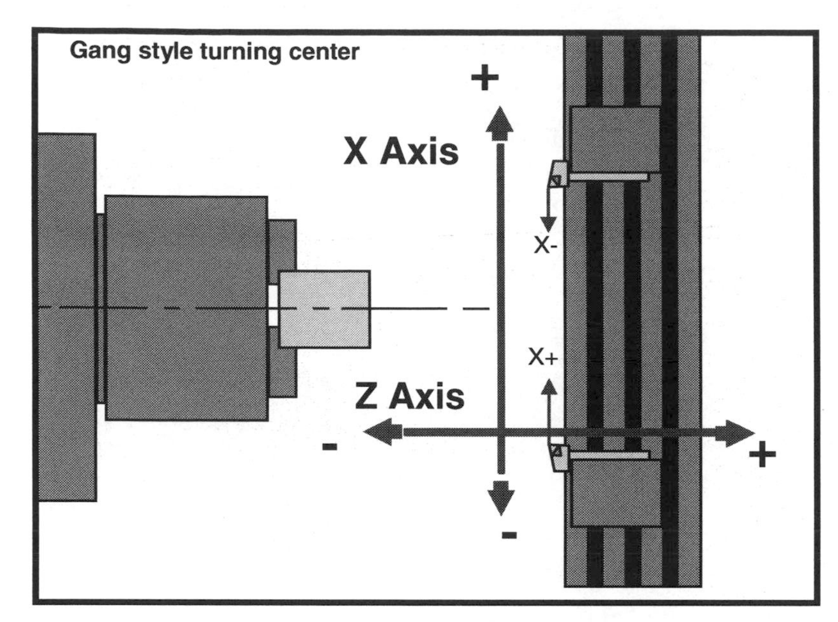

Figure 1.11 – Gang style turning center (axis polarity reflects tooling-table motion)

A gang style turning center is usually quite small (under 10 hp spindle motor) and is commonly used for bar work. Of course, the basic design of this machine eliminates the possibility for a tailstock, so shaft work cannot be done.

Notice that the entire table moves to form the X and Z axes. With X, this can be a little troublesome. For tools mounted on the far side of the spindle centerline, X+ will be as it is for all types of turning centers we have introduced (X+ is away from spindle centerline). But for tools on the near side of the spindle center, X+ is the direction toward the spindle center.

Swiss-type CNC turning centers (also called sliding headstock turning centers)

This machine is a little different. The tool (turret) only moves in the X axis for this type of machine. The tool remains stationary in the Z axis. Instead, the *workpiece* moves in the Z axis. Figure 1.12 shows one.

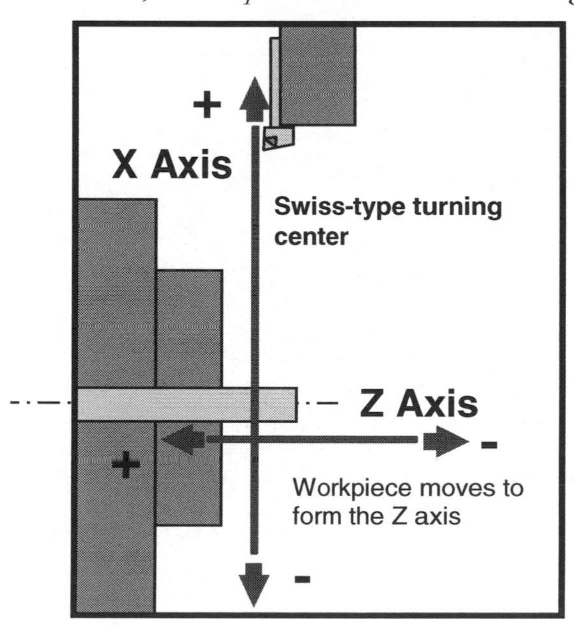

Figure 1.12 – Swiss type CNC turning center

While at first glance the Z axis may appear to be reversed as compared to the machines we've shown so far, remember that it is the workpiece that is moving in the Z axis. This kind of machine always uses a bar feeder.

Since machining will always occur very close to the workholding device, machining will be very well supported. This allows the machining of very small diameter, yet quite long workpieces. (Think of a needle valve.)

A reminder about this text – Once again, we will be using the universal style slant bed turning center for all example programs shown in this text. When appropriate, we will be pointing out special considerations for the other types of turning centers we have just shown.

Programmable functions of turning centers

A true CNC turning center will allow you to control just about any of its functions from within a program. There should be very little operator intervention during a CNC cycle. Here we list some common functions that can be programmed on all *true* turning centers. While we do show the related CNC words used to command these functions, our intention here is not to teach programming commands (yet). It is to simply make you aware of the kinds of things a programmer can control through a program.

Spindle

The spindle of all turning centers can be programmed in at least three ways, activation (start/stop), direction (forward/reverse), and speed (in either surface feet/meters per minute or revolutions per minute). Many turning centers additionally provide multiple power ranges (like the transmission of an automobile).

Spindle speed

You can precisely control how fast the spindle of a turning center rotates. An *S word* is used for this purpose. There are two ways to specify spindle speed. When the spindle is in rpm mode, an S word of S500 specifies a speed of 500 revolutions per minute (rpm). When the spindle is in *constant surface speed mode*, an S word of S500 specifies a speed of 500 surface feet per minute (sfm), assuming you are working in the inch measurement system. (If you work in the Metric measurement system, S500 will specify 500 meters per minute when in constant surface speed mode.)

We will describe the two spindle speed modes – as well as how to determine how and when to use them – in Lesson Two.

Spindle activation and direction

You can also control which direction the spindle rotates – *forward or reverse*. The forward direction is used for right hand tooling (when machining occurs toward the workholding device). It will appear as counter-clockwise when viewed from in front of the machine. The reverse direction is used for left hand tooling and will appear as clockwise when viewed from in front of the spindle.

Three *M codes* control spindle activation. M03 turns the spindle on in the forward direction (used with right-hand tools). M04 turns the spindle on in a reverse direction (for left-hand tools). M05 turns the spindle off.

What is an M code?

M codes control many of the programmable functions of a turning center. In many cases, you can think of them as being like programmable on/off switches. "M" stands for "miscellaneous" or "machine" function.

M codes are created by the machine tool builder, and will often vary from one turning center to another. Here we show some common M codes, but you must look in your machine tool builder's programming manual to find the complete list of M codes for a given turning center.

These M codes don't vary:

M00: Program stop
M01: Optional stop
M03: Spindle on (forward)
M04: Spindle on (reverse)
M05: Spindle off
M08: Flood coolant on
M09: Coolant off
M30: End of program

These M codes vary:

M___: Low spindle range
M___: High spindle range
M___: Tailstock quill forward
M___: Tailstock quill reverse
M___: Automatic door open
M___: Automatic door close
M___: Chip conveyor on
M___: Chip conveyor off
M___: ________________
M___: ________________
M___: ________________
M___: ________________
M___: ________________

Spindle range

Many, especially larger turning centers, have two or more spindle ranges. Spindle ranges are like the gears in an automobile transmission. Generally speaking, lower ranges are used for power – higher ranges are used for speed. With most turning centers, spindle range selection is done with M codes. While the specific M code numbers for spindle range selection will vary from one machine tool builder to another, many turning center use **M41** to select the low range and **M42** to select the high range. We'll use these two M codes (**M41**: low and **M42**: high) to specify spindle range selection throughout this text.

Turning centers vary when it comes to what will actually happen when the spindle range is changed. Some, especially older machines use a mechanical gearbox that must be engaged for the low range. These machines commonly require that the spindle be stopped during the range change. While the spindle stoppage, range change, and spindle restart will occur automatically, these machines can take from three to ten seconds or more to change ranges.

Some, especially newer machines have spindle motors with multiple windings. Two or more sets of windings within the motor itself control range selection. With these machines, range changing is almost instantaneous – and the spindle does not have to stop when the spindle range is changed.

It is important to know the power characteristics for the turning center/s you will be working with in order to make the correct spindle range selection for a given machining operation. Every turning center manufacturer will provide a power curve chart like the one shown in Figure 1.13 to document a machine's spindle power characteristics. You will normally find this power curve chart in the machine tool builder's operation or programming manual.

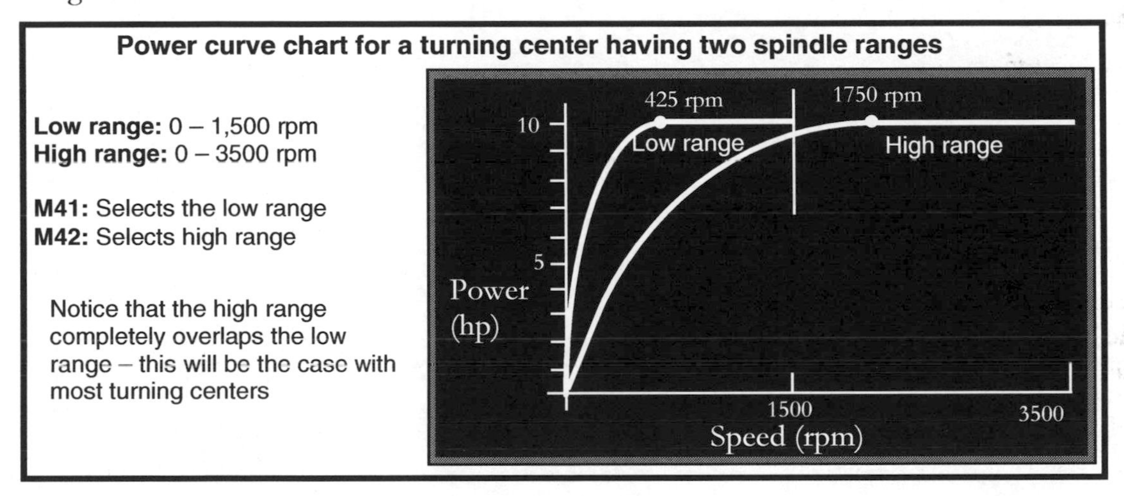

Figure 1.13 – An example spindle power curve chart

With this example spindle power curve chart, notice that the low range runs from zero to fifteen hundred rpm and achieves full power at 425 rpm. The high range runs from zero to thirty-five hundred rpm, completely overlapping the low range. Full power is achieved in the high range at 1,750 rpm.

Again, knowing your machine's spindle power characteristics is important for selecting the appropriate spindle range. We give a general rule-of-thumb for spindle range selection: *perform roughing operations in the low range and finishing operations in the high range.*

While this is a good rule of thumb, there are times when it isn't correct. Consider, for example, machining a steel workpiece that is less than 1.0 inch in diameter. Say the cutting tool manufacturer recommends a speed of 600 sfm for the rough turning operation. The formula to calculate rpm is as follows:

Rpm = 3.82 times speed in sfm divided by the diameter to be machined

In this case, the *slowest* speed needed for the roughing operation will be 2,292 rpm (3.82 times 600 sfm divided by 1.0). In this case, since full power is achieved at 1,750 rpm in the high range, and since the low range will peak out at 1,500 rpm – slowing the machining operation – the high range should be selected for this operation.

Feedrate

You know that all turning centers have at least two linear axes, X and Z. You also know that the cutting tool (for most turning center) moves along with these two axes. It is the motion of the cutting tool while it is in contact with the workpiece that causes machining to occur. It is important that the motion *rate* (how quickly the tool moves) be appropriate to the machining operation being performed. In CNC turning center terms, this motion rate is called *feedrate.*

An F word is used to specify feedrate. And like spindle speed, feedrate can be specified in two ways. It can be specified in *per minute* fashion or in *per revolution* fashion. As the names imply, when feedrate is specified in per minute fashion, it specifies how far the cutting tool will move during one minute. When feedrate is specified in per revolution fashion, it specifies how far the cutting tool will move during one spindle revolution.

Also as with spindle speed (at least in constant surface speed mode), feedrate specification is related to the measurement system you use. In the inch mode, feedrate is specified in either inches per minute (ipm) or inches per revolution (ipr). In metric mode, feedrate is specified in either millimeters per minute (mmpm) or millimeters per revolution (mmpr). Feedrate selection is discussed in much greater detail in Lesson Two. For now, we'll simply introduce the related words and give a few examples.

- F word – Feedrate specification
- G20 – Inch mode
- G21 – Metric mode
- G98 – Feed per minute mode
- G99 – Feed per revolution mode

Here are a few examples of feedrate specification:

```
N010 G20 G98 F4.0 (4.0 inches per minute)
N020 G20 G99 F0.015 (0.015 inches per revolution)
N030 G21 G98 F100.0 (100.0 millimeters per minute)
N040 G21 G99 F0.5 (0.5 millimeters per revolution)
```

What is a G code?

G codes are called *preparatory functions.* They prepare the machine for what is coming up – in the current command and possibly in up-coming commands. In many cases, they set *modes,* meaning once a G code is *instated* it will remain in effect until the mode is changed or cancelled.

Here we list a few common G codes, but don't worry if they don't make much sense yet. Upcoming discussions will clarify.

Common G codes:

G00: Rapid motion
G01: Straight line motion
G02: Circular motion (CW)
G03: Circular motion (CCW)
G04: Dwell
G20: Inch mode selection
G21: Metric mode selection
G28: Zero return command
G40: Cancel tool nose radius comp.
G41: Tool nose radius comp. left
G42: Tnr comp. right
G50: Spindle limiter
G70: Finishing cycle
G71: Rough turning cycle
G72: Rough facing cycle
G76: Threading cycle
G96: CSS mode
G97: RPM mode
G98: Feed per minute
G99: Feed per revolution

Look in your control manufacturer's manual for a full list of G codes.

Turret indexing (tool changing)

With the exception of gang style turning centers, all turning center introduced in this lesson have a turret into which cutting tools are placed (twin spindle turning centers have two turrets). Specific turret design will vary from one machine tool builder to another. All turning centers (even gang style turning centers) must provide a way to specify cutting tool selection. Figure 1.14 shows the turret of a typical turning center.

Figure 1.14 – A typical turning center turret

The turret shown in Figure 1.14 is currently holding turning tools (external machining tools) as well as boring tools (internal machining tools). Notice that the internal tools protrude quite a distance from the turret face (they stick out). This can cause interference problems with a large workpiece and/or the workholding device. You must always be concerned with the potential for interference problems as you choose the turret stations into which you place cutting tools.

With many turrets, for example, you cannot place a long drill or boring bar in a station that is adjacent to a turning tool that machines a small diameter (like a facing tool) without interference problems. As the turning tool faces a workpiece to center, the adjacent boring bar will be driven into the chuck or workpiece.

Turret station and offset selection

A T word specifies which cutting tool will be used. For turning centers that have a turret, the T word will actually cause the turret to index to the specified turret station. But there's a little more to the T word than turret index.

For most machines, the T word is a four-digit word. The first two digits specify the turret station and *geometry offset* to be used with the tool (geometry offsets assign *program zero* – and will be discussed in Lessons Seven and Twelve). The second two digits of the T word specify the *wear offset* to be used with the tool (wear offsets allow the operator to make minor adjustments – and will be discussed during Lesson Thirteen).

The command

```
N020 T0404
```

will cause these three things to occur:

- the turret to index to station number four (first two digits)
- geometry offset number four will be selected (first two digits)
- wear offset number four will be selected (second two digits)

Almost all current model turning centers have *bi-directional turrets*. That is, the turret can rotate in either direction. When a T word is given, most machines will cause the turret to automatically rotate in a direction that that provides the shortest rotational distance to the specified tool

With gang style turning centers, of course, there is no turret to index. Only two things will happen with the previous command: geometry and wear offset number four will be selected.

Again, offset use is the topic of future lessons (again, Lessons Seven, Twelve, and Thirteen). For now, just remember that most programmers will make the wear offset number the same number as the turret station number and geometry offset number.

Coolant

All turning centers allow programmable control of *flood coolant*. Coolant is commonly used to cool the workpiece during machining and to lubricate the machining operation. Two M codes are used to control coolant. Almost all turning center manufacturers use M08 to turn on flood coolant and M09 to turn it off.

Other possible programmable functions

The programmable functions introduced to this point are available for all current model turning centers. Those we list from this point are related only to certain machine types – or they are optional functions that are not supplied with all turning centers.

Tailstock

Turning centers that can perform shaft work (like universal style slant bed turning centers) are equipped with a tailstock. The tailstock is used to support the right end of a long workpiece during machining (the end opposite the workholding device). Most machinists would agree that when the length of the workpiece exceeds about three to four times its diameter, a tailstock should be used to provide support during machining.

Though most current model turning centers have *programmable* tailstocks, machine tool builders vary with how they cause the tailstock body and quill to move. Many use a series of M codes. M16 and M17 may be used to cause quill movement forward and reverse while M28 and M29 may be used to cause tailstock body movement forward and reverse.

Some machine tool builders actually cause tailstock motion by engaging the turret to the tailstock and pulling it along with the Z axis. If your machine has a programmable tailstock, you must reference your machine tool builder's programming manual to determine how it is programmed.

Programmable steady rest

Like a tailstock, this programmable function can be equipped with turning centers that perform shaft work. For extremely long workpieces, the tailstock by itself may not provide enough support. The chuck may provide ample support for one end and the tailstock for the other, but the workpiece will be unsupported in the middle. A programmable steady rest will provide support anywhere it is required over the length of the workpiece. This accessory is discussed in the Appendix (after Lesson Twenty-Three).

Bar feeders and chuck activation

In bar applications, the raw material comes in the form of a long bar (as long as fifteen feet). The bar is usually relatively small in diameter (under 2.0 inches or so). Most bar feeders do not require anything in the way of special programming commands since they apply a constant pressure to drive the bar through the spindle. Most bar fed turning centers use the collet chuck (or chuck jaw) activation to control the bar advance. When the chuck is opened, the bar automatically advances. Closing the chuck will secure the bar in place for machining. Two M codes control the chuck open and close process, but machine tool builders vary when it comes to *which* M code numbers they use. M15 might be used to open the chuck while M14 may be used to close it. Again, you'll have to reference your machine tool builder's programming manual to determine the M code numbers for chuck open/close.

Most machine tool builders recommend using a turret station to hold a bar-stop that will restrain the bar during its advance. To cause a bar advance, first the bar-stop is brought to within a small distance of the existing bar end (within about 0.100 inch). Next the chuck is opened (with M15 for instance) and the bar advances a small amount to contact the bar stop. The bar stop is then moved in the Z plus direction to advance the bar the appropriate amount (part length plus cut off amount plus facing stock amount). Finally, the chuck is closed (with M14 for example) and the bar stop is retracted. We explain the programming of bar feed applications in detail in the Appendix (after Lesson Twenty-Three).

Part catcher

In bar applications, a workpiece is cut off after every cycle. During the cut-of machining operation, a part catcher is commonly used to keep the workpiece from falling into the chip pan after being parted from the bar. A part catcher is commanded to swing into position just prior to cut-off operation. As the workpiece is cut off, it falls gently into the part catcher and exits the machining area. The part catcher is then retracted. Two M codes are commonly used to activate the part catcher, but again, machine tool builders vary with regard to *which* M codes they use. M18 may be used to advance the part catcher, M19 to retract it. If your turning center has a part catcher, reference your machine tool builder's programming manual to determine the related M codes.

Tool touch off probe

While this device may not be programmable, it is an extremely helpful accessory – one that is being equipped with more and more turning centers. While we won't be assuming that your turning center has one, we will include detailed information about how it is used. Once you understand how it can help CNC setup people and operators, you'll want to have one for all of your turning centers.

A tool touch off probe will help during the initial setup of a job – and during long production runs. During initial setup, it can dramatically facilitate the task of program zero assignment (discussed in Lessons Six and Seven). The setup person will first (manually) swing the tool touch off probe into its active position. They will then bring each cutting tool into contact with a stylus on the probe (actually twice). This will automatically align the cutting tool with the program zero point (again, program zero is discussed in Lessons Five, Six, and Seven).

During the production run, the tool touch off probe will help during dull tool replacement. The same techniques used during setup will be repeated when a dull tool is replaced.

Unfortunately, many existing turning centers are not equipped with a tool touch off probe, meaning you will have to learn other methods of performing the tasks they accomplish. Rest assured that we'll show alternatives to using a tool touch off probe in this text. Once you see both ways, you'll probably want all of your *future* turning centers equipped with tool touch off probes.

Automatic tool changing systems

As stated earlier, the turret of most turning centers can hold a maximum of twelve tools. This may not be enough tool stations for some applications – especially when the CNC user wants to keep all cutting tools in the machine – or when the machine is equipped with live tooling. For this reason, more and more machine tool builders are equipping their turning centers with an automatic tool changing device (similar to one found on a CNC *machining* center). This device automatically loads cutting tools into the turret of the turning center, minimizing tool loading time during setups while increasing the number of cutting tools that can be used by a program.

Exceptions to X axis

We mentioned this earlier, but we want to elaborate. Almost all turning centers manufacturers orient the X axis as we have shown – the X minus direction is turret motion toward the center of the spindle (getting smaller in diameter). However, there are at least two turning center manufacturers we know of (Mori Seiki and Nakamura Tome) that sometimes reverse X axis polarity. With these rather unique machines, X plus is motion as the turret moves *toward* the spindle center.

A quick fix

If you must work on a turning center that has the X axis reversed, it can be quite confusing, especially if your company also owns other turning centers with the X axis as we have shown. You can easily reverse the direction of the X axis by turning on a feature called *X axis mirror image* (a standard feature with most turning centers). By turning on X axis mirror image for those machines with the X axis reversed, you will make all machines in your shop consistent with one another – at least when it comes to X axis polarity.

If you do not turn on X axis mirror image for turning centers with which the X axis is reversed, all X coordinates must be designated with a minus sign (-). Also, circular motion words (G02 and G03 introduced in Lesson Ten) must be reversed – as must tool nose radius compensation commands (G41 and G42 discussed in Lesson Fourteen). Additionally, tool offset direction in the X axis will also be reversed (discussed in Lesson Thirteen). These inconsistencies make it quite difficult for people to work with both types of machines (with and without the X axis reversed).

Gang style turning centers with cutting tools on both sides of the spindle centerline

As discussed, a tooling table is used to hold cutting tools with this kind of machine. And cutting tools can be placed on the front or back side of the spindle centerline. For tools on the front side, X motion toward the spindle centerline is reversed (again, X plus is toward the spindle center).

This causes the same inconsistency just described. That is, the X axis polarity for any tool on the front side of the spindle centerline is reversed. Again, this causes the features circular motion (G02 and G03), tool nose radius compensation (G41 and G42), and all tool offsets to also be reversed. Additionally, all X coordinates for front-side tools must be programmed with X minus coordinates.

Again, if the feature *programmable mirror image for the X axis* is used (a standard feature with most gang type turning centers), the back turret and the front turret can be programmed in exactly the same manner. Simply turn on mirror image in the X axis before programming any motions for cutting tools on the front side of the spindle and turn it off before commanding motions for cutting tools on the back side of the spindle center.

Two G codes control turn programmable X axis mirror image on an off. Unfortunately, the specific G codes vary from one control model to another. With one popular control, G68 turns it on and G69 turns it off. You must reference your control manufacturer's programming manual to find the related G codes.

Center cutting axis

Some turning centers are equipped with an axis of motion that can only move along the spindle's centerline (parallel to the Z axis). The machine tool builder *Miyano* calls this axis the *B axis*. The purpose for this axis is to perform hole machining operations that are right on the spindle centerline. Operations like drilling, reaming, and tapping can be performed. Also, this motion can even be programmed during the activation of the other two axes (X and Z), meaning simultaneous machining can be accomplished which will minimize cycle time.

What else might be programmable?

While this text will acquaint you with the most common programmable functions of turning centers, you must be prepared for more. Other programmable devices that may be equipped on your turning center include chip conveyer, automatic door open and close, automatic loading devices, and a variety of other application-based accessories. If you have any of these functions, you must reference your machine tool builder's programming manual to learn how these special features are programmed.

> Check with an experienced person in your company or school to find out what other programmable features are available on the turning center you will be working with.

Key points for Lesson One:

- There are several types of CNC turning centers.
- With all types of CNC turning centers, X is the diameter controlling axis and X minus is the toward spindle center, Z is the length controlling axis and Z minus is toward the workholding device.
- X axis positions are specified in diameter.
- Some turning centers – those that have live tooling – can additionally perform machining-center-like machining operations).
- You must understand the functions of your turning center that can be programmed.
- Spindle can be controlled in at least three ways (activation, direction, and speed). Additionally, many turning centers have more than one spindle range.
- Spindle speed can be specified in rpm or in surface feet/meters per minute.
- Feedrate specifies the motion rate for machining operations.
- Feedrate can be specified in per revolution fashion or per minute fashion.
- Coolant can be activated to allow cooling and lubricating of the machining operation.
- Most turning centers have a turret in which cutting tools are placed.
- Machines in the United States allow the use of inch or metric mode.
- You must determine what else is programmable on your turning center/s.

Quiz

1) Which style of turning center is the most popular?

A) Universal slant bed style C) Twin spindle
B) Chucker style D) Gang style

2) Specify the correct axis letter for each moving component below (universal style slant bed turning center).

a) ____Diameter controlling axis
b) ____Length controlling axis
c) ____Rotation of the spindle (live tooling machines only)

3) Specify the correct axis letter and polarity for each tool motion below (C-frame style vertical machining center).

a) _____Tool motion bigger in diameter
b) _____Tool motion toward the chuck

4) Provide the CNC word or command needed to activate the following:

a) Feedrate of 0.015 ipr: ____________
b) Feedrate of 50.0 mmpm: ____________
c) Stop spindle: ____________
d) Turn on the flood coolant: ____________
e) Turn off the coolant: ____________
f) Index to station number seven: ____________
g) Select inch mode: ____________

Talk with experienced people in your company to learn more: Do your turning centers have more than one spindle range? If so, what are the speeds for each range? At what rpm does the spindle achieve maximum horsepower in each range? Do any of your machines have automatic doors? If so, what are the related M codes? How many cutting tools can your turning centers hold?

Answers: 1: A, 2a: X, 2b: Z, 2c: C, 3a: X+, 3b: Z-, 4a: G20 G99 F0.015, 4b: G21 G98 F50.0, 4c: M05, 4d: M08, 4e: M09, 4f: T0707, 4g: G20

STOP! Do practice exercise number one in the workbook.

Note: While this text includes several exercises and programming activities, we can also provide a separate workbook/answer book combination that will further confirm your understanding of presented material.

Lesson 2

Understanding Turning Center Speeds and Feeds

Speed and feed selection is one of the most important basic-machining-practice-skills a programmer must possess. Poor selection of spindle speed and feedrate can result in poor surface finish, scrapped parts, and dangerous situations. Even if speeds and feeds do allow acceptable workpieces to be machined – if they're not efficient – productivity will suffer.

You now know the basic configurations available for CNC turning centers. You know the main components, the directions of motion (axes), and the polarity for each axis. And you know that all turning centers have certain programmable functions – machine features that you can control from within a program. Two of the programmable functions introduced in Lesson One are spindle speed and feedrate. In Lesson Two, we're going to elaborate on these two important programmable functions.

As you know, spindle speed is the rotation rate of the machine's spindle (and workpiece). Feedrate is the motion rate of the cutting tool as it machines a workpiece. These two *cutting conditions* are extremely important to machining good workpieces. You will select a spindle speed and feedrate for every machining operation that must be performed.

Cutting tool manufacturers supply technical data including the recommendation of a cutting speed and feedrate for the cutting tools they supply. Recommendations are based on three important factors:

1) The machining operation to be performed
2) The material to be machined
3) The material of the cutting tool's cutting edge

The machining operation to be performed

This criterion determines the style of cutting tool that must be used to perform the machining operation. As the programmer, *you* will be the person making this decision. It requires that you draw upon your basic machining practice experience.

As stated in the Preface and Lesson One, turning centers can perform a wide variety of machining operations, and specific cutting tools are used to perform each operation. The specified spindle speed and feedrate must be appropriate to the cutting tool selection. Some cutting tools, like rough turning tools, perform very powerful machining operations and can remove a great deal of material from the workpiece per pass – while others, like small boring bars, perform lighter machining operations and can only remove a small amount of workpiece material per pass.

The material to be machined

This criterion determines (among other things) how quickly material can be removed from the workpiece. Soft materials, like aluminum can be machined faster than hard materials, like tool steel. So generally speaking, you'll use faster spindle speeds and feedrates for softer materials.

The material of the cutting tool's cutting edge

The material used for some cutting tools comprises the entire tool. A standard twist drill or end mill, for example, is made entirely of high speed steel (or cobalt, or some other material).

With other tools, like many turning tools, boring bars, grooving tools, and threading tools, the shank of the tool is made from one material (like steel), while the very cutting edge of the tool is made from another material (like carbide or ceramic). This lowers the cost of the cutting tool. With these kinds of cutting tools, only the cutting edge of the cutting tool is made from the (expensive) cutting tool material. The cutting edge component of the tool is called an insert. Figure 1.15 shows this kind of cutting tool.

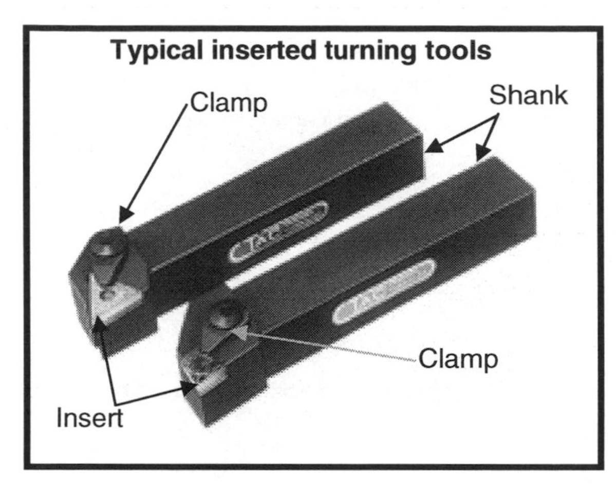

Figure 1.15 – Examples of cutting tools that use inserts

The cutting tool material also determines how quickly machining can be done. Generally speaking, the harder the cutting tool material, the faster machining can be done.

Based upon knowing these three criteria, spindle speed and feedrate selection is usually quite simple. You simply reference the cutting tool manufacturer's recommendations. These recommendations – along with other cutting condition recommendations like depth of cut – are usually provided in the tooling manufacturer's technical manuals – or they may be right in the sales catalog for the cutting tools.

- Speed will be recommended in **surface feet per minute** (sfm) if the data is given in inch, or **meters per minute** if the data is given in metric (most cutting conditions provided in the United States are given in inch). This is the amount of material (in feet or meters) that will pass by the tool's cutting edge in one minute.
- Feedrate will be recommended in **per revolution** fashion (either inches or millimeters per revolution – ipr or mmpr).

Say for example, you must finish turn 1018 cold-drawn steel with a finish turning tool that has a carbide insert. You look in the tooling manufacturer's technical handbook and find they recommend a speed of 700 surface feet per minute and a feedrate of 0.005 inches per revolution.

With conventional turning equipment like engine lathes and turret lathes, there is only *one way* to specify spindle speed (in rpm) and *one way* to select feedrate (in per revolution fashion). Since the speed for conventional lathes must be specified in rpm, a machinist must to convert the 700 sfm to the appropriate speed in rpm based on the diameter of the workpiece that is being machined. Here is the formula to do so:

Rpm = sfm times 3.82 divided by the diameter to be machined

If, for example, the workpiece diameter to be machined is 4.0 inches, the necessary speed for 1018 cold drawn steel (if it is to be run at 700 sfm) will be 668 rpm (700 times 3.82 divided by 4.0).

As you can imagine, calculating spindle speed in rpm can be pretty cumbersome, especially if several diameters must be machined by the same tool. And consider a facing operation. As soon as a facing tool begins moving, the diameter it is machining will change.

The two ways to select spindle speed

CNC turning centers allow you to specify spindle speed in *two* ways. You can do so in rpm, or you can specify spindle speed directly in surface feet per minute (or meters per minute if you are working in the metric mode). In our previous turning example, this means you can specify the spindle speed to be used as 700 sfm – eliminating the rpm calculation.

This second method of selecting spindle speed for CNC turning centers is called *constant surface speed* (css). As the name implies, constant surface speed will cause the machine to constantly update (change) the spindle speed in rpm to maintain the specified speed in surface feet per minute (or meters per minute if you work in the metric mode). As a cutting tool moves in the X axis (changing diameter), spindle speed in rpm will also change. As the cutting tool moves to a smaller diameter, speed in rpm will increase. As it moves to a larger diameter, spindle speed in rpm will decrease.

Two preparatory functions (G codes) specify which of the spindle speed modes (rpm or css) you want to use for speed selection.

G96 – select constant surface speed mode
G97 – select rpm mode

An S word specifies the actual speed. Three M codes are used for spindle activation.

M03 - Spindle on forward (right hand tools)
M04 - Spindle on reverse (left hand tools)
M05 - Spindle off

Here are two examples (working in the inch mode):

N050 G96 S500 M03 (Turn spindle on forward at 500 sfm)
N050 G97 S500 M04 (Turn spindle on reverse at 500 rpm)

As stated, constant surface speed mode (G96) will cause the machine to constantly and automatically update the spindle speed in rpm based on the current diameter the cutting tool is machining.

If for example, you have selected a speed of 500 sfm and the tool is currently machining a 3.5 in diameter, the machine will run the spindle at 546 rpm (3.82 times 500 divided by 3.5). There will be no need for you to perform the spindle rpm calculation – the machine does this for you.

If facing a workpiece (machining direction is X minus), the machine will constantly increase the spindle rpm as the facing operation occurs. See Figure 1.16 for a graphic illustration of how the diameter being machined changes during a facing pass– and the impact this has on spindle speed in rpm.

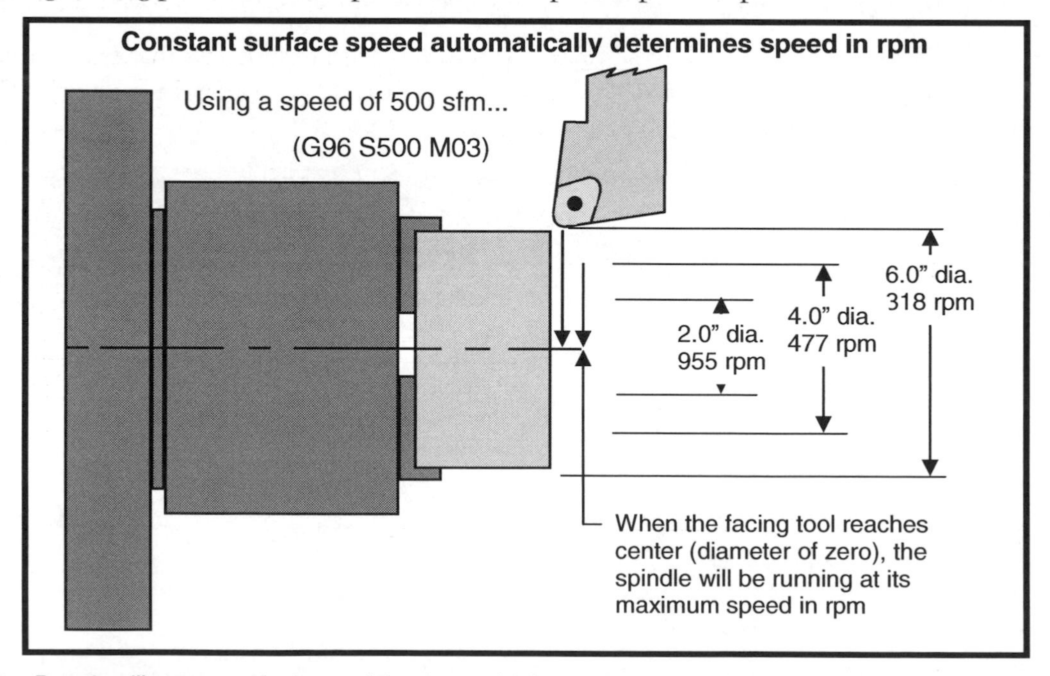

Figure 1.16 – Drawing illustrates a facing tool that is machining to the center of a workpiece

When to use constant surface speed mode

You should use constant surface speed mode (G96) whenever a single point cutting tool will be machining more than one diameter on the workpiece. Examples include rough and finish facing, rough and finish turning, rough and finish boring, necking (grooving), and cutting off (parting).

For these kinds of operations, constant surface speed provides three important benefits:

1) Constant surface speed simplifies programming. As you have seen, it eliminates the need for rpm calculations.

2) Since the appropriate rpm will be used as machined diameters change, the witness marks (finish) on the workpiece will be consistent from one surface to another. This is also related to the fact that feedrate will be specified in inches per revolution (ipr) or millimeters per revolution (mmpr). As the spindle speed changes, so does the feedrate per minute (feedrate selection is presented later in this lesson).

3) Since spindle speed in rpm will be correct during *all* machining operations, tool life will be extended to its maximum.

When to use rpm mode

There are three times when constant surface speed cannot be used – so spindle speed in rpm must be calculated and specified in the rpm mode (G97).

First, you must specify spindle speed in rpm for any cutting tool that machines right on the spindle's centerline. We call these tools *center-cutting tools*. Examples include drills, taps, and reamers. These tools machine a hole right in the center of the workpiece.

Again, these tools are sent right to the spindle's centerline (a diameter of zero). If you specify speed in the constant surface speed mode (G96) for a center cutting tool – even as just *one* surface foot per minute – the spindle will run at its maximum speed in rpm when a cutting tool is sent to a diameter of zero. (3.82 times one divided by zero is infinity).

When you machine with a center cutting tool, you must calculate and specify spindle speed in rpm. If for example, you must drill a 0.75 diameter hole and the drill manufacturer recommends a speed of 80 sfm based upon the material you are machining, the required speed will be 407 rpm (3.82 times 80 divided by 0.75).

Second, rpm mode must be used when chasing threads. The machine must perfectly synchronize the spindle speed with the feedrate motion during the multiple thread-passes required for machining a thread. With most machines, this cannot be done in the constant surface speed mode.

And third, if your machine has live tooling (introduced in Lesson One), spindle speed must always be specified in rpm when live tool are being used.

We'll mention one more time that some programmers elect to use the rpm mode even though constant surface speed *could* be used. When a turning tool or boring bar is machining but one diameter (or even several diameters that are close together), there is not much of an advantage to using constant surface speed – other than eliminating the need for the rpm calculation. So some programmers will calculate an rpm based upon the largest diameter being machined and program the operation in the rpm mode. See Figure 1.17 for an example.

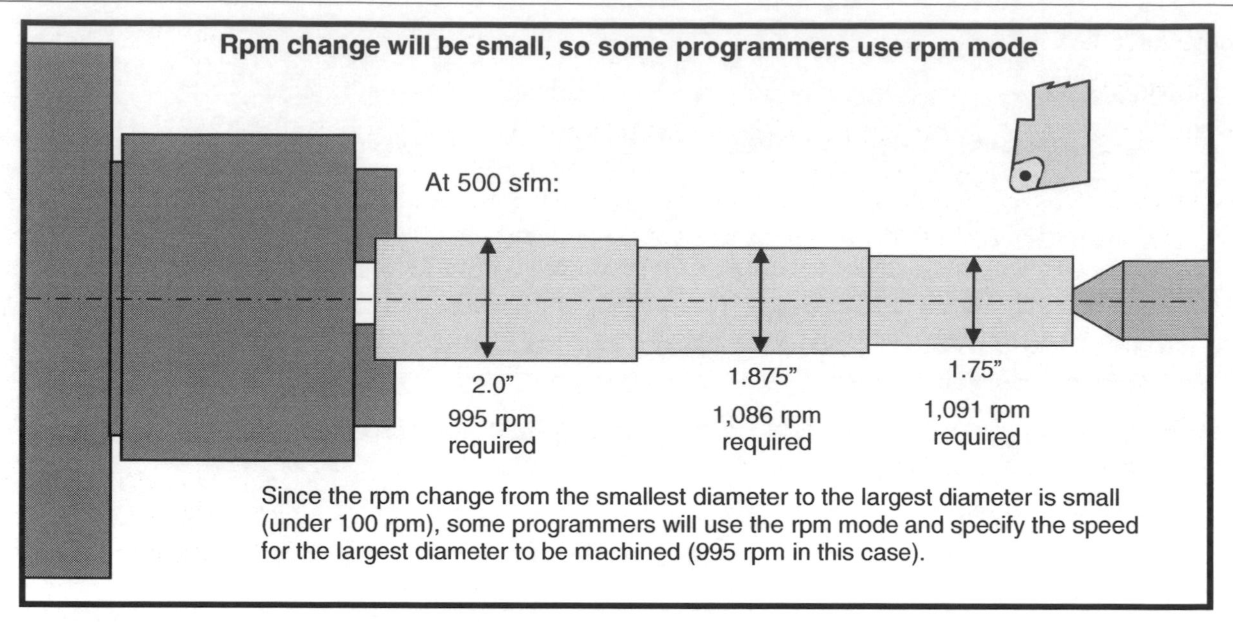

Figure 1.17 – With small diameter changes, rpm changes will also be small

How fast will the spindle be running when constant surface speed is used?

Say you're going to be rough turning a mild steel workpiece with a cutting tool that has a carbide insert. The cutting tool manufacturer recommends that you run the spindle at 500 sfm. In the program at the beginning of the tool, you give this command:

```
N050 G96 S500 M03
```

This command will start the spindle at 500 sfm (assuming the inch mode is being used) in the forward direction. But when the spindle starts, how fast will it be running in rpm?

Based upon the information just provided, you cannot answer this question. Prior to answering, you must know the diameter at which the cutting tool is currently positioned. And even then, you must perform the rpm calculation (3.82 times sfm divided by diameter).

For people that have experience running a conventional lathe, this can be a little unnerving. Machinists are accustomed to specifying spindle speed directly in rpm – so when the spindle starts on a conventional lathe, they will know precisely how fast the spindle will run in rpm. If you want to know the precise rpm at which the spindle will start with CNC turning centers (when specifying speed in the constant surface speed mode), you must perform the rpm calculation – and you must know the diameter position of the cutting tool in order to perform this calculation.

How fast can the spindle rotate?

As stated in Lesson One, you must reference the machine tool builder's documentation (commonly the programming manual) in order to determine your machine's spindle characteristics. Say, for example, you find that your turning center has two spindle ranges. The low range runs from 0 – 1,500 rpm. The high range runs from 0 – 5,000 rpm. This means, of course, that when the spindle is in the low range, it cannot run any faster than 1,500 rpm. When it is in the high range, it cannot run faster than 5,000 rpm.

When you specify spindle speed in the constant surface speed mode, the current spindle range (low or high) will determine the maximum spindle speed. Using the rpm calculation, the machine will attempt to run the spindle at the appropriate rpm. If the rpm calculation renders a speed in rpm that is higher than the maximum speed allowed in the current spindle range, the spindle will simply *peak out* at the maximum rpm of the spindle range and run at this speed. Consider these scenarios for the machine just described.

Scenario number one: Say you are rough turning a workpiece from an 8.0" diameter down to a 1.0" diameter. Based upon the cutting tool being used and workpiece material being machines, the cutting tool manufacturer recommends a speed of 800 sfm. Since this is a powerful machining operation, you select the low spindle range.

When rough turning begins, the spindle will be running at about 334 rpm (3.82 times 800 divided by 8.0). As the rough turning tool makes roughing passes, the spindle speed in rpm will increase. When the rough turning tool reaches 2.0" in diameter, the spindle will attempt to run at 1,528 rpm (3.82 times 800 divided by 2.0). Since the maximum rpm in the low spindle range for this machine is 1,500 rpm, the machine will not be able to achieve the appropriate rpm. It will peak out at 1,500 rpm. If you continue to rough turn the workpiece in the low spindle range, this rough turning pass – as well as the rest of the roughing passes – will be performed at 1,500 rpm.

Scenario number two: Whenever you face a workpiece to center (a common machining operation), the spindle will run up to the maximum speed of the current spindle range. For our example machine, this means it will run up to 1,500 rpm if the low range is selected or 5,000 rpm if the high range is selected.

For small, perfectly round workpieces, this will be acceptable. The workpiece will run true in the spindle all the way up to the machine's maximum speed – even in the high range. But you must exercise extreme caution when workpieces are larger – and especially when they are not perfectly round.

Castings, for example, are notorious for being out-of-round. An out-of-round workpiece will wobble in the workholding device when the spindle rotates. The faster the spindle runs, the more machine vibration this wobbling will cause. Of course, wobbling is caused by the fact that the workpiece is not truly concentric with the spindle – and it will place stress on the workholding device used to secure the workpiece.

If this stress is excessive, the workpiece will actually be released by the workholding device. This makes for a very dangerous situation. A workpiece that is rotating at a very high rate will be bouncing around inside the machine – and could actually come right through the door of the machine.

When you must machine large workpieces that are not truly round, you must be very careful not to allow the spindle to reach a speed in rpm that causes the machine to vibrate. A test can be made during the machine's setup to determine this maximum spindle speed. Once you know how fast the spindle/workpiece can safely rotate, you can specify a spindle limiting command in the program that will keep the spindle from exceeding this speed.

The maximum spindle speed test: The setup person will load a workpiece and start the spindle at a very slow rpm. They will continue to increase the spindle speed in small increments until they *start* to feel vibration. The speed at which the machine begins to vibrate will be reduced by about twenty percent – and will be the maximum speed for this workpiece while it is in its rough state.

How to specify a maximum speed for the constant surface speed mode

Here is a way to limit the maximum speed in rpm that the spindle can achieve. In essence, you will be superceding the maximum rpm of each spindle range. The spindle limiter is specified with a G50 command. The command

N055 G50 S2000 (Limit spindle speed to 2,000 rpm)

specifies that the spindle will not be allowed to exceed 2,000 rpm, even if the constant surface speed mode is being used and the machine has calculated a speed that is greater than 2,000 rpm. If this command is specified at the beginning of the program for our example machine (maximum speed in the high range is 5,000 rpm), the spindle will not be allowed to run faster than 2,000 rpm, even if the high range is selected. If, after specifying the spindle limiter shown above in line N055, you program a facing tool to face to center (zero diameter) in the high spindle range, the spindle will peak out when it reaches 2,000 rpm.

A potential limitation of constant surface speed

While constant surface speed is an extremely important programming feature, we must point one potential limitation. If it is not efficiently programmed, it can increase program execution (cycle) time. The reason for this has to do with the fact that a turning center's spindle cannot instantaneously respond to speed changes in rpm. It takes time for the machine to respond to spindle rpm changes.

How much time rpm changes take for a given turning center is based on several factors, including machine size, horsepower, size & weight of the work holding device, and weight of the workpiece being machined. Generally speaking, the bigger the machine, the more the time it takes for the spindle to respond to rpm changes.

One way to determine spindle response time is to actually measure it with a stop-watch. Say you do so and find these characteristics for one of your machines:

- 0-1,000 rpm takes 2 seconds
- 0-2,000 rpm takes 4 seconds
- 0-3,000 rpm takes 6 seconds
- 0-4,000 rpm takes 8 seconds

Say your program has a turret index position of 8.0 inches in diameter (the X axis). This position provides ample clearance for safely indexing the turret. When each tool is finished machining, it is sent to this position. If the machine is in the constant surface speed mode (and assuming the workpiece is smaller than 8.0 inches in diameter), the spindle will slow down during this motion. If the speed is 600 sfm, for example, the spindle will be running at 286 rpm whenever a cutting tool is at this 8.0 inch diameter turret index position.

When you command a cutting tool to approach a workpiece in the X axis, the spindle speed in rpm will increase accordingly. Say the tool is approaching to a 0.75 in diameter. At 600 sfm, the spindle will increase to 3,056 rpm. For the example machine just shown, this will take six seconds.

It's likely that the approach movement will occur much faster than the spindle response, meaning the machine will pause for about five seconds, while the spindle gets up to speed. The reverse will happen during each tool's retract to the 8.0 inch diameter turret index position (slowing back to 286 rpm). For each of these approach/retract motions, about five seconds will be added to program execution time (for our example machine). This is the reason why constant surface speed can waste cycle time if not efficiently programmed. We'll show how to efficiently program constant surface speed in Lesson Sixteen.

The two ways to specify feedrate

As with spindle speed, there are two ways to specify feedrate for CNC turning centers. As mentioned in Lesson One, feedrate can be specified in per revolution fashion (inches or millimeters per revolution) or in per minute fashion (inches or millimeters per minute). Again, two G codes are used to specify which feedrate mode will be used.

G98 – feed per minute
G99 – feed per revolution

When you power up on a CNC turning center, the feed per revolution mode (G99) is automatically selected. (By the way, CNC words that are automatically selected at power-up are called *initialized* words.) This means that if you do not include a feedrate mode specifying G code in a program, the machine will assume the feed per revolution mode. You'll notice that many of the example programs provided in this text make this assumption (they don't include a G99).

Also as mentioned in Lesson One, an F word actually specifies feedrate. Here are two examples (assuming you are working in the inch mode).

```
N060 G98 F30.0 (Feedrate of 30.0 ipm)
N060 G99 F0.012 (Feedrate of 0.012 ipr)
```

When to use the feed per revolution mode

We recommend using the feed per revolution feedrate mode (G99) for almost all machining operations you perform on CNC turning centers. In the per revolution feedrate mode, feedrate specifies how far the cutting tool will move during one spindle revolution.

Per revolution feedrate mode is especially helpful with cutting tools that use the constant surface speed mode (like rough and finish turning tools, rough and finish facing tools, rough and finish boring tools, and grooving tools). As the spindle changes speed in rpm based upon the current diameter position of the cutting tool, the feedrate in inches or millimeters per minute will also change. This will cause witness marks (finish) to be consistent for all surfaces being machined.

Even if you're using the rpm mode (possibly for a drill that machines a hole in the center of the workpiece), it is always easier to specify feedrate in per revolution fashion. Again, most cutting tool manufacturers recommend feedrates for their cutting tools in per revolution fashion.

When to use the feed per minute feedrate mode

Frankly speaking, about *the only time we recommend using the feed per minute mode is when you must cause a controlled motion with the spindle stopped.* If the spindle is stopped, of course, the axes will not move regardless of how large a feedrate is specified in the per revolution mode.

If your turning center has a bar feeder, for example, and if the bar feed operation requires the spindle to be stopped prior to feeding a bar, the feed per minute mode must be used for the bar advance motion (bar feeder programming is shown in the Appendix which is after Lesson Twenty-Three).

If your turning center is equipped with live tooling, operations performed with live tools require that the machine be in the live tooling mode (not the normal turning mode). In live tooling mode, the machine's main spindle is off (at least from a turning operation standpoint), meaning that feedrate for live tools must be programmed in per minute fashion.

Calculating feedrate per minute – To calculate feedrate in per minute fashion, multiply the desired feedrate per revolution times the previously calculated spindle speed in rpm. For example, say you must use the live tooling mode and drill a 0.5 diameter hole. For the material you must machine, the drill manufacturer recommends a speed of 80 sfm and a feedrate of 0.008 ipr. First, calculate the speed in rpm – 3.82 times 80 divided by 0.5 – or 611 rpm. Now multiply 611 times 0.008 – which renders a per-minute feedrate of 4.88 ipm.

Again, any time you must cause a feedrate motion when the spindle is stopped, you must use the feed per minute mode. One other example is when performing a light broaching operation.

Some programmers do prefer to use the feed per minute mode for operations when the feedrate will remain consistent in feed per minute for the entire machining operation (even though feed per revolution could be used). When drilling a hole, for example, once the feed per minute is calculated, it will work for the entire drilling operation.

An example of speed and feed usage

Figure 1.18 provides an illustration of what happens when you use the constant surface speed spindle mode with the per revolution feedrate mode to perform machining operations. Notice that spindle rpm changes with workpiece diameter. So does feedrate in inches per minute change with changes in spindle rpm.

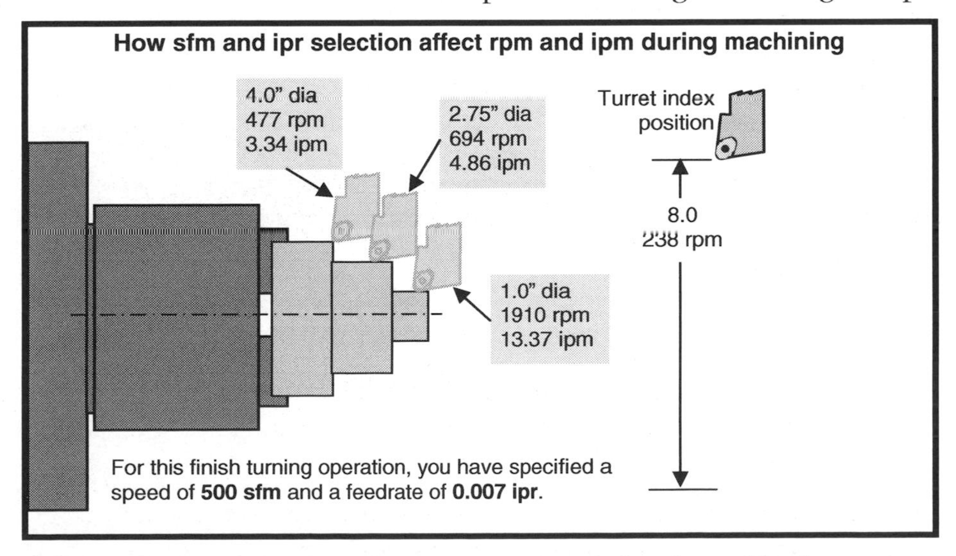

Figure 1.18 – How spindle speeds and feedrates change as a workpiece is machined

As you study Figure 1.18, notice how many speed and feed calculations the machine is making – all based upon your spindle speed selection in surface feet per minute and feedrate selection in inches per revolution.

Key points for Lesson Two:

- With CNC turning centers, there are two ways to specify spindle speed (constant surface speed and rpm) and two ways to specify feedrate (feed per revolution and feed per minute).
- Constant surface speed mode (G96) should be used with single point turning tools that machine more than one diameter during the machining operation. Rpm mode (G97) must be used for tools that machine on center (like drills) and for threading operations.
- Constant surface speed mode will cause the machine to automatically select the appropriate spindle rpm based upon the specified speed in surface feet/meters per minute and the current diameter position of the cutting tool.
- Per revolution feedrate mode (G99) is initialized (automatically selected at power-up) and should be used for almost all machining operations. Feed per minute mode (G98) should only be used for feedrate movements when the main spindle is stopped.
- Using per revolution feedrate mode will ensure that witness marks (surface finish) on the workpiece will be consistent for all surfaces being machined.
- Rpm = 3.82 times sfm divided by the diameter being machined.
- Ipm = previously calculated rpm times feedrate per revolution.

Quiz

1) Which spindle speed mode should you select for cutting tools that machine more than one diameter?

A) Constant surface speed mode B) Rpm mode

2) Which feedrate mode should you use for almost all machining operations?

A) Per revolution mode B) Per minute mode

3) Provide a CNC command that will start the spindle at 400 sfm in the forward direction.

____

4) Provide a CNC command that will start the spindle at 500 rpm in the reverse direction.

____

Answers: 1: A, 2: A, 3: G96 S400 M03, 4: G97 S500 M04

STOP! Do practice exercise number two in the workbook.

Lesson 3

General Flow Of The CNC Process

The tasks of programming, setup, and operation are but three of the things that must be done in order to get a CNC job up and running. It really helps to understand how these tasks fit into the bigger picture of a company's manufacturing environment.

CNC machine tools are being used by all sorts of companies. Indeed, if a company manufacturers anything, it's likely that they're using at least some CNC machine tools. With the diversity of applications, there comes diversity in what is expected of CNC people. Understanding where your company fits in to this diverse group should help you understand what will be expected of you.

Companies that use CNC turning centers

There are many factors that contribute to how a CNC-using company applies its CNC turning centers. These factors include (among others) lot sizes, lead times, percentage of new jobs, closeness of tolerances held, materials machined, and company type. The most important of these factors is company type.

When it comes right down to it, there are only four types of companies that use CNC machine tools:

- Product producing companies – get revenue from the sale of a product
- Workpiece producing companies – (also called job-shops or contract-shops) get revenue from the sale of production workpieces to product producing companies
- Tooling producing companies – get revenue from the sale of manufacturing tooling (fixtures, jigs, molds, dies, gauges, cutting tools, etc.) to product producing and workpiece producing companies
- Prototype producing companies – get revenue from the sale of prototypes to product producing companies

There are also overlaps in company type. For example, some product producing companies have a tool-room in which CNC machine tools are used – or they may have a research and development department that produces prototypes. Some workpiece producing or tooling producing companies have a product of their own.

While there will be exceptions to what we say here, some pretty good generalizations can be made based upon the company type alone, especially when it comes to what CNC people will be doing.

Product producing companies tend to have more resources than workpiece producing, tooling producing, and prototype producing companies. Since their profit is one step removed from manufacturing (a product won't come to market unless the company can make a profit), they tend to engineer all facets of the manufacturing environment. For this reason, they commonly break up the tasks related to CNC machine tool usage. People will specialize in the tasks they perform.

This will maximize machine tool utilization. It ensures that machines are running for as high a percentage of time as possible. Many CNC tasks will be performed while the machine is running production (like programming for upcoming jobs, gathering components needed for future setups, and assembling cutting tools, among others). So while a CNC operator is running a job on the CNC machine, other people are getting ready to run the next (and other upcoming) jobs. Again, this minimizes the amount of time that the CNC machine is down between production runs.

On the other hand, workpiece producing, tooling producing, and prototype producing companies tend to charge an hourly rate for machine usage time. Their resources will be much smaller – and they will require much more of their CNC people. One person may be responsible for all of the tasks related to running a

CNC job. They will sacrifice machine tool utilization to some extent in order to run jobs with a small number of CNC people. This means machines may sometimes sit idle while the person responsible for the machine performs all or some of the tasks related to CNC machine usage.

What will you be doing?

This text includes enough information to help you fully master CNC turning center usage, and you'll probably want to learn it all. But if you work for a product producing company, it's likely that you'll only have to master a portion of what we present in this text. If you work for a workpiece producing, tooling producing, or prototype producing company, you'll probably have to master just about everything shown.

Common job titles in the CNC environment include:

- Process engineer – This person (often the CNC programmer) determines which machine will be used to produce the workpiece and develops the sequence of machining operations, or *process*
- Tooling engineer – This person (often the CNC programmer) designs any special workholding devices (fixtures) needed for a job and determines which cutting tools will be used to machine the workpiece
- CNC programmer – This person creates the CNC program and develops the setup and production run documentation used by people that actually run the job on the CNC machine
- CNC setup person – This person gets the CNC machine ready to run a job
- Inspector – Once a setup is made and the first workpiece is run, this person checks the workpiece to confirm that it is within specifications
- CNC operator – Once a setup is completed and the first workpiece passes inspection, this person runs out the job
- CNC helper – This person does as much as possible to help in the shop – gathering components (fixtures, cutting tools, inserts, etc.) needed by others in the CNC environment, cleaning machines, performing basic preventive maintenance, and in general, doing whatever it takes to minimize work for other CNC people
- Tool crib attendant – This person gathers and assembles tooling components for upcoming jobs and supplies perishable tooling (like inserts) to CNC operators currently running jobs

> Ask an experienced person in your company what your CNC-related responsibilities will be.

Flow of the CNC process

Here we give an example of how a typical shop will handle a CNC job. By no means will this flow be the same for all shops. As stated, workpiece producing companies tend to have one or two people handling all steps in this flow. Product producing companies will usually break these steps up into small tasks, each to be handled by a different person. We present the related tasks in the approximate order that jobs are run.

Study the workpiece drawing

All questions about how a workpiece must be machined are answered by the workpiece drawing. Everyone involved with a job must get acquainted with the workpiece to be produced.

Decision is made as to which CNC machine to use

If the company has more than one CNC turning center, there may be some question as to which one should be used to produce a given workpiece. Based on the required number of workpieces (lot size), the accuracy required of the workpiece, the material to be machined, the required surface finish, and the shop loading on any one CNC turning center (among other things), the decision is made as to which CNC turning center to use.

The machining process is developed

If more than one machining operation must be performed by the CNC machine, someone with machining practice background (commonly the programmer) must come up with a sequence of machining operations to be used to machine the part. The program will follow this machining process. It is during this step that cutting tools needed for the job will be determined. Cutting conditions (speeds, feeds, and depths-of-cut) must also be calculated.

Tooling is ordered and checked

Based on a previously developed process, the required tooling must be obtained. This includes workholding tooling, like chuck jaws or collets, as well as cutting tools.

The program is developed

In this step, the programmer codes the program into a language that the CNC machine can understand. This step is, of course, one major focus of this text. We present six Key Concepts to help you understand this step.

Setup and production run documentation is made

Part of the programmer's responsibility is to make it clear as to how the setup is to be made at the machine – as well as how the production run is to be completed. Drawings are commonly made to describe the work holding setup. A cutting tool list is made, specifying the components needed for each tool, the offsets to be used for each tool, and the turret station number into which each tool will be loaded. Production run documentation commonly includes workpiece loading instructions, and instructions related to holding size on the workpiece during the production run. Truly, anything that helps a setup person or operator should be documented.

Program is loaded into the CNC control's memory

Once the program is prepared, there are two common methods used to load it into the control's memory. One way is for someone to physically type it through the keyboard and display screen of the control panel. Frankly speaking, this is a rather cumbersome way to get the program into the control's memory. For one thing, the keyboard of the control panel is quite difficult with which to work. The keyboard is not oriented in a logical way (most are not like the keyboard of a computer) and usually the control panel itself is mounted vertically, resulting in the person typing the program experiencing fatigue. In many cases, the machine will sit idle while the program is being typed. Even a relatively short program will take time to enter. A CNC machine makes a very expensive typewriter!

A more popular way to enter programs into the control's memory is to use another device for typing programs. A personal computer is commonly used for this purpose. The software used for this computer application resembles a common word processor. In fact, most word processor software programs can be used for the purpose of typing CNC programs. Once the program is entered through the computer's keyboard, it can be saved on the computer's hard drive. When the program is needed at the machine, it can be quickly transmitted to the machine from the computer. This transmission takes place almost instantaneously, even for lengthy programs. The software, computer, and cabling used to transfer CNC programs to and from the CNC machine are called a distributive numerical control (DNC) system.

Again, a computer makes it much easier for the person typing the program. They can sit in a comfortable environment to type the program. Most importantly, the program can be typed *while* the machine continues to machine workpieces for the current production run, meaning almost no machine time is wasted while the program is loaded.

The setup is made

Before the program can be run, the setup must be made. Using the set-up instructions (commonly prepared by the programmer), the setup person makes the workholding setup. Cutting tools are assembled and loaded into the proper turret stations. Measurements must also be made to determine the position of program zero for each tool. And certain program zero assigning and cutting tool-related offsets must be entered.

The program is cautiously verified

It is very rare that a new program requires no modification at the machine. Even if the programmer does a good job programming the required cutting tool motions, there will almost always be some optimizing that is necessary to minimize program execution time (especially for larger lots).

Production is run

At this point the machine is turned over to a CNC operator to complete the production run. While the programmer's and setup person's job could be considered finished at this point, there may be some long term problems that do not present themselves until several workpieces are run. For example, cutting tool wear may be excessive. Tools may have to be replaced more often than the company would like. In this case, the speeds and feeds in the program may have to be adjusted.

Corrected version of the program is stored for future use

Many workpieces that run on CNC machines are run on a regular basis, especially in product producing companies. At some future date it will probably be necessary to run the workpiece again. If changes have been made during a new program's verification, it will be necessary to transmit the corrected CNC program back to the program storage device (DNC system). If this step is not done, the program will have to be verified again the next time the job is run.

Key points for Lesson Three:

- There are four types of companies that use CNC machine tools.
- The type of company dramatically affects what is expected of CNC people.
- You must know what you will be expected to do in the CNC environment.
- It helps to understand the bigger picture of how a job is processed through a CNC using company.

Talk to experienced people in your company...

... to learn more about how your company operates. If you work for a CNC-using company, here are some questions you'll want to get answered.

1) What kind of company do you work for: product producing, workpiece producing, tooling producing, or prototype producing?

2) Who are the programmers? Who does setups? Is there a tool crib? If so, who's the attendant? Do setup people inspect their own workpieces or is there a special inspector that inspects the first workpiece? What will *you* be doing?

3) Who develops the machining process for jobs run on CNC machines? Is it the CNC programmer?

4) Do programmers write programs manually or do they use some kind of computer aided manufacturing (CAM) system?

5) Does the company have a distributive numerical control (DNC) system? If so, where is it and how is it used?

6) How often are new programs run? Does the setup person verify programs alone, or does the programmer come to the machine to help?

STOP! Do practice exercise number three in the workbook.

Lesson 4

Visualizing The Execution Of A CNC Program

A CNC programmer must possess the ability to visualize the movements a CNC machine will make as it executes a program. The better a person can visualize what the turning center will be doing, the easier it will be to prepare a workable CNC program.

Once again, we stress the importance of basic machining practice as it applies to CNC turning center usage. A machinist that has experience running a conventional lathe will have seen machining operations taking place many times. While this experience by itself does not guarantee the ability to *visualize* a machining operation (seeing the machining operation taking place in your mind), it dramatically simplifies the task of learning how to visualize a CNC program's execution.

When a machinist prepares to machine a workpiece on a conventional lathe, they will have all related components needed for the job right in front of them. The machine, cutting tools, workholding setup (chuck), and print are ready for immediate use. It is highly unlikely that the machinist will make a basic mistake like forgetting to start the spindle before trying to machine the workpiece.

On the other hand, a CNC turning center programmer will be writing the program with only the workpiece drawing to reference. No tooling – no machine – and no workholding setup will be available to them. For this reason, a programmer must be able to visualize just exactly what will happen during the execution of the program – and this can sometimes be difficult, since this visualization must take place in the programmer's mind. A beginning programmer will be prone to forget certain things – sometimes very basic things (like turning the spindle on prior to machining the workpiece).

In this lesson, we will acquaint you with those things a programmer must be able to visualize. We will also show the first (elementary) program example to stress the importance of visualization.

Visualization is necessary to develop a set of instructions

Consider the visualization it takes to write a set of travel instructions to get someone from the airport to your company. Before you can write the instructions, you must be able to visualize the path from the airport to your company. If you cannot visualize the path, you can't write the instructions. Worse, if you think you can visualize the path (but you're wrong), you'll write incorrect instructions and the person following your instructions will get lost!

In similar fashion, if you cannot visualize the path a cutting tool will take as it machines a workpiece, you cannot write the CNC commands that drive the cutting tool through this path.

Airport Dr.
N
Highland St.
March Ave.
Lance Dr.
Elm St.
Company

1) Take airport exit to Highland Dr. Turn left.
2) Take Highland Dr. 4 mi to Elm St. Turn right.
3) Take Elm St. 1 mi to March Ave. Turn left.
4) Take March Ave 2 mi to Lance Dr. Turn right.
5) Take Lance Dr. 1/2 mi to company (on right).

Program make-up

Like the sentences making up a set of travel instructions, a CNC program is made up of *commands* (also called blocks). Within each command are *words*. Each word is made up of a *letter address* (N, X, Z, T, etc.) and a *numerical value*. Figure 1.19 shows the beginning of a CNC program that stresses program make-up.

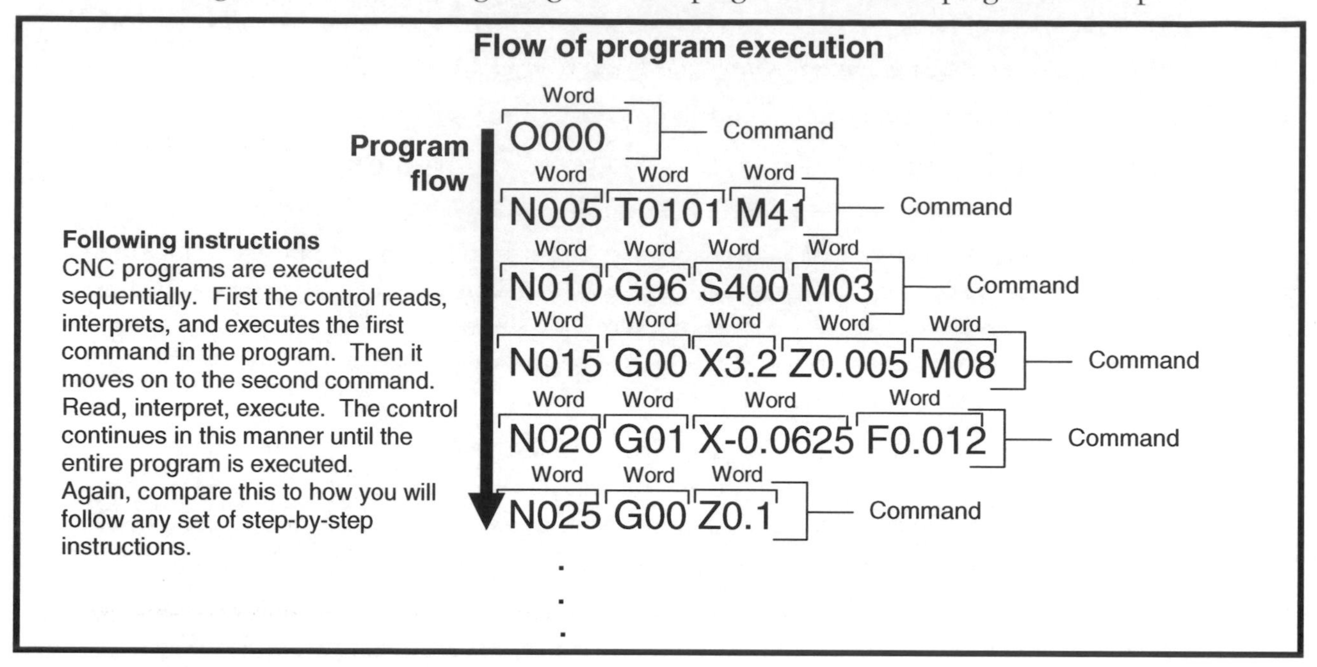

Figure 1.19 The flow of a CNC program

Method of program execution

You can also compare writing a CNC program to giving a set of step-by-step assembly instructions. For example, maybe you have just purchased a bookcase that requires assembly. The instructions you receive will be in sequential order. You will perform step number one before proceeding to step number two. Each step will include a sentence (or paragraph) explaining what it is you are supposed to do at the current time. As you follow each step, performing the given procedure, you are one step closer to finishing.

In like fashion, a CNC program will also be executed in sequential order. The CNC control will read, interpret, and execute the first command in the program. It will then go on to the next command. Read, interpret, execute. The control will continue this process for the balance of the program.

Keep in mind that the CNC control will execute each command *explicitly*. Compare this to the set of instructions for assembling the bookcase. In the set of assembly instructions, the manufacturer may be rather vague as to what it is you are supposed to do in a given step. They may also assume certain things. They may assume, for example, that you have a screwdriver and that you know how to use it. Assembly instructions can be so vague that they may be open to interpretation. A CNC control, by comparison, will make no assumptions. Each command will be very explicit - and any one CNC command will have only one resultant machine action or set of actions.

An example of program execution

To stress the sequential order of execution, and the visualization that is necessary to write programs, let's look at a very simple turning center example. We will first show the steps a machinist will perform to machine a very simple workpiece on a conventional lathe. Then we will show the equivalent CNC program that will perform the same machining operation on a CNC turning center.

Figure 1.20 shows the drawing for this machining operation. In this example, a machinist will be turning a 3.0 inch diameter down to 2.875. We're assuming the workpiece is already in the chuck and the cutting tool is in the tool holder.

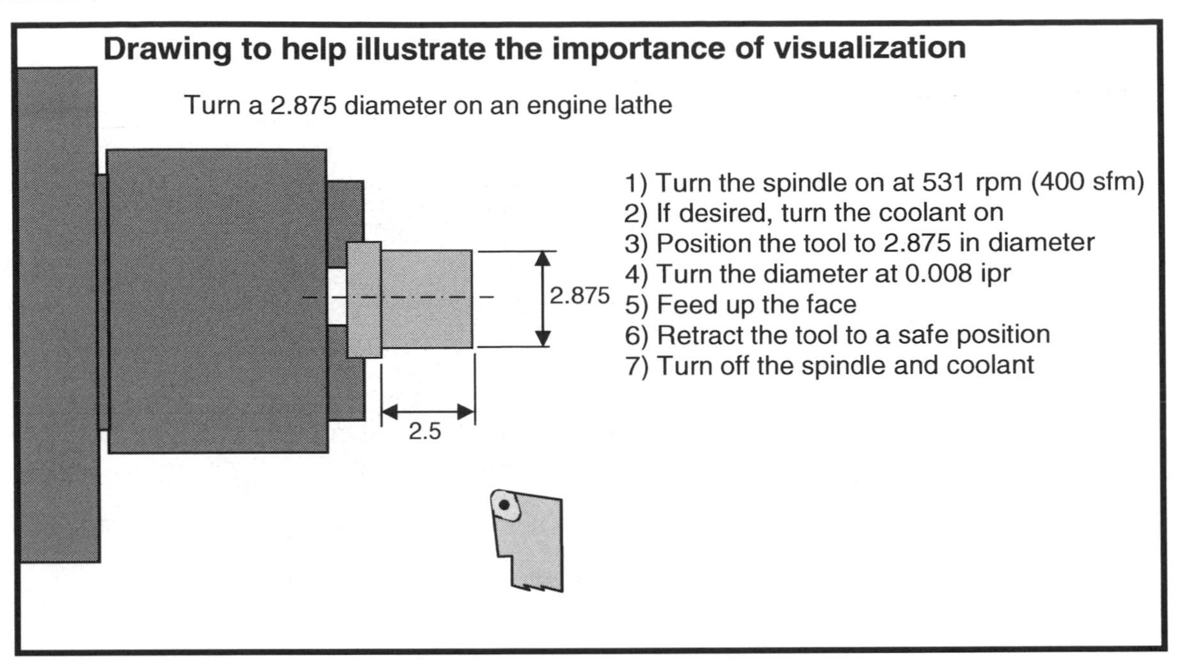

Figure 1.20 Drawing for example illustrating program execution flow

Figure 1.20 includes a step-by-step procedure a machinist will follow to machine the workpiece. Admittedly, this is a very simple machining operation. But remember, you must be able to see each step happening in your mind if you are going to be able to write a CNC program.

The CNC program to machine the 2.875 diameter

Now here is a CNC program to turn the diameter on a CNC turning center. Again, we assume that the workpiece is in the chuck and the cutting tool is in the turret (station number one) when this program is executed.

```
O0001 (Program number)
N005 T0101 M41 (Index the turret to station number one and select the low spindle range)
N010 G96 S400 M03 (Start the spindle at 400 sfm in the forward direction)
N015 G00 X2.875 Z0.1 M08 (Position the tool flush with the diameter to machine and clear of the
  workpiece in Z, turn on the coolant)
N020 G99 G01 Z-2.5 F0.008 (Machine the 2.875 diameter at 0.008 ipr)
N025 X3.1 (Feed up the face)
N030 G00 X8.0 Z6.0 (Retract to a safe position)
N035 M30 (End of program command – this command stops the spindle and turns off the coolant)
```

Though the actual commands in this program probably don't make much sense, the messages within parentheses should nicely clarify what is happening in each command. (By the way, you can include messages within parentheses in your own programs – though in actual programs they must be in upper case characters).

Though this example is a complete and workable program, our intention is not to teach the programming words being used (yet). Instead, we want to stress two things. First is the sequential order by which the program will be executed. The control will execute line number N005 before moving on to line number N010. Then line number N015. And so on —until the entire program is activated.

Second is the visualization that is necessary in order to write programs. For this example, you must be able to visualize (see in your mind) the turning tool machining the workpiece before you can write this program.

Indeed, if you cannot visualize how a tool will move as it performs a machining operation, you will not be able to write the related CNC commands. For those machining operations with which you are not familiar, you will (at the very least) need the help of an experienced machinist.

Sequence numbers

Look at the example program again. Notice that each command of this program begins with an *N word.* N words are called *sequence numbers.* They are command identification numbers. For the most part, the CNC control will actually ignore them. They are just in the program to help you keep track of commands in the program. They really help in this text to point out specific things about a particular command – they confirm that you're looking at the command that's being discussed.

For longer programs, sequence numbers help you confirm that you are truly looking at the desired command (many commands look alike). This is extremely important when you must make a modification to the program.

Again, the CNC control ignores sequence numbers. Sequence numbers can be left out of the program if you desire. Some programmers eliminate sequence numbers, for example, if the control's memory capacity is small. But we recommend that you include them in your programs – and place them in a logical order.

Throughout this text, every command of every program will have a sequence number. And you'll notice that, for the most part, we skip five numbers between sequence numbers. That is, we'll use N005, then N010, then N015, and so on. We recommend using this technique so you can add commands to your programs and still start them with sequence numbers.

A note about decimal point programming

As you have seen in the previous example program, certain CNC words allow a decimal point to be included within a numerical value. All current CNC controls allow a decimal point to be placed in CNC words which require *real numbers* (numbers that require a portion of a whole number). But most current controls do not allow a decimal point in words that always require integer values (whole numbers). All axis words (like X and Z) often require real numbers, so a decimal point is allowed. Spindle speed is specified with an integer value, so you are not allowed to include a decimal point in an S word.

Beginning programmers have the tendency to leave out needed decimal points. You must remember to include a decimal point within each word that allows a decimal point. If you do not, some very strange things can happen. Here's why.

Older controls (over thirty years old) do not allow a decimal point in any CNC word. These old controls require a *fixed format* for all real numbers needed in the program. Trailing zeros are required for these older controls to imply where the decimal point should be placed. For example, an X word of 5.0 in will be specified as "X50000" (note the four trailing zeros) if decimal point programming is not allowed – assuming the inch mode is being used.

Newer controls are *backward compatible.* This means programs written for older controls can still run in current controls. If a current CNC control sees a word that allows a decimal point, but the decimal point is not specified in the word, it will simply place the decimal point automatically – four places to the left of the right-most digit (in the inch mode).

Here's an example that illustrates the kind of mistake you can make. Say you intend to specify an X word of 3.0 in. The correct way to designate this movement is "X3.0" or "X3.". But say you make the mistake of omitting the decimal point. You specify "X3", leaving out the decimal point. In this case, the control will incorrectly interpret the X word. Instead of taking this word as 3.0 inches, the control will place the decimal point four places to the left of the right-most digit. In this case, "X3" will be taken as X0.0003, not 3.0 inches.

Concentrate on overcoming this tendency for making mistakes of omission. Get in the habit of including a decimal point within those words that allow it. These letter address words include: F for feedrate, I, J, & K, for circular motion commands, R for radius, and X, Y & Z for axis movements. Words that do not allow a decimal point and must be programmed as integer values include: N, G, H, D, L, M, S, T, O, and P.

A decimal point tip

When programming whole numbers in CNC words that allow a decimal point, be sure to carry the value out to the first zero after the decimal point. For example, write

X4.0

Instead of

X4.

This will force you to write/type the decimal point. In similar fashion, when specifying values under one, begin the value with the zero to the left of the decimal point. For example, write

X0.375

instead of

X.375

Again, this will force you to write/type the decimal point.

Other mistakes of omission

Knowing the mistakes beginners are prone to make may help you avoid them. Beginners tend to forget things in their programs. As stated, they forget to program a decimal point. They forget to turn on the spindle. They forget to turn on the coolant. They forget to drill a hole before rough boring in it. You will have to concentrate very hard on avoiding mistakes of omission. We're hoping that if you know about this tendency, that it will help you avoid it.

Modal words

Many CNC words are *modal*. This means that the CNC word remains in effect until changed or canceled. In the previous example program, notice the G01 command in line N020. This happens to be a linear motion instating word. The positioning movement that results from line N020 will be done at a feedrate of 0.008 inches per revolution. The very next (line N025) makes a movement up the face X to approach a position slightly off the workpiece. But notice that there is no motion type specification in this command. This movement will still occur in the linear motion mode (and at 0.008 ipr) since G01 is modal. Modal words do not have to be repeated in every command.

Initialized words

Certain CNC words are *initialized*. This means the CNC control will automatically instate these words at power-up. For example, most machines used in the United States will be initialized in the inch mode (again, the instating word is G20). If a company will be using the inch mode exclusively, they can depend upon the CNC control to be in this state at all times (they'll never select the metric mode), meaning there is no need to actually include a G20 in the program.

Letter O or number zero?

As you look at the example program in Figure 1.11, you see several zeros. But the *only* letter O in this program is the very first character (which happens to be the letter address that specifies a program number). All other characters that look like the letter O are actually *zeros*.

For example, notice the spindle-on-forward M code in line N010 (M03). This is specified as M-zero-three, not M-oh-three. Again, the character in the middle is a number zero, not a letter O. This is very important. The control will not behave properly if you use a letter O here. One very common beginner's mistake is to use the upper-case letter O instead of a number zero.

Another beginner's mistake is to use the upper case I or the lower case L (l) to specify the number one (1). A CNC machine makes this distinction. You must use the number one (1) for this character.

Word order in a command

The order in which CNC words appear in a command has no bearing upon how the command will be executed. For example, the command

N050 G00 X1.5 Z0.1 M08

will be executed in exactly the same way as

N050 M08 X1.5 G00 Z0.1

Key points for Lesson Four:

- You must be able to visualize the motions cutting tools will make as they machine workpieces.
- CNC programs are executed in sequential order, command by command.
- Commands are made up of words.
- Words are comprised of a letter address and a numerical value.
- You've seen your first complete CNC program. Though it was short, you've seen the sequential order by which CNC programs are executed – and you should understand the importance of being able to visualize a program's execution.
- We've introduced some program-structure-related points, including decimal point programming, modal words, and initialized words.
- You've seen some of the mistakes beginners are prone to making – like mistakes of omission and using an upper case letter O instead of a number zero.

Quiz

1) You must have basic machining practice experience in order to be able to visualize cutting tool motions.

True False

2) A CNC program is executed
a) Randomly
b) Sequentially
c) Right-to-left
d) None of the above

3) A decimal point must be included in all CNC words.

True False

4) A modal word
a) is not allowed
b) Is automatically instated at power up
c) remains in effect until changed or cancelled
d) only takes effect in the command in which it is included

4) An initialized word
a) is not allowed
b) Is automatically instated at power up
c) remains in effect until changed or cancelled
d) only takes effect in the command in which it is included

Talk with experienced people in your company to learn more: What kinds of machining operations does your company perform on its CNC turning centers? Ask to watch the execution of existing programs to see the motions for these machining operations.

Answers: 1: True, 2: b, 3: False, 4: c, 5: b

STOP! Do practice exercise number four in the workbook.

Lesson 5

Program Zero And The Rectangular Coordinate System

A programmer must be able to specify positions through which cutting tools will move as they machine a workpiece. The easiest way to do this is to specify each position relative to a common origin point called program zero.

You know that turning centers have two linear axes – X and Z. You also know these axes move and that they have a polarity (plus versus minus). In order to machine a workpiece in the desired manner, each axis must, of course, be moved in a controlled manner. One of the ways you must be able to control each axis is with precise *positioning control.*

In the early days of NC (before CNC, over forty years ago), a programmer was required to specify drive motor rotation in order to cause axis motion. This meant they had to know how many rotations of an axis drive motor equated to the desired amount of linear motion for the moving component of the axis (turret). As you can imagine, this was extremely difficult – it was not logical. There is no relationship between drive motor rotation and motion of the moving component. Today, thanks to program zero and the rectangular coordinate system, specifying positions through which cutting tools will move is much easier.

The rectangular coordinate system has an origin point that we'll be calling *program zero.* It allows you to specify all positions (we'll be calling *coordinates*) from this central location. As a programmer, *you* will be choosing the location for program zero – and if you choose it wisely, many of the coordinates you will use in the program will come directly from your workpiece drawing, meaning the number and difficulty of calculations required for your program can be reduced.

Graph analogy

We use a simple graph to help you understand the rectangular coordinate system as it applies to CNC. Since everyone has had to interpret a graph at one time or another, we should be able to easily relate what you already know to CNC coordinates. Figure 1.21 is a graph showing a company's productivity for last year.

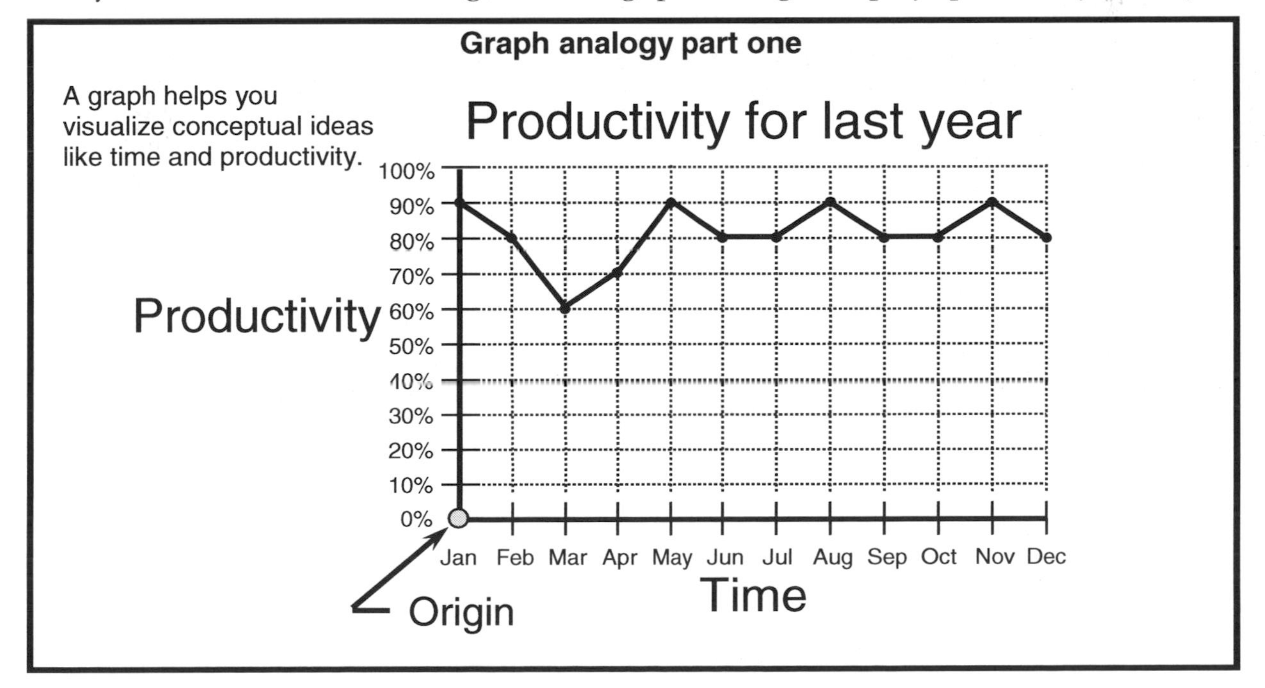

Figure 1.21 Graph example used to illustrate rectangular coordinate system

In Figure 1.21, the *horizontal base line* represents time. The *increment* of the time baseline is specified in months. One whole year is the *range*, given from January through December. The *vertical base line* represents productivity. The *increment* for this base line is specified in ten percent intervals and ranges from 0% to 100% productivity.

In order for a person to make this graph, they must have the productivity data for last year. They will plot a point along the vertical line corresponding to January (the vertical base line in this case) and along the horizontal line corresponding to the percentage of productivity (90% in our case). This plotting of points will be repeated for every month of the year. Once all of the points are plotted, a line or curve can be passed through each point to show anyone at a glance how the company did last year.

A graph is amazingly similar to the rectangular coordinate system used with CNC. Look at Figure 1.22.

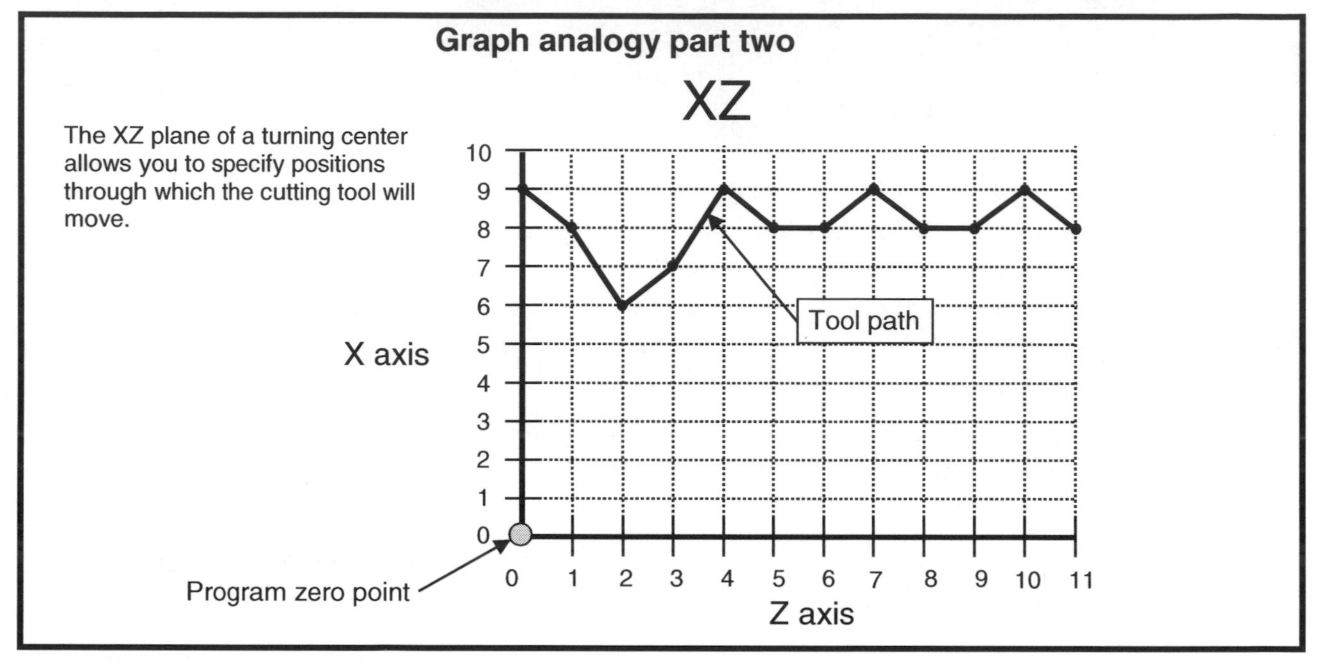

Figure 1.22 – The coordinate system of a turning center (XZ plane)

For CNC turning centers, the *horizontal base line* represents the Z axis. The *vertical base line* represents the X axis. The *increment* of each base line is given in linear measurement. If you work in the inch mode (which we use throughout this text), each increment is given in inches. The smallest increment is 0.0001 inch, meaning each axis has a very fine grid. If you work in the metric mode, each increment will be in millimeters. In the metric mode, the smallest increment is 0.001 mm. The *range* for each axis is the amount of travel in the axis (from one over-travel limit to the other).

A metric advantage – Again, since most people in the US are accustomed to the inch mode, we use it for all examples in this text. However, you should know that there is an accuracy advantage with the metric mode. The advantage has to do with the least input increment – or resolution – for each axis. As stated, in the metric mode the least input increment is 0.001 mm, which is less than half of 0.0001 inch (0.001 mm is actually 0.000039 in). Think of it this way: A ten inch long linear axis has 100,000 programmable positions in the inch mode. The same ten inch long linear axis has 254,000 programmable positions in the metric mode!

More about polarity

In the graph example shown in Figure 1.21, notice that all points are plotted after January and above 0% productivity. The area up and to the right of the two base lines is called a *quadrant*. This particular quadrant is quadrant number one. The person creating the productivity graph intentionally planned for coordinates to fall in quadrant number one in order to make it easy to read the graph.

From Lesson One, you know that each machine axis has a *polarity*. As the turret (and cutting tool) moves toward the workholding device in the Z axis, it is moving in the Z minus direction. As it moves toward the spindle center in the X axis, it is moving in the X minus direction (for almost all turning centers).

The rectangular coordinate system makes determining the polarity of coordinates used in a program *very* simple. From a programming standpoint, you will specify polarity for a given coordinate by determining whether a coordinate position is above or below program zero in X – or to the left or right of program zero in Z. Figure 1.23 shows how.

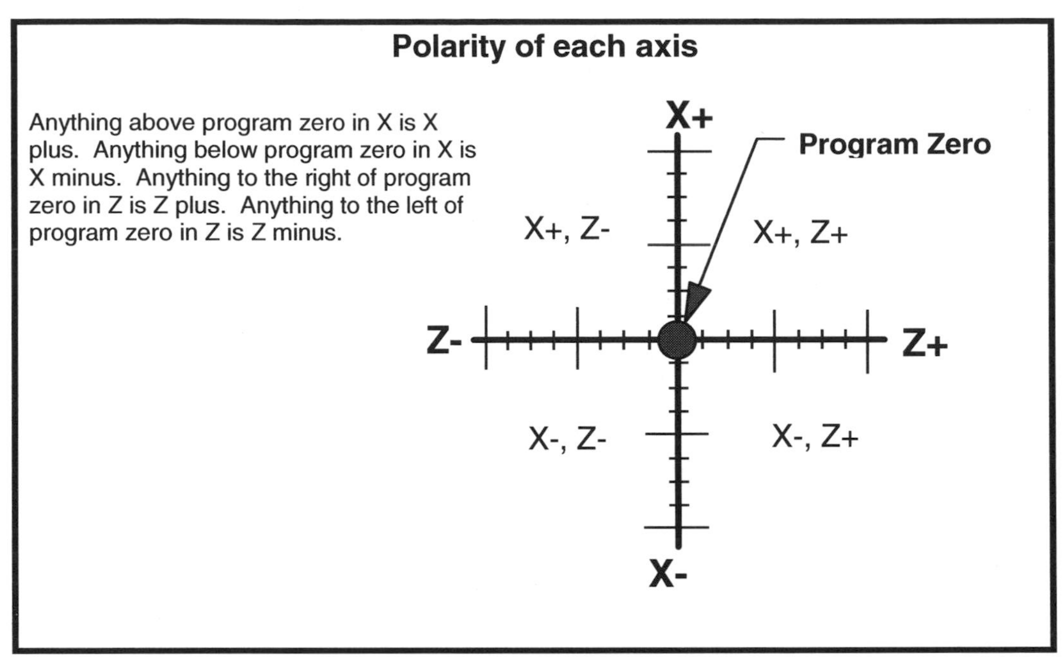

Figure 1.23 – How you determine polarity for coordinates in your program

It just so happens that all CNC machines will *assume* that a coordinate is positive (plus) unless a minus sign is specified. The CNC word: **X2.0**, for example, specifies a position along the X axis of *positive* two inches. Some controls will actually generate an alarm if the plus sign is included within the word, meaning you must let the control assume positive values. Only include a polarity sign if it is negative (-).

Remember an X coordinate specifies a diameter position. A diameter of 1.5" will be specified as **X1.5**.

Wisely choosing the program zero point location

As the programmer, *you* determine the program zero point location for every program you write. Frankly speaking, program zero could be placed in just about *any* location. As long as the coordinates used in your program are specified from the program zero point, the program will function properly. Though this is the case, the *wise* selection of the program zero point will make programming much easier. It may also make it easier for the setup person.

In X

Always place the X axis program zero point at the spindle/workpiece center in the X axis. This will allow you to specify all X coordinates as diameter positions.

In Z

For the most part, we recommend placing the Z axis program zero point at the end of the workpiece from which the workpiece is dimensioned (there is an exception to this recommendation that we will discuss in Lesson Six). Since most turned workpieces are dimensioned from the end you'll be machining, this means program zero in Z will be the right end of the finished workpiece.

Figure 1.24 shows an example of program zero placement based upon these recommendations.

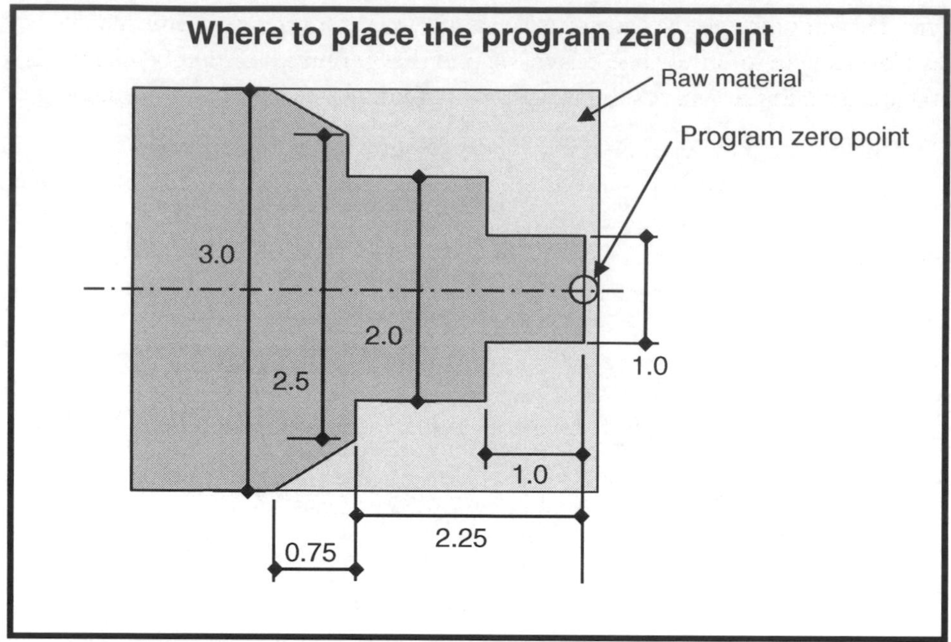

Figure 1.24 – An example of program zero placement

With this workpiece, notice that the Z related dimensions begin from the right side of the finished workpiece – so this is where we've placed the program zero point in Z. Since most coordinates used in the program will be to the left of the Z axis program zero point, the polarity for most Z coordinates will be negative.

For the X axis, again, always place the program zero point in the center of the workpiece. Based upon this placement of program zero in X and Z, Figure 1.25 shows the coordinates that will be needed to finish turn this workpiece (we're not showing any coordinates needed for rough turning).

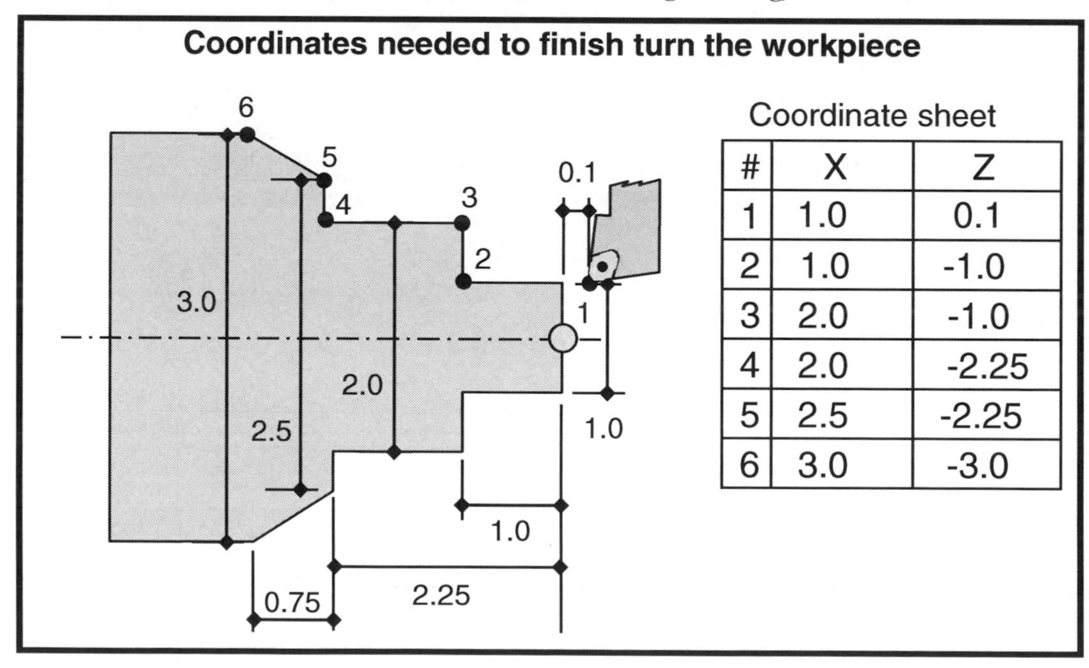

#	X	Z
1	1.0	0.1
2	1.0	-1.0
3	2.0	-1.0
4	2.0	-2.25
5	2.5	-2.25
6	3.0	-3.0

Figure 1.25 – Coordinates needed for finish turning

Notice that all coordinates for points one through six on the coordinate sheet are specified *from the program zero point*. For the Z coordinates, since points 2 through 6 are to the left of the program zero point, they are negative values. When program zero is placed at the right end of the finished workpiece in Z, the only time you'll have a positive Z coordinate (to the right of program zero) is when you specify an approach position (like point number one in Figure 1.25).

With program zero placed at the center of the spindle almost all X coordinates will be positive – and again, the machine assumes a coordinate to be positive unless the minus sign is used. The only time you'll need a negative X coordinate is when the tool passes the spindle centerline – as is required when facing a workpiece to center. Figure 1.26 shows why.

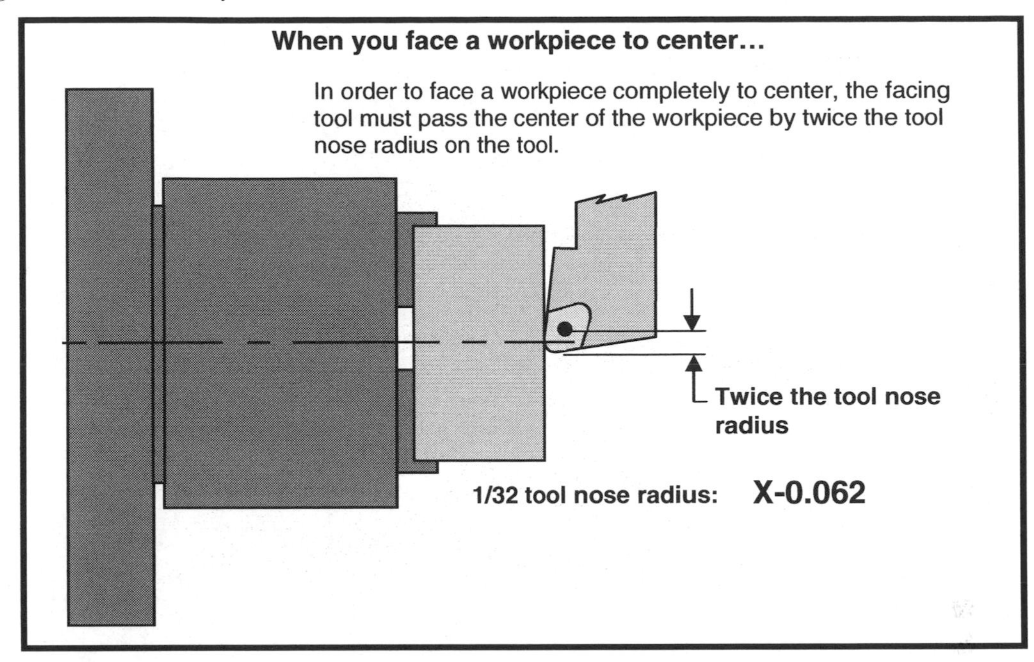

Figure 1.26 – Facing a workpiece completely to center requires a negative X coordinate

You will be programming the very tip of each cutting tool (in both axes). For this reason, you must send the extreme tip of a facing tool past center by twice the tool nose radius to ensure that the face gets completely machined. At a position of **X0**, there will still be material to be machined (in the form of a small point).

Once again, all X coordinates are specified in diameter. With the drawing shown in Figure 1.25, all X axis-related dimensions happen to be given directly in diameter, making it very easy to determine X coordinates needed in the program. It will not always be quite this simple. Consider, for example, the drawing shown in Figure 1.27.

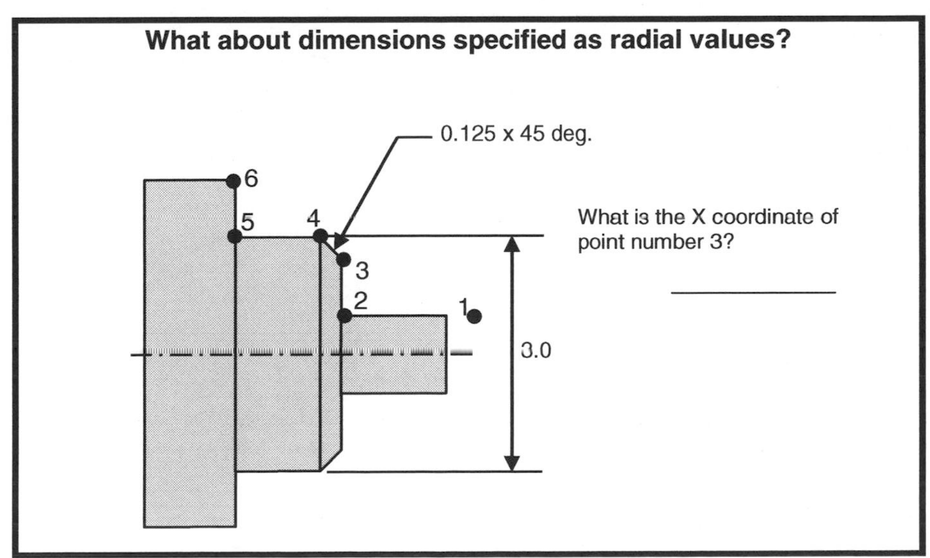

Figure 1.27 – Drawing that has an X axis dimension specified as radial value

Notice that the design engineer has specified a chamfer on the middle diameter of this workpiece. The chamfer size is 0.125 (1/8") by 45 degrees. This means the distance from point 3 to point 4 along the X axis (on the side) is 0.125. A chamfer (or radius) is a *radial dimension.* The X coordinate for point 3, of course, must be the *diameter* of the workpiece at point number 3. To calculate this diameter, you must double the chamfer size (two times 0.125 is 0.25) and subtract the result from the 3.0 inch diameter. The X coordinate for point 3 is 2.75.

A common beginner's mistake is not to double the chamfer or radius size before adding or subtracting it from the closest diameter. For point 3 in Figure 1.27, they will incorrectly specify point number 3 as 2.875 instead of 2.75. This will result in the angle of the chamfer being considerably less than 45 degrees.

Figure 1.28 shows another example, this time for a fillet radius.

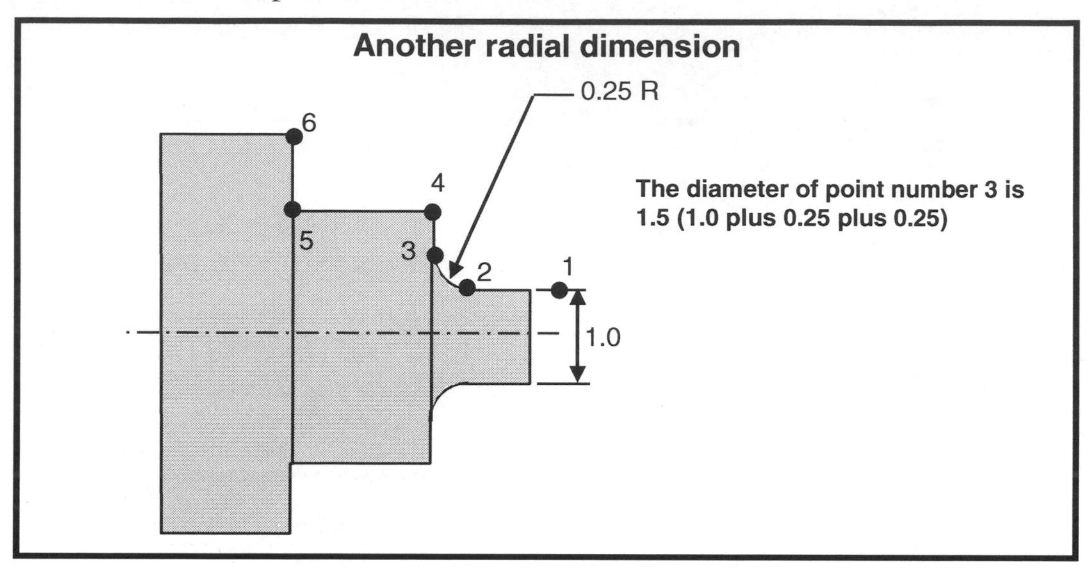

Figure 1.28 – another example of a radial dimension

The diameter of point 3 (X1.5) is determined by doubling the 0.25 radius (a radial dimension on the drawing) and adding the result to the 1.0 diameter.

What about tolerances?

Almost all dimensions shown in this text do not specify a tolerance. But as you know – design engineers specify a tolerance for *every* dimension on a blueprint.

Which value do you program?
Specify the mean value for every dimension (coordinate) you use in your program. The mean value is, of course, right in the middle of the tolerance band. With some tolerances, it is more difficult to determine the mean value than with others.

Plus/minus tolerance: 4.0 +/- 0.002
The mean value is specified right in the dimension. 4.0 is the mean value of the dimension above.

Uneven tolerance: 4.0 +0.003, -0.001
The mean value must be calculated. Divide the overall tolerance (0.004 above) by two (0.002) and subtract it from the high limit (4.003 is the high limit). The mean value of the dimension is 4.001.

High and low limit specified: 4.004 / 4.001
The mean value must be calculated. Again, divide the overall tolerance (0.003 above) by two (0.0015) and subtract it from the high limit (4.004 is the high limit). The mean value of the dimension is 4.0025.

Practice calculating coordinates

Instructions: Mark the location of program zero based upon the way this workpiece is drawn. Then fill in the coordinate sheet below.

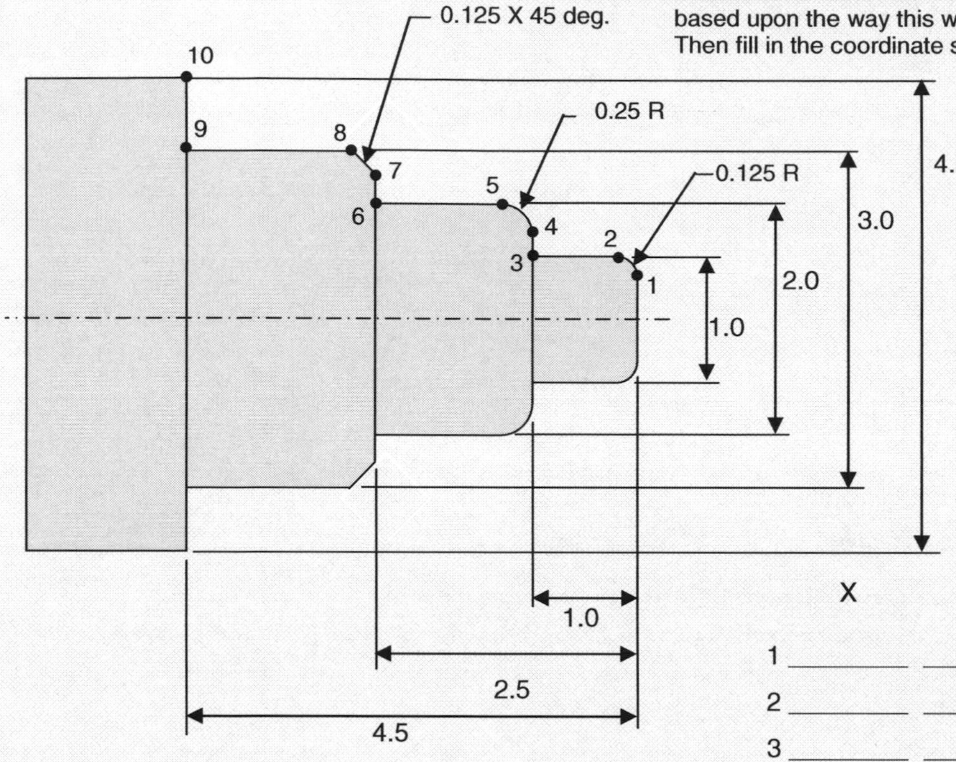
0.125 X 45 deg.
0.25 R
0.125 R
10
9
8
7
6
5
4
3
2
1
4.5
3.0
2.0
1.0
1.0
2.5
4.5

	X	Z
1	________	________
2	________	________
3	________	________
4	________	________
5	________	________
6	________	________
7	________	________
8	________	________
9	________	________
10	________	________

Answers:

Program zero is at the center of the workpiece in X and at the right end of the finished workpiece in Z.

1: X0.75 Z0
2: X1.0 Z-0.125
3: X1.0 Z-1.0
4: X1.5 Z-1.0
5: X2.0 Z-1.25
6: X2.0 Z-2.5
7: X2.75 Z-2.5
8: X3.0 Z-2.625
9: X3.0 Z-4.5
10: X4.5 Z-4.5

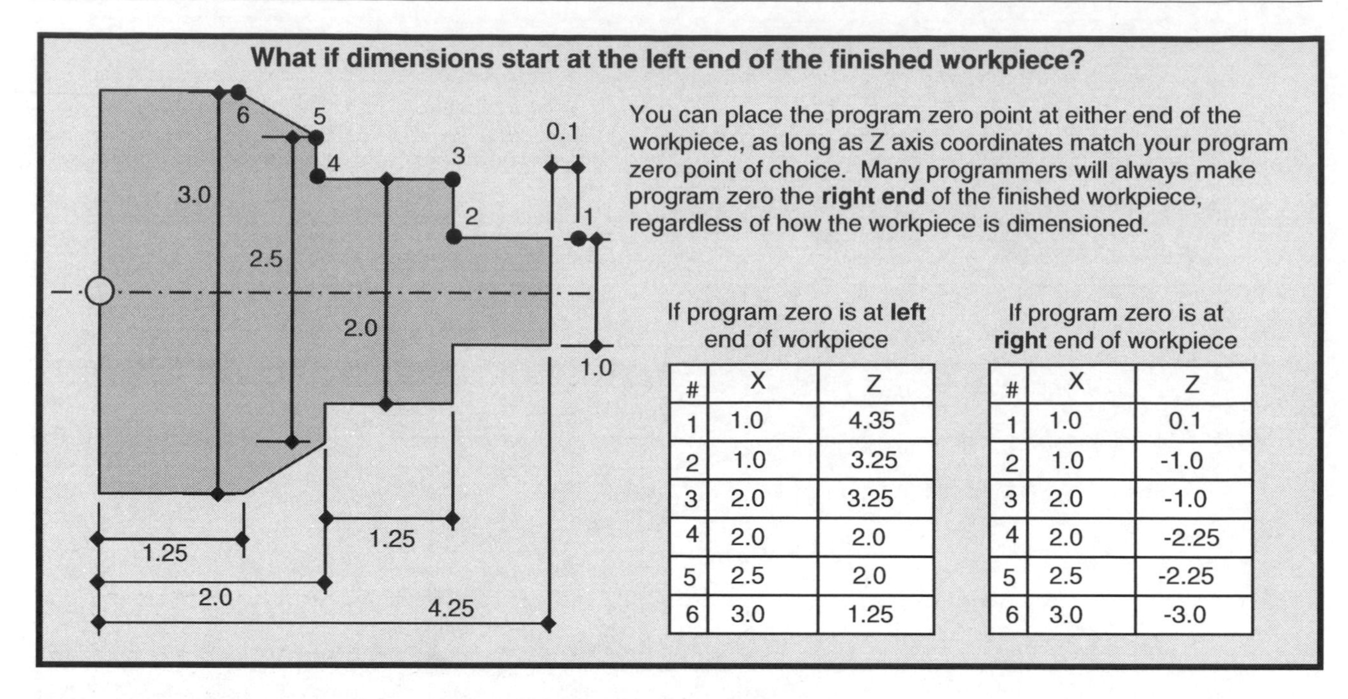

You can place the program zero point at either end of the workpiece, as long as Z axis coordinates match your program zero point of choice. Many programmers will always make program zero the **right end** of the finished workpiece, regardless of how the workpiece is dimensioned.

If program zero is at **left** end of workpiece

#	X	Z
1	1.0	4.35
2	1.0	3.25
3	2.0	3.25
4	2.0	2.0
5	2.5	2.0
6	3.0	1.25

If program zero is at **right** end of workpiece

#	X	Z
1	1.0	0.1
2	1.0	-1.0
3	2.0	-1.0
4	2.0	-2.25
5	2.5	-2.25
6	3.0	-3.0

Absolute versus incremental positioning movements

Though we have not actually said so yet, when you specify coordinates from the program zero point (as we have just introduced), it is called *absolute positioning.* The letter addresses X and Z (for most turning centers) is used to specify absolute positioning. So when you specify a position as

X1.0 Z0.1

the machine will know that the point of reference for this position is the program zero point.

All positions we've shown in this lesson so far has been with absolute positioning. There is, however, another way to specify positioning. It is called *incremental positioning.*

With incremental positioning, the point of reference for the movement is the cutting tool's current position. You specify the amount of motion you want in an incremental positioning movement. While at first this may sound easy, it actually complicates positioning movements – especially in the X axis.

With most machines, two different letter addresses are used to specify incremental positioning. The letter address U is used to specify incremental motions in the X axis. Since the X axis is specified in diameter, letter address U specifies a diameter increase or decrease from the cutting tool's current diameter. The letter address W is used to specify incremental movements in the Z axis. For example,

U1.0 W-1.0

will cause the cutting tool to increase in diameter by 1.0 inch (an actual movement of 0.5 inch along the X axis) and move in the negative Z direction by 1.0 inch.

Frankly speaking, it is quite difficult to follow a series of motion commands that are specified with incremental positioning. And by the way, if a mistake is made in a given incremental positioning command, all movements from the point of the mistake will also be incorrect. If the same mistake is made with absolute positioning, only one command will be incorrect.

We urge beginners to work exclusively with absolute positioning. While we'll show one good application for incremental positioning much later (in Lesson Twenty-One), almost all examples shown in this text will use absolute positioning.

Another way to specify absolute and incremental positioning

Some CNC controls (especially those placed on machines made in the United States) use two G codes to specify the positioning method. For these machines, the absolute positioning mode is specified by a G90 word. Once a G90 is specified, all coordinates are taken to be from the program zero point. G91 is used to specify incremental positioning mode. Once a G91 is specified, all coordinates will be taken from the cutting tool's current position.

When these two G codes are used to specify the positioning method, the letter addresses X and Z are used for both positioning modes. Consider these example commands:

> G90 X1.0 Z0.1 (Move to a 1.0 diameter and a position 0.1 from program zero in Z – point of reference is program zero)
>
> G91 X1.0 Z0.1 (Move 1.0 larger in diameter in X [actual move will be 0.5] and plus in the Z axis by 0.1 inch – point of reference is the cutting tool's current position)

Any series of motions can be commanded in either the absolute or incremental mode. Look at Figure 1.29.

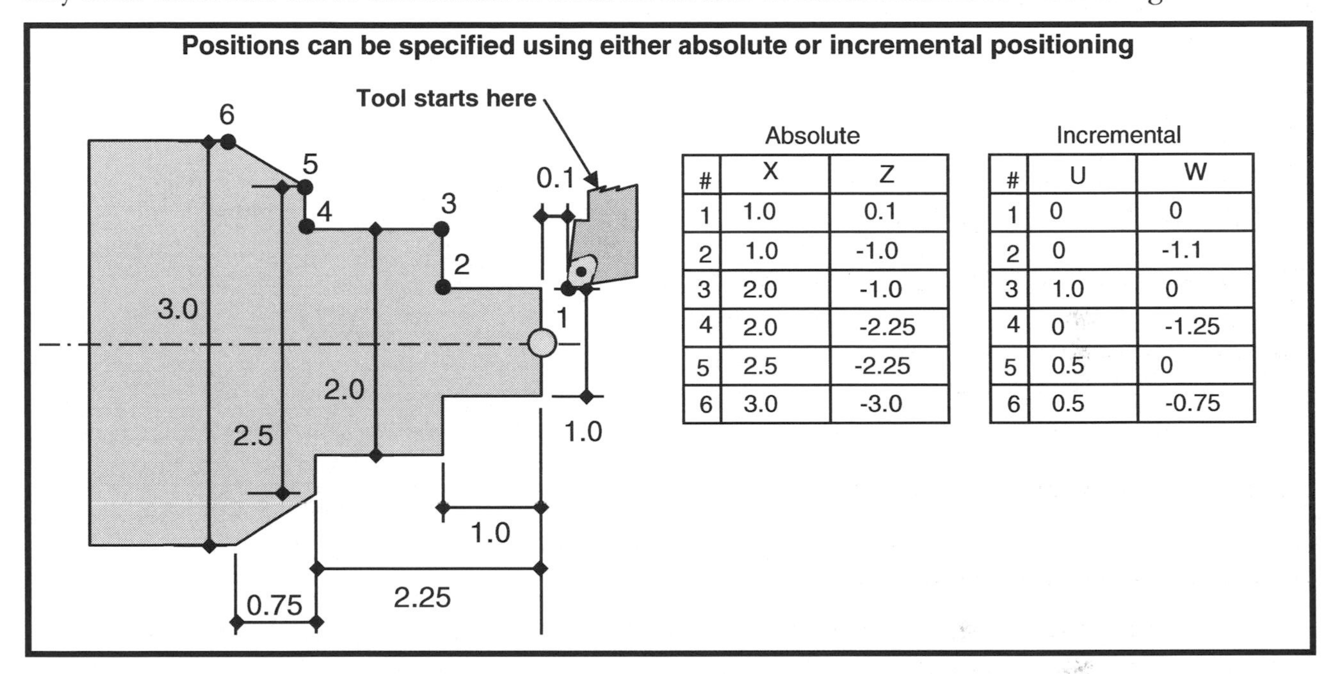

Absolute

#	X	Z
1	1.0	0.1
2	1.0	-1.0
3	2.0	-1.0
4	2.0	-2.25
5	2.5	-2.25
6	3.0	-3.0

Incremental

#	U	W
1	0	0
2	0	-1.1
3	1.0	0
4	0	-1.25
5	0.5	0
6	0.5	-0.75

Figure 1.29 – Movements can be specified using either absolute or incremental positioning

As you can see, absolute positioning makes more sense. Coordinates often match print dimensions – but even when they don't – the point of reference for each position is the same – the program zero point. Incremental positioning doesn't make much sense. Positions are nothing more than a whole series of disjointed movements, each taken from the tool's previous position. And in the X axis, many dimensions must be doubled since you must specify a diameter increase or decrease.

A decimal point reminder

As stated in Lesson Four, coordinates (X and Z) require real numbers. That is, values that often include a portion of a whole number. You must remember to include a decimal point with coordinates, even when you are specifying a whole number value. For example, if you want to specify an X coordinate of three inches, you must specify

> X3.0

not

> X3

If you specify "X3", the control will use the fixed format for the X word. It will automatically place the decimal point in a position four places to the left of the right-most digit (in inch mode). The value "X3" will be taken as X0.0003, and not X3.0.

We repeat our suggestion about how to specify any real number value:
When programming whole numbers in CNC words that allow a decimal point (real number values), be sure to carry the value out to the first zero after the decimal point. For example, use

X4.0

Instead of

X4.

This will force you to write/type the decimal point. In similar fashion, when specifying values under one, begin the value with the zero to the left of the decimal point. For example, use

X0.375

instead of

X.375

Again, this will force you to write/type the decimal point.

Most programmers prefer to use decimal point programming for obvious reasons. It doesn't make much sense to program using the fixed format. But be ready for some computer aided manufacturing (CAM) systems that automatically output G code level CNC programs in fixed format – without the decimal point. They can be a little tough to interpret.

Key points for Lesson Five:

- You must be able to specify positions (coordinates) within CNC programs.
- You know that the rectangular coordinate system of a CNC machine is very similar to that used for a graph.
- In CNC terms, the origin for a coordinate system is called the program zero point.
- From a programmer's viewpoint, polarity for each positioning movement is based upon the commanded position's relationship to program zero.
- The program zero point location is determined based upon how the workpiece drawing is dimensioned.
- For the X axis, coordinates must be specified in diameter for most turning centers.
- When you specify coordinates relative to program zero, you're working with absolute positioning.

STOP! Do practice exercise number five in the workbook.

Lesson 6

Determining Program Zero Assignment Values

The programmer chooses the program zero point location. But the CNC turning center must also be told where program zero is located – it must be able to move cutting tools as instructed by the program. This means the setup person will probably be quite involved with program zero assignment.

You know from Lesson Five that the program zero point is the origin for your program. All coordinates specified in your program are taken from program zero (when using absolute positioning). You also know that the program zero point is determined based upon how the print is dimensioned. The program zero point is always placed at the center of the workpiece in the X axis (since X-related dimensions are specified on the print in *diameter*) and usually at the right end of the finished workpiece in the Z axis.

You must understand that just because you want the program zero point to be in a certain location, doesn't mean the CNC turning center is automatically going to know where it is. A conscious effort must be made to *assign* program zero. Program zero assignment is the task of telling the CNC turning center where the program zero point is located for each cutting tool. You can *think of this task as marrying your program to the workholding setup* that is made on the machine.

Much of what we present in Lessons Six and Seven is more related to setup than it is to programming. However, a CNC programmer must be able to *instruct and direct* setup people, providing instructions related to how a given setup must be made. This means they must understand as much about setups as setup people – this includes an understanding of how program zero is assigned.

The programmer must, of course, tell the setup person where the program zero point is located. This is commonly done in the setup documentation – on a *setup sheet* that provides directions for making the setup. For the purpose of this text, we'll say that the program zero point will always be placed at the center of the workpiece in the X axis and the right end of the finished workpiece in the Z axis – as we recommend in Lesson Five.

Program zero must be assigned independently for *each* cutting tool

A wide variety of cutting tools can be used on CNC turning centers. Figure 1.30 shows some of the most common cutting tools used in the normal turning mode (not considering live tools).

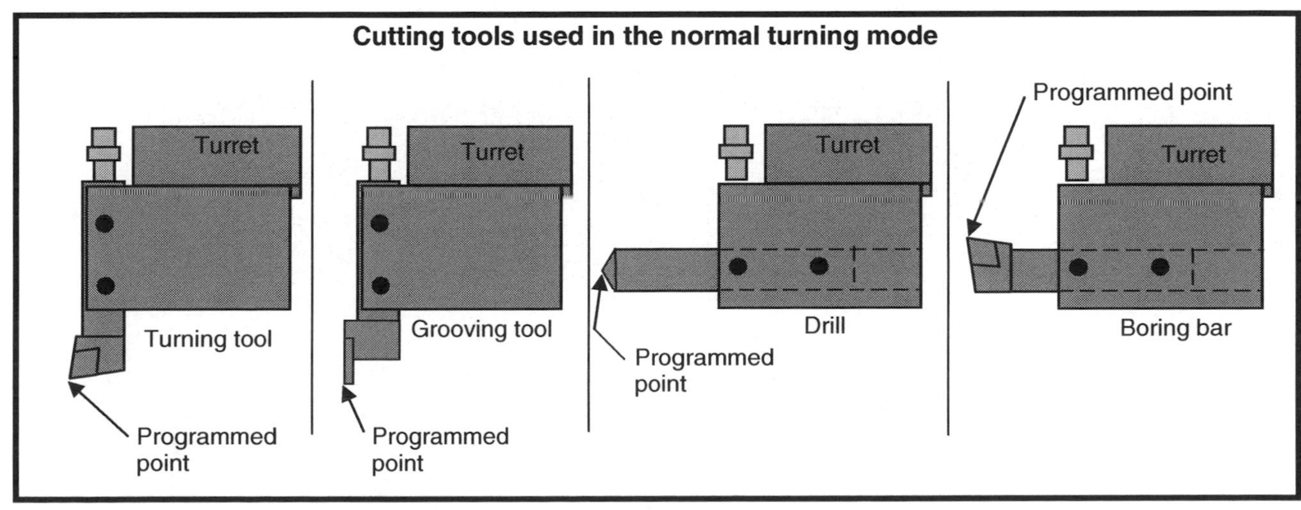

Figure 1.30 – Common cutting tools used in the normal turning mode

Additionally, machines that are equipped with live tooling can also use rotating tools. Figure 1.31 shows two end mills – one in an X axis tool holder and the other in a Z axis tool holder.

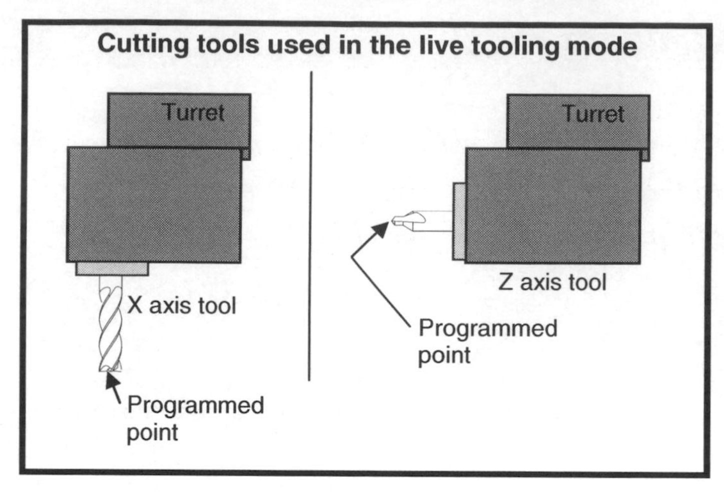

Figure 1.31 – Cutting tools used in the live tooling mode

As you look at Figures 1.30 and 1.31, notice that the point (or tip) of each cutting tool is in a different position. A program zero assignment for a turning tool will not be correct for a boring bar (or any other tool shown). Figure 1.32 more clearly illustrates this be super-imposing two cutting tools in the same turret station.

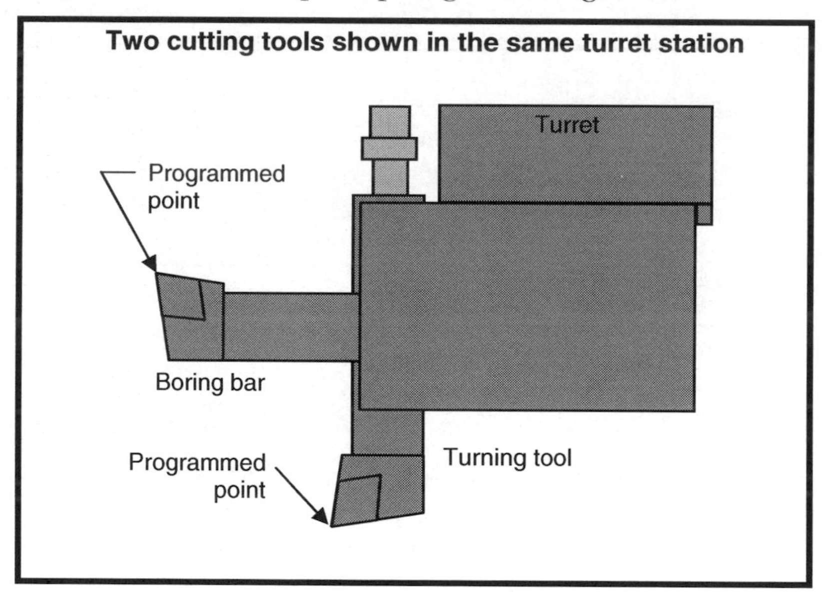

Figure 1.32 – The tool tip is in a different position for each style of cutting tool

Again, the programmed point for a turning tool is in a different location than it is for a boring bar – or a drill – or a grooving tool – or a threading tool – or any live tool. Yet the turning center must be able to correctly position any of these cutting tools to the coordinates you specify in a program – and of course these coordinates are specified relative to a central program zero point. This means the machine must be told – by one means or another – where each cutting tool is placed in the machine relative to the program zero point. *Each cutting tool will require its own program zero assignment.*

Even for very similar cutting tools a separate program zero assignment is necessary. Consider two identical turning tools being used in a job. One is being used for rough turning and the other is being used for finish turning. If the same program zero assignment is used for both tools, even a tiny deviation in the position of the programmed point (especially for the finishing tool) will cause the cutting tool to incorrectly machine the workpiece – probably resulting in a scrap workpiece. We'll discuss this in greater detail a little later. For now, the main point is: *each cutting tool requires a separate program zero assignment.*

Understanding program zero assignment values

Again, a CNC turning center must be told where each cutting tool is located relative to program zero. For the most part, *program zero assignment values specify the distances between the tool tip and the program zero point in each axis.* These distances are determined when the machine is positioned at the *zero return position* in each axis.

What is the zero return position?

The zero return position (also called the *home position* and *machine zero*), is a precise and consistent reference position in each axis. Many turning center manufacturers place the zero return position close to the plus over-travel limits in each axis. (An exception to this statement is with long-bed turning centers – with these machines the Z axis zero return position may be placed near the center of the Z axis.) Figure 1.33 shows an example.

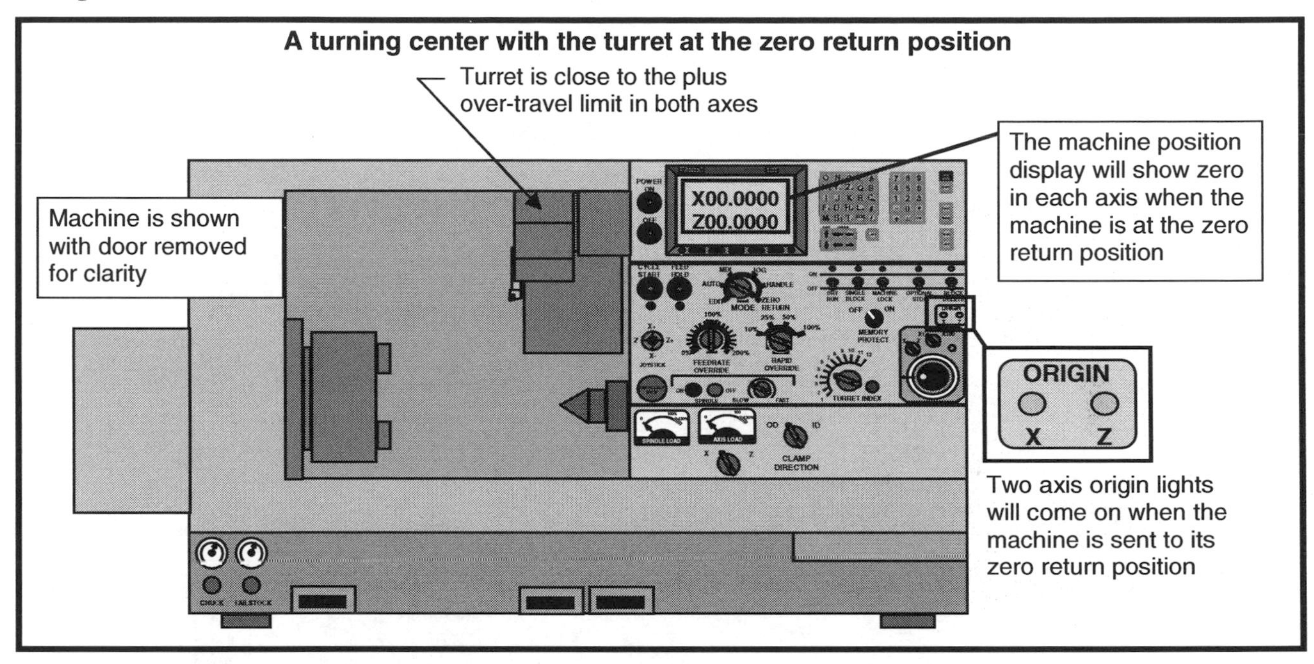

Figure 1.33 – A turning center with the turret resting at the zero return position

The actual procedure used to send each axis to its zero return position is shown in Lesson Twenty-Seven. For now, we simply want to make it clear that the zero return position is the point-of-reference for program zero assignment values.

Program zero assignment values

The X axis program zero assignment value is the diameter a cutting tool will machine while the turret (and selected cutting tool) is at the X axis zero return position. This is true for all cutting tools and regardless of which method is used to assign program zero.

The Z axis program zero assignment value will vary based upon the method of program zero assignment being used. With two of the methods we show (not our favorite methods), the Z axis program zero assignment value is the distance between the tool tip (with the turret at the Z axis zero return position) and the program zero point in the Z axis.

Figure 1.34 shows the program zero assignment values for this method of program zero assignment.

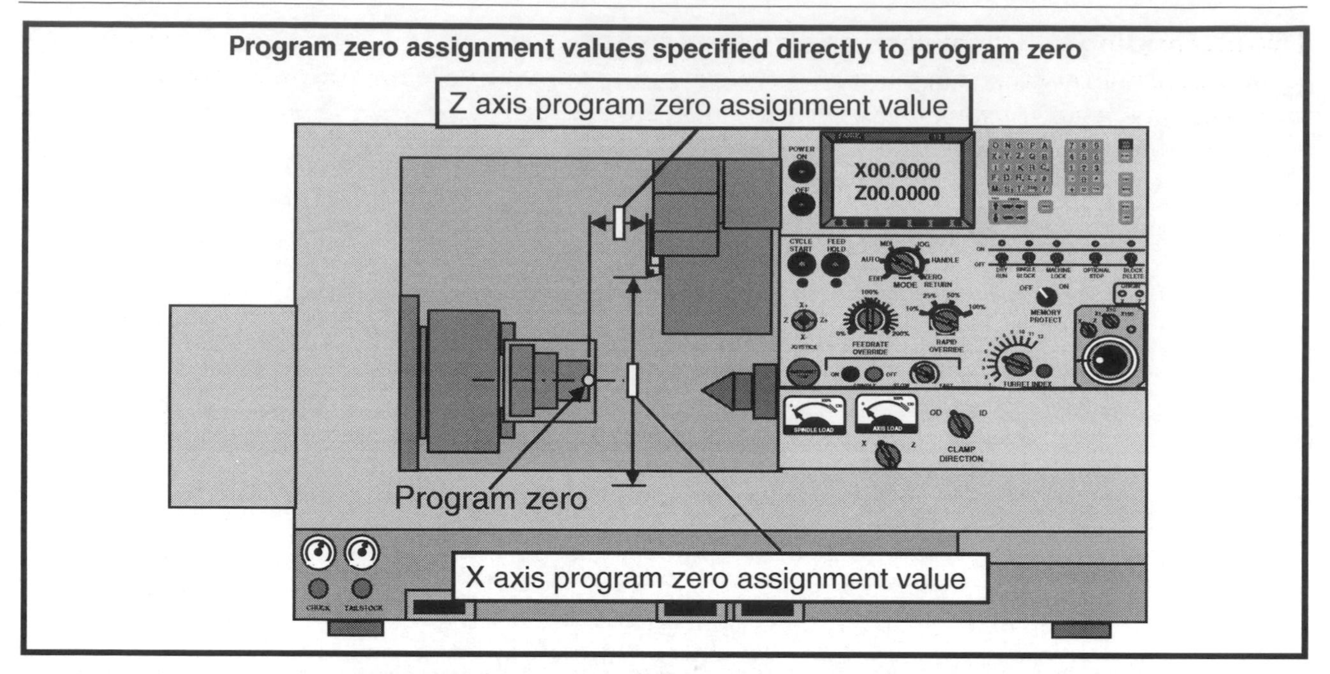

Figure 1.34 – Program zero assignment values specified directly to the program zero point in each axis

With the two other methods (our preferred methods), the Z axis program zero assignment value for each tool is the distance between the tool tip (while the turret is at the Z axis zero return position) and the *chuck face*. A special work shift value tells the machine how far it is from the chuck face to the program zero point. Figure 1.35 shows this.

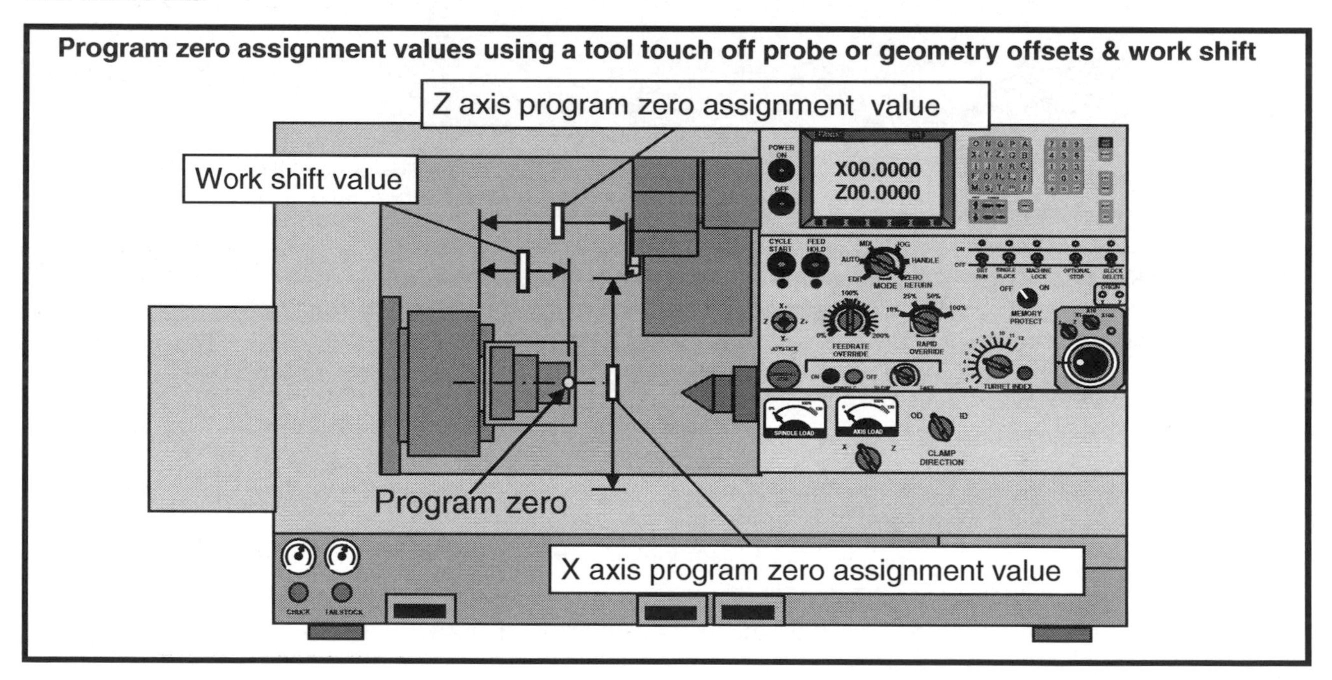

Figure 1.35 – Z axis program zero assignment values are specified to the chuck face and a work shift value is used.

We prefer the method shown in Figure 1.35 because it minimizes work from setup to setup. Any tool remaining in the turret from one job to the next will require but one program zero assignment – when it is used for the first time. That is, a program zero assignment done for a tool in one job will still be correct for it in the next job – and any upcoming job. The work shift value must be changed, of course, if the workpiece length changes from job to job.

With the program zero assignment values shown in Figure 1.34, if the workpiece length changes from one job to the next (as it almost always does), the Z axis program zero assignment value for *every* tool must be determined in *every* job. As you can imagine, this results in duplication of effort.

Regardless of which method of program zero assignment is being used, program zero assignment values must be determined – and they will be entered into the control during setup – into the program or into something called *geometry offsets.*

How do you determine program zero assignment values?

In Lesson Seven, we show the four most popular ways to assign program zero on turning centers. These methods vary based upon machine age and whether or not the turning center has a tool touch off probe. Every method of program zero assignment requires program zero assignment values to be determined. The actual techniques used to determine program zero assignment values also vary based upon the method used, and whether or not your turning center has a feature called *measure function.*

The **measure function** is an optional feature for turning centers that have *geometry offsets.* Most machine tool builders include it with their package of standard features – so if your machine has geometry offsets, it is likely to have the measure function. This function facilitates the calculation and entry of program zero assignment values – and is discussed during Lesson Seven.

Figure 1.36 shows one way to determine program zero assignment values. Frankly speaking, this method is rather crude. It is only used for older machines that *do not* have geometry offsets – when program zero must be assigned in the program. With it, we're simply using the turning center as a measuring device – and we're using the *relative position display screen* to help measure each program zero assignment value.

Again, this method is rather crude – and current model turning centers don't require this method. But it should nicely show what program zero assignment values represent. Note that this procedure can be easily modified for use with work shift. Instead of facing a workpiece to Z0 prior to setting the Z axis relative position display to zero (steps five and six in Figure 1.36), the tool tip can be brought flush with the face of the chuck.

Other (easier) methods for determining program zero assignment values are shown in Lesson Seven.

Must you actually measure program zero assignment values at the machine?

In Lesson Seven, we present the four most popular ways to assign program zero on turning centers. In order from most preferred to least preferred, they are:

- Using a tool touch off probe
- Using geometry offsets with work shift
- Using geometry offsets without work shift
- In the program with G50

Several factors, like age of the machine and available features determine which method will be used for a given machine.

Only the last method (in the program with G50) requires that you measure program zero assignment values using the techniques shown in Figure 1.36. With all other methods of assigning program zero, these techniques are simplified.

Rather than show all of the alternatives here in Lesson Six, we present them in Lesson Seven as we discuss each program zero assignment method in detail. This will allow you to concentrate on only the method/s of program zero assignment you must use.

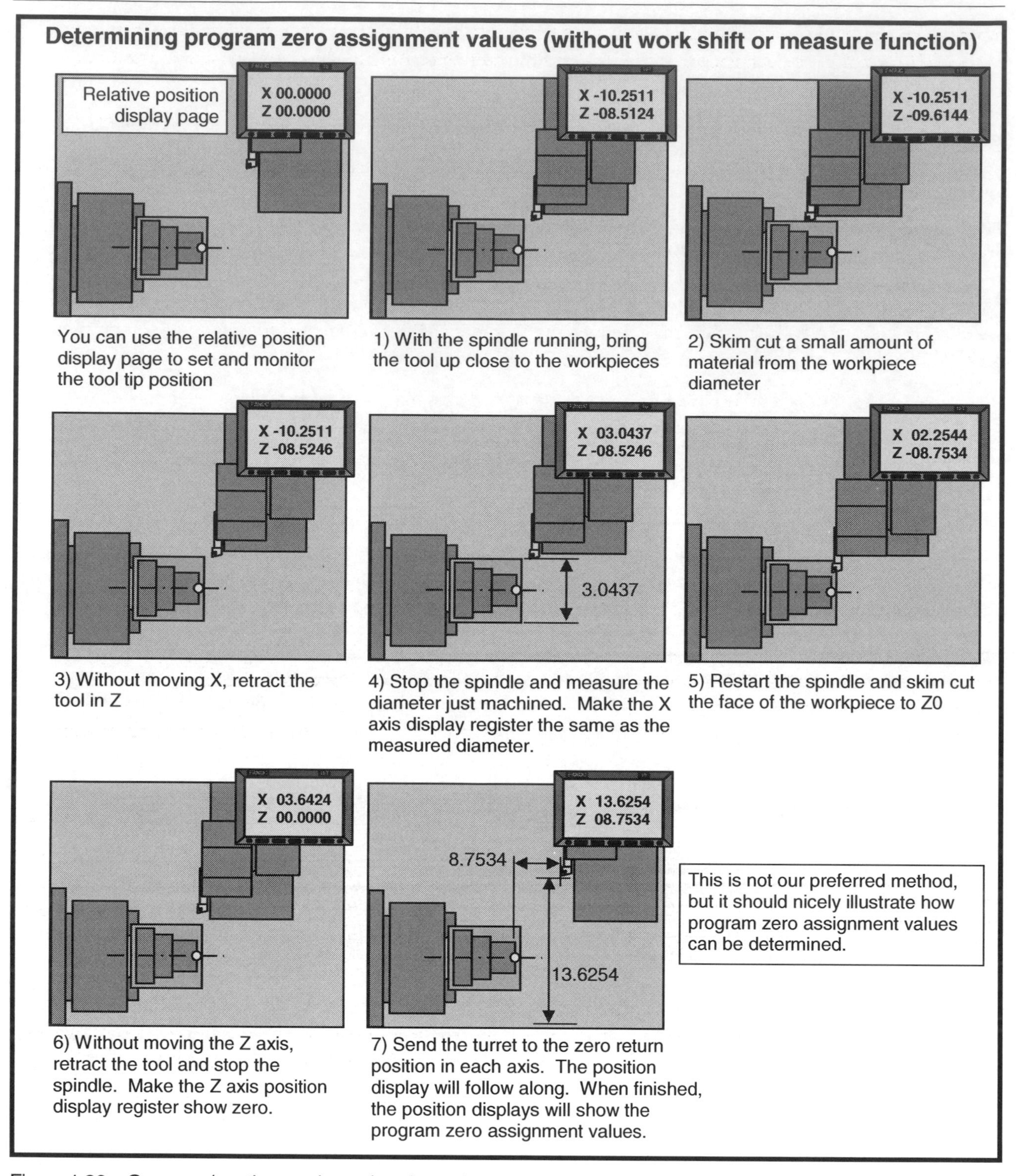

Figure 1.36 – One way (a rather crude way) to determine program zero assignment values

Other kinds of cutting tools

To this point, we've only addressed program zero assignment values for turning tools. Similar techniques can be used for just about any kind of single point tool (threading tool, grooving tool, and boring bar).

For other cutting tools that are held in a turning tool holder (like a grooving tool or threading tool), most setup people will use the machined diameter and face shown in Figure 1.36 and simply touch the tool tip to these surfaces (not actually skim cutting as is shown in Figure 1.36).

Center cutting tools, like drills and taps present a special problem for determining X axis program zero assignment values. (Note that determining Z axis program zero assignment values is the same as it is for any turning tool.) Any center cutting tool is held in an internal tool holder. This kind of holder has a round hole into which the cutting tool is placed. If the shank of the cutting tool is not the same diameter as this hole, a bushing will be used as a spacer. The X axis program zero assignment value is the diameter of the tool tip (centerline of the internal tool holder) while the turret is at the X axis zero return position, as Figure 1.37 shows.

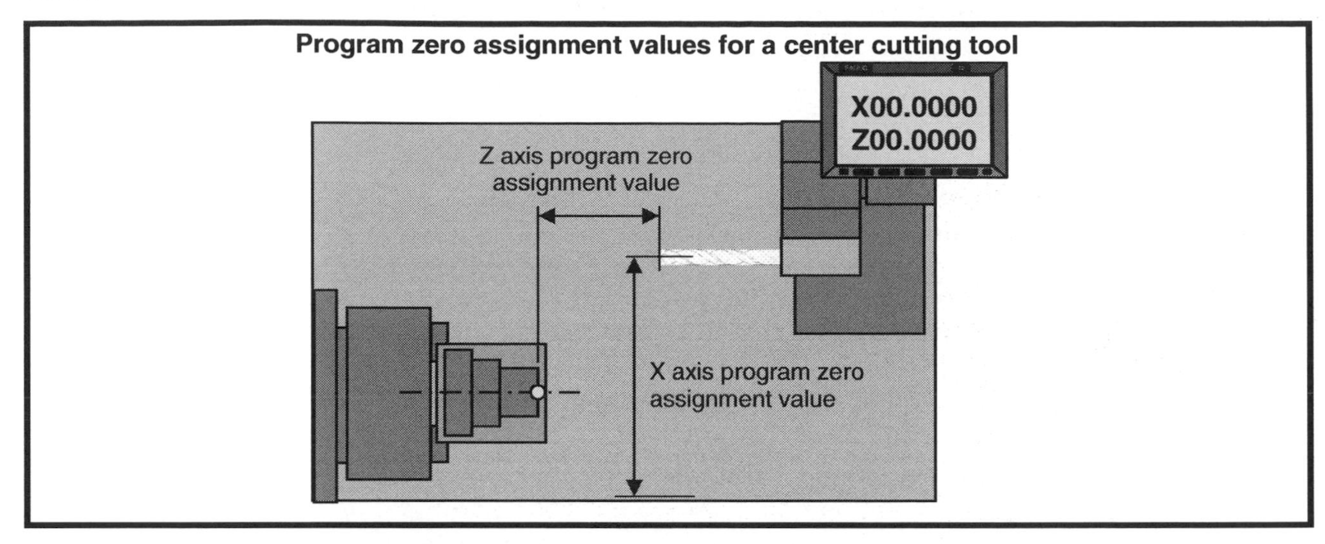

Figure 1.37 – Program zero assignment values for a center cutting tool (drill, tap, reamer, etc.)

Since the X axis program zero assignment value is based upon the centerline of the internal tool holder, it will remain the same for *any* center cutting tool. That is, every center cutting tool will have the same X axis program zero assignment value, regardless of which turret station it is in.

Because this value is so consistent, some machine tool builders actually publish it. But if your machine tool builder has not published it, rest assured that it is relatively easy to measure.

First, mount a magnetic base dial indicator on the face of the chuck. Second, place an empty internal tool holder in a turret station. Third, manually position the (empty) internal tool holder up close to the dial indicator. Fourth, using the dial indicator, adjust the X axis until the center of the internal tool holder is in line with on the spindle centerline. Fifth, make the X axis relative position display show zero. Finally, send the X axis to its zero return position. The position display will follow along. When the X axis reaches the zero return position, the X axis relative position display will show the X axis program zero assignment value for any drill, tap, reamer, or any center cutting tool. Write this value down. You'll need it often.

Boring bars present yet another problem for X axis program zero assignment. While you can use the same techniques as are used for an external tool (like a turning tool), if you are going to skim cut a diameter – like you do for a turning tool – a hole must exist in the workpiece. If you are machining a piece of tubing, this will not be a problem. But if you're working with a solid slug, it is quite inconvenient to machine a hole in a piece of raw material just so you can determine X axis program zero assignment values for boring bars.

We offer a short-cut. Most boring bar manufacturers specify the distance from the tool tip to the centerline of the boring bar in their technical documentation, as Figure 1.38 shows. Simply double this value and add it to the (previously determined) program zero assignment value for any center cutting tool.

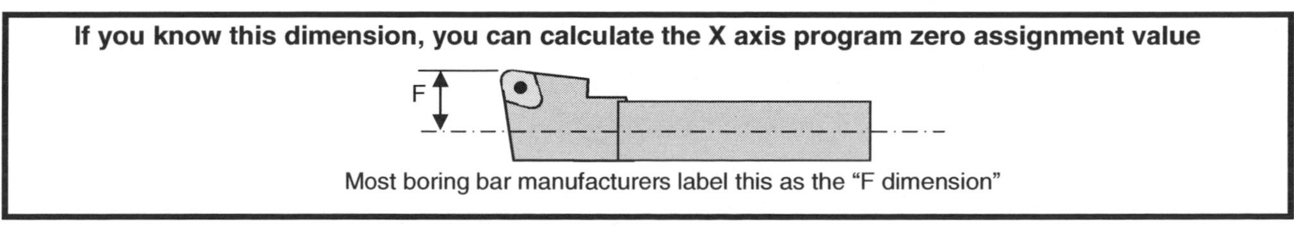

Figure 1.38 – Most boring bar manufacturers publish this value

Will you really have to determine all of these values?

Unless you're working with a very old turning center, you won't be determining program zero assignment values with the techniques shown in this lesson. However, it is quite important to know what these values represent. In Lesson Seven, we present several simplified methods.

Other considerations that affect program zero assignment values

It is difficult to relate all of the implications of program zero assignment values before you have some experience with the machine. Armed with what you know so far and what is presented in Lesson Seven, you will be able to determine program zero assignment values and get the machine up and running production. However, you may find yourself repeating program zero assignment measurements from one setup to the next, repeating trial machining operations during setup (and whenever tools are replaced during a production run), and in general, not using the machine in a very efficient manner.

Unfortunately, if this is your first exposure to program zero assignment values, you may need to get some experience before much of what we present here will make much sense. If you're becoming at all confused, skip this presentation until you gain some experience with the machine.

How accurate are the program zero assignment values?

Regardless of how you determine program zero assignment values, they are determined using *static* measurements. There will always be a small deviation between what you measure during setup and the dynamics of machining a workpiece. That is, you can *perfectly* measure program zero assignment values for a given tool, yet that workpiece surfaces machined by the tool will still deviate from programmed sizes. This deviation is caused by the dynamics of the cutting operation. Some tools will tend to push the workpiece away (referred to as *tool pressure*) while others tend to dig into the workpiece.

Say for example, you will be finish turning a 3.0 inch diameter on a medium carbon steel workpiece. You measure the X axis program zero assignment value during setup using the techniques shown in Figure 1.36. During this process, you skim cut a small amount of material from the diameter of a raw workpiece – just enough to clean up. You measure this diameter to help determine the program zero assignment value for the tool. And again, you do so *perfectly*.

When you skim cut to measure program zero assignment values, the cutting tool will not be using the same cutting conditions (speed, feedrate, and depth-of-cut) as it will during the machining of an actual workpiece. Cutting conditions during the machining of a workpiece will be much more aggressive. This difference in cutting conditions will cause increased tool pressure. The tool will tend to push the workpiece away slightly – and the 3.0 inch will probably come out slightly oversize. The harder the material and the smaller the rake angle on the tool, the greater will be the tool pressure. With medium carbon steel and a negative rake took, the 3.0 inch diameter might come out to 3.001 inch or more.

While a 0.001 inch diameter deviation may not sound like much, most CNC turning center users are holding tolerances that are smaller than 0.001 inch (0.025 mm) on a regular basis.

In this scenario, at least tool pressure will cause the work surface to have *excess stock*. Re-running the tool (after an adjustment) may make it possible to save the workpiece. (Experience machinists may be questioning this since there will be *another* difference in tool pressure when the cutting tool attempts to machine just 0.001 inch from the 3.0 inch diameter– it's likely the cutting tool will machine much more than 0.001 inch. You may still end up scrapping the workpiece.)

There are also situations when the difference in cutting conditions will cause a cutting tool to dig in and machine too much stock. Cutting tools with a large positive rake angle have this tendency, especially when the workpiece is not well supported. In the case of the 0.001 inch overall diameter tolerance, if the deviation is more than 0.001 inch in diameter, the workpiece will be scrapped.

And again, we're assuming that program zero assignment values have been perfectly measured – and that they have been perfectly entered into the machine. Mistakes, of course, will also result in deviations (machining too much or tool little material from the workpiece).

Trial machining
When you are concerned that a cutting tool's program zero assignment values are not accurate enough to machine workpiece surfaces within their tolerance bands, you can use *trial machining* (test cutting) techniques. These techniques can be used the very first time any cutting tool is used to finish machine the workpiece. Trial machining is not necessary for less critical machining operations (like any roughing operation). Trial machining is discussed in great detail during Lessons Eleven through Thirteen.

Generally speaking, trial machining involves making an adjustment prior to machining with a tool for the first time. The trial machining adjustment will cause the cutting tool to leave some additional material on all surfaces it machines (usually 0.01 to 0.02 inch [0.25 to 0.50 mm]). After the cutting tool machines under the influence of this trial machining adjustment, the setup person will stop the machine and measure the machined surfaces. Based upon the measurement, they will know how much to adjust in order to make the cutting tool machine precisely to size. After making this second adjustment, they will re-run the cutting tool.

While each CNC person will develop a feel for when trial machining is required, as you get started with a CNC turning center we recommend that you trial machine whenever a cutting tool must machine surfaces having tolerances smaller than about +/-0.005 inch. For many applications, this includes all finishing operations.

An advantage of tool touch off probes
If calibrated properly, a tool touch off-probe will eliminate the need for trial machining. The tool touch-off probing system can take into consideration the small deviation between the static measurement taken during setup and the dynamics of the cutting operation. This is a very important benefit – one that can save countless hours of setup time – time that has traditionally been required for trial machining purposes.

Tool wear
All cutting tools will wear during their lives. With some tools, like turning tools and boring bars, tool wear will cause machined surfaces to change. External surfaces (like turned diameters) will *grow* – internal diameters (like bored holes) will *shrink*. If small tolerances must be held, it is likely that the operator will have to make adjustments *during the tool's life*. We call these adjustments *sizing adjustments*.

Consider, for example, a 3.0 inch turned (external) diameter that has a plus or minus 0.0005 inch tolerance (specified as 3.0 +/-0.0005). When the setup person runs the first workpiece, say they *target* the mean value of this tolerance band (3.0) and when the job is turned over to a CNC operator to start running production, this diameter is coming out precisely to 3.0 inches in diameter.

As the finish turning tool that machines this 3.0 inch diameter continues to machine workpieces, it will eventually begin to show signs of wear. After fifty workpieces have been machined, for example, say the operator measures the 3.0 inch diameter and it comes out to 3.0002. This 0.0002 deviation is, of course, caused by tool wear – a small amount of material (0.0001 inch in this case) has worn away from the tool's cutting edge.

At this point, the tool is still machining properly, and there isn't a need for an adjustment (yet). But after fifty more workpieces have been machined, the operator measures the 3.0 inch diameter and it comes out to 3.0004. It is still within the tolerance band (it's a good workpiece), but this diameter is coming dangerously close to its high limit (anything larger than 3.0005 inch will be too big). If this trend continues, the tool will eventually begin machining this diameter too large – making scrap workpieces.

Now the operator will need to make an adjustment (we show how during Lesson Thirteen) to make this tool continue machining the 3.0 inch diameter correctly. Indeed, this kind of adjustment may have to be made several times – depending upon how small a tolerance must be held – before the tool is completely worn out and in need of replacement.

Replacing dull tools during a production run
Though many cutting tools used on turning centers are inserted tools (using carbide inserts) and it may be possible to perfectly replace them, there are three more points you must understand about cutting tools.

First, and as stated, all cutting tools will wear during the course of their lives. As tools wear, they tend to leave more material on the surfaces they machine (external diameters grow, internal diameters shrink). If small tolerances are to be held, it's likely that the CNC operator will need to make adjustments to allow for tool wear during the cutting tool's life.

When it comes time to replace a dull tool, the operator must reset the adjustment to its original value. But it can be difficult to remember just *how much* adjustment has been made during the tool's life. And of course, if no adjustment is made for the new insert, the new tool will remove too much stock when it machines for the first time (scrapping the workpieces).

Second, the skill level of the CNC operator has a lot to do with how repeatable insert replacement will be. Some operators can replace inserts in exactly the same position while others cannot (practice is required to master the skill of perfectly replacing inserts).

And third, the tolerance of the insert itself may not be sufficient to allow perfect placement. Inserts may not be perfectly the same size from one to the next. For these reasons, it may be necessary to repeat trial machining during dull tool replacement for tight tolerance work.

This is yet another advantage of a tool touch off probe. If the tool touch off probe is used after dull tool replacement (as it is during setup), and if it is calibrated to allow for the deviation discussed earlier, there will be no need to make another adjustment after replacing a dull tool.

Using cutting tools from one job to the next

If you use our preferred methods of program zero assignment (shown in Lesson Seven), you will not have to measure program zero assignment values for cutting tools used in the most recent production run. Considering how many tools stay in the turret from one job to the next, this will dramatically reduce setup time.

As you go from one production run to the next, the tools used in the previous job will continue machining correctly (with current settings). The only exception to this statement is if cutting conditions (speeds, feeds, depth of cut, workpiece material, etc.) change substantially from one production run to another.

Key points for Lesson six:

- Program zero assignment values must be determined before program zero can be assigned.
- Each cutting tool requires its own set of program zero assignment values.
- Program zero assignment values are the distances between the program zero point and the machine's zero return position.
- The work shift value can be used in the Z axis to keep Z axis program zero assignment values consistent for tools used in consecutive jobs.
- One way to determine program zero assignment values is to actually measure them during setup at the machine.
- We've shown how program zero assignment values can be determined for external (turning) tools, center cutting tools, and internal (boring) tools.

Talk to experienced people in your company...

...to learn more about how program zero assignment values are determined.

1) How do setup people determine program zero assignment values?

2) Do any of your machines have a tool touch off probe?

3) Are geometry offsets used to assign program zero? Do setup people use the work shift value?

4) Ask to watch setup people measuring program zero assignment values.

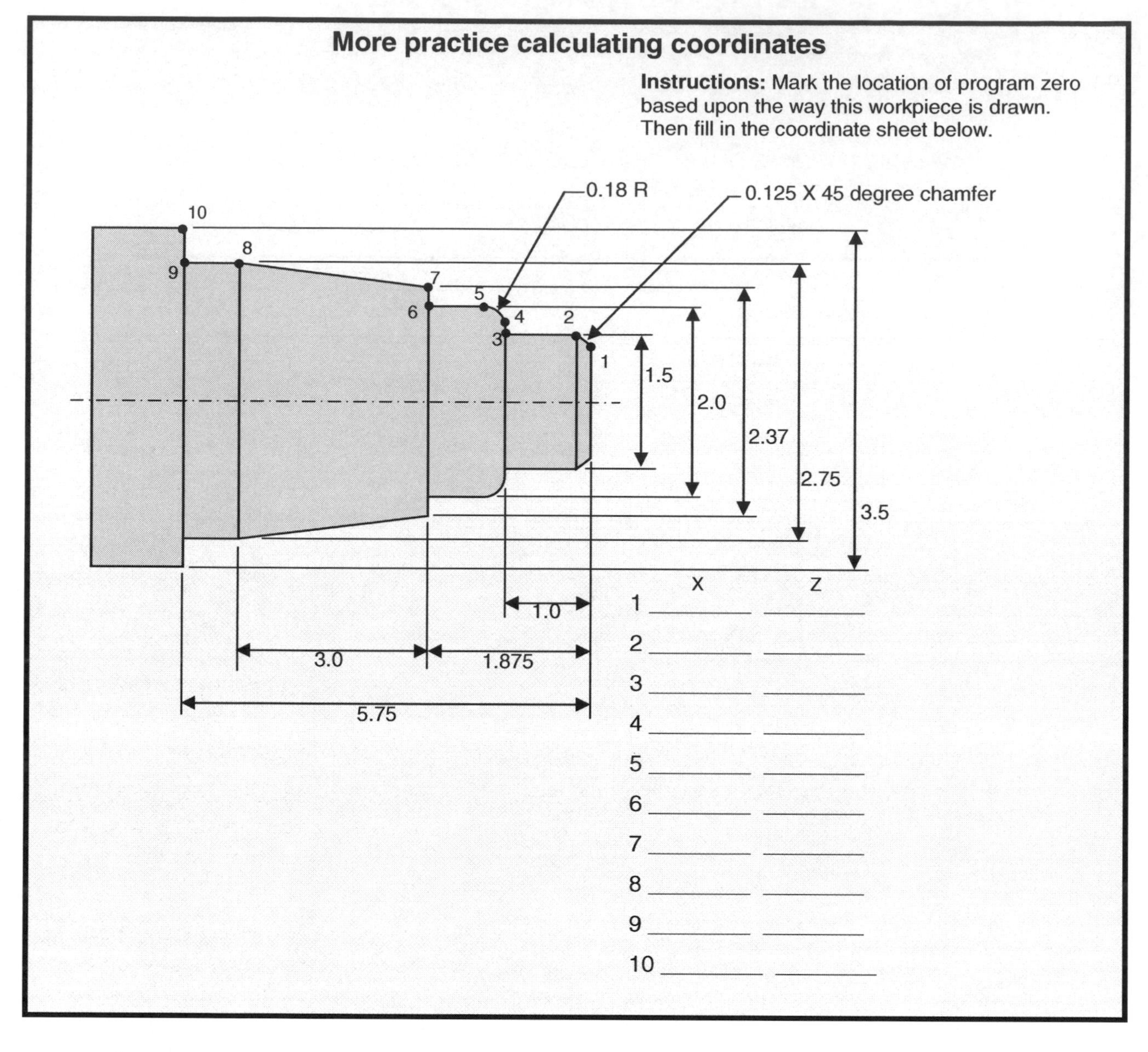

Answers:

Program zero is the upper right corner of the workpiece.

1: X1.25 Z0
2: X1.5 Z-0.125
3: X1.5 Z-1.0
4: X1.64 Z-1.0
5: X2.0 Z-1.18
6: X2.0 Z-1.875
7: X2.37 Z-1.875
8: X2.75 Z-4.875
9: X2.75 Z-5.75
10: X3.5 Z-5.75

STOP! Do practice exercise number six in the workbook.

Lesson 7

Four Ways To Assign Program Zero

Once program zero assignment values are determined as shown in Lesson Six, you must assign program zero. The most popular way to do so is with geometry offsets.

Like Lesson Six, this lesson has more to do with setup than with programming. Again, programmers must know enough about making setups to instruct setup people. This includes knowing how to assign program zero.

You know that cutting tools used in turning centers vary from type to type. You know that each will require its own program zero assignment. You know that program zero assignment values are used to assign program zero. You know that program zero assignment values are the distances between the tool tip and the program zero point (or the face of the chuck if the work shift value is used). These values are determined while the machine is at its zero return position. And you have seen one rather crude way to determine program zero assignment values – manually measuring them at the machine during setup.

In this lesson we're going to show how to *assign* program zero. The most popular and efficient way to do so is with *geometry offsets.*

Understanding geometry offsets

Offsets are storage registers within the control. They are used to store numerical values (numbers) – and will not be used until they are *invoked* by the program. In general, offsets are used to separate certain *tooling related values* from the CNC program – keeping you from having to know them when a program is written.

Geometry offsets are the registers used to assign program zero. In them, you'll be placing program zero assignment values. Turning center controls come with at least sixteen sets of geometry offsets (most come with thirty-two). One set of geometry offsets is used per tool. Since most turning center turrets can hold no more than twelve cutting tools, there will always be more than enough geometry offsets.

Of the four ways we show to assign program zero, three of them use geometry offsets. And frankly speaking, you should use geometry offsets to assign program zero unless your machine does not have them – as will be the case with older machines (machines made before about 1987). Figure 1.39 shows the first display screen page of geometry offsets.

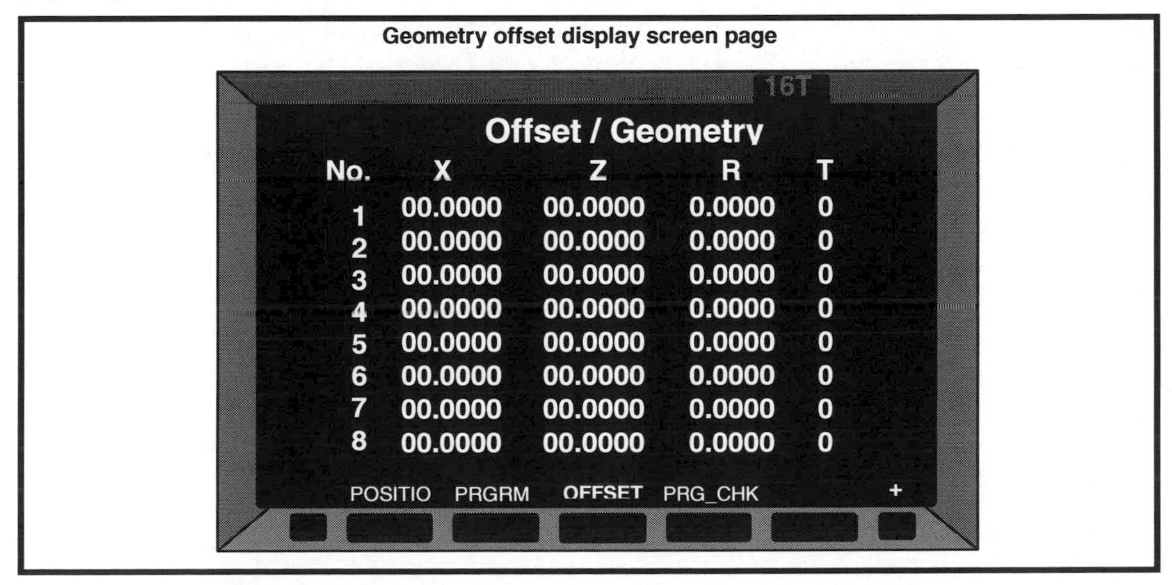

Figure 1.39 – First page of geometry offsets

With this particular control model, the first eight geometry offsets are shown on page one. By pressing the page down button on the control panel, other pages of geometry offsets can be displayed. Notice that each geometry offset has four registers (X, Z, R, and T). Only the X and Z registers are used for program zero assignment. We'll discuss the R and T registers in Lesson Fourteen.

Say you have measured the program zero assignment values for every cutting tool to be used in a job using techniques shown in Figure 1.36 (Lesson Six). With the program zero assignment values at hand, they can be simply entered into geometry offset registers. However, geometry offsets do have a *polarity* (plus versus minus). They specify the distance *from* the tool tip (while the turret is at the zero return position) *to* the program zero point. If the zero return position is close to the plus over-travel limit for each axis (as it is with most turning centers), geometry offset values always will be negative (-). Figure 1.40 shows the geometry offsets page again, this time after program zero assignment values have been entered.

Geometry offset display screen page

16T

Offset / Geometry

No.	X	Z	R	T
1	-12.2328	-14.2423	0.0000	0
2	-12.3864	-14.2358	0.0000	0
3	-15.2373	-10.2335	0.0000	0
4	-16.2931	-10.2847	0.0000	0
5	-16.3857	-11.8373	0.0000	0
6	00.0000	00.0000	0.0000	0
7	00.0000	00.0000	0.0000	0
8	00.0000	00.0000	0.0000	0

POSITION PROGRAM **OFFSET** PRG_CHK +

Figure 1.40 – Fixture offsets page after program zero assignment values are entered.

Again, the values placed in geometry offsets are simply the program zero assignment values needed for each cutting tool. In Figure 1.40, each X register represents the diameter of the tool tip while the machine is at the X axis zero return position (as negative values). And this will be true regardless of which method of program zero assignment is used. As you go from job to job, X axis geometry offset values will remain the same for any tools that remain in the turret.

The Z axis registers of Figure 1.40 represent the distances in Z from the tool tip while the machine is at the Z axis zero return position to the Z axis program zero point on the workpiece. When the next job must be run, all Z axis geometry offsets will change if the workpiece to be machined in the next job is longer or shorter than it is in the last job. As stated in Lesson Six, this can result in much duplicated effort. We'll be presenting two methods for program zero assignment that eliminate this duplication of effort – and in general – simplify the task of program zero assignment. These will be the first two methods shown in this lesson.

How geometry offsets are instated

Again, each cutting tool requires its own program zero assignment (geometry offset). From within a program, a four digit *T word* is used to instate geometry offsets. Actually the T word does three things.

- The first two digits of the T word specify the turret station and cause the turret to index
- The first two digits of the T word also specify and instate the geometry offset
- The second two digits of the T word specify and instate the wear offset

Consider the T word:

T0101

The first two digits command the turret to index to station one and instate geometry offset number one. The second two digits instate wear offset number one. Most programmers will keep things simple and use the same wear offset number as station number and geometry offset number (T0202, T0404, T0606, etc.).

When the machine executes the command

N030 T0505

it will index the turret to station five and instate geometry offset number five. It will also instate wear offset number five. The machine will look in geometry offset number five to find the related program zero assignment values. With this information, it can correctly position the cutting tool to the coordinates specified in the program and machine the workpiece properly. More is presented about programming with geometry offsets during Lesson Twelve.

The four most common ways to assign program zero

Actually, there are *many* ways to assign program zero with CNC turning centers. These methods vary based upon the age of the turning center as well as whether or not the turning center has a tool touch off probe (sometimes called a quick-setter or tool setter). Program zero assignment methods can even vary based upon personal preferences of setup people.

While you will have to be prepared for variations, we show the four most popular methods for program zero assignment in order from most preferable to least preferable:

- **With a tool touch off probe** (use this method if your machine has a tool touch off probe)
- **With geometry offsets using work shift** (use this method if your machine doesn't have a tool touch off probe but it does have the geometry offset and work shift features)
- **With geometry offsets without work shift** (use this method if your machine has neither a tool touch off probe nor the work shift function, yet does have the geometry offset feature)
- **With G50 in the program** (use this method if your machine doesn't have a tool touch off probe or geometry offsets – older machines)

Admittedly, this may get a little confusing. You'll have to check with an experienced person in your company to determine which method/s of program zero assignment your company uses. If you're unsure – or if you don't currently work with for a CNC-using company, please study the first two program zero assignment methods shown (with a tool touch off probe and with geometry offsets & work shift). They tend to be the most common methods used with *current* CNC turning centers.

If your company has a variety of different CNC turning centers ranging in age from very old to very new, you may find that all four of these methods are being used in your company (only one method for a given machine, of course). In this case, you'll eventually have to learn them all. But to get started, you might want to limit your studies to just one or two of these methods – we'd recommend starting with the first one shown (with the tool touch off probe) and then work your way through the list as you master each method.

You may even find that setup people are using a different method of program zero assignment all together. This often happens when setup people have no formal training – when they are left on their own to figure out how program zero assignment should be done. You may have to adapt to these methods – or try to convince them that methods shown in this text may be better. But if you can't convince them, you'll have to adapt to their methods. And if you truly understand what is presented in Lessons Six and Seven, you should be able to adapt to just about any method of program zero assignment.

Using a tool touch off probe to assign program zero – also called a *tool setter* (your 1st choice)

This is our most preferred method for two reasons. First, the tool touch off probe provides several advantages:

- It makes program zero assignment very easy
- It eliminates the need for calculations
- It eliminates the possibility for entry mistakes
- It eliminates the need for trial machining (during setup and when dull tools are replaced)
- It automatically clears the wear offset

Second, since the chuck face is used as the point of reference for geometry offset entries, and since the chuck face position does not change from job to job (in most companies), geometry offsets measured for a cutting tool used in one setup will still work in the next setup. That is, *cutting tools need only be measured once* – when they are first placed in the turret.

A tool touch off probe requires the machine to have geometry offsets. So if your machine has a tool touch off probe, you can rest assured that it also has geometry offsets. Also, machine tool builders will additionally supply the work shift function when their machines are equipped with tool touch off probes. Figure 1.37 shows how geometry offsets and the work shift value will be specified when a tool touch off probe is used.

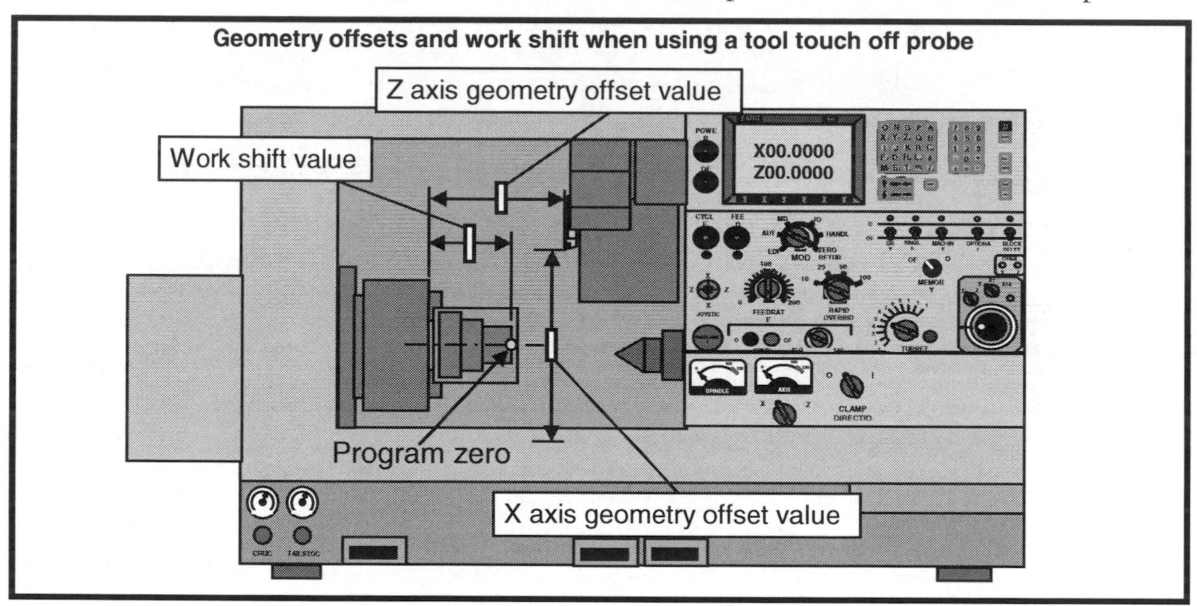

Figure 1.37 – Geometry offset and work shift value use when a tool touch off probe is used

The X axis geometry offset value is the diameter the cutting tool will machine when at the X axis zero return position. **The Z axis geometry offset value** is the distance between the tool tip and the *chuck face* along the Z axis. If you have a tool touch off probe, the X and Z geometry offset values will be automatically determined and entered each time you use the probe.

The work shift value is the distance between the chuck face and the Z axis program zero point on the workpiece. We've shown the Z axis program zero point at the right end of the finished workpiece, but it could, of course, be at the *left* end of the finished workpiece based upon how the workpiece is dimensioned on the print.

A tool touch off probe will make it quite simple to quickly and accurately assign program zero for single point tools (turning tools, boring bars, threading tools, grooving tools, etc.). With most machines, first the setup person or operator manually brings the tool touch off probe into the probing position (though some machines do allow probe activation to be done with a push-button or to be programmed). The act of bringing the probe into position also turns it on.

Prior to using the tool touch off probe, most turning centers additionally require the setup person to call up the geometry offset display screen page and bring the cursor to the geometry offset number of the cutting tool for which program zero is being assigned. Note that some turning centers do not require this. These machines constantly monitor which turret station is active so they can determine which geometry offset must be entered when a stylus position is touched. This, of course, simplifies the program zero assignment process.

Stylus use

The tool touch off probe has four stylus positions. Each cutting tool will require that two stylus positions be touched. You must determine *which two* stylus positions must be used for a given style of cutting tool.

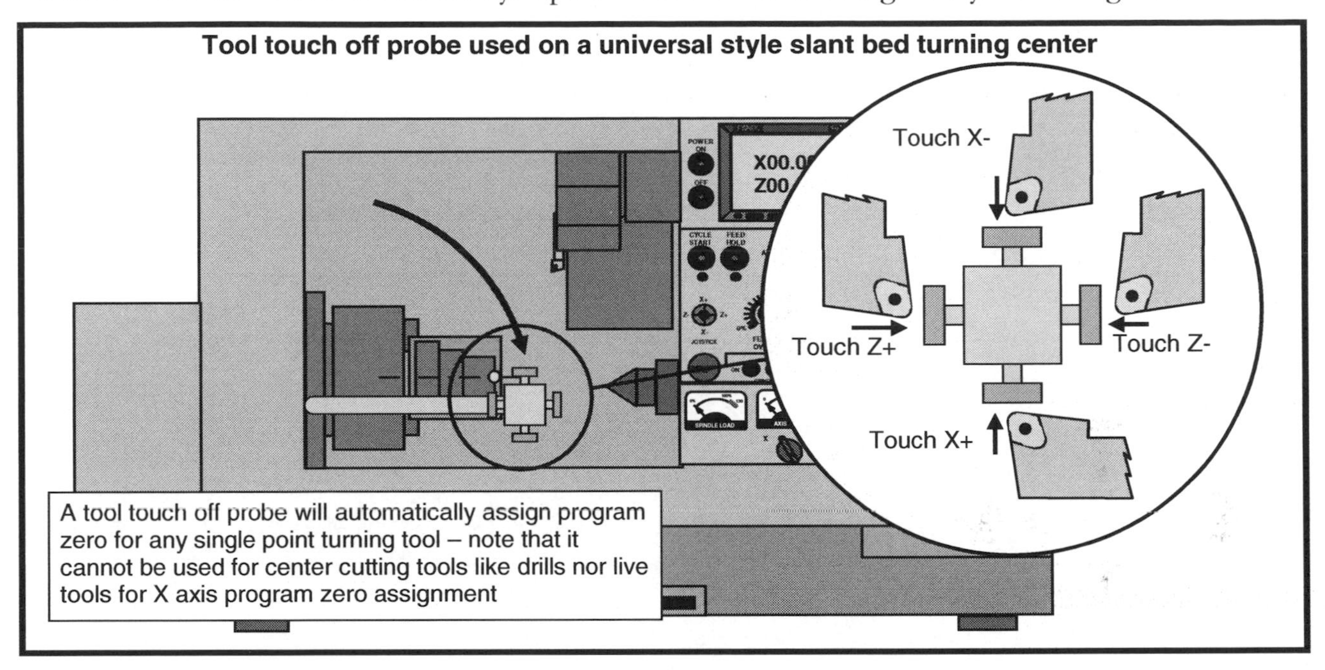

Figure 1.38 – A tool touch off probe

The upper stylus position (labeled Touch X- in Figure 1.38) will be used for any external turning tool, including rough and finish turning tools, grooving tools, and threading tools. When this stylus position is touched, the machine will automatically enter the cutting tool's X axis geometry offset value.

The lower stylus position (labeled Touch X+ in Figure 1.38) will be used for any internal turning tool, including rough and finish boring bars, internal grooving tools, and internal threading tools. Note that it will *not* be used for center cutting tools, like drills and taps. When this stylus position is touched, the machine will automatically enter the cutting tool's X axis geometry offset value.

The right stylus position (labeled Touch Z- in Figure 1.38) will be used for any tool that machines toward the chuck, including rough and finish turning tools, grooving tools, threading tools, rough and finish boring bars, internal grooving tools, and internal threading tools. It will also be used for center cutting tools, like drills and taps. When this stylus position is touched, the machine will automatically enter the cutting tool's Z axis geometry offset value.

The left stylus position (labeled Touch Z+ in Figure 1.38) will be used for any back turning tool, including back turning tools and (if possible) back boring bars. When this stylus position is touched, the machine will automatically enter the cutting tool's Z axis geometry offset value.

Again, each cutting tool requires two stylus positions to be touched. An external turning tool, for example, will be touched to the upper stylus position (for X) and to the right stylus position (for Z). A boring bar will be touched to the lower stylus position (for X) and to the right stylus position (for Z).

What about center cutting tools?

There is one style of cutting tool for which the tool touch off probe can provide only limited help – a center cutting tool (drill, tap, reamer, etc.). This includes any live tooling cutting tool. The tool touch off probe can only determine the Z axis geometry offset value for this kind of tool – by touching the right stylus position (labeled Touch Z- in Figure 1.38). It is not possible for a tool touch off probe to determine the X axis geometry value for this kind of tool. But as stated in Lesson Six, the X axis program zero assignment value (geometry offset X value) for any center cutting tool will be consistent. Once you determine it for one center cutting tool (as is shown in Lesson Six), it will remain the same for *any* center cutting tool – in any turret station.

Procedure to use a tool touch off probe

While you must be prepared for variations from one machine to another, here is the general procedure to use a tool touch off probe.

- 1) Bring the tool touch off probe into the probing position (this activates the probe).
- 2) Be sure the turret is in a position that allows safe indexing (without interference). Jog the axes to a safe index position if required. The procedure to jog the axes is shown in Lesson Twenty-Seven.
- 3) Index the turret to the desired station number. Some machines allow manual turret indexing. Others require manual data input (MDI) to be used. We show both procedures to index the turret in Lesson Twenty-Seven.
- 4) Carefully jog the tool up close to the stylus position to be touched. The procedure to jog the axes is shown in Lesson Twenty-Seven.
- 5) Select the geometry offset display screen page and bring the cursor to the geometry offset number to be set (the geometry offset number is the same as tool station number).
- 6) Most turning centers require that each stylus position be touched at a consistent feedrate. For these machines, you must select the appropriate jog feedrate using the jog feedrate switch. (Unfortunately, machines vary with regard to *what feedrate* is used, so you must reference you machine tool builder's instruction manual for more information.)
- 7) Jog the tool tip into the stylus. When the stylus is touched, motion will stop and the related geometry offset value will be automatically entered.
- 8) Repeat for the cutting tool's other stylus position. Again, each cutting tool requires that *two* stylus positions be touched – one for the X axis and the other for the Z axis. The only exception to this is center cutting tool –for which the X axis geometry offset value cannot be determined with the tool touch off probe.
- 9) Repeat steps two through eight for all other new cutting tools used in the job. Note that the geometry offset values for cutting tools that remain in the turret from the last job will still be correct for the next job. These cutting tools need not be probed again.

Figure 1.39 shows an example procedure for using the tool touch off probe with an external turning tool (like a finish turning tool).

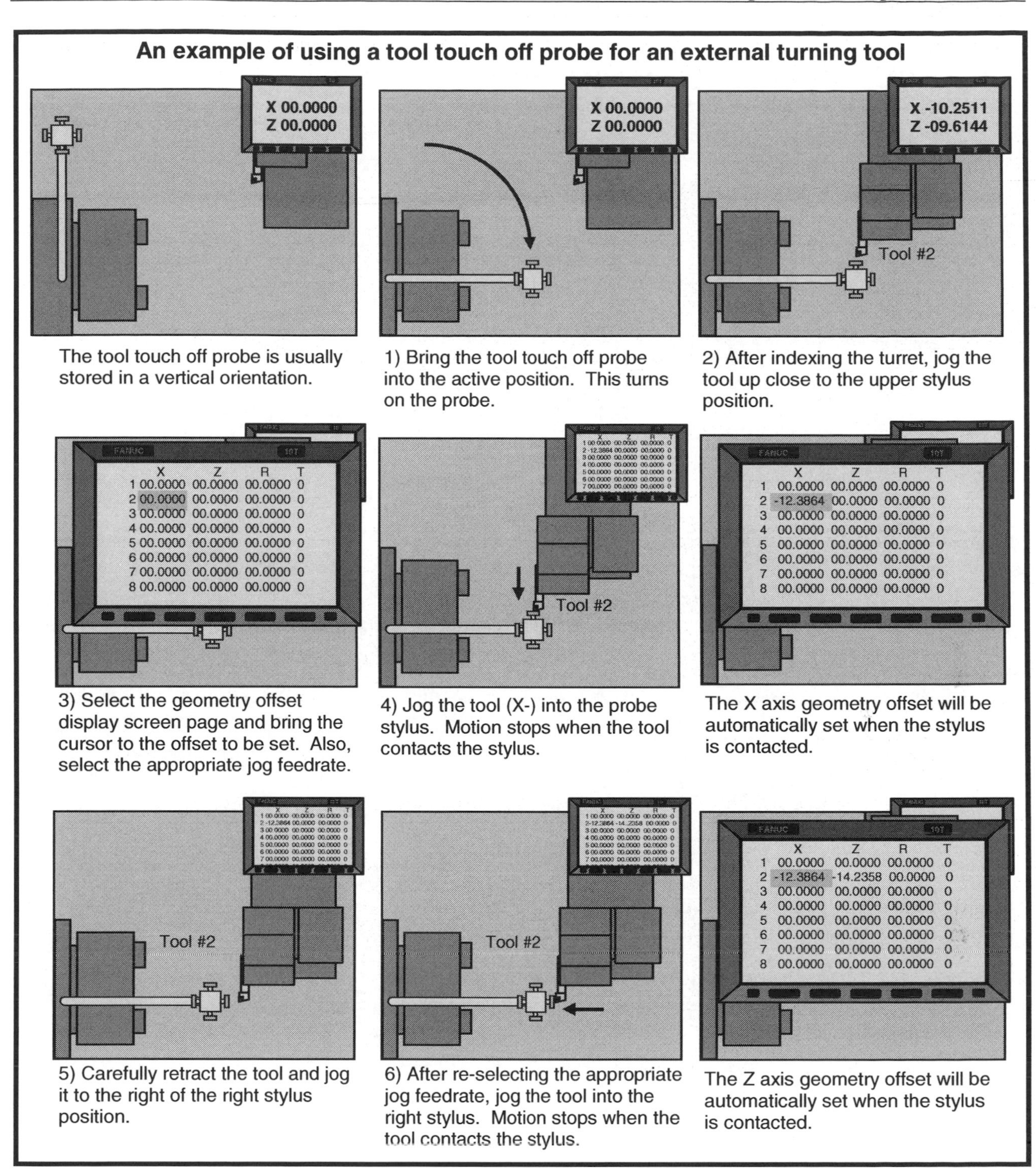

Figure 1.39 – Using the tool touch off probe with an external turning tool

As you can see from the example procedure shown in Figure 1.39, the X or Z geometry offset for the tool will be automatically entered when a cutting tool contacts a stylus position on the tool touch off probe.

While our procedure is pretty specific, remember that machines vary. With some machines, for example, the tool touch off probe can be activated by a push button or specified with a program command (step one in Figure 1.39). It may not be necessary to select the geometry offset number to be set (step three in Figure 1.39). Additionally, the recommended feedrate used for jogging the tool into a stylus position will vary – you must reference your machine tool builder's operation manual to find it. (What is most important is that you use a *consistent* feedrate from one time you jog a tool into a stylus position to the next.)

How does it work?

There's nothing magical going on. The machine will simply calculate and enter geometry offset values. Remember what the X and Z geometry offset values represent. The geometry offset X value is the diameter of the tool tip when the X axis is at its zero return position (though it is a negative value in the geometry offset). The Z axis geometry offset is the distance from the tool tip with the Z axis at its zero return position to the chuck face (again, a negative value).

Part of calibrating a tool touch off probe (which most CNC setup people and operators are *not* required to do) is telling the machine how far it is from the upper and lower probe stylus positions to the spindle center. With this information, and since the machine can always tell how far each axis is from its zero return position, the machine can determine the cutting tool's X axis geometry offset value whenever the upper or lower stylus positions are contacted.

For the Z axis, part of calibrating the tool touch off probe is telling the machine how far it is from the left and right stylus positions to the chuck face. With this information – along with the current position of the Z axis – the machine can determine the cutting tool's Z axis geometry offset value whenever the left or right stylus positions are contacted.

What about the wear offset for a cutting tool that has been probed?

Admittedly, we have not discussed *wear offsets* yet (they are the subject of Lesson Thirteen). Wear offsets are used during a cutting tool's life to adjust for small workpiece size changes that occur as cutting tools wear. As a cutting tool machines more and more workpieces, a small amount of material will be worn away from its cutting edge. A wear offset adjustment tells the machine to machine a little more material, which compensates for tool wear – and keeps the cutting tool machining properly.

When a stylus position is touched, you know that the related geometry offset value is automatically entered. It is important to know that contacting a stylus position also causes the cutting tool's *wear offset* to be cleared (set to zero). This is important since if the wear offset is not cleared, any wear offset that carries over from the previous tool will cause the new cutting tool to machine too much material. It is likely to scrap the next workpiece.

In the example shown in Figure 1.38, for example, step four requires that tool number two be jogged into the upper stylus position. When the tool tip contacts the upper stylus, geometry offset number two's X register value will be automatically set. Though we don't show it in Figure 1.38, *wear offset* number two's X register value will be set to zero.

The same is true for the Z axis. In step six of Figure 1.38, we touch the tool tip into the right stylus position. This automatically sets geometry offset number two's Z register value. It also clears wear offset number two's Z register value.

What about tool pressure?

As you know from Lesson Six, tool pressure is the tendency for a cutting tool to make the workpiece push away from the cutting edge during a machining operation. It tends to make external diameters slightly larger than intended and internal diameters slightly smaller than intended. When tolerances are small, this can cause problems when machining the first workpiece – and it is one of the reasons why *trial machining* must be done (trial machining is introduced in Lesson Six).

If properly calibrated, most tool touch off probes can account for tool pressure – meaning when they're used, the cutting tool will machine the workpiece perfectly (or at least within its tolerance band) on its first try. This eliminates the need for trial machining.

You may remember from Lesson Six that another reason why trial machining is required is that geometry offset values may be incorrectly measured and/or entered. This is an important concern when geometry offset values are *manually* measured and/or entered. But since a tool touch off probe automates the process of geometry offset value measurement and entry, it effectively eliminates this reason for trial machining. *Setup people and operators that use tool touch off probes never need to trial machine (as long as the tool touch off probe is properly calibrated).*

Determining and entering the work shift value

Figure 1.39 shows the work shift display screen page for a popular Fanuc control model.

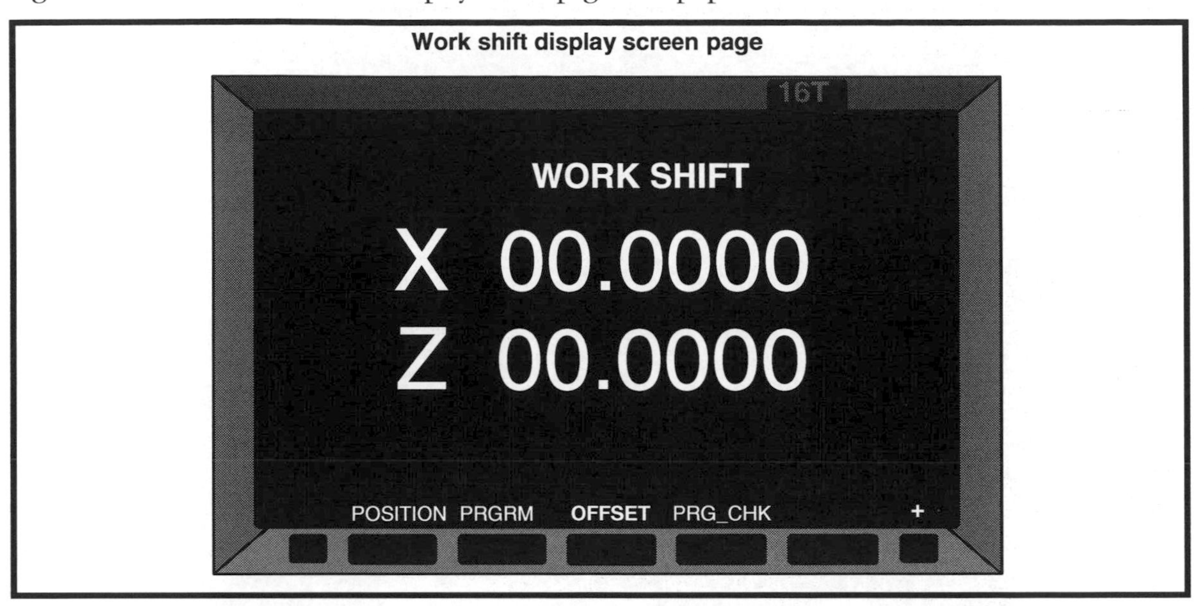

Figure 1.39 – Work shift page

The X and Z values on this page specify the distances from the center of the spindle (in X) and the face of the chuck (in Z) to the program zero point to be used by the program. In the X axis, the work shift will *always* be zero (program zero in X is always placed at the spindle center). In Z, the work shift value will be the length of the workpiece plus the jaw length (if program zero in Z is placed at the right end of the finished workpiece). With most machines, the polarity for this value is positive.

For almost all turned workpieces, both ends must be machined. For the first end (first operation), the work shift value is not terribly critical. In most cases, all that is important is that the raw material (stock) be balanced on both ends of the workpiece. That is, the stock on the front face will be the same as the stock on the back face. Figure 1.40 shows an example.

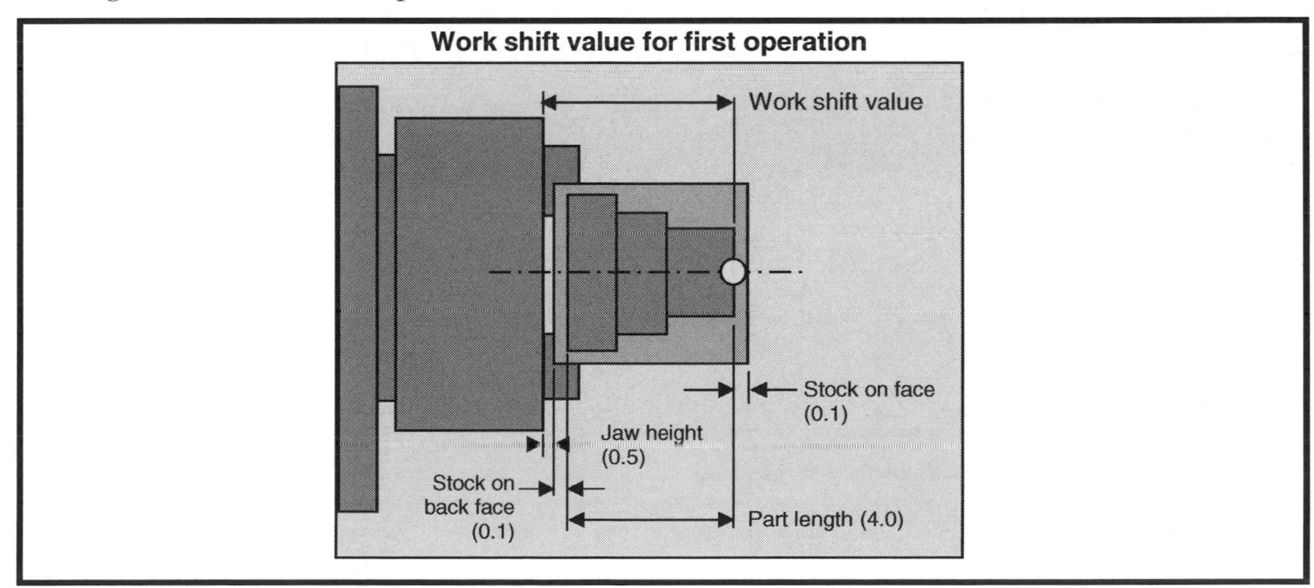

Figure 1.40 – The work shift value

Say, for example, the finished length of the finished workpiece is 4.0 inches. The raw material (bar) has been cut to a length of 4.2 inches. In this case, most setup people will machine 0.1 inch of material from the first operation – leaving another 0.1 inch of material to be machined from the other face in the second operation.

One way to determine the work shift value is to measure it. For this first operation, the setup person might use a caliper or scale (or even a tape measure) to measure the work shift value. Again, it is not very critical if they are simply balancing stock for first and second operations.

This value can also be calculated if you know the jaw height. The jaw height can, of course, be measured with a caliper or (more precisely) with a depth gauge. In Figure 1.40, we're showing that the jaw height is 0.5 inches. For the 4.0 inch workpiece that has been cut to 4.2 inches long, the setup person will add 4.1 to the 0.5 inch jaw height in order to come up with the Z axis work shift value. This entered as shown in Figure 1.41.

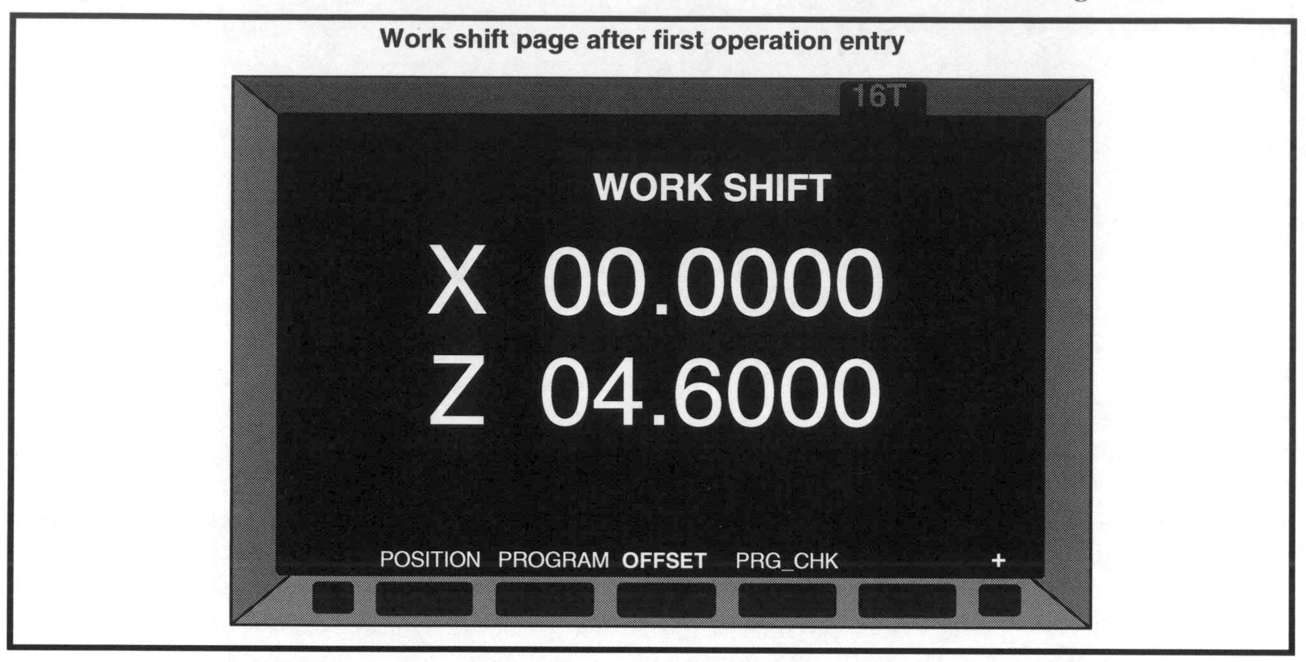

Figure 1.41 – Work shift entry for first operation

The Z axis work shift value is much more critical when it comes time to machine the *other* end of the workpiece (the second operation). This value will, of course, control the overall length of the finished workpiece – and depending upon the tolerance for the overall workpiece length, it might be pretty critical indeed.

Figure 1.42 shows the workholding setup for the second operation.

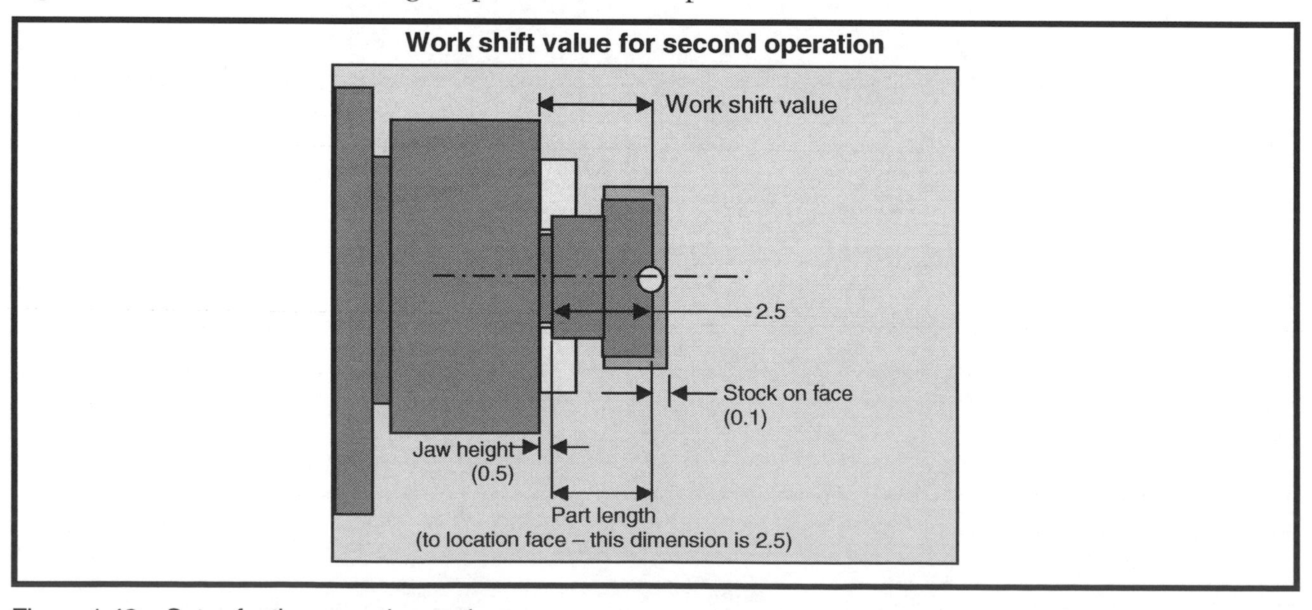

Figure 1.42 – Setup for the second operation

Notice that the jaws have been bored to hold on the middle diameter – and that they locate on the face of this diameter. The jaw height is still 0.5 inches (measured with a depth gauge). The distance from the locating face to the finished end of the workpiece is 2.5 (this dimension can be found on or calculated from the workpiece drawing). The work shift value for this operation will be Z3.0, as is shown in Figure 1.43.

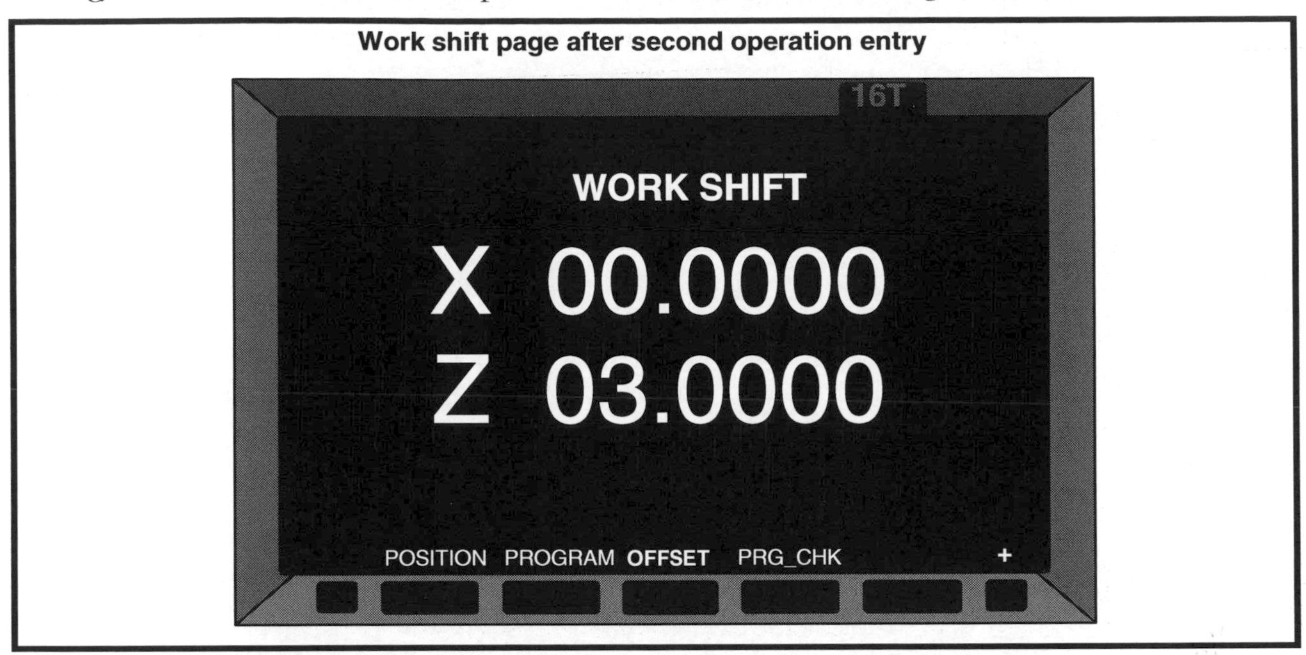

Figure 1.43 – Work shift value for second operation

A note about the polarity of the work shift value

We've been showing the polarity of the Z axis work shift value as *from* the chuck face *to* program zero. And of course, this will always result in a positive work shift value. You must be prepared, however, for control models that reverse this polarity. With these machines, the Z axis work shift value will be entered as a negative value. Ask an experienced setup person which polarity is used for work shift values.

Using geometry offsets with work shift to assign program zero (your 2nd choice)

This is second-most preferred method. It should be used for turning centers that do not have a tool touch off probe, but do have geometry offsets and the work shift function. Advantages include:

- Eliminates the need for calculations
- Minimizes the possibility for entry mistakes (if measure function is used)
- Automatically clears the wear offset

Additionally, since the chuck face is used as the point of reference for geometry offset entries, and since the chuck face position does not change from job to job (in most companies), geometry offsets measured for a cutting tool used in one setup will still work in the next setup. That is, *cutting tools need only be measured once* – when they are first placed in the turret.

While tool touch off probes are becoming very popular, not all turning centers are equipped with them. Yet most current model Fanuc-controlled turning centers that do not have a tool touch off probe *are equipped with geometry offsets and the work shift value.*

These features are used in exactly the same way as they are on machines with tool touch off probes. In fact, the work shift value is determined and entered in exactly the same manner as it is when a tool touch off probe is used (see the presentation just given about the work shift value – we will not repeat it). The only difference is that more work is required to determine and enter geometry offset values.

Though geometry offsets and the work shift function are used in the same way as they are when using a tool touch off probe, remember that a tool touch off probe can also be calibrated to allow for tool pressure. And with a tool touch off probe, the tasks of measuring geometry offset values and entering them are automated. A tool touch off probe eliminates the need for trial machining.

This is not the case when the turning center does *not* have a tool touch off probe. With the methods we show from this point on, there is no (feasible) way to calibrate for tool pressure – and you *might* make a mistake with the measurement and/or entry of geometry offset values. So you must use trial machining techniques with cutting tools that must machine within small tolerances bands.

With this second method of program zero assignment, the **X axis geometry offset value** is the diameter of the tool tip while the X axis is at its zero return position. The **Z axis geometry offset value** is the distance from the tool tip at the Z axis zero return position to the chuck face. And the Z axis **work shift value** is the distance between the chuck face and the program zero point for the program in Z (the X axis work shift value is always zero). See the next illustration.

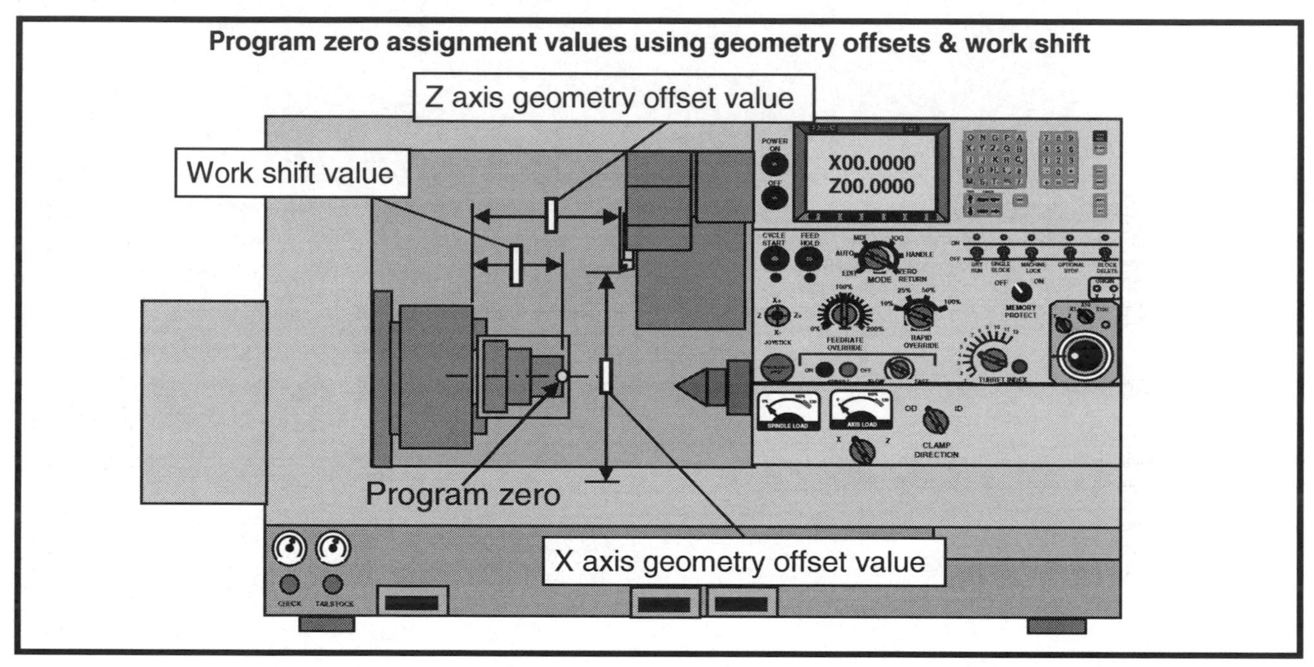

Instead of using a tool touch off probe to measure and enter geometry offset values, you will be doing so in a more manual manner. While most turning centers are equipped with a helpful *measure function* to simplify the task, you'll have to physically skim cut a sample workpiece and measure the machined surfaces during the task of program zero assignment.

Understanding the measure function

The measure function can be thought of as a poor man's tool touch off probe. With a tool touch off probe, the machine is calibrated to know the distance from each stylus position to the spindle center in X and the chuck face in Z. The machine also keeps track of axis position (in each axis) relative to the zero return position. So when a stylus position is contacted, the machine can automatically calculate and enter the related geometry offset value for the tool.

In similar function, the measure function will cause the machine to calculate and enter geometry offset values. But instead of contacting a stylus position, *you'll be telling the machine a cutting tool's current position* – relative to spindle center in X and the chuck face in Z. Based on this information, the machine will calculate and enter the related geometry offset values. Figure 1.44 shows an example of the procedure.

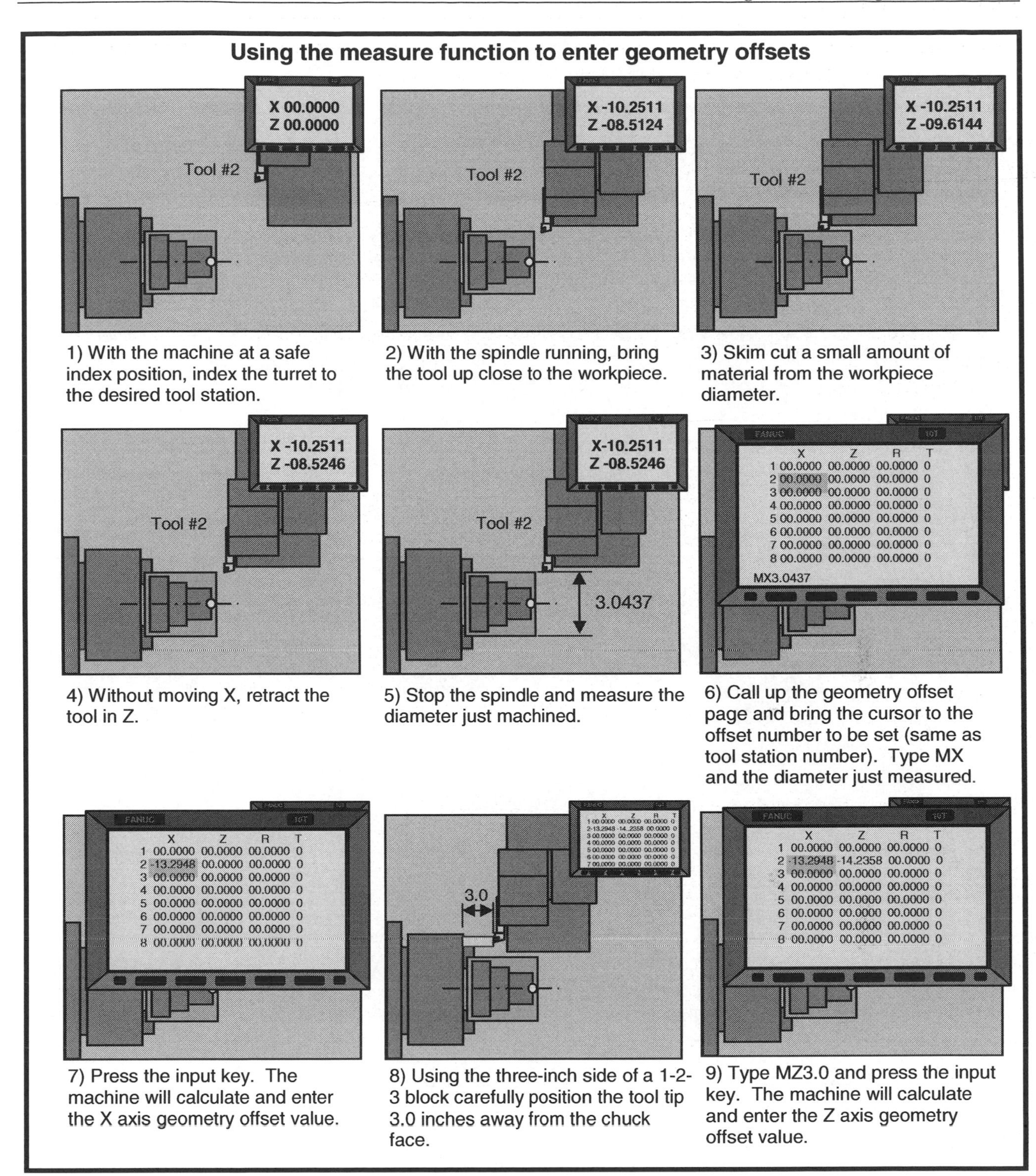

Figure 1.44 – Procedure to use the measure function to enter geometry offset values

In step five of Figure 1.44, notice that the tool is at a known diameter of 3.0437 (we've just machined this diameter and measured it). In step six and with the cursor on the related geometry offset number (number two in our case), you type MX3.0437. (MX stands for measure X.) Since the machine constantly knows the current X axis position relative to the zero return position, it can calculate and enter the X axis geometry value for this tool.

In similar fashion, step eight of Figure 1.44 brings the tool tip to a known position relative to the chuck face. We've used the three-inch side of a 1-2-3 block to keep from scuffing the chuck face – and note that some turning centers don't have enough Z axis travel to bring the tool tip up to the chuck face. (You can, of course, use any block of known length.) When typing and entering **MZ3.0**, the machine can calculate and enter the Z axis geometry offset value for this tool. MZ stands for measure Z. The three inch value is used because the tool is currently three inches from the chuck face (if you use a four inch block, you type **MZ4.0**, of course).

What if my machine doesn't have the measure function?

Without the measure function, your job is more difficult. While geometry offset values still represent the tool tip diameter (in X) and the distance to the chuck face (in Z), you must use the machine's relative position display screen page and actually measure & enter geometry offset values yourself. Figure 1.45 shows how.

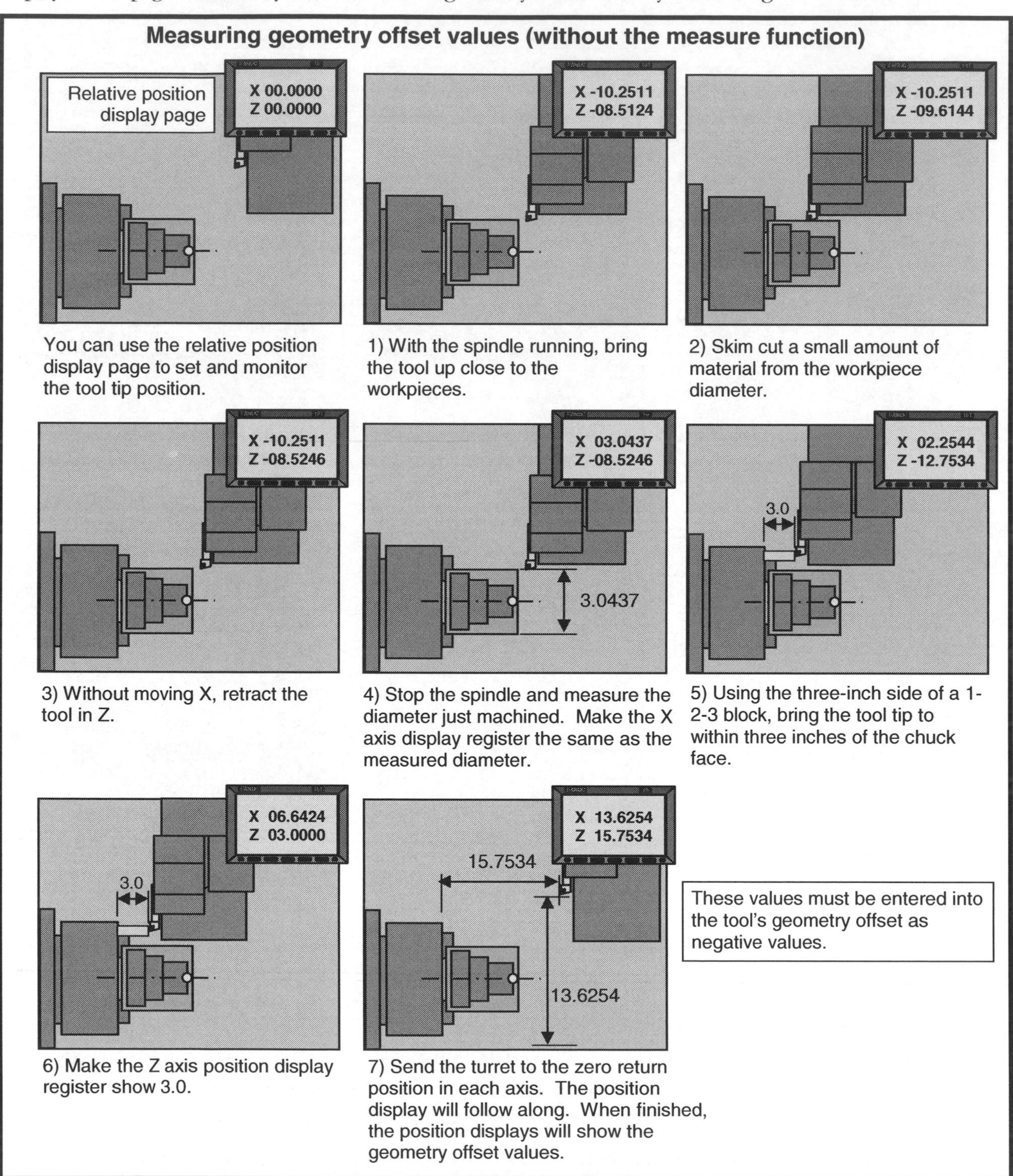

Figure 1.45 –Measuring geometry offset values (without the measure function)

As you can see in Figure 1.45, the relative position display page makes it possible to measure geometry offset values – but it is not nearly as easy to determine geometry offset values as it is when using the measure function. Additionally, once you have determined geometry offset values, you must enter them on the geometry offset page (as negative entries). This opens the door to entry mistakes. And of course, you must still complete the task of program zero assignment by determining and entering the work shift value.

The procedure to determine and enter the work shift value is shown after the discussion of how the tool touch off probe is used – and will not be repeated.

Using geometry offsets without work shift to assign program zero (your 3rd choice)

Some (older) turning centers have geometry offsets, but do not have the work shift function. With these machines, each Z axis geometry offset value is the distance from the tool tip at the Z axis zero return position to the Z axis program zero point (not to the face of the chuck). Figure 1.46 shows this.

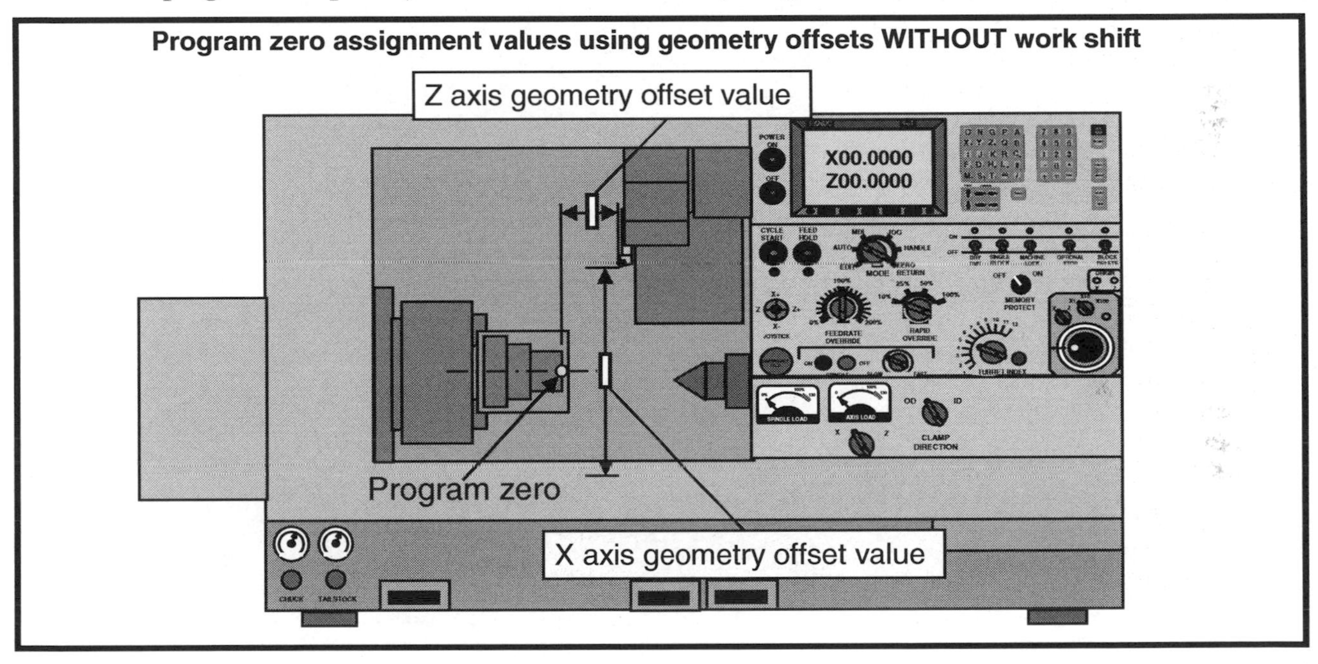

Figure 1.46 – Geometry offset values when work shift value is not used.

This eliminates the important advantage of the work shift function. As you go from job to job, Z axis geometry offsets must be determined for every cutting tool being used – even for cutting tools used in the most recent job (we're assuming workpiece length changes from job to job). So, Z axis *geometry offsets values must be determined and entered for every cutting tool in every job.*

Note that it is possible that your machine does have the work shift function, but the setup people in your company don't know how it can help – meaning they are using geometry offsets without work shift. In this case, be sure to show them this text! Work shift can eliminate much duplicated effort.

As introduced previously, the *measure function* can be used to help with the calculation and entry of geometry offsets. Figure 1.47 shows the procedure.

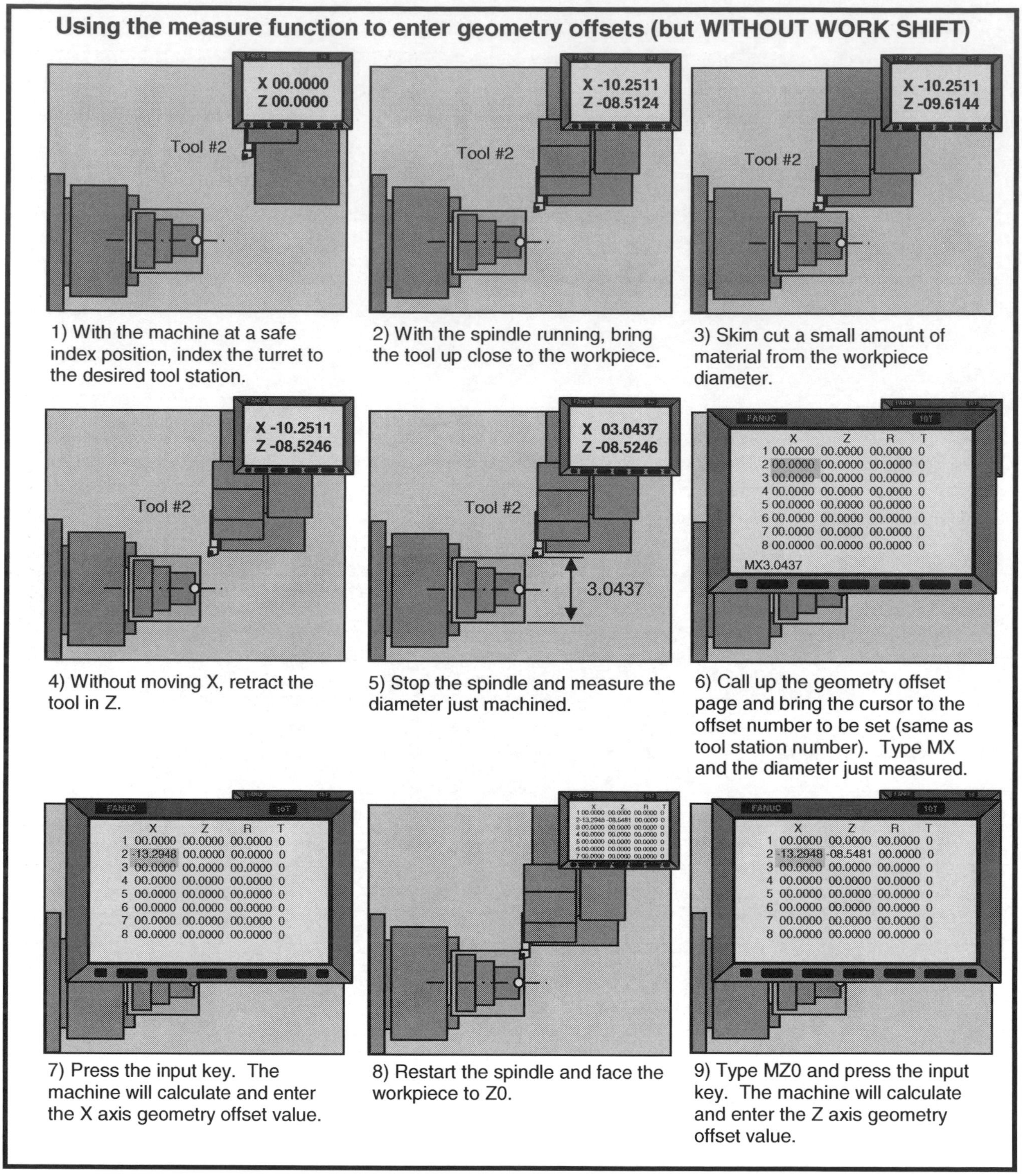

Using the measure function to enter geometry offsets (but WITHOUT WORK SHIFT)

1) With the machine at a safe index position, index the turret to the desired tool station.

2) With the spindle running, bring the tool up close to the workpiece.

3) Skim cut a small amount of material from the workpiece diameter.

4) Without moving X, retract the tool in Z.

5) Stop the spindle and measure the diameter just machined.

6) Call up the geometry offset page and bring the cursor to the offset number to be set (same as tool station number). Type MX and the diameter just measured.

7) Press the input key. The machine will calculate and enter the X axis geometry offset value.

8) Restart the spindle and face the workpiece to Z0.

9) Type MZ0 and press the input key. The machine will calculate and enter the Z axis geometry offset value.

Figure 1.47 – Procedure to measure geometry offset values when not using the work shift value

In step eight of Figure 1.47, notice that the cutting tool is positioned to the program zero point in the Z axis (Z0). When **MZ0** is entered in step nine, the control will automatically calculate and enter the distance from the tool tip at the Z axis zero return position to the Z axis program zero point.

What if my machine doesn't have the measure function?

Without the measure function, your job is more difficult. While geometry offset values still represent the tool tip diameter (in X) and the distance to the program zero surface on the workpiece (in Z), you must use the machine's relative position display screen page and actually measure & enter geometry offset values yourself. Figure 1.48 shows how.

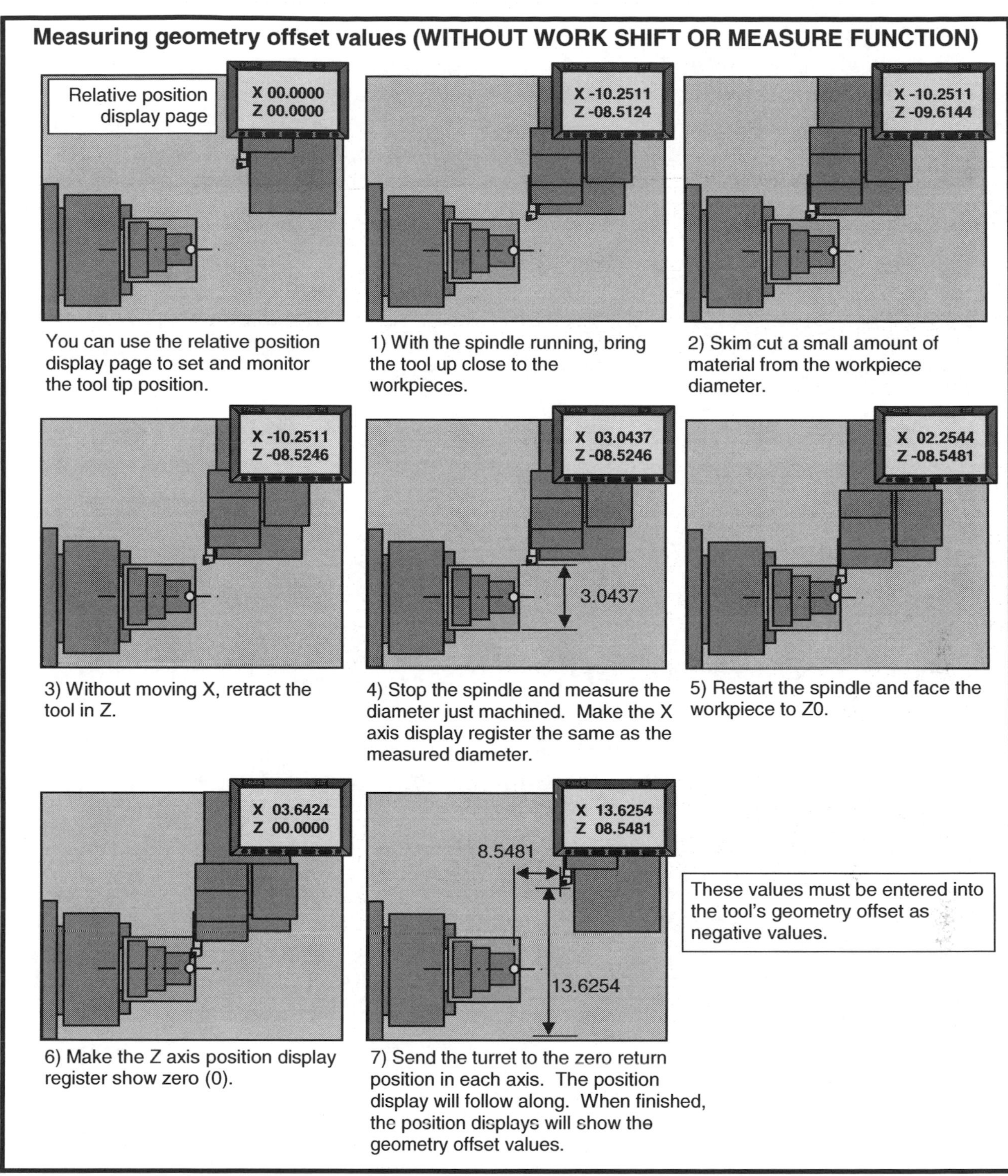

Figure 1.48 – Procedure to measure geometry offsets without the work shift value or the measure function

Assigning program zero in the program with G50 or G92 (your last resort)

Very old Fanuc-controlled turning centers (machines made before about 1987) do not have geometry offsets. With these machines, a command in the program is used to assign program zero – the G50 command (note that some turning centers use G92 for the same purpose – you must ask an experienced programmer which G code is used to assign program zero). While newer machines do have a use for G50 (as a spindle limiter), G50 used for spindle limiting has nothing to do with program zero assignment – it is introduced in Lesson Two.

You should only assign program zero in the program if your machine does not have geometry offsets. We have seen some setup people that continue to use G50 to assign program zero even on newer machines that have geometry offsets.

They may, for example, be trying to maintain compatible methods among several machines. Maybe the company owns some older machines that require program zero to be assigned in the program along with some newer machines that have geometry offsets.

While maintaining compatible methods from machine to machine is desirable, assigning program zero with geometry offsets provides *many* advantages over doing so in the program with G50. Using G50 to assign program zero on a machine that has geometry offsets is like limiting yourself to keyboard entry with desktop computers that have a mouse. Advantages of geometry offsets over G50 include (among others):

Enhanced safety
The most important advantage of using geometry offsets (over assigning program zero with G50 in the program) is that the machine need not be in a special position before the program can be activated. The turret can be in any position that allows safe indexing and approach. In essence, the machine cannot be out of position when geometry offsets are used. This eliminates a lot of crashes.

Enhanced efficiency
With geometry offsets, the turret need not be sent to a special pre-determined position prior to indexing. Again, the turret can be in any location as long as it can index without interfering with the workpiece or chuck. This means the turret index position can be closer to the workpiece when it indexes – which minimizes cycle time.

Better ease-of-use
As you have seen, assigning program zero with geometry offsets is quite easy – especially with the work shift value and the measure function. No such features are available when you assign program zero in the program with G50. You must measure program zero assignment values for every tool in every setup. And once they are measured, *you must edit the CNC program in at least two places (more likely four places) for each tool.*

Re-running tools
It is often necessary to re-run cutting tools in a program (when verifying programs and when trial machining, for instance). It is much easier to restart a program when geometry offsets are used to assign program zero than it is when program zero is assigned in the program.

Let us repeat the main point: ***You should only use G50 to assign program zero when your machine does not have geometry offsets.***

Determining the program zero assignment values for each tool

As with any other method of program zero assignment, you must determine program zero assignment values when program zero is assigned in the program. Figure 1.49 shows the related values.

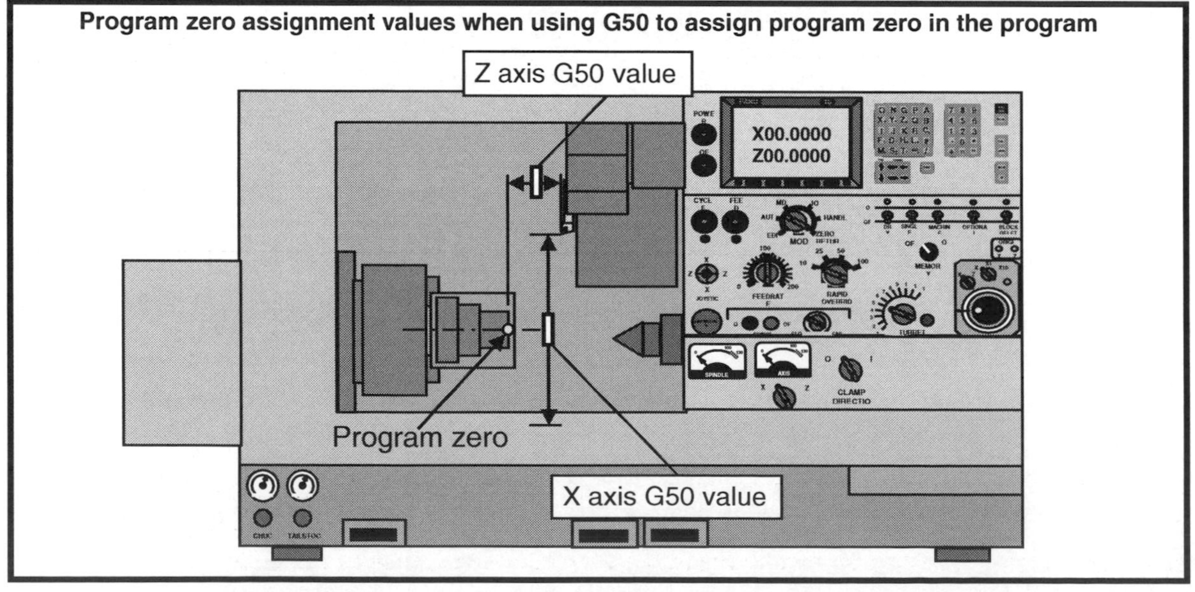

Figure 1.49 – Program zero assignment values when G50 is used to assign program zero in the program

Determining program zero assignment values is identical to doing so when using geometry offsets without the measure function or the work shift value. Figure 1.50 shows the procedure for a turning tool.

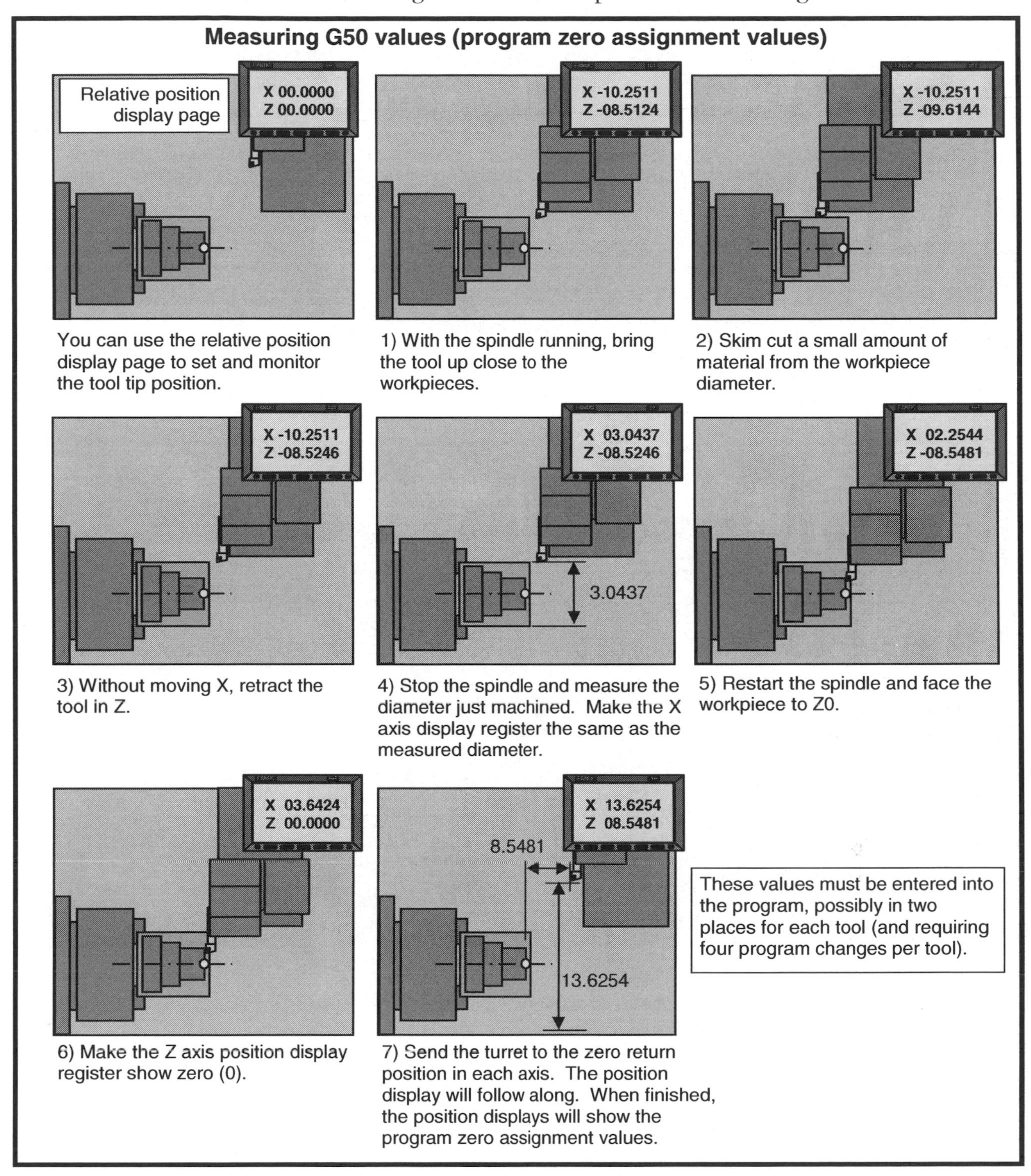

Figure 1.50 – Determining program zero assignment values when using G50 to assign program zero

Programming format when program zero is assigned in the program

A program zero assignment command – the G50 command – assigns the program zero point (again, some turning centers use G92 for this purpose). Since each cutting tool has its own program zero assignment values, *each tool will require its own G50 command.* This command is usually the very first command for each tool. The program zero assignment values (shown in Figure 1.50) will be used in G50 commands.

The polarity for program zero assignment values used in a G50 command is taken *from* the program zero point *to* the tool tip in each axis. The X axis G50 value will always be plus. If the Z axis zero return position is at the Z axis plus over travel limit, the Z axis G50 value will also always be plus. The only time you'll have a negative G50 value is in the Z axis, when machining a very long workpiece and the tool tip is on the negative side of the program zero point when the machine is at its starting position.

Say for example, a turning tool's X axis program zero assignment value has been found to be 13.6254. Its Z axis program zero assignment value has been found to be 8.5481. (Remember, these are the distances between the program zero point (center of the workpiece in X and the right end of the finished workpiece in Z) and the tool tip while the machine is resting at its zero return position. In this case, the command

```
G50 X13.6254 Z8.5481
```

will be used to assign program zero for this tool. Notice that we simply specify the program zero assignment values within the G50 command.

The major limitation of assigning program zero in the program is that the machine *absolutely, positively* must be in the correct position (the zero return position is our recommended starting position) before the program can be run. This is a common source of crashes (damaging collisions). If the turret is not where it's supposed to be, the control will simply *believe* the G50 program zero assignment values when the program is run. And it's likely that since the zero return position is close to the plus over-travel limits on many machines, if the turret has been moved, it's going to be *closer* to the workpiece in X and/or Z. The resultant approach motion will be longer than it should be, and again, this will cause a crash.

For this reason, it is a good idea to program the machine to go to the zero return position prior to the G50 command, at least for the first tool in the program. If the machine is out of position, it will go to the zero return position prior to executing the G50 command. A G28 command is used to send the machine to its zero return position. Though we won't explain the details of how G28 works at this point (G28 is explained in Lessons Fifteen and Sixteen), the commands

```
N005 G28 U0 W0 (Send both axes to the zero return position)
N010 G50 X13.6254 Z8.5481 (Assign program zero)
```

will first send the machine to its zero return position and then assign program zero for the tool.

Here's a more complete example. Say the setup person measures the program zero point for a facing tool and finds them to be 10.2033 in X and 12.4933 in Z. Here is a short program that faces a 6.0 in diameter workpiece and stresses how program zero assignment is done in the program.

Program:

```
O0002 (Program number)
N003 G28 U0 W0 (Ensure that the machine is in the proper starting position)
N005 G50 X10.2033 Z12.4933 (Assign program zero)
N010 G00 T0101 (Index to station number one)
N015 G96 S300 M03 (Turn spindle on CW at 300 SFM)
N020 G00 X6.2 Z0 M08 (Rapid into position, turn on coolant)
N025 G01 X-.062 F.01 (Face workpiece slightly passed center)
N030 G00 Z.1 (Rapid away in Z)
N035 G28 U0 W0 T0 (Return to starting point [zero return position], cancel offset)
N040 M01 (Optional stop)
N045 .
N050 . (Program continues)
```

In line N005, the G50 command tells the machine the tool's current location (while the machine is resting at the zero return position). The G28 command in line N003 makes sure the machine is truly at the zero return position when the G50 command is executed. From this point, the machine knows where the program zero point is located so it can move the cutting tool in the absolute positioning mode accordingly. In line N035, after the facing operation and at the end of the tool, the tool is commanded to go back to its starting position (the zero return position).

This program assumes you want the zero return position to be the program's starting point. Since the zero return position is usually close to the plus over-travel limit in each axis, the machine must move a lengthy distance to approach the workpiece with each cutting tool. And this lengthy motion must be made as the tool retracts for a turret index (tool change). For most machines, the zero return position doesn't make a very *efficient* program starting position.

There is a way to change the program's starting position, making it closer to the workpiece. But doing so makes program zero assignment even *more* difficult, and creates more potential for operator error. If you're just starting out, and if you must assign program zero in the program, we urge you to use the zero return position as the program's starting position.

Using the zero return position as the starting position for your programs allows you to simply include a G28 command at the end of each tool to send the machine to its zero return position. It also makes it easy for the operator to know when the machine is in its proper starting position, since two axis origin lights will come on when the machine is sent to its zero return position. If both axis origin lights are both on, the machine is in the proper location to start the program. If one or both are off, the machine is out of position. While this may not be very efficient, programming is kept simple and operation is kept safe and easy (at least as simple, safe, and easy as it can be when program zero is assigned in the program).

Making the machine more efficient when assigning program zero in the program

As stated, the zero return position isn't a very efficient tool change position for most turning centers. Consider, for example, a machine having a long bed with 40 inches of Z axis travel. If the machine's zero return position is close to the Z plus over-travel limits, and if you're running a short workpiece on a long-bed machine, each tool will have a time consuming (wasteful) approach and retract movement.

Also as stated, you should only consider trying to optimize approach and retract movements once you've gained some experience and confidence with the machine. But it is possible and desirable to reduce these wasteful motions, and your savings will be greatest with larger lot sizes. Here is a general procedure you can follow to change the tool change position from the zero return position to a location closer to the workpiece. Note that you can use this procedure for both the X and Z axes, but it is usually the Z axis that has its zero return position furthest from the workpiece. Usually reducing the approach/retract distance for only the Z axis will be sufficient to reduce program execution time.

1) Measure the program zero assignment values just as described in Figure 1.50 - All the way back to the zero return position in X and Z. Write down these values for each tool.

2) Come up with the distance from the zero return position to the safe index point - With the machine resting at the zero return position, first zero the U and W values on the relative position display screen page. Next, manually jog the turret to a position closer to the workpiece, but that still allows the turret to be safely indexed (you may want to manually index the turret all the way around to confirm that there won't be interference). When you're satisfied with your new starting position the U and W values will be showing the distances from the zero return position to your safe index point. To keep calculations as simple as possible, we recommend rounding these values to the next *smaller* whole number. Again, you can modify the program's starting position for the X axis, but it is usually the Z axis approach and retract motions that are the most wasteful. You may wish to keep the X axis zero return position as the program's starting position in X to keep things as simple as possible.

3) Subtract the value measured in step two from all program zero assignment values measured in step one - Say you determined in step two that the turret could be moved 18 inches from the zero return position closer (minus) to the workpiece in the Z axis. Your safe index position will be 18 inches from the zero return position. In this case, subtract 18 inches from all Z axis program zero assignment values (one for each tool) measured in step one. The resultant values will be used in each G50 command within the program.

4) Modify all retract motions within the program - You can no longer use the simple G28 command for the purpose of retracting tools to the tool change position (zero return will no longer be the tool change position). For this reason, all retract movements must reflect your new tool change position. Your positioning movement must specify the same position as is used in the G50 command. If, as in our previous example, the modified G50 command is

```
N010 G50 X10.3857 Z9.2843
```

the retract motion to the tool change position will be

```
N095 G00 X10.3857 Z9.2843 T0
```

(instead of the simpler G28 U0 W0 T0) This must be done for each tool's retract motion.

5) Be sure to send the machine to this tool change position prior to running the program - In our example, the operator must move the machine 18 inches in the Z minus direction from the zero return position before the program can be run. They will not, of course, have to do this once the machine is in production since the program will leave the machine where it started (the retract motion command for the last tool will leave the machine at its tool change position). But considering how often the operator must move the machine to this special program starting position (when there's a mishap, after power up, when rerunning tools, and several other times), it can be tedious and error prone to manually cause this movement.

For this reason, many programmers include a special program segment to send the machine to its planned starting position. The commands:

```
.
.
N455 M30 (Normal program ending command)
N999 G28 U0 W0 (Note special sequence number that is easy to remember)
N1000 G00 W-18.0 (Incrementally move 18 inches in the Z minus direction)
N1001 M30 (End of program command)
```

When the operator wishes to send the machine to its planned starting position, they'll scan to sequence N999 (an easily remembered sequence number) and run the program from there. The machine will go to the zero return position and then move to by the appropriate amount to the planned starting position (18 inches in our case).

6) With the machine resting at the planned starting position, zero the values on the relative display screen page - Since the zero return position is no longer being used to change tools, the operator will no longer be able to monitor the axis origin lights to tell if the machine is in the proper starting position. One trick that overcomes this problem is to use the relative display screen values for the same purpose. If the U and W values are zero (after performing this step), the machine is in its planned starting position. If for any reason these values are not zero, the machine is out of position.

As you can see, modifying the program's starting position is not easy – and it is error prone. Again, you should only attempt it after you have gained experience and confidence with your machine.

Key points for Lesson Seven:

- There are many ways to assign program zero. We show the four most popular ways in order of most preferable to least preferable.

- If your machine has a tool touch off probe, use it to assign program zero – it provides many advantages.
- If your machine has geometry offsets, use them to assign program zero.
- Use the measure function and work shift value with geometry offsets to minimize work and duplication of effort.
- Ask an experience setup person what the polarity is for the work shift value for your machine/s.
- Only assign program zero in the program (with G50) if your machine does not have geometry offsets.

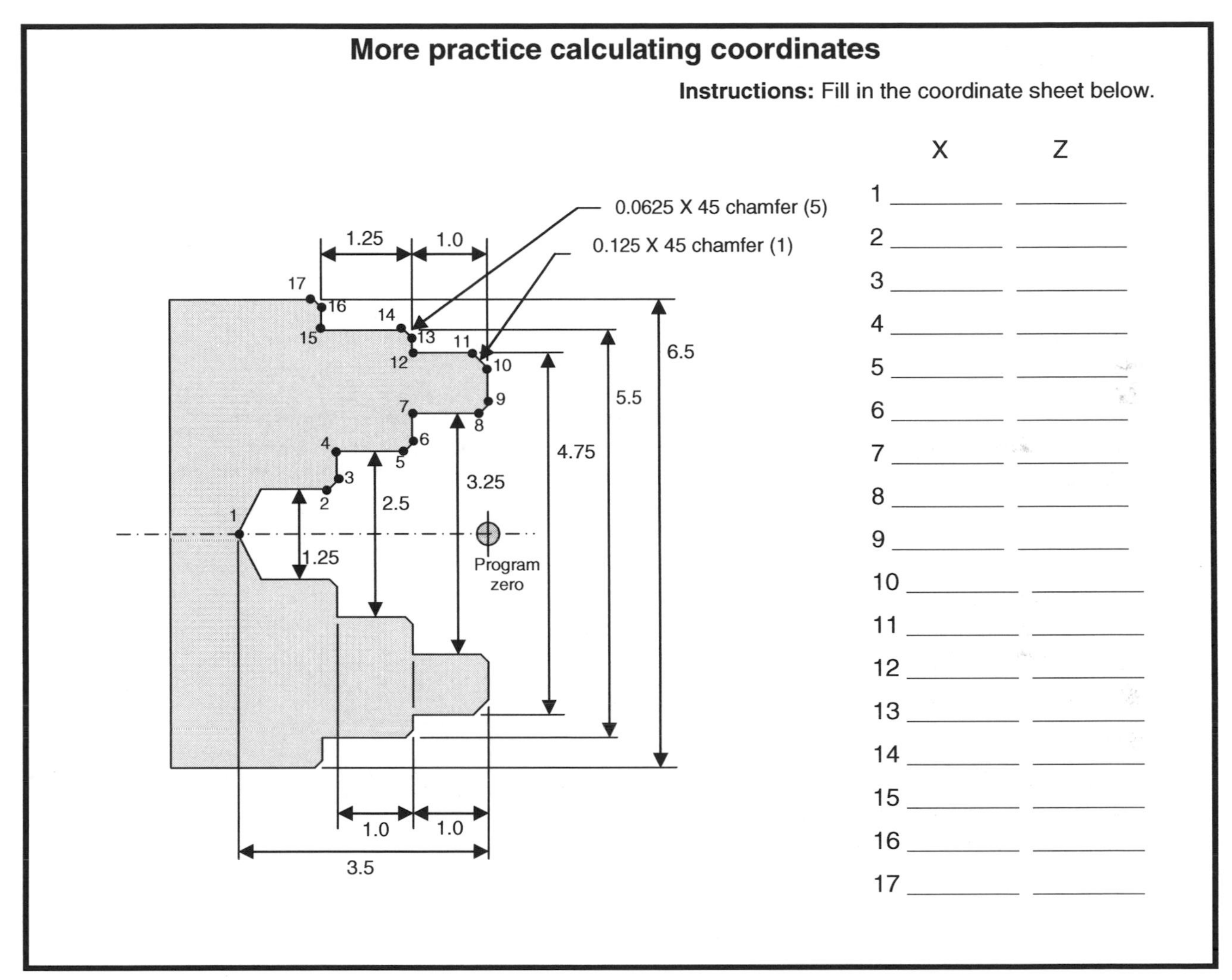

Answers:

1: X0 Z-3.5
2: X1.25 Z-2.0625
3: X1.375 Z-2.0
4: X2.5 Z-2.0
5: X2.5 Z-1.0625
6: X2.625 Z-1.0
7: X3.25 Z-1.0
8: X3.25 Z-0.0625
9: X3.375 Z0
10: X4.5 Z0
11: X4.75 Z-0.125
12: X4.75 Z-1.0
13: X5.375 Z-1.0
14: X5.5 Z-1.0625
15: X5.5 Z-2.25
16: X6.375 Z-2.25
17: X6.5 Z-2.3125

Talk to experienced people in your company...

... to learn more about how program zero is assigned on your company's turning centers. If you work for a CNC-using company, here are questions you'll want to get answered.

1) Do any of your company's machines have a tool touch off probe? If so, is it being used? Is it also being used when dull tools are replaced? Ask to see it demonstrated.

2) For machines without a tool touch off probe, do they have geometry offsets? Do they have the work shift value? Do they have the measure function? Are these features being used as is shown in this text? Ask to see program zero assignment demonstrated.

3) If the work shift value is used with any of your machines, what is its polarity (plus or minus)?

4) Does your company own any older machines that don't have geometry offsets? Is program zero being assigned with G50? Ask to see program zero assignment demonstrated.

5) Are any setup people in your company using a method of program zero assignment that we have not shown in this text? If so, why? Ask to have this method explained and demonstrated.

STOP! Do practice exercise number seven in the workbook.

Lesson 8

Introduction To Programming Words

As you know, all CNC words include a letter address and a numerical value. The letter address identifies the word type. You should be able to quickly recognize the most common CNC word types.

You know that CNC programs are made up of commands – and that commands are made up of words. Words are categorized into types – and each word type has a special meaning to the machine. Each word type is designated by a *letter address*. You already know a few the letter addresses, like N for sequence number, G for preparatory function, X and Z for axis designations, S for spindle speed, F for feedrate, and M for miscellaneous (or machine) functions. In this lesson, we're going to introduce the rest of the word types.

If you are a beginner looking at the word types for the first time, you may want to read this section a few times to get better acquainted with these word types. Note that we are *not* asking you to memorize the word types – just to get familiar with them. In Key Concept Number Five – program formatting – we will provide you with a way to remember each word's function.

Also, this lesson is only intended to *introduce* each word, not to give you an in-depth description. When appropriate, we'll point you to the lessons that discuss the word type in more detail.

You will find that certain words are seldom used, meaning you will have little or no need for them. Other words are constantly used, and you will soon have them memorized.

Some CNC words have more than one function, depending on commanded format. We will be showing you the primary (most common) function of the word next to the "A" description and the secondary uses for the word next to the "B" and "C" descriptions.

Once you've seen a word type a few times, it should not be too difficult to remember its function. Again, most word types are aptly named with an easy-to-remember letter address. Additionally, only about fifty words are used consistently, so try to look at learning to program a CNC turning center as like learning a foreign language that contains only fifty words.

As you continue with this text, *use this lesson as a reference*. If you come across a word or word type you don't recognize, remember to come back to this lesson. You've probably already noticed that we provide a quick reference for CNC words on the inside front and back cover of this text, but information in this lesson is a little more detailed.

Words allowing a decimal point

Current CNC controls allow you to include a decimal point in those words that are used to specify *real numbers* (values that require a portion of a whole number). You must remember to include a decimal point with these words or the control will revert to the fixed format for the word (as discussed in lessons three and four). Word types that allow a decimal point include:

X, Y, Z, U, W, C, I, K, F, E, and R

Certain CNC words are used to specify *integers* (whole numbers). These word types do *not* allow a decimal point:

O, N, G, P, L, S, T, and M

O

This is the word most CNC controls use for a **program number**. All machines discussed in this text allow you to store multiple programs in the memory of the control. The program number allows the specification of an individual program currently stored in the control. You will be assigning every program a program a number from 0001 through 9999 (O0001 through O9999). The O word will be the *very first word* in the program. A decimal point is not allowed with the O word. Program numbers are discussed in more detail in Lesson Fifteen.

N

This word specifies a **sequence number**. Sequence numbers are used to identify commands in a CNC program. They allow you to organize each command in the program by number. This allows easier modification of the program at the machine. Sequence numbers are not required to be in any particular order and can even repeat in the program. Actually, they need not be in the program at all. But for the sake of organization purposes, we recommend that you include them in the program and place them in an understandable order. We use them for all example programs in this text. For most examples, we skip five numbers for each sequence number (going by fives). This allows extra commands to be added to the program while still using sequence numbers. A decimal point is not allowed with the N word. Sequence numbers are introduced in Lesson Four and discussed in more detail in Lessons Fifteen and Sixteen.

G

This word specifies a **preparatory function**. Preparatory functions prepare the machine for what is coming – in the current command – and often in upcoming commands. They set modes (though some G codes are non-modal). There are many G words, but only a few are used on a consistent basis. For a list of all G codes, see the list at the end of this lesson. A decimal point is not allowed with most G words. But do note that with Fanuc controls, some G codes do use a decimal point (like G84.1 for rigid tapping). A decimal point used in this fashion is simply part of the G code's designation – and is not the true application for a decimal point. Preparatory functions are introduced in Lesson One and are discussed often throughout this text.

X

A. The primary use for the X word is to designate coordinates in the absolute positioning mode along the X axis. That is, it is the **X axis designator**. The X word allows a decimal point. An X position of 10 inches will be specified X10.0. Axis designators are introduced in Lesson One and discussed often during this text.

B. The secondary use for the X word is that it can be used to specify **dwell time** in seconds in a dwell command (G04). Dwell commands are used to make axis motion (for all axes) pause for a specified length of time. The dwell command is discussed in Lesson Twenty-Three.

C. X is also used to specify the **major or minor diameter of a thread** with the multiple repetitive cycle for threading (discussed in Lesson Twenty).

Y

For turning centers equipped with a Y axis (not many turning centers have this function) this letter address specifies **absolute positioning along the Y axis**. A decimal point is allowed with the Y word.

Z

A. The primary use for the Z word is as the **Z axis designator in the absolute positioning mode**. The Z word allows a decimal point.

B. The secondary use for the Z word is that it has some **special applications with certain multiple repetitive cycles** (multiple repetitive cycles are discussed in Lessons Eighteen through Twenty).

U

A. The primary use for letter address U is to specify movements using **incremental positioning along the X axis**. Incremental positioning is introduced in Lesson Five. It is also discussed in Lesson Twenty-One. U specifies a *diameter change*. The command U1.0 will cause the cutting tool to increase by one inch (0.5 inch of actual motion). A decimal point is allowed with the U word.

B. Letter address U is also used for a **special purpose with multiple repetitive cycles** (discussed in Lessons Eighteen through Twenty).

C. Letter address U can also be used within a dwell command (G04) to specify **dwell time** (in seconds) for a dwell command. For example the command G04 U0.5 specifies a 0.5 second dwell. The dwell command is presented in Lesson Twenty-Three.

W

Letter address W is used to specify movements using **incremental positioning along the Z axis**. Incremental positioning is introduced in Lesson Five. It is also discussed in Lesson Twenty-One. The W word allows a decimal point.

C

A. Most turning centers that are equipped with live tooling also have rotary axis within the spindle (though some have a simple indexer). Turning centers with live tooling can perform machining operations that are normally associated with machining center (milling) work. If the turning center has a rotary axis, this letter address is used to specify angular position along the C axis. That is, letter address C is the **C axis designator**. With a C axis, machining can occur during rotation. A decimal point is allowed with the C word. The rotary axis is introduced in Lesson One and is discussed in detail in the Appendix (after Lesson Twenty-Three).

B. For live-tooling turning centers equipped with a one-degree indexer within the spindle (instead of a rotary axis) the C word is the **indexer activator**. You are not allowed to use a decimal point with the C word when it is used to activate an indexer. C50 specifies a fifty degree index. With an indexer, machining can only occur *after* the indexer rotates (not during rotation).

R

The letter address R is used to specify the **radius of a circular move**. The R word allows a decimal point. Circular motion is discussed in Lesson Ten.

I & K

A. I and K words can be used to specify the **arc center point of a circular motion**. While they are still effective, we strongly recommend that beginners concentrate on using the R *word* to specify the arc size in a circular move because it is much easier. I and K allow a decimal point. Circular motion (including the use of directional vectors) is discussed in Lesson Ten.

B. Letter address I is also used to specify the **amount of taper in a taper thread** when the threading multiple repetitive cycle is used. The threading multiple repetitive cycle is discussed in Lesson Twenty.

F

The F word specifies **feedrate** – which is the motion rate for machining operations. It is used with straight line and circular motion commands (see G01, G02, and G03 in Lesson Ten), along with any other interpolation types equipped with your turning center/s. Feedrate is affected by the currently instated measurement system mode (inch or metric). Inch mode is specified with G20 and metric mode is specified with G21. Feedrate can be specified in two ways, feed per revolution and feed per minute. Feedrate specification is introduced in Lesson Two and discussed throughout this text.

E

The letter address E is used to specify the **pitch of a thread** during a threading command. Threading commands are discussed in Lessons Seventeen and Twenty. While an F word can also be used to specify the pitch of the thread, the F word allows only four places of accuracy. With the E word, up to six digits of accuracy are allowed. For example, with a twelve threads per inch thread, the pitch is 0.0833333333... (1/12). With the F word this pitch can only be specified to four places (as F0.0833). With the E word six places of accuracy can be specified (as E0.083333).

S

A. The letter address S specifies a **spindle speed**. There are two ways to specify spindle speed – in constant surface speed (surface feet per minute or meters per minute, depending upon measurement system mode) and in rpm. Spindle speed selection is discussed in Lesson Two.

B. An S word specified within a G50 command can be used to designate the **maximum rpm** allowed when in constant surface speed mode. The spindle limiter is introduced in Lesson Two and discussed again in Lessons Fifteen and Sixteen.

T

The letter address T is a four digit value and designates three different functions on most machines. The first two digits of the T word specify the **turret station to which the turret will index**. For machines with geometry offsets, the first two digits also **instate the geometry offset number**. The second two digits of the T word **specify the wear offset number** to be used with the tool. For example, the command T0101 rotates the turret to station number one, instates geometry offset number one, and instates wear offset number one. The T word is introduced in Lesson One. It is also discussed in Lesson Seven. And it is presented in Lessons Eleven through Fourteen. The T word does not allow a decimal point.

M

An M word specifies a **miscellaneous function** (also called a **machine function**). You can think of M words as programmable on/off switches that control functions like coolant and spindle activation. For a list of all M words, see the list at the end of this lesson. Note that machine tool builders will select their own set of M words. While there are many standard M word numbers, you must consult your own machine tool builder's programming manual to find the exact list for your particular machine/s. A decimal point is not allowed with the M word. Miscellaneous functions are introduced in Lesson One and discussed numerous times throughout this text.

P

A. The P word can be used to specify **dwell time** in seconds for a dwell command (G04). Dwell commands are used to make axis motion (for all axes) pause for a specified length of time. A dwell time of three seconds is specified as P3000 (a decimal point is *not* allowed). Note the fixed format of the P word. There are three places to the right of the automatically placed decimal point position. Other examples: P2500 is 2.5 seconds, P500 is 0.5 second, and P10000 is 10 seconds. Note that the X word can also be used to specify the time for a dwell command – and since it allows a decimal point – most programmers prefer using it instead of the P word. Again, a decimal point is not allowed with the P word. The dwell command is discussed in Lesson Twenty-Three.

B. The secondary use for the P word is with sub-programming to specify the **subprogram program** number of the program to be called. A decimal point is not allowed with the P word. Sub-programming is discussed in Lesson Twenty-One.

C. P is also used with multiple repetitive cycles to specify the **starting command of a finish pass designation**. Multiple repetitive cycles are discussed in Lessons Eighteen through Twenty.

Q

This letter address is used with multiple repetitive cycles to specify the **ending command of a finish pass designation**. Multiple repetitive cycles are discussed in Lessons Eighteen through Twenty.

L

The L word is used with sub-programming to specify the **number of executions** for the subprogram. A decimal point is not allowed with the L word. Subprograms are discussed in Lesson Twenty-One.

EOB (end of block character)

EOB stands for **end-of-block**. Though it is not actually a CNC word, it is an important part of a CNC program. It is called a *command terminator*. On a turning center's display screen, it usually appears as a semicolon (;). When you type a program using a computer, it is automatically inserted at the end of each command when you press the *Enter key* (though it will not appear on the computer display). Though it is automatically inserted into your program when typing programs on a computer, it is not automatically entered if you type (or modify) programs through the keyboard and display of a CNC turning center. A special key on the control's keyboard labeled EOB must be pressed to insert the end-of-block character into the program. This must be done at the end of every command. Program structure is discussed in Lessons Fifteen and Sixteen.

/ (slash code)

This is called the **block delete** word (also called **optional block skip**). It is the slash character on your keyboard (under the question mark on most keyboards). It works in conjunction with an on/off switch on the control panel (labeled block delete or optional block skip). If the switch is on when the machine reads the slash code, the machine will ignore the words to the right of the slash code. If the switch is off, the machine will execute the words in the normal manner. Block delete, including several applications for this feature, is discussed in Twenty-Three.

G and M codes

Here we list most of the G and M codes that can be used in programming, providing little more than the name for each word. Rest assured that the most often used G and M codes are discussed in detail in this text. You can find documentation for lesser used G and M codes in your control manufacturer's programming manual.

G codes

As you know, G codes specify preparatory functions. They prepare the machine for what is to come – in the current command – and possibly in up-coming commands. They set modes.

G code limitation:

With most controls, *only three compatible G codes are allowed per command.* If you exceed this limitation, most controls will *not* generate an alarm. They will simply execute the last three G codes in the command – ignoring those prior to the last three. For example, in the command

G99 G40 G20 G96 (Select feed per revolution mode, cancel tool nose radius compensation, select metric mode, and select constant surface speed mode)

only the G40, G20, and G96 codes will be executed. The G99 code will be ignored. If needed in a program, these four G codes must be broken into two commands.

By *compatible* G codes, we mean G codes that work together. For example, in the command

G20 G40 G00 (select the inch mode, cancel tool nose radius compensation, and rapid mode)

all G codes are compatible. They work together. But you cannot, of course, specify G20 and G21 (selecting inch and metric mode) in the same command.

Option G codes

Some of the G codes in the up-coming list are *option G codes.* It is impossible to tell whether a given option G code is included in your control or not by just looking at our list since most machine tool builders include a standard package of options when they purchase controls from the control manufacturer. Our list shows what one popular control manufacturer specifies as options (Fanuc). Probably when your company purchased the turning center from your machine tool builder, other (option) G codes came with the machine. If there is any question as to whether your turning center has any a particular option G code, you can perform a simple test at the machine to find out if the G code is available to you (or you can call your builder to find out if the G code was included).

To make the test for an option G code, simply command the G code in the MDI mode (techniques given in Lesson Twenty-Seven). You need not even specify the correct format for the G word. If you receive the alarm Unusable G code or G code not available, your machine does not have the G code. If you receive no alarm or if the alarm is related to the format of the G code, the option G code should be available for you to use.

What does initialized mean?

In the up-coming list, we also state whether the G code is *initialized*, meaning whether it is automatically instated the machine's power is turned on.

What does modal mean?

Most G codes are *modal*, meaning once they are invoked, they remain in effect until they are changed or cancelled. This means you do not have to keep repeating modal G codes in every command. There are some G codes that are non-modal (also referred to as *one-shot G codes*). Non-modal G codes only have an effect on the command in which they are included.

The most popular G codes

As you look at the entire list, it may seem a little intimidating. Admittedly, there are a lot of G codes. Here are the most popular G codes (about thirty), along with where they are discussed in this text.

G00, G01, G02, G03: motion types – discussed in Lesson Ten

G04: dwell command – discussed in Lesson Twenty-Three

G20, G21: inch and metric mode – discussed in Lessons One, Five, and Fifteen

G28: zero return command – introduced in Lesson Five and discussed in more detail in lessons fourteen and fifteen
G40, G41, G42: tool nose radius compensation – discussed in Lesson Fourteen
G50: spindle limiter – discussed in Lessons Two, Fifteen, and Sixteen
G70: finishing multiple repetitive cycle – discussed in Lessons Eighteen and Nineteen
G71: rough turning and boring multiple repetitive cycle – discussed in Lesson Eighteen
G72: rough facing multiple repetitive cycle – discussed in Lesson Nineteen
G76: threading multiple repetitive cycle – discussed in Lesson Twenty
G96: constant surface speed spindle mode – discussed in Lesson Two
G97: rpm spindle mode – discussed in Lesson Two
G98: per minute feedrate mode – discussed in Lesson Two
G99: per revolution feedrate mode – discussed in Lesson Two

G code	Description	Status	Initialized
G00	Rapid motion	std	yes
G01	Straight line cutting motion	std	no
G02	Circular motion (CW)	std	no
G03	Circular motion (CCW)	std	no
G04	Dwell command	std	no
G09	Exact stop check command	option	no
G10	Offset setting by program command	option	no
G112.1	Polar coordinate interpolation	option	no
G113.	Cancel polar coordinate interpolation	option	yes
G20	Inch mode	std	yes (for most US machines)
G21	Metric mode	std	no (for most US machines)
G22	Safety zone setting	option	no
G23	Safety zone cancel	option	yes
G27	Zero return check	std	no
G28	Zero return command	std	no
G29	Return from zero return	std	no
G30	Second reference point command	option	no
G31	Skip cutting command	option	no
G32	Basic thread cutting command	std	no
G34	Variable lead thread cutting	option	no
G36	Automatic tool compensation in X	option	no
G40	Tool nose radius compensation	std	yes
G41	Tool nose radius compensation left	std	no
G42	Tool nose radius compensation right	std	no
G50	Program zero designation command	std	no
G50.1	Mirror image	option	no
G51.1	Mirror image cancel	option	yes
G52	Local coordinate system	std	no
G54-G59	Work coordinate systems 1-6	std	no
G61	Exact stop check mode	std	no
G64	Normal cutting mode	std	yes

G code	Description	Status	Initialized
G65	Custom macro call command	option	no
G66	Custom macro modal call	option	no
G67	Custom macro modal call cancel	option	yes
G70	Finishing cycle	std	no
G71	Rough turning and boring cycle	std	no
G72	Rough facing cycle	std	no
G73	Pattern repeating cycle	std	no
G74	Peck drilling cycle	std	no
G75	Grooving cycle	std	no
G76	Treading cycle	std	no
G90	One pass turning cycle	std	no
G94	One pass facing cycle	std	no
G92	One pass threading cycle	std	no
G96	Constant surface speed mode	std	no
G97	RPM speed mode	std	yes
G98	Feed per minute mode	std	no
G99	Feed per revolution mode	std	yes

Common M codes used on a CNC turning center

Again, machine tool builders vary with regard to the M codes they provide – so this is just a partial list. You must reference your machine tool builder's programming manual for a complete list of the M codes available for a particular CNC turning center.

Note that some CNC control models (like many supplied by Fanuc) allow but one M code per command. If you include more than one M code, it's hard to predict what will happen for all machines. Some machines will stop executing the program while others will generate an alarm. Yet others will continue executing the program, ignoring all but the last M code in the command.

M CODE	DESCRIPTION
M00	Program stop
M01	Optional stop
M02	End of program (does not rewind memory for most machines – use M30)
M03	Spindle on in a forward direction
M04	Spindle on in a reverse direction
M05	Spindle stop
M08	Flood coolant on
M09	Coolant off
M30	End of program (rewinds memory)
M41	Low spindle range
M42	High spindle range
M98	Sub program call
M99	End of sub program

Other M Codes for your machine (found in your machine tool builder's manuals)

____ ______________________________

____ ______________________________

____ ______________________________

____ ______________________________

____ ______________________________

____ ______________________________

____ ______________________________

____ ______________________________

Key points for Lesson Eight:

- While there are many CNC words available to programmers, there are only about fifty words that are used on a regular basis. Look at learning CNC programming as like learning a foreign language that has only fifty words.
- The letter addresses for many CNC word types are easy to associate with their usage (F for feedrate, T for turret, S for spindle, etc.). But other word types are more difficult to remember.
- You've been exposed to all of the word types used with CNC turning center programming in this lesson – as well as where (in this text) you can find more information about the most often-used words.
- There are only about thirty G words used on a regular basis.
- Only three compatible G codes are allowed per command with most CNC controls.
- Only one M code is allowed per command with some CNC controls (including Fanuc).
- You must reference your machine tool builders programming manual to find the complete list of M codes for your turning center/s.

STOP! Do practice exercise number eight in the workbook.

Key points: Key Concept one, Know your machine from a programmer's viewpoint

We've tried to provide a good foundation on which you can build your understanding of CNC turning center usage. Let's recap the most important points from Key Concept Number One.

Lesson One – Machine configurations:

- There are several types of CNC turning centers.
- With all types of CNC turning centers, X is the diameter controlling axis and X minus is the toward spindle center, Z is the length controlling axis and Z minus is toward the workholding device.
- X axis positions are specified in diameter.
- Some turning centers – those that have live tooling – can additionally perform machining-center-like machining operations).
- You must understand the functions of your turning center that can be programmed.
- Spindle can be controlled in at least three ways (activation, direction, and speed). Additionally, many turning centers have more than one spindle range.
- Spindle speed can be specified in rpm or in surface feet/meters per minute.
- Feedrate specifies the motion rate for machining operations.
- Feedrate can be specified in per revolution fashion or per minute fashion.
- Coolant can be activated to allow cooling and lubricating of the machining operation.
- Most turning centers have a turret in which cutting tools are placed.
- Machines in the United States allow the use of inch or metric mode.
- You must determine what else is programmable on your turning center/s.

Lesson Two – Understanding turning center speeds and feeds:

- With CNC turning centers there are two ways to specify spindle speed (constant surface speed and rpm) and two ways to specify feedrate (feed per revolution and feed per minute).
- Constant surface speed mode (G96) should be used with single point turning tools that machine more than one diameter during the machining operation. Rpm mode (G97) must be used for tools that machine on center (like drills) and for threading operations.
- Constant surface speed mode will cause the machine to automatically select the appropriate spindle rpm based upon the specified speed in surface feet/meters per minute and the current diameter position of the cutting tool.
- Per revolution feedrate mode (G99) is initialized (automatically selected at power-up) and should be used for almost all machining operations. Feed per minute mode (G98) should only be used for feedrate movements when the spindle is stopped.
- Using per revolution feedrate mode will ensure that witness marks (surface finish) on the workpiece will be consistent for all surfaces being machined.
- Rpm = 3.82 times sfm divided by the diameter being machined.
- Ipm = previously calculated rpm times feedrate per revolution.

Lesson Three – General flow of the CNC process:

- There are four types of companies that use CNC machine tools.
- The type of company dramatically affects what is expected of CNC people.
- You must know what you will be expected to do in the CNC environment.
- It helps to understand the bigger picture of how a job is processed through a CNC using company.

Lesson Four – Visualizing program execution:

- You must be able to visualize the motions cutting tools will make as they machine workpieces.
- CNC programs are executed in sequential order, command by command.
- Commands are made up of words.

- Words are comprised of a letter address and a numerical value.
- You've seen your first complete CNC program. Though it was short, you've seen the sequential order by which CNC programs are executed – and you should understand the importance of being able to visualize a program's execution.
- We've introduced some program-structure-related points, including decimal point programming, modal words, and initialized words.
- You've seen some of the mistakes beginners are prone to making – like mistakes of omission and using an upper case letter O instead of a number zero.

Lesson Five – Program zero and the rectangular coordinate system:

- You must be able to specify positions (coordinates) within CNC programs.
- You know that the rectangular coordinate system of a CNC machine is very similar to that used for a graph.
- In CNC terms, the origin for a coordinate system is called the program zero point.
- From a programmer's viewpoint, polarity for each positioning movement is based upon the commanded position's relationship to program zero.
- The program zero point location is determined based upon how the workpiece drawing is dimensioned.
- For the X axis, coordinates must be specified in diameter for most turning centers.
- When you specify coordinates relative to program zero, you're working with absolute positioning.

Lesson Six – Determining program zero assignment values o:

- Program zero assignment values must be determined before program zero can be assigned.
- Each cutting tool requires its own set of program zero assignment values.
- Program zero assignment values are the distances between the program zero point and the machine's zero return position.
- The work shift value can be used in the Z axis to keep Z axis program zero assignment values consistent for tools used in consecutive jobs.
- One way to determine program zero assignment values is to actually measure them during setup at the machine.
- We've shown how program zero assignment values can be determined for external (turning) tools, center cutting tools, and internal (boring) tools.

Lesson Seven – Four ways to assign program zero:

- There are many ways to assign program zero. We show the four most popular ways in order of most preferable to least preferable.
- If your machine has a tool touch off probe, use it to assign program zero – it provides many advantages.
- If your machine has geometry offsets, use them to assign program zero.
- Use the measure function and work shift value with geometry offsets to minimize work and duplication of effort.
- Ask an experience setup person what the polarity is for the work shift value for your machine/s.
- Only assign program zero in the program (with G50) if your machine does not have geometry offsets.

Lesson Eight – Introduction to CNC words:

- While there are many CNC words available to programmers, there are only about fifty words that are used on a regular basis. Look at learning CNC programming as like learning a foreign language that has only fifty words.

- The letter addresses for many CNC word types are easy to associate with their usage (F for feedrate, T for turret, S for spindle, etc.). But other word types are more difficult to remember.
- You've been exposed to all of the word types used with CNC turning center programming in this lesson – as well as where (in this text) you can find more information about the most often-used words.
- There are only about thirty G words used on a regular basis.
- Only three compatible G codes are allowed per command with most CNC controls.
- Only one M code is allowed per command with some CNC controls (including Fanuc).
- You must reference your machine tool builders programming manual to find the complete list of M codes for your turning center/s.

You Must Prepare To Write Programs

While this Key Concept does not involve any programming words or commands, it is among the most important of the Key Concepts. The better prepared you are to write a CNC program, the easier it will be to develop a workable program.

Key Concept Number Two is made up of but one lesson:

8: Preparation for programming

Frankly speaking, this Key Concept can be applied to any CNC related task, including setup and operation. Truly, the better prepared you are to perform a task, the easier it will be to correctly complete the task. When it comes to making setups, for example, if you have all of the needed components to make the setup at hand (jaws, cutting tools, cutting tool components, program, documentation, etc.), making the setup will be much easier and go faster. So *gathering needed components* is always a great preparation step to perform.

The same goes for completing a production run. If the CNC operator has all needed components (raw material, inserts for dull tool replacement, a place to store completed workpieces, gauging tools, etc.), they will be able to smoothly complete the production run.

In this Key Concept, however, we're going to limit our discussion to preparation steps that you can perform to get ready to write a CNC program. Remember, the actual task of *writing* a CNC program is but part of what you must do. Certain things must be done *before* you're adequately prepared to write the program.

Preparation for programming is especially important for entry-level programmers. For the first few programs you write, you will have trouble enough remembering the various CNC words – remembering how to structure the program correctly – and in general – you'll have trouble getting familiar with the entire programming process. The task of programming is infinitely more complicated if you are not truly ready to write the program in the first place.

Preparation and time

Without adequate preparation, writing a CNC program can be compared to working on a jigsaw puzzle. A person doing the puzzle has no idea where each individual piece will eventually fit. The worker makes a guess and attempts to fit the pieces together. Since the worker has no idea as to whether pieces will fit together, it is next to impossible to predict how long it will take to finish the puzzle.

In similar fashion, if you attempt to write a CNC program without adequate preparation, you will have a tendency to piece-meal the program together in much the same way as a person doing a jig saw puzzle. You will not be sure that anything will work until it is tried. The program may be half finished before it becomes obvious that something is seriously wrong with the process. Worse, the program may be completed and being verified on the CNC turning center before some critical error is found.

CNC machine time is much more expensive than your own time. There is no excuse for wasting precious machine time for something as avoidable as a lack of preparation.

You can also liken the preparation that is required to write a CNC program to the preparation needed for giving a speech. The better prepared the speaker, the easier it will be to make the presentation, and the more effective the speech will be. Truly, the speaker must think through the entire presentation (probably several times) before the speech can be presented. Similarly, the CNC programmer must think through the entire CNC process, setup, and program before the program can be written.

With adequate preparation, writing the program will be *much* easier. Most experienced programmers will agree that the actual task of writing a CNC program is the *easy part* of the programming process. The real work is done in the preparation stages. If preparation is done properly, writing the program will be a simple matter of translating what you want the machine to do (from English) into the language a CNC machine can understand and execute.

Though preparation is so very important, it is amazing to see how many so-called *expert* programmers muddle through the writing of a program without preparing at all. While an experienced person may be able to write workable programs for simple applications with minimal preparation (and even then they only gain this ability through trial-and-error practice), even the so-called expert programmer have problems with more complex programs.

To think that you are saving time by not preparing to write a CNC program can be a terrible mistake. In reality, you usually *add time* to programming and program-verification time if you do not prepare. The period of time saved by not preparing will be quickly lost when you consider the problems caused by lack of preparation. And again, waste is compounded if the added time caused by lack-of-preparation results in lost machine time.

Preparation and safety

Wasted time is but one of the symptoms of poor preparation. Indeed, it may be the least severe one. Poor preparation will often result in all kinds of mistakes.

A CNC turning center will follow a CNC program's instructions *to the letter*. While the control may go into an alarm state if it cannot recognize a given word or command, it will give absolutely no special consideration to motion mistakes. Indeed, a turning center cannot detect motion mistakes. The level of problem encountered because of motion mistakes ranges from minor to catastrophic.

Minor motion mistakes usually do not result in any damage to the machine or tooling, and the operator is not exposed to a dangerous situation. However, the workpiece will not be correctly machined. For example, say you intend to drill a hole at 0.015 inches per revolution with a feedrate word of F0.015 in the drilling command. But, you place the decimal point in the wrong location. You specify a feedrate of F.0015 instead of F0.015. In this case, the machine is being told to run the drill at a much slower feedrate than you intend. No damage to the tool or machine will result – but machining will take much longer than it should.

This mistake will be more serious, of course, if the programmer incorrectly specifies a feedrate of F0.150 instead of F0.015. This time the tool will probably break – and the workpiece may be damaged. If the machine does not have a fully enclosed work area, debris flying out of the work area could injure the operator.

Catastrophic mistakes can result in damage to the machine and injury to the operator. For example, say you intend to position a cutting tool at the machine's *rapid rate* (the machine's fastest motion rate – which for some machines is well over 1,000 inches per minute). The correct position for your command is Z0.1, which is 0.100 inch away from the workpiece. But you make a mistake and include a minus sign for this motion word (Z-0.1 instead of Z0.1). The tool is being told to crash into the part (at rapid). Depending on what kind of tooling is being used, this will, at the very least, cause the tool to break. Worse, the workpiece could be pushed out of the chuck and be thrown from the machine (maybe even through the machine's door). Possibly, if the setup is very sturdy and the tool is very rigid, damage to the machine's way system and/or axis drive system will result. If the tool breaks and parts fly out of the work area, the operator could be injured.

We're not saying these things to scare you. There are program verification procedures that, if followed, almost guarantee that no crash will ever occur, regardless of how severe the positioning mistake. (These verification techniques are shown in Lessons Twenty-Four and Twenty-Eight.) However, as the programmer of a CNC turning center, you must constantly recognize the potential for dangerous situations.

Due to this constant potential for personal injury and/or damage to the machine tool, everyone involved with CNC equipment must treat the turning center with great respect. You must be constantly aware of this possible danger and do everything you can to avoid dangerous situations. Being adequately prepared to write a CNC program is the single most important thing you can do to achieve this end.

We cannot overstress the importance of preparation. Just as the well-prepared speaker is less apt to make mistakes during his or her presentation, so will the well-prepared CNC programmer be less apt to make mistakes while writing a CNC program.

Typical mistakes

It helps to know the mistakes a beginning programmer is prone to make. Knowing these tendencies should help you avoid them. There are actually four categories of mistakes you'll be prone to making.

Syntax mistakes

Syntax mistakes are mistakes related to the basic structure of your program. CNC programming is extremely structured. If a command is encountered that the machine cannot process, it will generate an alarm. Since a beginning programmer is still getting acquainted with the various CNC words and commands, they are very prone to making syntax mistakes.

For example, a beginner might intend to give the command **G96** to instate the constant surface speed mode. But by mistake the command **G6** is given. Since most CNC turning centers do not have a function commanded by **G6**, a syntax alarm will be generated. That is, the machine will go into *alarm state*, halting the program's execution and placing an alarm message on the display screen to help you diagnose the alarm. For this reason, syntax mistakes are usually quite easy to find and diagnose since the message will tell you the reason for the alarm. Other examples of syntax mistakes include:

N055 G02 X3.0 Z-2.0 F4.0 (Missing arc radius designator [R] for circular move)
N055 T02 (Incorrect tool station designator [T02 should be T0202])
N055 G98 P1000 (Incorrect call to subprogram [G98 should be M98])

Motion mistakes

Motion mistakes can be harder to find and diagnose since no alarm will be sounded. They can also be much more severe since the control will perform the *incorrect* motions, regardless of how severe the consequences.

Minor motion mistakes may not cause any damage, but can be very difficult to catch – even when cautious program verification procedures are followed. For example, if a small mistake of less then 0.010 inch is made in a cutting command, everything will appear to be just fine throughout all program verification procedures. Yet when the first workpiece is machined, it will not pass inspection.

More severe motion mistakes can cause more severe damage (and personal injury), but they are usually easier to find during the program's verification. An incorrect positioning movement of more than about 0.5 in is usually pretty easy to spot.

Motion mistakes are commonly caused by miscalculations (incorrect arithmetic) or transposition of numbers. For example, the programmer may have meant to specify X1.125, but instead specified X1.215. As stated, the control will interpret and execute the transposed word, regardless of how severe the consequences.

Mistakes of omission

This is the kind of mistake a beginner is most prone to making. Beginners tend to forget things. They forget to turn the spindle on. They forget to turn the coolant on. They forget to include a feedrate word. They forget to drill a hole before boring it. They forget to include a decimal point in CNC words that require it. Unfortunately, only experience will help you spot your own tendencies when it comes to mistakes of omission. As you write and verify your first few programs, you'll likely find things you have forgotten to do.

Process mistakes

Again, you must have a good understanding of basic machining practice. Since turning centers are multi-tool machines, you must be able to develop a *sequence of machining operations* by which a workpiece can be properly machined (the process). Even a poorly written CNC program can eventually be made to work if the process is good. However, even a perfect program (one that does exactly what the programmer intends) will fail if the process is bad. Beginning programmers who may be a little weak in their processing skills should seek the help of experienced machinists when developing processes. Here is an example of a poor process. See if you can spot the two mistakes before reading on to the next paragraph.

1) Rough face and turn (roughing tool)

2) Finish face and turn (finishing tool)

3) Drill hole in center of workpiece (2.0" drill)

4) Rough bore (1.25" boring bar)

5) Finish bore (1.25" boring bar)

While processing is a *very* subjective topic, most experienced machinists will agree that this process breaks an important rule of basic machining practice: *"You should rough everything before you finish anything."* Prior to finish facing and turning (step two in the process above), the hole should be drilled and rough bored (steps three and four).

As with most powerful machining operations, it is likely that the workpiece will shift a small amount during the drilling and rough boring in this process. If it shifts after the finish facing and turning but before finish boring, the external diameters will not be concentric with the internal diameters. And again, even if the program is written perfectly (machining with this poor process), the workpiece will not pass inspection.

Here is the corrected process:

1) Rough face and turn (roughing tool)

2) Drill hole in center of workpiece (2.0" drill)

3) Rough bore (1.25" boring bar)

4) Finish bore (1.25" boring bar)

5) Finish face and turn (finishing tool)

Again, if you feel at all weak in your processing skills, be sure to show your intended process to an experienced person. This will confirm that what you intend to do will work.

Lesson 9

Preparation Steps For Programming

Any complex project can be simplified by breaking it down into small pieces. This can make seemingly insurmountable tasks much easier to handle. CNC turning center programming is no exception. Learning how to break up this complex task will be the primary focus of Lesson Nine.

As you now know, preparation will make programming easier, safer, and less error-prone. Now let's look at some specific steps you can perform to prepare to write CNC programs.

Prepare the machining process

Process sheets, also called routing sheets, are used by most manufacturing companies to specify the sequence of machining operations that must be performed on a workpiece during the manufacturing process. The person who actually prepares the process sheet must, of course, have a good understanding of machining practice, and must be well acquainted with the various machine tools the company owns. This person determines the best way to produce the workpiece in the most efficient and inexpensive possible way, given the company's available resources.

In most manufacturing companies, this involves *routing* the workpiece through a series of different machine tools and processes. Each machine tool along the way will perform only those operations the process planner intends, as specified on the routing sheet. This commonly means that some non-CNC machine tools are needed to complete a given workpiece. For example, the workpiece shown in practice exercise at the end of Lesson Seven might have the following routing sheet.

Op. #:	Operation:	Machine:
10	Procure material	Vender ID #12322
20	Cut bar stock to 5.7" long	Cut off saw
30	Clean and de-burr	Cleaning tanks
40	Face, turn, drill, and bore	CNC turning center
50	Clean and de-burr	Cleaning tanks
60	Plate with nickel	Finishing tanks

Note that only operation 40 of this process requires a CNC machine tool. When a CNC machine is involved in the process plan, often the CNC machine will be required to perform several machining operations on the workpiece (as is the case in our example). As you know, CNC turning centers are multi-tool machines, meaning several tools can be used by one program. In some companies, the process sheet will clearly specify the order of machining operations that must be performed by the CNC machine. However, the vast majority of process planners do not get so specific with their routing sheets. Instead, the *sequence of machining operations to be performed on the CNC machine is left completely to the CNC programmer.* If the programmer must develop the machining order, of course, they must possess a good understanding of basic machining practice.

Regardless of *who* does the planning, the step-by-step machining order required to machine a workpiece on the CNC machine must be developed before the CNC program can be written. With a simple process, an experienced programmer may elect to develop the process as the program is being written. While some experienced programmers have the ability to do this, beginning programmers will find it necessary to plan *and document* the machining process first.

Developing your machining process before the program is written will serve several purposes. First (as stated), it will allow you to check the process for errors in basic machining practice before writing the program. You will be forced to think through the entire process *before* the first CNC command is written. You will have the opportunity to spot a problem with the process that will be difficult to spot and/or repair if the process is developed while the program is being written.

Second, developing the machining process prior to writing the program allows you *concentrate on your machining practice skills separate from your CNC programming skills.* While developing the machining process, you concentrate on the machining practice in order to develop a workable process. Your mind is occupied only with the task at hand. While programming, you concentrate on programming skills, translating your process into a language that the CNC machine can understand.

Third, if you spot an error in your thinking, you will be able correct the mistake *before* sending the program out to the machine. Indeed, you should spot the mistake before you write the program. Remember, there is never an excuse for wasting machine time for something as basic as inadequate preparation.

Fourth, developing the machining process prior to programming will provide you with *documentation.* You'll be able to use your process as a checklist while programming to ensure that you don't forget a machining operation. And anyone that must work on your program in the future will be able to quickly familiarize themselves with your machining process if they have this documentation. Figure 2.1 shows an example planning form that you can use to develop and document your machining process.

Part no.:	**Date:**	**Machining Process Planning Form**				
Part name:	**Programmer:**					
Machine:	**Material:**					
Seq.	**Operation description**	**Tool**	**Station**	**Speed**	**Feed**	**Note**

Figure 2.1 Example sequence of operations form

Notice how well this form allows you to document the process your program will use. Months or years after a CNC program is developed, there may be a need to revise it. If the person doing the revision can view the completed process planning form shown in Figure 2.1, it will be much easier to make the necessary changes.

The last reason we will give to plan the process first is to simply help you remember the operations to perform during programming. Remember, beginners tend to make mistakes of omission. You will have enough to think about when it comes to remembering the various commands needed in the program. The process planning form can be used as a kind of step-by-step set of instructions by which you machine the workpiece.

It can be used as a check-list. Without this form, you will be prone to omitting important machining operations from the CNC program.

Develop the needed cutting conditions

Before a program can be completed, cutting conditions must be determined for the various cutting tools used in the program. All cutting tools will need a *spindle speed*. As you know from Lesson Two, any cutting tool that machines a variety of diameters (like turning tools, facing tools, boring bars, and grooving tools) will use the constant surface speed mode. If you work in the inch measurement system, this speed will be specified in surface feet per minute (sfm). If you work in the Metric system, this speed will be specified in meters per minute (mpm). Speed in sfm or mpm is specified right in the cutting tool manufacturer's technical manual or catalog – based upon the cutting tool material to be used and the workpiece material to be machined.

As you also know, center cutting tools (like drills, taps, and reamers) and threading tools require speed to be specified in rpm – meaning you'll have to calculate this speed based upon the recommended speed in sfm (or mpm).

You must also determine a *feedrate* for each cutting tool. With turning centers, the feedrate for most cutting tools is specified in per revolution fashion – inches per revolution (ipr) if you work in the inch mode, or millimeters per revolution (mpr) if you work in the metric mode. Like spindle speed, you'll find the recommended feedrate for a given cutting tool in the cutting tool manufacturer's technical documentation or catalog.

For roughing tools (like rough facing, rough turning, and rough boring tools), you must also determine a *depth-of-cut* for the tool – as well as how much *finishing stock* you will leave for the finishing tool. You must also determine whether or not to use *coolant* based upon the workpiece and cutting tool material.

It is important to come up with the various cutting conditions needed in the program while developing the machining process – and again – *before* you write the program. This will keep you from having to break out of your train of thought while programming. If you use a planning form like the one in Figure 2.1, you will be able to document the speeds and feeds needed for programming right on the form. For speeds, remember to include what kind of speed value you're using (sfm, mpm, or rpm).

Here are a few terms and formulae used when calculating cutting conditions:

sfm (surface feet per minute) – this is the linear amount of material that will pass by the cutting tool's cutting edge/s during one minute. You find sfm recommended in reference books related to cutting conditions like the tooling manufacturers' technical manuals (possibly the cutting tool manufacturer's catalog).

rpm (revolutions per minute) – this is the number of spindle revolutions that occur in one minute.

ipr (inches per revolution) – for any cutting tool, this is the distance a cutting tool will move during one revolution of the spindle. For most cutting tools, this value can be found in reference handbooks related to cutting conditions.

Ipm (inches per minute feedrate) – this is the desired amount of cutting motion during one minute. Note that since turning centers can be programmed directly in inches (or millimeters) per revolution, you won't need to determine the inches per minute feedrate for most cutting tools.

Some formulae:

rpm = 3.82 * sfm / workpiece or cutting tool diameter (in inches)

Ipm = ipr * rpm

pitch (for threading or tapping) = 1 / number of threads per inch

The data provided by most cutting tool manufacturers includes the speed (in sfm) and feedrate (in ipr). This information is based upon the cutting tool material (high speed steel, carbide, ceramic, etc.) and the workpiece material (mild steel, medium carbon steel, high carbon steel, stainless steel, aluminum, etc.). When appropriate, cutting tool manufacturers will also specify whether or not you should use coolant – as well as the recommended depth of cut for a roughing tool. In some cases, they will even provide recommendations about how the cutting tool should move as it machines the workpiece.

For most turning tools, you will specify speed in surface feet (or meters) per minute (G96 mode) and feedrate in inches (or millimeters) per revolution (G99 mode). For center cutting tools, you specify spindle speed in rpm (G97 mode) and feedrate in inches (or millimeters) per revolution (again, G99 mode).

An example

Say you must machine a mild steel workpiece with this machining process:

Operation:	**Tool**:
1) Rough face and turn	Rough face and turn tool (carbide insert)
2) Drill hole in center of workpiece	2.0" drill (carbide insert drill)
3) Rough bore	1.25" boring bar (carbide insert)
4) Finish bore	1.25" boring bar (carbide insert)
5) Finish face and turn	Finish face and turn tool (carbide insert)

You must first determine some important information about your cutting tools, including what they are made of (cutting edge material). In our case, all of the cutting tools are using carbide inserts. The manufacturer of these inserts (Kennemetal, Sandvik, Valenite, etc.) will recommend speed and feed for each insert in their technical manual or catalog. They will do so based upon the insert *grade* (tooling manufacturer's name for the insert), the material to be machined (mild steel in our example), and the machining operation to be performed (rough turn, finish turn, drill, etc.). Expected tool life may also be specified.

Note that the recommendations made by cutting tool manufacturers are just that – *recommendations*. They commonly assume that you have made a very rigid and secure setup and that cutting tools are capable of machining at their maximum capabilities – which may not always be the case. You may elect to reduce these recommendations by ten to twenty percent until you have the job up and running.

We'll say that all of the carbide inserts used in this job are made by the same manufacturer – meaning you'll need only reference manual to come up with cutting conditions recommendations. For most tools – especially turning tools, boring bars, and grooving tools, determining speeds and feeds is quite simple.

Roughing tools

For the first machining operation (rough face and turn), say you look up this insert grade in the cutting tool manufacturer's documentation. Based upon the fact that you are using this insert for rough turning mild steel, they recommend using a speed of 450 sfm and a feedrate of 0.012 ipr. They also recommend using a maximum depth-of-cut (again, when rough turning with this insert) of 0.25 inch – assuming your machine is capable of such a large depth-of-cut. And for finishing stock, they recommend leaving 0.040 inch of material on all diameter surfaces (on the side – 0.080 inches on all diameters) and 0.005 on all faces.

Similar information will be provided for the inset used to rough bore the workpiece (operation number three in the process above.

As you can see, cutting conditions are pretty much spelled out for you for roughing tools. For the rough turning tool, you'll use the constant surface speed mode (G96) and specify the speed with the command

N005 G96 S450 M03 (Start spindle forward at 450 sfm)

For feedrate, you'll use the feed per revolution mode (G99) and specify feedrate with the command

N010 G99 F0.012 (Note that G99 is initialized – some programmers elect to leave the G99 word out of the program)

As you develop the *tool path* for this tool, you'll use a 0.25 inch depth-of-cut – again, assuming your machine is capable of such a powerful rough cut. And for finishing, you will leave 0.040 inch of material on all diameter surfaces and 0.005 inch of material on all faces. For diameters, this means leaving 0.080 inch of material on each diameter (0.040 inch on the side).

Drilling

For the second machining operation in the process above (drilling the 2.0 inch hole in the center of the workpiece), you look in the cutting tool manufacturer's manual or catalog and see that they recommend a speed of 500 sfm and a feedrate of 0.008 ipr when using this carbide insert to drill into mild steel. Since this is a center cutting tool, of course, speed must be specified in rpm (not sfm). So, using the formula shown above, you calculate the required speed in rpm (3.82 times 500 divided by 2.0 is 955 rpm). You'll use this command in the program for starting the spindle:

N090 G97 S955 M03 (Start spindle forward at 955 rpm)

For feedrate, you'll use this command:

N095 G99 F0.008 (Again, note that G99 is initialized – some programmers elect to leave the G99 word out of the program)

Finishing tools

For the fourth and fifth machining operations in the process above (finish boring and finish turning), say the cutting tool manufacturer recommends a speed of 550 sfm and a feedrate of 0.006 ipr when finishing mild steel using this insert – and based upon the surface finish requirement for the workpiece. You'll be using these commands for starting the spindle and selecting the feedrate.

N150 G96 S550 M03 (Start spindle forward at 550 sfm)

N155 G99 F0.006 (Select a feedrate of 0.006 ipr)

Chasing threads

You'll find similar recommendations for other tool types, including reamers and grooving tools. There is one more machining operation we need to mention. *Chasing a thread* requires a threading tool to make several passes over the thread. Each pass makes the thread deeper – until the complete thread depth is reached. While cutting tool manufacturers will specify a speed in surface feet per minute this speed must be converted to rpm (threading cannot be done in the constant surface speed mode). If, for example, the threading tool manufacturer recommends a speed of 400 sfm (based upon the material you are threading), and the thread is of a 3.0 inch diameter, the required speed will be 509 rpm (3.82 times 400 divided by 3.0).

Also, cutting tool manufacturers cannot recommend a feedrate for threading. The feedrate is related to the pitch of the thread. Thread pitch is the distance for crest to crest and is equal to one divided by the number of threads per inch. A thread specified as

2.0-16

has a diameter of 2.0 inches and 16 threads per inch. The pitch will be 0.0625 (1.0 divided by 16). The feedrate needed to produce this thread is F0.0625.

Note that for metric threads, pitch is specified as part of the thread's designation. A metric thread specified as

60-1.25

has a diameter of 60mm and a pitch of 1.25mm.

Cutting conditions can be subjective

Many programmers, indeed many machinists, choose feeds and speeds using a *seat-of-their-pants* approach. They do so by watching and listening to the cutting operation – and manually overriding the programmed spindle speed and feedrate until the machining operation *looks and sounds right.* Admittedly, expert machinists can do this pretty well, and it's hard to argue with success. But the majority of people we see using these techniques don't even come close to using efficient cutting conditions. Instead, cutting conditions they use tend to be well under the cutting tool manufacturer's recommendations. While many factors determine how quickly a cutting tool can machine (like rigidity of the setup, length (overhang) of the cutting tool, and sharpness of the cutting tool), we urge beginners to at least *start with* the recommendations made by the manufacturer of the cutting tools they use.

Do the required math and mark-up the print

As stated in the Preface of this text, the word *numerical* in computer numerical control implies a strong emphasis on numbers and math. Most college curriculums related to CNC do require a strong math background. However, most forms of CNC equipment require less math than you might think. Many CNC turning center programs can be completely prepared solely with simple addition and subtraction. Every practice exercise you've worked on so far, for example, requires nothing more hat addition and subtraction. A basic knowledge of right angle trigonometry is also helpful, but not always mandatory.

While there are times when a manual programmer must apply trigonometry, a relatively simple chart can be used to determine the formulae related to solving right-angle trigonometry problems. This makes it relatively easy to solve trig problems, even for a person who knows little about trigonometry. The formulae shown in Figure 2.2, for example, offer the solutions you will need to solve almost all trig problems related to CNC programming. While a serious CNC programmer will eventually want to thoroughly learn trigonometry, it will not be mandatory at the start of your programming career.

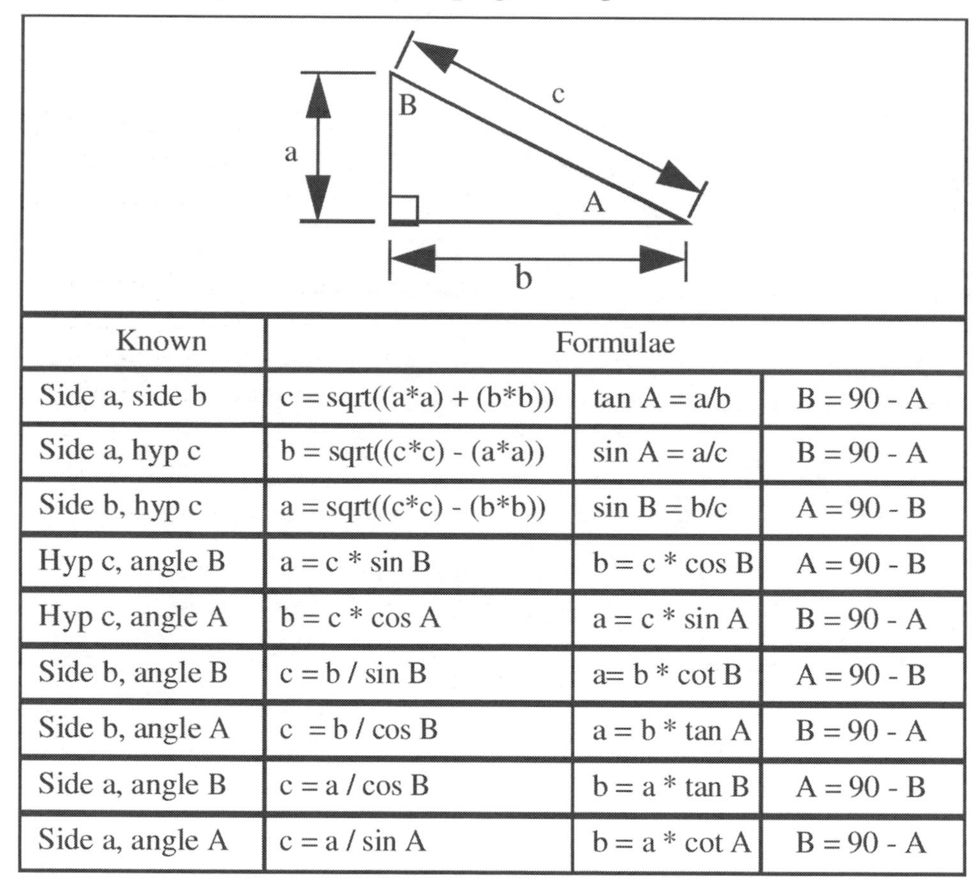

Known	Formulae		
Side a, side b	c = sqrt((a*a) + (b*b))	tan A = a/b	B = 90 - A
Side a, hyp c	b = sqrt((c*c) - (a*a))	sin A = a/c	B = 90 - A
Side b, hyp c	a = sqrt((c*c) - (b*b))	sin B = b/c	A = 90 - B
Hyp c, angle B	a = c * sin B	b = c * cos B	A = 90 - B
Hyp c, angle A	b = c * cos A	a = c * sin A	B = 90 - A
Side b, angle B	c = b / sin B	a= b * cot B	A = 90 - B
Side b, angle A	c = b / cos B	a = b * tan A	B = 90 - A
Side a, angle B	c = a / cos B	b = a * tan B	A = 90 - B
Side a, angle A	c = a / sin A	b = a * cot A	B = 90 - A

Figure 2.2 – Chart to help you solve right-angle trigonometry problems

The trick to using the chart in Figure 2.2 is finding the appropriate formula needed to solve your problem. It is based upon knowing *two things* about the triangle in question. If you know any two things about a right-angle triangle, this chart will help you determine anything else you need to know about it.

In drawing shown at the top of Figure 2.2, notice that angular specifications are given with A and B (the third angle is, of course, ninety degrees). Lengths of the triangle sides are given with a, b, and c (c being the hypotenuse, the longest side of the triangle).

Look at the drawing in Figure 2.3 for an example.

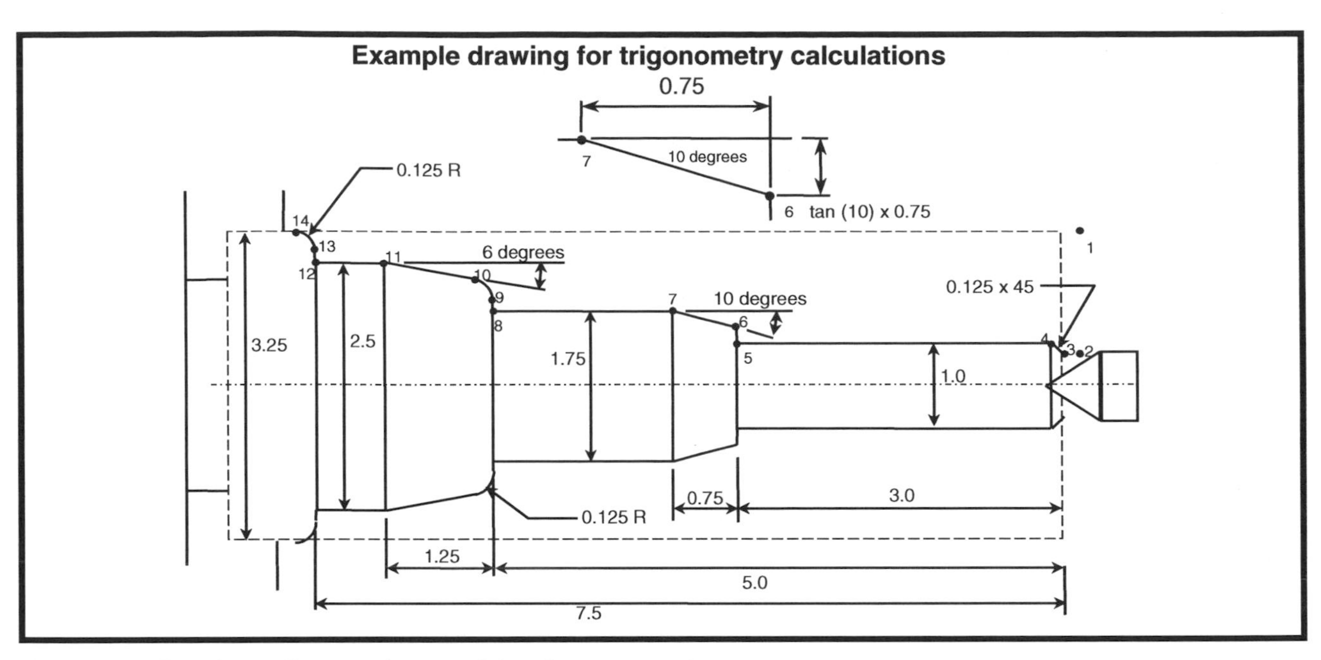

Figure 2.3 – Drawing to illustrate the use of the trigonometry chart

You can use simple addition and subtraction to calculate most of the points labeled on this print (points 1, 2, 3, 4, 5, 7, 8, 11, 12, 13, and 14). But for points labeled 6, 9, and 10, trigonometry must be used to determine at least one of the coordinates (X and/or Z). Let's begin with the simpler of the two problems – point number 6.

Assuming that program zero is placed at the right end of the finished workpiece, the Z coordinate of point 6 is, of course Z-3.0. This coordinate is specified right on the drawing. But the X coordinate requires right angle trigonometry. See the blow-up drawing at the top of Figure 2.3. The taper angle on this diameter is specified as 10 degrees. The length of the taper is specified as 0.75 inch. Based upon these two known values, we need to calculate the distance from point 6 to 7 along the X axis. The blow-up drawing nicely shows the related triangle.

We know the values for side b and angle A. Based upon these two known values, we need to determine the length of side a. As you look at the chart in Figure 2.2, which formula will you use to calculate the length of side a?

While this may take some study, you should eventually agree that the seventh row down specifies side b and angle A as known values. The middle column shows the formula for side a (side a = length of side b times the tangent of angle A). With an electronic calculator that includes trig functions, you find that the tangent of ten degrees is 0.17632. When this value is multiplied times 0.75, the result is 0.1322. So the distance from point 6 to point 7 along the X axis is 0.1322. But since X coordinates are specified in diameter, this value must be doubled and then subtracted from the diameter of point 7 (1.75) in order to come up with the diameter of point 6. 1.75 minus 0.2644 is 1.4856. So the diameter of point 6 is 1.4856.

Now look at the taper from points nine through eleven. This taper includes a radius that is tangent to the previous face as well as the taper (a very common problem for turning center programmers). Admittedly, finding the correct triangles to solve in this example is much more difficult. But this is about as tough as it gets, so if you can follow this step-by-step procedure, you should be prepared to calculate points required for any radius-to-taper situation. Here is an enlarged drawing that shows the related triangles.

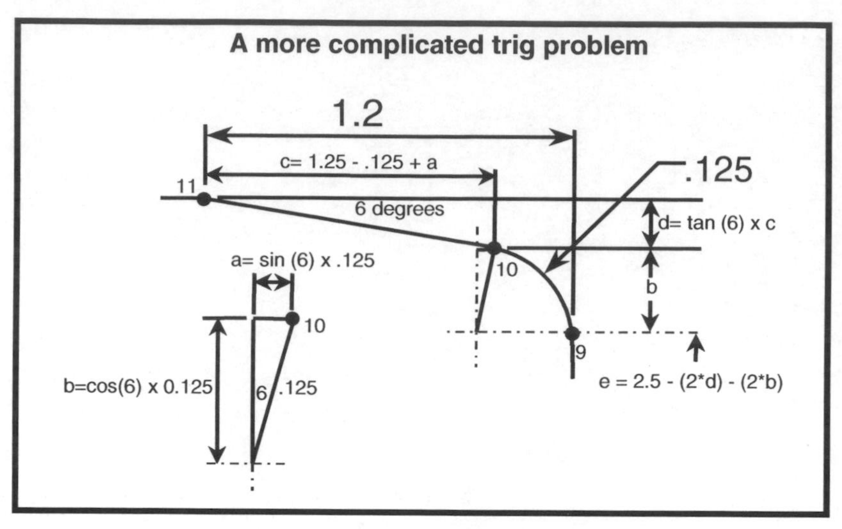

Drawing for solving radius-to-taper problems (work from a through e)

1) Calculate the short side of the small triangle (small letter a in our example)
Do this by taking the sine of the taper angle (six degrees in our case) times the radius size. The sine of 6 degrees is 0.1045. When multiplied times 0.125 (the size of the radius), the length of small side a is 0.0130.

2) Calculate the long side of the small triangle (small letter b in our example)
Do this by taking the cosine of the taper angle (again, six degrees) times the radius size. The cosine of 6 degrees is 0.9945. When multiplied times 0.125 (the size of the radius), the length of long side b is 0.1243.

3) Calculate the long side of the large triangle (small letter c in our example)
Do this by subtracting the previously calculated short side of the small triangle (small a) from the radius size and subtract the result from the taper length. In our example this renders 1.138 (1.25 - 0.125 + 0.0130)

4) Calculate the short side of the large triangle (small letter d in our example)
Do this by multiplying the tangent of the taper angle times the previously calculated short side (c) value. The tangent of six degrees is 0.1051. When multiplied times the previously calculated long side of the large triangle (small letter c, which is 1.138 in our case), the result is 0.1196.

5) Calculate the diameter of the beginning point of the radius (e, the X coordinate of point nine in our case)
Do this by first adding the previously calculated long side of the small triangle (b) plus the short side of the large triangle (d). You must double the result, since we're looking for a diameter value, and then subtract it from the diameter at which the taper begins (2.5 in our case). This renders 2.013 (2.5 - [2 times 0.1196] - [2 times0.1243]) for our example.

6) Calculate the diameter of the radius ending point (X of point ten in our case)
Do this by subtracting twice the previously calculated short side of the large triangle (d) from the diameter from which the taper begins (2.5 in our case). This renders a value of 2.2608 (2.5 - [2*.1196]).

7) Calculate the Z coordinate for point number ten
Do this by subtracting the previously calculated long side of the large triangle (c) from the Z position of point number eleven. Point number eleven's Z coordinate is 6.25 and when 1.138 (the value of c) is subtracted from it, the result is 5.112.

Admittedly, solving trig problems can be quite challenging, even with relatively simple right-angle trigonometry. This is one of the reasons why computer aided manufacturing (CAM) systems are so popular.

Other ways to come up with coordinates

The more complex your workpiece drawings are, the more difficult it will be to calculate coordinates. While manual programmers get pretty good at the related math, there are some alternatives. Here are a few suggestions.

Do you have an electronic file of the workpiece drawing? – In most companies, a computer aided design (CAD) system is used to draw the workpiece. The design engineer saves the drawing in a special file format (commonly called iges or dxf). If you have access to this file – and the CAD system with which it is created, you can use this file to learn anything you must about the coordinates needed for programming. For the workpiece shown in Figure 2.3, for example, if you have the electronic file, you can call it up using the CAD system and easily determine the coordinates for points 6, 9, and 10. You must, of course, understand how to use the CAD system to do so – or you can ask a person who is experienced with the CAD system to determine these coordinates for you.

Purchase an inexpensive CAD system – If you don't have access to the electronic drawing (this is commonly true in contract shops), remember that you can use an inexpensive CAD system and draw the workpiece yourself (using the workpiece drawing). Some inexpensive CAD systems (well under $100.00) are relatively powerful – yet easy to learn and use. While you'll have to learn to use it, the CAD system will eliminate the need to perform tedious and error-prone manual calculations.

Does your company own a computer aided manufacturing (CAM) systems? – Many companies that machine complex workpieces on a regular basis have CAM systems. These systems do more than calculate coordinates. They actually create the entire CNC program.

Do you really need a CAM system? – While we may be a little ahead of ourselves, you will find that if you can determine the coordinates needed for finishing the workpiece (finish turning, finish facing, or finish boring), it is quite easy to program roughing operations using a feature called *multiple repetitive cycles*. These cycles are presented in Key Concept Six. In many cases, these features rival what can be done with a CAM system. Again, if you can command the motions required for finishing, the machine will do all the roughing for you – based *on but one CNC command*. In many companies, this eliminates the need to purchase an elaborate CAM system – at least for the company's CNC turning centers.

Marking up the print

The programmer assigned to prepare a program must be given a copy of the blueprint. In most companies, the programmer can do whatever they wish to the print if it helps with programming. This working copy of the print is usually kept with the program's documentation (not used for manufacturing).

Depending on the complexity of the workpiece to be produced, interpreting the print can range from quite simple to very difficult. Once the programmer studies the print and understands the machining operations that must be performed, they should mark up the print to make programming simpler, and to allow others having to work on the program in the future to understand what they've done. The purpose for doing so is to clarify as much as possible. If you've been doing the practice exercises, you've already seen some of the marking-up techniques we recommend.

The first print marking technique we suggest is to take high-lighting pen with a bright color and mark those workpiece surfaces that require machining by your program. For complicated workpieces, especially cast or forged workpieces, it is not uncommon that only a small portion of the workpiece must be machined. Using the routing sheet and/or your own knowledge of what must be done on the workpiece, you can isolate just what must be machined for all to see.

Next, mark up the location of the program zero point you'll be using for coordinate calculations. Draw a dot and make a note as to its position.

It's also wise to sketch in the work holding device (commonly chuck jaws). This also helps you when planning the setup, especially if you're going to have interference problems when machining surfaces coming close to the workholding device. If a work support device (like a tailstock or steady-rest) is going to be used, be sure to sketch it on the drawing as well.

And if it's feasible, mark up your tool paths right on your working copy of the print. Depending upon how cluttered your print is becoming, you may have to make separate sketches for illustrating tool paths. With tool path points decided, number each point for helping with coordinate calculations.

Doing the math

How dimensions are described on the part print will determine how much math is required for your program. Some companies have their design engineers supplying all drawings with datum surface dimensioning. When this technique is used, each dimension on the print will be taken from only one position in each axis. While this may reduce the need to do calculations, it does not completely eliminate the need for doing math. You'll see why in just a bit.

Of course, not all design engineers use datum surface dimensioning techniques – indeed most do not. You will probably be expected to do a great deal of math (commonly addition and subtraction) in order to determine X and Z coordinates required in the CNC program. Regardless of how much math is required to determine program coordinates, it is wise to perform these calculations *before* you write the program. This will keep you from having to break out of your train-of-thought to calculate coordinates as you write the program.

If you have been doing the practice exercises in this text, you've already done some coordinate calculations. You have also seen our method of using a *coordinate sheet* for documenting coordinates needed in a program. This method involves numbering each position on the print with a point (like connect the dots) and placing each point's X and Z values on a separate coordinate sheet.

However, our practice exercises have only required that you calculate positions that are *right on workpiece surfaces.* And we've provided all of the points for you to calculate. In reality, *not all coordinates* required in the program will actually be on the workpiece (for clearance positions and when finishing stock must be allowed with a roughing tool, for example). And you'll be on your own to determine which positions are required in the program. Once again, this brings up the need to understand basic machining practice. You must be able to plan each cutting tool's movement path (commonly called the tool path) in order to plot the points needed for your program.

Figure 2.4 shows the cross section of a workpiece that must be programmed. We'll use it to illustrate the coordinates that will be needed in the program.

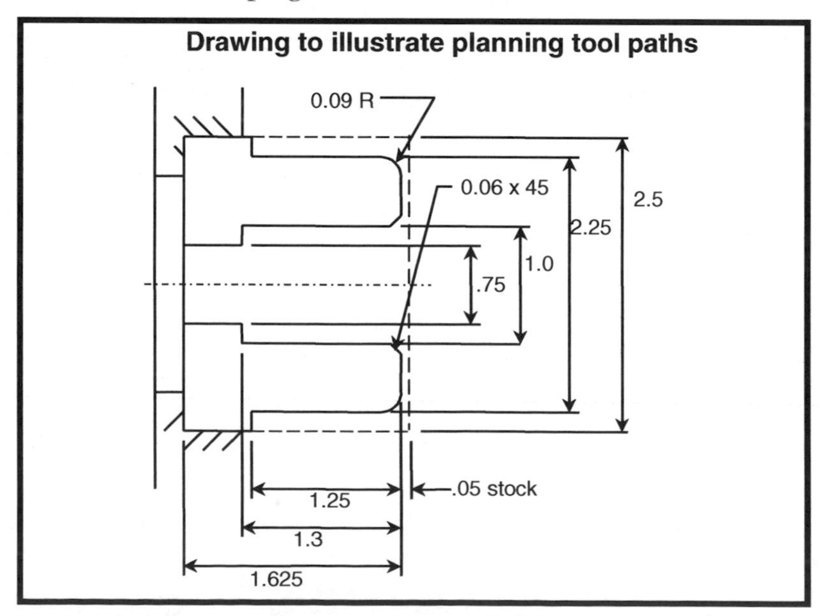

Figure 2.4 – Drawing to illustrate the planning of tool paths

Here is the process we will use to machine the workpiece.

Description	Tool	Stat.	Feedrate	Spindle speed
1) Rough face and turn	80 degree diamond tool	1	0.012 ipr	400 sfm
2) Drill through	3/4" drill	2	0.010 ipr	350 rpm
3) Rough bore	5/8" boring bar	3	0.007 ipr	400 sfm
4) Finish bore	5/8" boring bar	4	0.004 ipr	450 sfm
5) Finish face and turn	80 degree diamond tool	5	0.007 ipr	500 sfm

Figure 2.5 shows a series of tool paths needed to machine the workpiece. We're showing one tool path per tool. And – we're only showing the upper half of the workpiece in each example (above center), but you should be able to see what's going on.

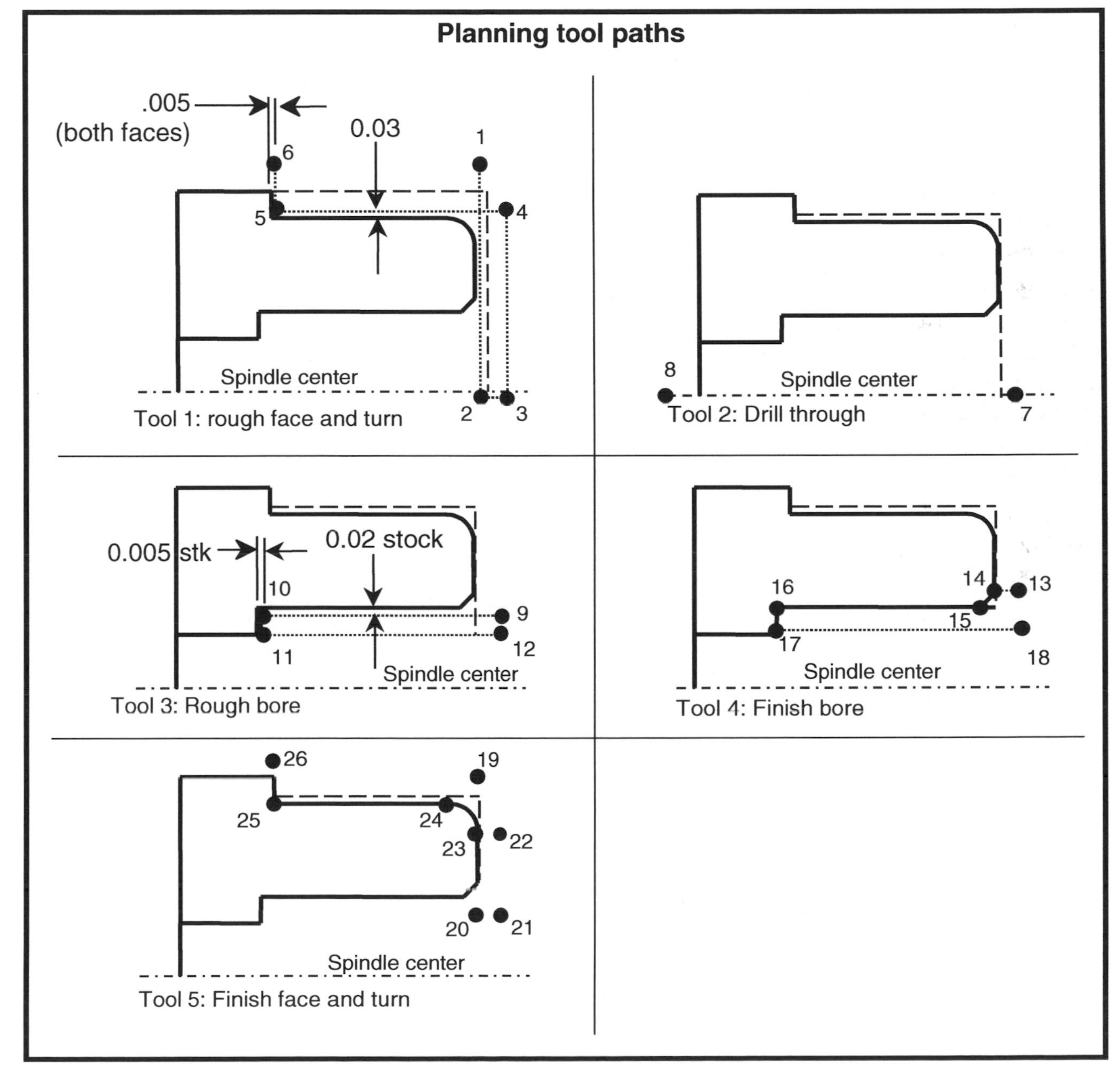

Figure 2.5 – Tool path planning for example

For each cutting tool, we've planned the path it must take in order to properly machine the workpiece. And again, *you* will have to do this for every cutting tool you program. We've placed a dot next to every location that the cutting tool must pass through and numbered it. While it helps if the numbering is in order, it is not completely mandatory. At this point, we're simply coming up with *all* of the needed points.

Notice that we've also included things in each tool path drawing that are not on the print. For the roughing tools (tools one and three), for example, we've specified how much finishing stock will be left on each surface.

Also, there are some things in each tool path that are *not* labeled. Point number one, for example, appears to be above the rough stock diameter – but by how much? The same is true for points 3, 4, 7, 9, 11, 12, 13, 18, 19, 20, 21 and 22. Each of these points is at a *clearance position* (either approaching the workpiece or retracting from it). Most programmers will use an approach/retract distance of 0.1 inch (2.5 mm) – which is the approach/retract distance we use throughout this text.

Unfortunately, the drawings for many turned workpieces are quite small and cluttered. You may not be able to adequately draw your tool paths right on your working copy of the print. You may have to make a series of special sketches (like those shown above) to help with your tool path documentation. Beginners have a tendency to skip this step if it seems too time-consuming. But without a very clear understanding of the points needed in your program, you'll become confused with your tool paths when you try to write the program.

While it's a little off the subject of doing math, again, we'll mention again the programming features that minimize the need to come up with a special set of coordinates for roughing operations. *Multiple repetitive cycles*, discussed in Key Concept Six, help with roughing operations and truly minimize the number of points for which coordinates must be calculated. But as the programmer, you'll be responsible for determining which coordinates will be needed in your program – and for calculating them before you write the program.

Once all points are plotted, a simple coordinate sheet will help you document your math calculations. Here is a completed coordinate sheet for the previous example.

#	X	Z
1	2.7	0.005
2	-0.062	0.005
3	-0.062	0.1
4	2.31	0.1
5	2.31	-1.245
6	2.7	-1.245
7	0	0.1
8	0	-1.87
9	0.96	0.1
10	0.96	-1.295
11	0.75	-1.295
12	0.75	0.1
13	1.12	0.1
14	1.12	0
15	1.0	-0.06
16	1.0	-1.3
17	0.75	-1.3
18	0.75	0.1
19	2.45	0
20	0.8	0
21	0.8	0.1
22	2.07	0.1
23	2.07	0
24	2.25	-0.09
25	2.25	-1.25

26 2.7 -1.25

Completed coordinate sheet

Check the required tooling

The next preparation step is related to the cutting- and workholding- tools used by your program. Tooling problems can cause even a perfectly written CNC program to fail. You can avoid production delays by considering potential tooling problems while you prepare for programming. Again, there is no excuse for machine down-time for as avoidable a reason as poor preparation.

First, you will need to confirm that the various cutting tools to be used by your program are *available*. You may incorrectly assume, for instance, that common tools are always in your company's inventory. It is always wise to double-check, even with relatively common tools. And always be sure your company has lesser-used tools in stock.

If a given tool is not in stock, of course, it must be ordered. Or you may have to make do with what your company does have available. If this is the case, your entire machining process may have to be changed (another reason to check *before* you begin programming).

Second, you must confirm that cutting tools are capable of machining the workpiece attribute/s you intend. Every insert used in turning tools and boring bars has a certain *geometry* that controls what the tool can and cannot do. Figure 2.6 shows a potential problem.

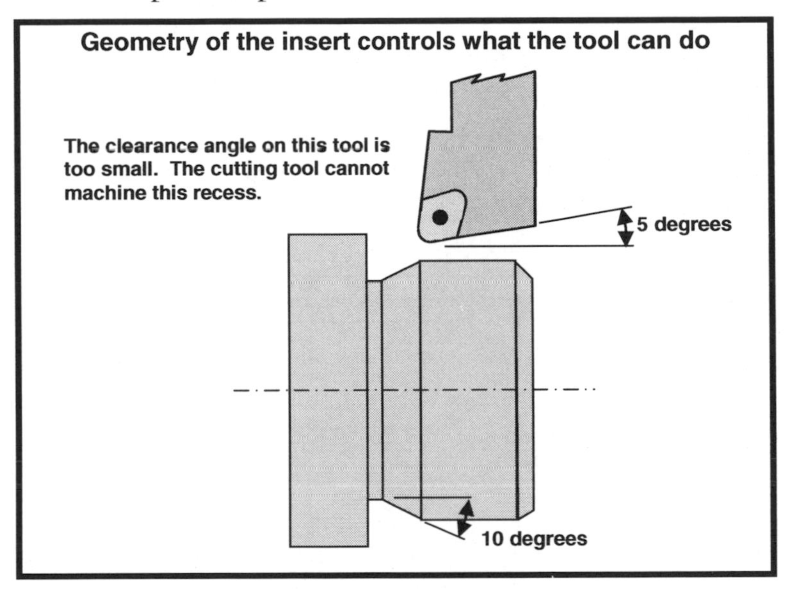

Figure 2.6 – A recess that cannot be machined with the intended turning tool

Notice that the cutting tool shown in Figure 2.6 has only a 5 degree clearance angle. It cannot machine a recess having more than a 5 degree angle.

Another tooling concern is the potential for *interference*. For the process given for the workpiece shown in Figure 2.4, for instance, the two boring bars must *fit* into the 3/4" drilled hole. We've selected 5/8" diameter bars, but we must confirm that they can machine the 1.3" deep face down to 0.75" in diameter without hitting on the back side of the boring bar. The only way to do this is to confirm (from the cutting tool drawing in the cutting tool manufacturer's catalog) that the distance from the tip of the boring bar to back of the boring bar is 0.75 inch or less, as is shown in Figure 2.7.

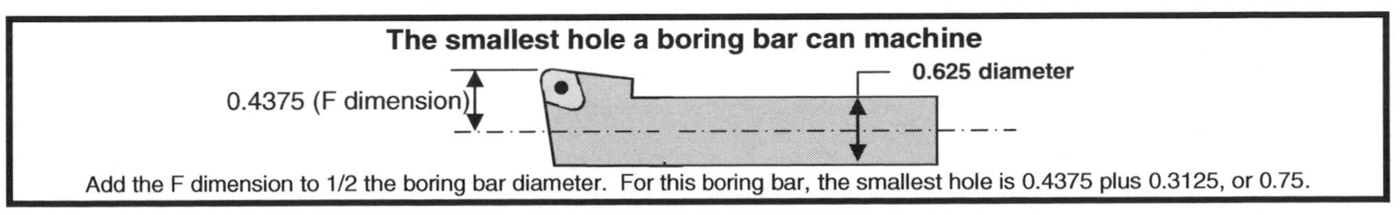

Figure 2.7 – How to determine the smallest hole a boring bar can machine

Yet another interference problem of which you must be constantly concerned has to do with adjacent turret stations. With many turning centers, you cannot place internal cutting tools (drills, boring bars, etc.) in turret stations that are adjacent to turning tools – or to other internal cutting tools. As a facing tool faces a workpiece to center, the adjacent drill or boring bar may collide with the chuck – or even the workpiece itself. The smaller the turret and the more tools it can hold – the more likely it will be that you must skip turret stations with internal cutting tools.

Plan the work holding set-up

CNC programmers are responsible for designing the setup used for machining. Once again, this emphasizes the importance of basic machining practice. You must plan the workholding setup as well as specify the cutting tools and their intended turret station numbers. We recommend planning the setup prior to developing the CNC program. Doing so will help you avoid wasting programming time should you discover a setup related problem during the planning of the setup. Additionally, many programming considerations depend upon how the setup is made.

We recommend using a setup sheet like the one shown in Figure 2.8 to help you plan and document the setup. While it has some limitations upon which you may wish to improve, this kind of setup sheet makes it easy for the setup person to understand how you intend the setup to be made. Mistakes made during setup, of course, can lead to dangerous situations. Anything you can do to clarify will minimize the potential for mistakes.

Part no.:	**Date:**	**Setup Sheet**
Part name:	**Programmer:**	
Machine:	**Program no:**	

Stat.	**Tool description**	**Offset**	**Insert**	**Instructions:**
				Sketch of setup:
Chuck:				
Clamping pressure:				
Other notes:				

Figure 2.8 – An example universal setup sheet

For extremely complicated setups, the programmer may not be the only person involved with planning setups. A tool designer may be responsible for the design of the setup. But the programmer is still commonly responsible for providing the setup person with setup *instructions*.

This universal setup sheet does not include a detailed list of components for the cutting tools used in the program. Many companies do include a more complete tool list, possibly on a separate page. In similar fashion, this setup sheet does not include a complete list of the workholding tools (specifically which top tooling jaws should be used). Again, many companies do include this kind of information so someone (probably *other* than the setup person) can be gathering all needed components even before the setup is made.

You should plan the setup and create the setup sheet *before* you write the CNC program. How the setup is made, of course, has a dramatic impact on how the program must be written. While you write the program, you'll need to know the position of workholding and work support devices in the setup so you can make sure that cutting tools don't hit them during their motions. When it comes to cutting tools, you'll need to know where they are placed in the turret (station numbers). *Without having planed the setup, you really can't write the CNC program.*

Other documentation needed for the job

The programmer is responsible for providing all documentation needed for the jobs they program. And documentation should be aimed at the lowest skill level of people performing the related tasks. As with setup documentation, we recommend that *all* documentation related to a job be done *prior to* writing the program. You never know when you'll come across something that affects the way the program must be written.

Production run documentation

With longer production runs, many companies turn the machine over to another person, the CNC operator, once the setup is made and the first workpiece passes inspection. In some cases, more than one operator will complete the job (first and second shift operators, for example). Production run documentation should be aimed at the CNC operator, providing instructions for how to load and unload workpieces, how often to take sampling measurements, approximately how long cutting tools will last before they must be replaced, and a list of all perishable tools (inserts, twist drills, taps, etc.) used in the job so they can be quickly found and replaced when they get dull.

Program listing

Most companies supply a printed copy of the program with the documentation that goes out to the setup person and operator who will be running the job. This will help them make modifications to the program if they are required.

Is it all worth it?

You have seen that there is a great amount of work involved with *preparing* to write a CNC program. Again, experienced programmers will agree that this is where the *real work* lies. Once preparation is done, you will have a clear understanding of what the program must do. There will be no questions left to ponder while you write the program – and writing the program will be relatively easy – especially after you have written a few programs. Also, the work you do in preparation to write programs will pay special dividends when it comes time to run your program on the machine – there will be fewer mistakes to find and correct.

Key points for Lesson Eight:

- The better prepared you are, the easier it will be to write the program – and the fewer mistakes you will make.
- There is no excuse for wasting machine time for something avoidable as lack of preparation.
- Beginners are prone to making mistakes in four categories: syntax mistakes, motion mistakes, mistakes of omission, and process mistakes.
- First, study and mark up the print to become familiar with what must be done.
- Second develop a machining process, including cutting tools and cutting conditions to be used.
- Third, do the math calculations, coming up with all coordinates needed in the program.
- Fourth, check that cutting tools are available and capable of performing their machining operations.
- Fifth, plan the workholding setup.
- Sixth, create any other documentation that is required for the job (like production run documentation).

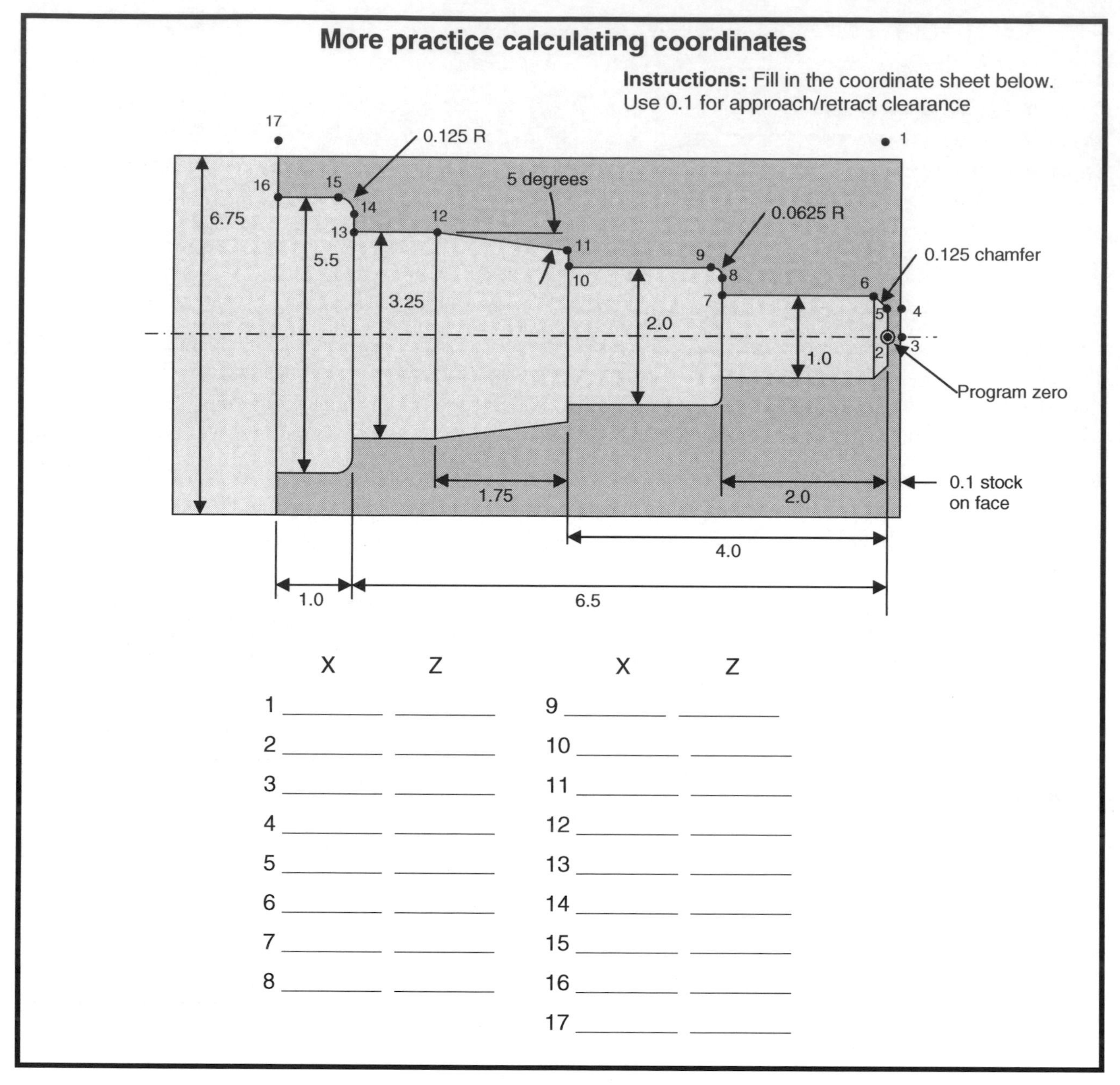

Answers:

1: X6.95 Z0
2: X-0.0625 Z0
3: X-0.0625 Z0.1
4: X0.75 Z0.1
5: X0.75 Z0
6: X1.0 Z-0.125
7: X1.0 Z-2.0
8: X1.875 Z-2.0
9: X2.0 Z-2.0625
10: X2.0 Z-4.0
11: X2.9438 Z-4.0
12: X3.25 Z-5.75
13: X3.25 Z-6.5
14: X5.25 Z-6.5
15: X5.5 Z-6.625
16: X5.5 Z-7.5
17: X6.95 Z-7.5

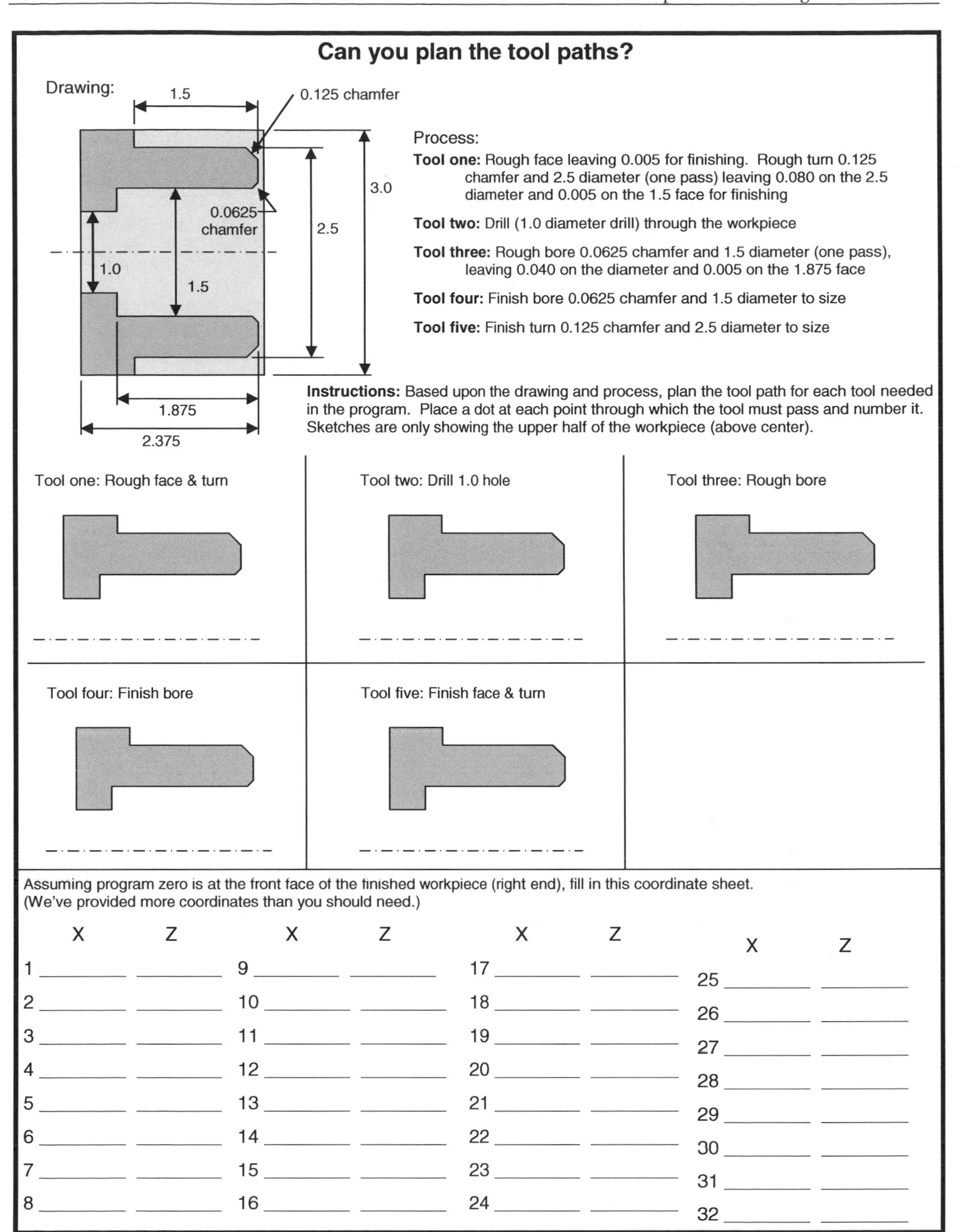

Can you plan the tool paths?

Process:

Tool one: Rough face leaving 0.005 for finishing. Rough turn 0.125 chamfer and 2.5 diameter (one pass) leaving 0.080 on the 2.5 diameter and 0.005 on the 1.5 face for finishing

Tool two: Drill (1.0 diameter drill) through the workpiece

Tool three: Rough bore 0.0625 chamfer and 1.5 diameter (one pass), leaving 0.040 on the diameter and 0.005 on the 1.875 face

Tool four: Finish bore 0.0625 chamfer and 1.5 diameter to size

Tool five: Finish turn 0.125 chamfer and 2.5 diameter to size

Instructions: Based upon the drawing and process, plan the tool path for each tool needed in the program. Place a dot at each point through which the tool must pass and number it. Sketches are only showing the upper half of the workpiece (above center).

Tool one: Rough face & turn

Tool two: Drill 1.0 hole

Tool three: Rough bore

Tool four: Finish bore

Tool five: Finish face & turn

Assuming program zero is at the front face of the finished workpiece (right end), fill in this coordinate sheet. (We've provided more coordinates than you should need.)

	X	Z		X	Z		X	Z		X	Z
1			9			17			25		
2			10			18			26		
3			11			19			27		
4			12			20			28		
5			13			21			29		
6			14			22			30		
7			15			23			31		
8			16			24			32		

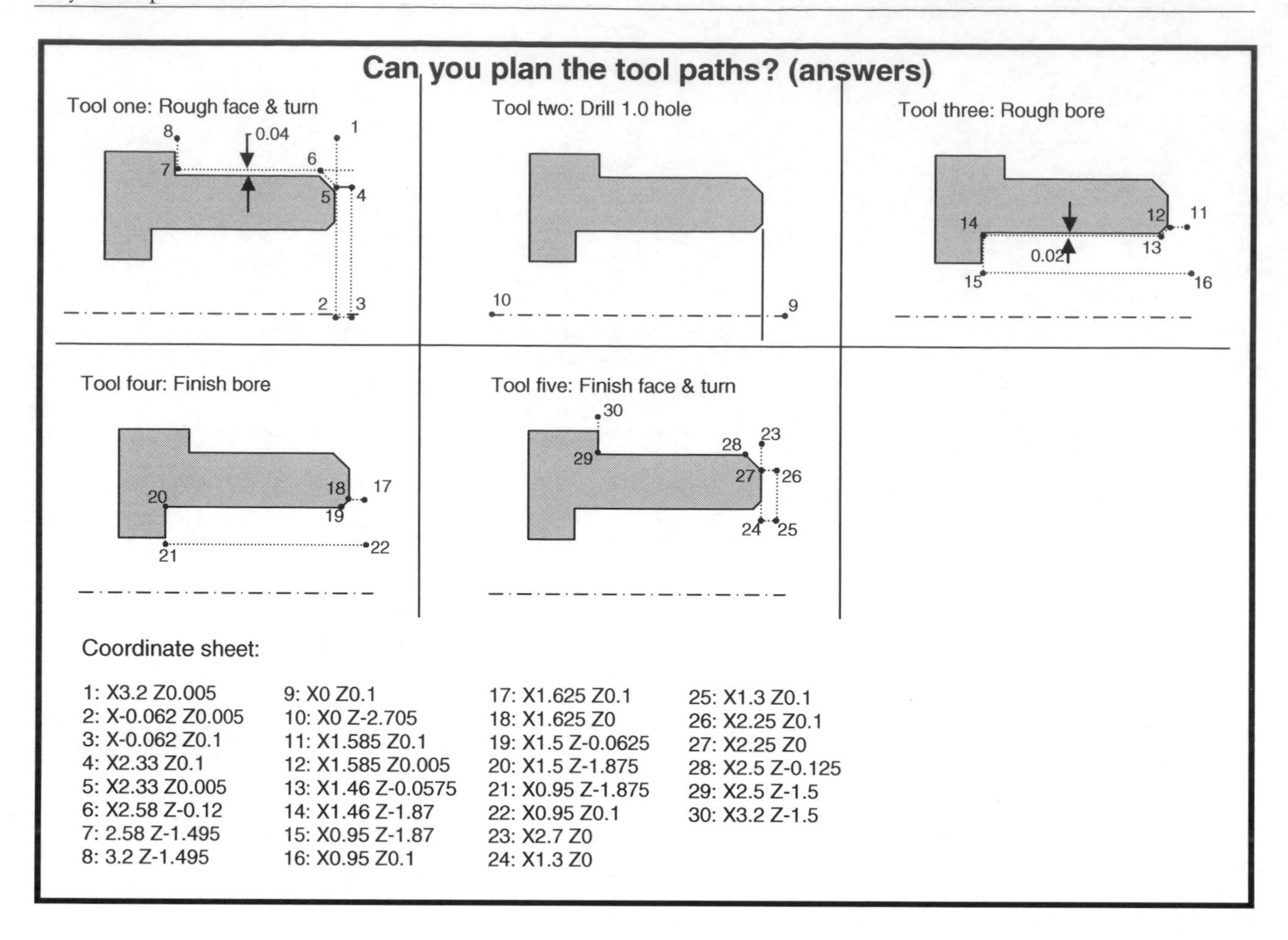

STOP! Do practice exercise number nine in the workbook.

Understand The Motion Types

Motion control is at the heart of any CNC machine tool. CNC turning centers have at least three ways that motion can be commanded. Understanding the motion types you can use in a program will be the focus of Key Concept Number Three.

Key Concept Number Three is another one-lesson Key Concept:

10: Motion types

You know that all CNC turning centers have at least two axes (X and Z). You also know that you can specify positions (coordinates) along each axis relative to the program zero point. These coordinates will be positions through which your cutting tools will move. In Key Concept Number Three, we're going to discuss the ways to specify *how* cutting tools move from one position to another.

What is interpolation?

If a single linear axis is moving (X or Z), the motion will of course, be along a perfectly straight line. For example, look at Figure 3.1.

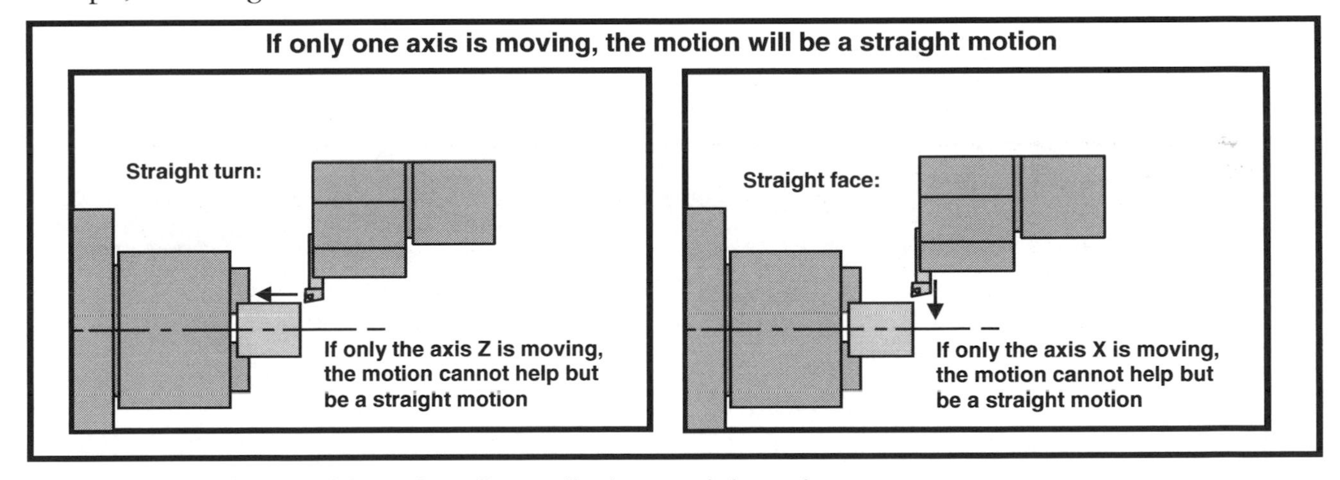

Figure 3.1 – A perfectly straight motion will occur if only one axis is moving

When turning a straight diameter on the workpiece (left view in Figure 3.1), only the Z axis is moving. And since the Z axis is a *linear* axis, this will force the motion to be perfectly straight – and parallel to the Z axis. In like fashion, when facing a workpiece (right view), only the X axis is moving – and again – this motion will be perfectly straight – parallel to the X axis.

But sometimes a straight surface on a workpiece is *tapered*. Additionally, there will be times when a cutting tool must machine a radius. This will require that *both* the X and Z axes move in a controlled manner, as shown in Figure 3.2.

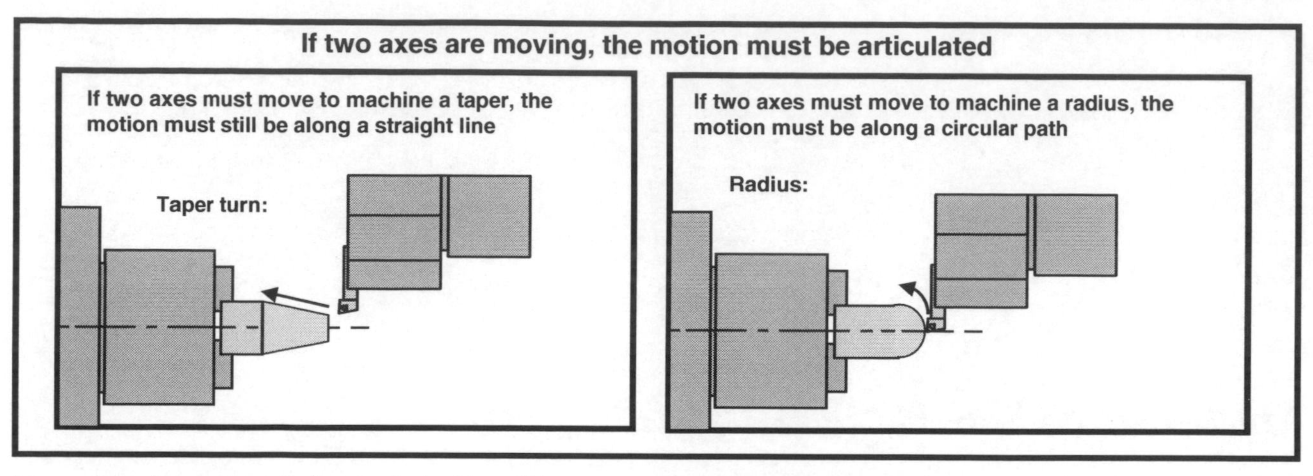

Figure 3.2 – Examples of when both the X and Z axes move in a controlled manner

The machine must *articulate* the motions shown in Figure 3.2. In CNC terms, this kind of articulation is called *interpolation.*

Look at Figure 3.3.

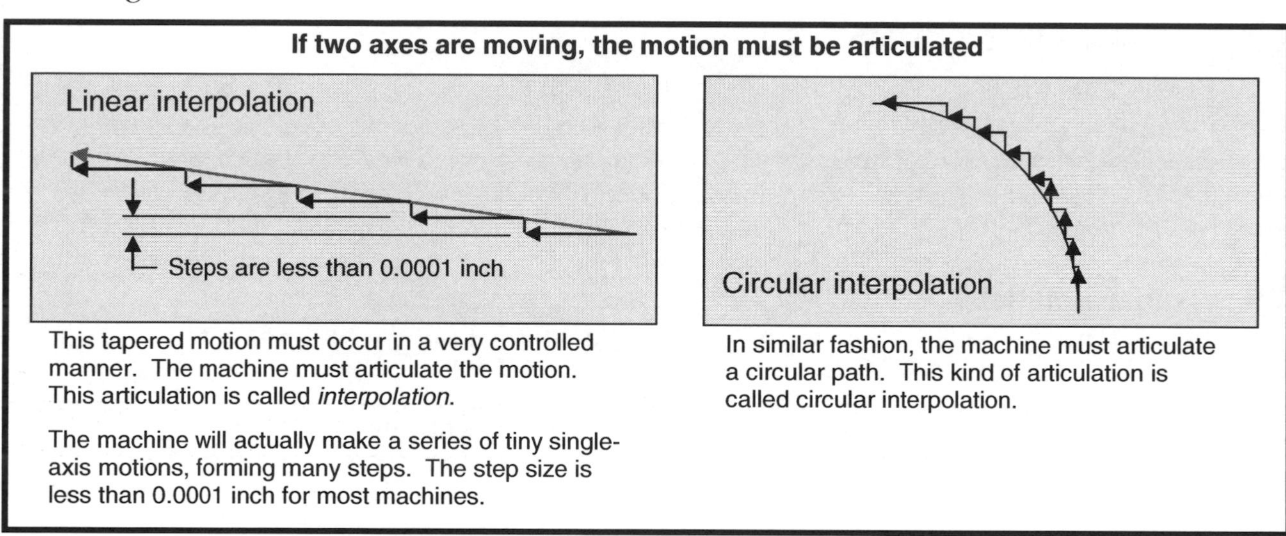

Figure 3.3 – The meaning of interpolation

As you can see, the machine will actually break these two-axis motions in to a series of very tiny single-axis steps. The step size will be under 0.0001 inch for most machines. These steps are so small that you will not be able to see them – or measure them – with most measuring devices. For all intents-and-purposes, all machined surfaces will appear to be perfectly straight (or circular) and without steps.

(By the way, the step size during interpolation is referred to as the machine's *resolution.* With many CNC machines, it is equal to the least-input-increment available in the measurement system you are using – 0.0001 inches in the inch mode or 0.001 millimeters in the metric mode. The smaller the step size, the finer will be machine's resolution – and the more precisely it will follow your commanded motions.)

Turning center manufacturers offer a variety of *interpolation types*, based upon what their customers will be doing with their machines. *All* offer at least the two interpolation types shown above – *linear interpolation* (also called straight-line motion) and *circular interpolation* (also called circular motion). Additionally, all turning centers come with a third motion type, called *rapid motion* – though we don't consider rapid motion to be a true form of interpolation since most machines do not articulate the tool's path during rapid motion.

Depending upon your company's specific needs, there may be other interpolation types available with your turning centers. However, these additional interpolation types will only be used for special applications. If, for example, your machine has live tooling, and if you'll be performing milling operations around the outside diameter of the workpiece (like milling flats), you'll need an interpolation type called *polar coordinate interpolation* (we discuss polar coordinate interpolation in the Appendix – after Lesson Twenty-Three). Again, special interpolation types are only used in special applications. The bulk of your programming will not require them.

Actually, it should be refreshing to know that there are *only* three common ways to cause axis motion – rapid, straight-line, and circular. Just about every motion a CNC turning center makes can be divided into one of these three categories. Once you master these three motion commands, you will be able to generate the movements required to machine a workpiece. Figure 3.4 illustrates them.

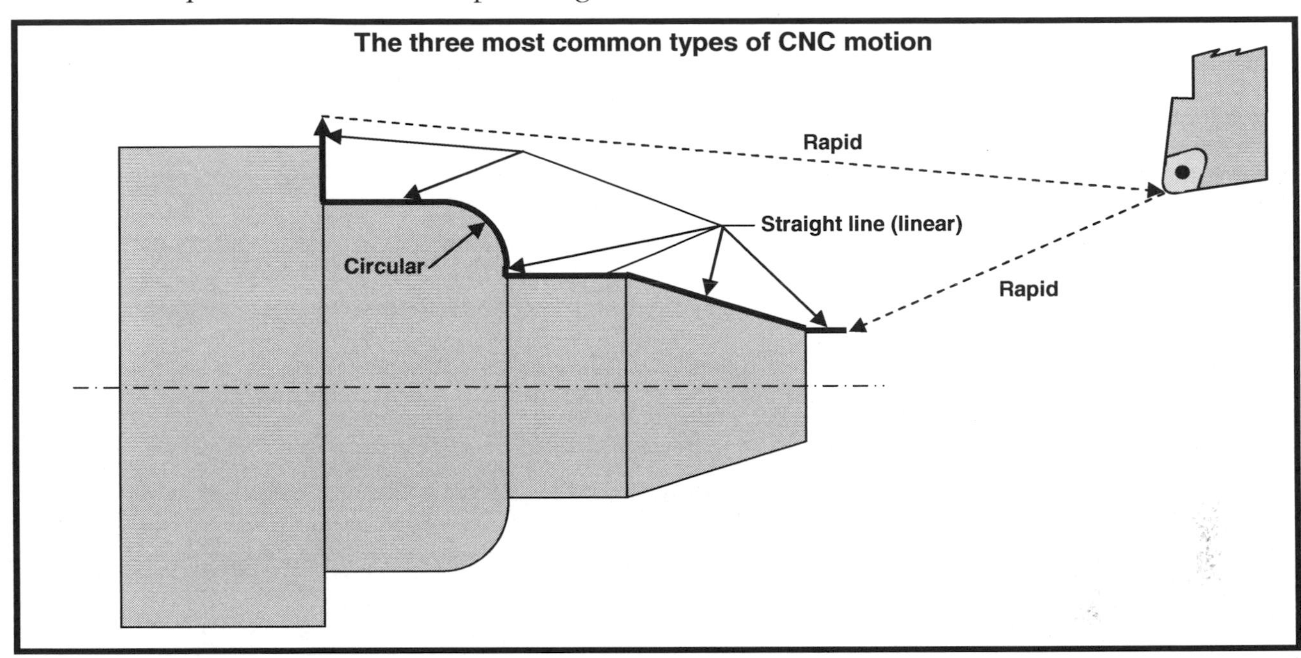

Figure 3.4 – The three kinds of motion available with all CNC turning centers

As you will see in Lesson Ten, it is actually quite easy to specify motion commands within a CNC program. In general, each motion will require you to specify the kind of motion (rapid, straight line, or circular) along with the motion's end point (coordinates at the end of the motion). Linear motion and circular motion additionally require that you specify the motion rate (feedrate) for the motion. Circular motion additionally requires that you specify the size of the arc being generated in the motion.

Lesson 10

Programming The Three Most Basic Motion Types

There are only three motion types used in CNC turning center programs on a regular basis – rapid, straight-line, and circular motion. You must understand how they are commanded.

While it helps to understand how the turning center will interpolate motion, it is not as important as *knowing how to specify motion commands in a program.* This is the focus of Lesson Ten. Let's begin our discussion by showing those things that all motion types share in common.

Motion commonalties

All motion types share five things in common:

First, they are all modal, meaning a motion type will remain in effect until it is changed. If more than one consecutive movement of the same type must be made, you need only include the motion type G code in the *first* command of the series of movements.

Second, each motion type command requires the *end point* of the motion. The control will assume the tool is positioned at the starting point of the motion prior to the motion command. Think of motion commands that form a tool path as being like a series of connect-the-dots.

Third, all motion commands are affected by whether or not you specify coordinates with absolute or incremental positioning. With most turning center controls, X and Z are used to specify absolute positioning. U and W are used to specify incremental positioning. As stated in Lesson Five, you should concentrate on specifying coordinates with absolute positioning (X and Z).

Fourth, each motion command requires *only* the moving axes. If specifying a motion in only one axis, only one axis specification (X or Z) need be included in the motion command. *Axes that are not moving can be (and should be) left out of the command.*

Fifth, leading zeros can be left out of the G codes related to motion types. This means the actual G codes used to instate the motion types can be programmed in one of two ways. G00 and G0 (stated G zero-zero and G zero) mean exactly the same thing to the control, as do G01 and G1, G02 and G2, G03 and G3. All examples in this text do include the leading zero.

Understanding the programmed point of each cutting tool

In order to generate appropriate motions for cutting tools, you must know the location on each cutting tool that is being programmed – which we're calling the *programmed point.* In some cases, you will have to modify program coordinates to compensate for certain attributes of cutting tools.

For example, hole-machining tools like drills, taps, and reamers will have a certain amount of *lead* that must be compensated when Z axis (depth) coordinates are calculated. You can calculate the lead for a twist drill (118 degree point angle) by multiplying 0.3 times the drill diameter. This lead must be added to the hole's depth when plunging through holes. See Figure 3.5.

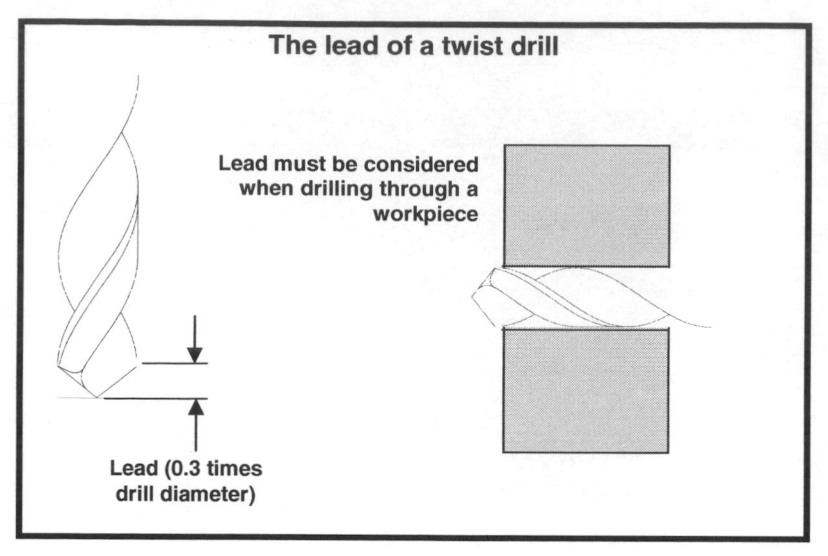

Figure 3.5 – The lead of a twist drill

Unfortunately, cutting tool manufacturers vary when it comes to *how much* lead there will be on certain hole-machining tools. A tap, for example, may have two, three, or four imperfect crests on the end of the tap. Reamer lead will vary with reamer size. You'll have to be quite familiar with the hole-machining tools your company uses in order to correctly compensate for lead when programming hole depths.

Single point tools, like turning tools, boring bars, and grooving tools will always have a small radius on the cutting edge. For the most part, you'll be programming the *extreme edges* of these tools in the X and Z axes. Figure 3.6 shows this.

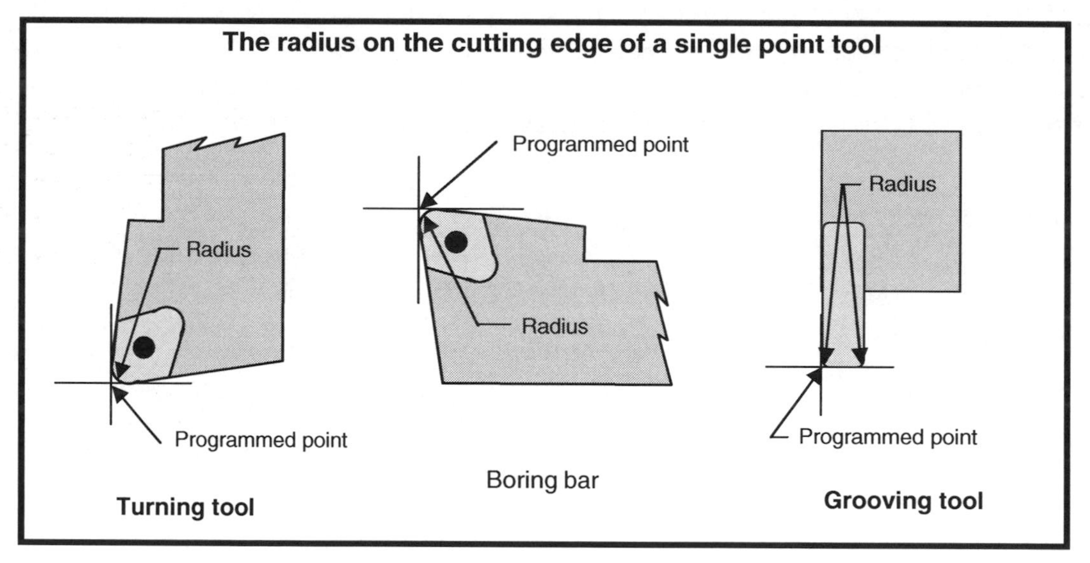

Figure 3.6 – All single point tools have a small radius on the cutting edge

There will be times when you'll have to compensate for a cutting tool's radius as you calculate coordinates for your program. One time is when facing a workpiece to center, as Figure 3.7 demonstrates.

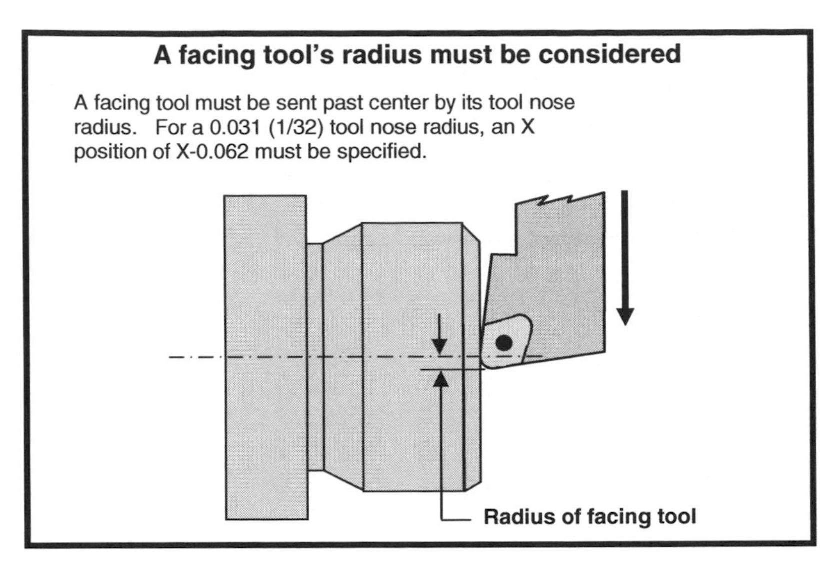

Figure 3.7 – Compensating for a facing tool's radius

If you simply bring the programmed point of the facing tool to the workpiece center (X0), the tool will not completely machine the face. There will be a small nub left in the center of the workpiece. To eliminate the nub, you must program an X coordinate that is slightly past center. The facing tool must be sent to a negative X position that is twice the tool nose radius. If using a tool with a 0.031 (1/32) tool nose radius, the end point for the facing operation must be X-0.062 (remember that X is specified in diameter).

Actually, there is a feature called *tool nose radius compensation* used to deal with many of the problems caused by the tool nose radius. Tool nose radius compensation will keep the cutting edge of the cutting tool flush-with (tangent-to) the work surface at all times. This dramatically minimizes the number of compensated coordinates you must calculate. Tool nose radius compensation is presented in Lesson Fourteen.

There is yet another time when you must calculate compensated coordinates based on tooling criteria. It has to do with grooving tools. Most programmers will program the *leading edge* of grooving tools (the side facing the chuck), as is shown in Figure 3.6. Again, you program the front (left) side of the groove and it machines the left side of the groove. If the groove is of the same width as the grooving tool, the *right* side of the grooving tool will machine the right side of the groove as the left side of the grooving tool machines the left side of the groove. Figure 3.8 shows the tool path needed for grooving tools.

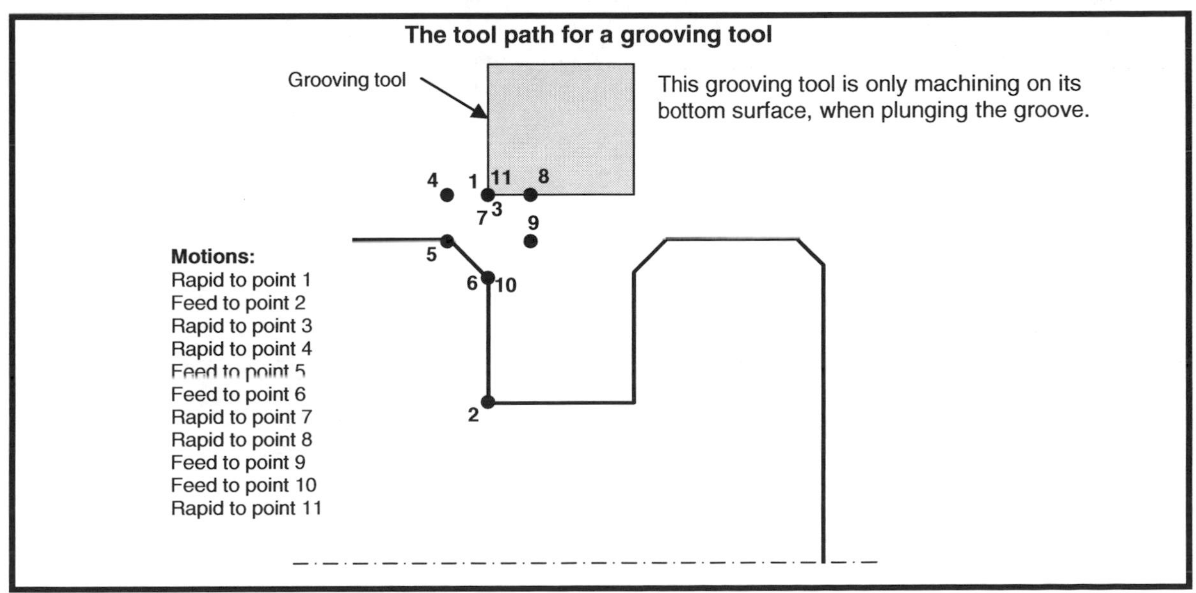

Figure 3.8 – The programmed point for a grooving tool

One last time we show that you must compensate for tooling during motion calculations has to do with threading. Though it is not a location on the cutting edge itself, most programmers program the *extreme leading edge* of the threading tool in the Z axis. Figure 3.9 shows this.

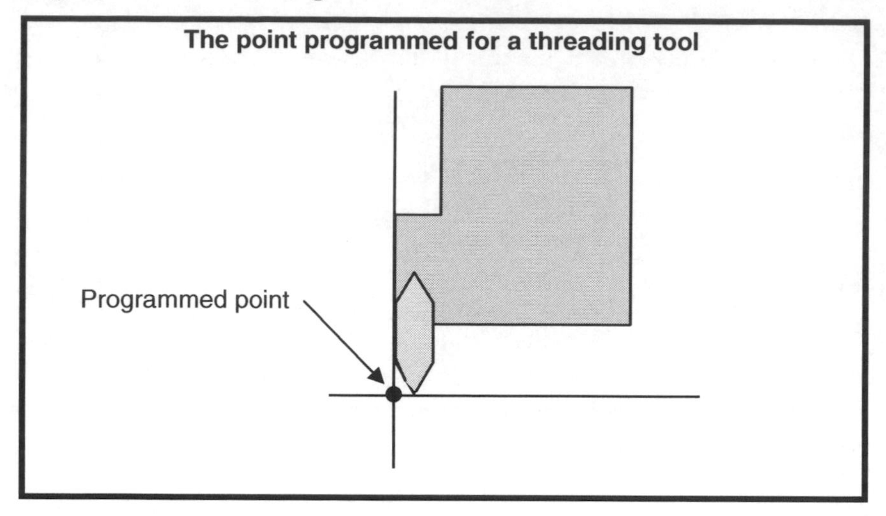

Figure 3.9 – The programmed point for a threading tool

Though some programmers do program the point of the threading tool insert, using the insert's leading edge as the programmed point will eliminate the possibility of interference when threading up to a shoulder. You can program a Z axis position quite close to the shoulder without fear of having the tool contact the shoulder.

On the other hand, if you program the point of the threading tool, calculating the Z end point will be more difficult and worrisome. You must know, of course, the distance between the tip of the threading tool and the leading edge of the threading tool. And if the setup person changes insert sizes (going to a bigger insert) a collision with the shoulder is likely to occur.

Figure 3.10 provides a summary, showing the programmed point for the most popular types of cutting tools used on turning centers.

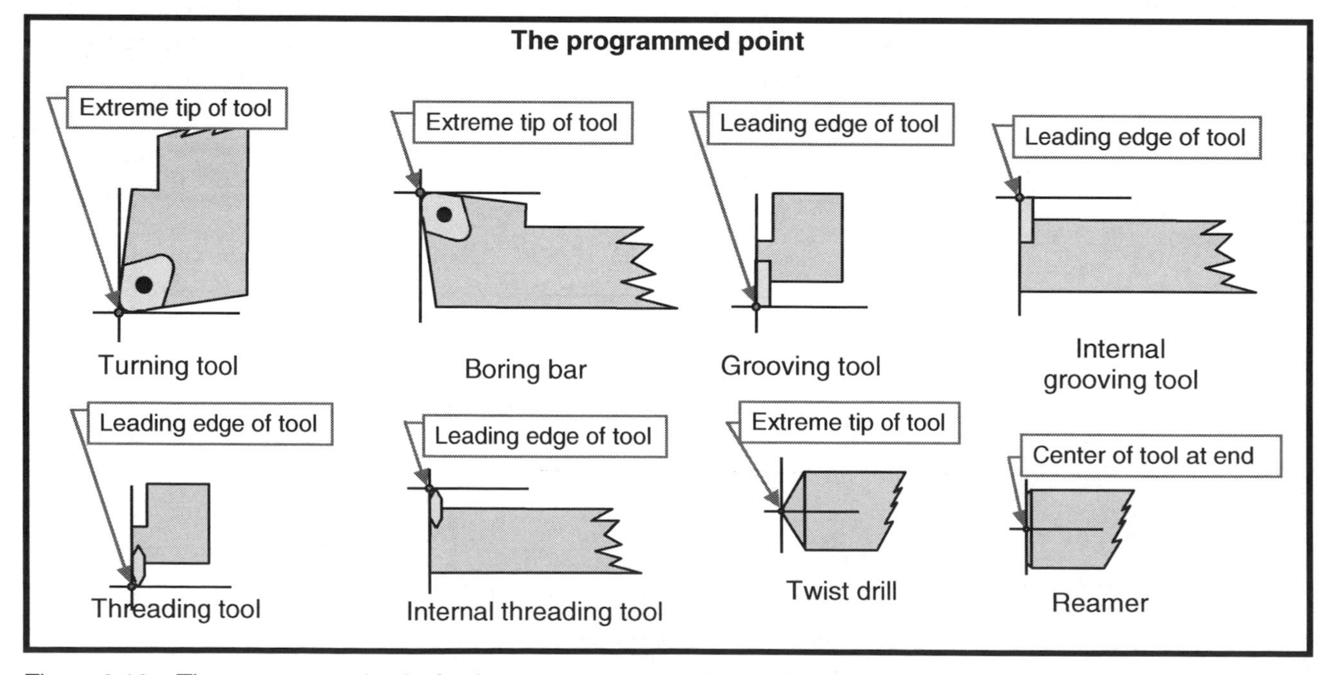

Figure 3.10 – The programmed point for the most common tools used on turning centers

G00 Rapid motion (also called positioning)

Rapid motion is commanded by G00 (or G0), and will cause the axes to move at the fastest possible rate. Rapid motion is used whenever the tool will not be machining during the motion. A good rule-of-thumb is: *if the tool won't be cutting, it should be moving at rapid.* Rapid motions are used for approaching the workpiece, making motions internal to the machining operation that are not cutting (like moving into position for another roughing pass), and when retracting the tool to the tool change position.

Current model turning centers have extremely fast rapid rates. Some can rapid at rates over 2,000 inches per minute. Even older turning centers have pretty fast rapid rates (at least 300 ipm). With these extremely fast rapid rates, you must be very careful with rapid movements. Turning center manufacturers offer two features that allow you to override the machine's rapid rate.

First, the *rapid override* function will allow you to slow the rapid rate at any time. However, machine tool builders vary with regard to how much control they provide with the rapid override switch. Some allow you to slow the machine in rather large increments. With one popular style, you'll have a four-position switch that allows 100%, 50%, 25%, and 10% of the rapid rate. While this is nice, 10% of rapid can still be quite fast. For example, if your turning center has a 1,200 ipm rapid rate, you'll only be able to slow the rapid rate to 120 ipm (10% of 1,200). This is still too fast to catch rapid motion mistakes.

Other turning center manufacturers provide more of a rheostat for rapid override control. With this kind of machine, your rapid override function will allow you to slow motion to a crawl.

Second, the *dry run* function of most machines allows you to use a special multi-position switch to override *all* motions the machine makes – including rapid motions. This function lets you slow motion to a crawl, but is not as handy as the second rapid override function mentioned above. We present more about how you take control of machine motions for the purpose of program verification in Lessons Twenty-Four and Twenty-Eight. For now, rest assured that you'll be able to keep the machine from moving at its very fast (and scary) rapid rate while you're getting familiar with the machine.

Again, the G code G00 (or G0) is used to specify rapid motion. Since all motion types are modal, you need only include a G00 in the first of consecutive rapid motion commands. You also include the motion's end point in X and/or Z in a G00 command.

If both axes will be moving in a rapid motion, it is important to know how the motion will be made. With most machines, each axis will move at its rapid rate until it reaches its destination. If the motions aren't equal distances, or if each axis has a different rapid rate (some turning centers have a slower rapid rate for the X axis), one axis will reach its destination before the other. This means that straight motion will not occur. The motion will appear as a *dog-leg* or *hockey stick.*

Figure 3.11 illustrates the points made about G00.

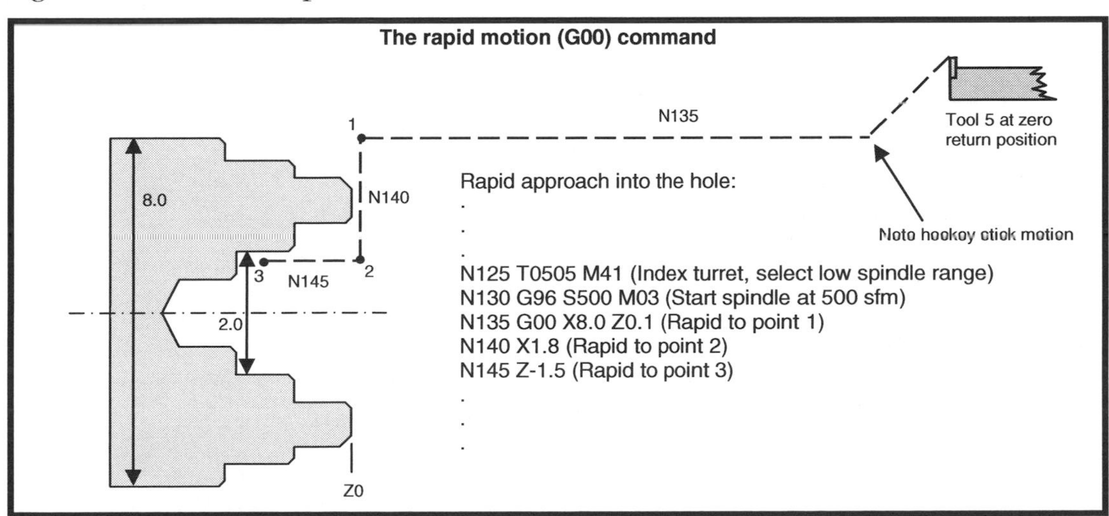

Figure 3.11 - Rapid motion example

Notice that the first approach movement requires a two-axis motion (line N135). Since the X axis doesn't have to move as far as the Z axis, the X axis will reach its destination first. The Z axis will continue moving even after the X axis stops. Notice how this motion resembles a hockey stick (hence our odd name). From this original approach position, the internal grooving tool is brought to a diameter that is smaller than the current hole size. Then it's sent into the hole. These last two motions are single axis movements.

Since G00 is modal, there is no need to include a G00 in line N140 or N145. While it wouldn't hurt to do so, we recommend leaving out redundant words for two reasons. First, you can minimize program length, and given the need to conserve memory space even with current model controls, this is an important consideration. Second, and more importantly, including redundant words and commands will open the door to making mistakes as you write and/or type the program.

In line N140, since only X is moving, there is no need to include a Z value in this command. In line N145, only the Z axis is moving – so there is no need to include an X value in this command. Again, doing so wouldn't hurt anything, but we recommend leaving out redundant words.

What is a safe approach distance?

During approach, most programmers like to keep the tool 0.1 inch away from *qualified surfaces* on the workpiece before they begin the machining operation. By qualified surfaces, we mean surfaces that are not varying by more than a few thousandths of an inch. Examples of qualified surfaces include the outside diameter (rough stock diameter) for cold drawn steel bars and surfaces that have been previously machined.

A good rule-of-thumb is to *make the approach distance at least ten times the amount that a surface will vary from part to part.* If the surface may vary by as much as 0.01 inch, an approach distance of 0.1 inch should be used.

An acceptable approach distance has a lot to do with the quality of the surface being approached. If approaching a surface that is prone to variation from one workpiece to the next (castings are notorious for this kind of variation), most programmers will keep the tool further away (say 0.25 inch) during approach movements.

When a surface is truly qualified (possibly it is machined by another tool in the program), even a 0.1 inch approach distance may be excessive. Keep in mind that the motion made *after* the approach will be done at the machining feedrate. The tool will be *cutting air* until it reaches the workpiece. For this reason, experienced programmers will often reduce rapid approach distances when approaching qualified surfaces (from 0.1 inch to 0.05 inch or even less).

One perfect example of when reducing rapid approach distance is safe (and recommended) is when a cutting tool will be facing a workpiece – and then rough turning it. Consider these commands. Though we have not introduced straight line motion (G01) yet, you should be able to understand the point being made.

```
O0001 (Program number)
(Rough face and turn tool)
N005 T0101 M41 (Index to tool one, select low spindle range)
N010 G96 S500 M03 (Start spindle at 500 sfm in the forward direction)
N015 G00 X6.2 Z0.005 M08 (Approach a 6.0 inch diameter workpiece, turn coolant on)
N020 G01 X-0.062 F0.012 (Face the workpiece past center)
N025 G00 Z0.1 (Rapid away from face)
N030 X5.75 (Rapid up to first diameter to rough turn)
N035 G01 Z-1.995 (First rough turning pass)
N040 X6.2 (Feed up face)
N045 G00 Z0.1 (Rapid back for next rough turning pass)
N050 X5.5 (Rapid to next rough turn diameter)
N055 G01 Z-1.995 (Second rough turning pass)
```

.

.

.

In line **N020**, we face the workpiece. The cutting tool is flush with the face just machined at the completion of this command. In line **N025**, the tool is retracted to a position of **Z0.1**, which is 0.095 inch away from the surface just machined (note that the tool faces to **Z0.005**, leaving 0.005 on the face to be machined). So after line **N025**, the tool is 0.095 away from the face just machined.

In line **N030**, the tool is sent to a diameter for its first rough turning pass. It is still 0.095 away from the face in Z. In line **N035**, the tool makes its first rough turning pass. It must feed (at 0.012 ipr) a distance of 0.095 before it cuts anything. The same is true in line **N055**, when the tool makes its second rough turning pass. It will do the same for as many rough turning passes as it must make.

Our point is that since the cutting tool has qualified the face, an approach distance of 0.095 is excessive. In this example, reducing the approach distance to, say, 0.050 will cut the time needed for approaching almost in half (some programmers will reduce the approach distance even more for this scenario).

Throughout this text, we use a rapid approach distance of 0.1 inch (2.5 mm). And we recommend that beginners use this approach distance for qualified surfaces and 0.25 or more for non-qualified surfaces (like cast surfaces). But as lot sizes grow, remember the effect that excessive approach distance can have on program execution time.

What about feed-off distance?

The same point applies to cutting motions that feed the tool *off* the workpiece. In line **N040** of the program just shown, the tool is feeding up the face and clear of the 6.0 stock diameter. By specifying an end point in line **N040** of **X6.2**, we're telling the machine to move the tool to a position that is 0.1 inch away from the diameter. That is, we've fed the tool off the workpiece by 0.1 inch. Again, some programmers will consider this to be excessive. If the 6.0 inch diameter is not varying by more than 0.010 inch or so, using an X position of **X6.05** in line **N040** will be sufficient to have the tool clear the workpiece.

This can be especially important when machining internal diameters with a boring bar. You must be careful not to allow the back side of the boring bar to contact the workpiece as it feeds off the smallest diameter being machined. Reducing feed-off distance in this case will not only save time, it will minimize the potential for a collision.

G01 linear interpolation (straight line motion)

Linear interpolation, commanded by G01 (or G1), will cause the cutting tool to move along a straight path at a specified feedrate. Linear interpolation is needed for two purposes. First, it's necessary whenever a straight surface must be machined (turn, face, chamfer, taper, drilling a hole, etc.). Since most workpieces have many straight surfaces, there will be many G01 commands in your programs.

Second, there will be times when you'll want the machine to move along a straight path at a controlled feedrate even though you're not cutting anything. A bar feeding turning center will require this kind of motion whenever advancing bar stock.

All turning centers have a limitation when it comes to *maximum feedrate.* Though it's not heavily publicized, the maximum feedrate for most turning centers is *about half the rapid rate.* If the machine can rapid at 800 ipm, for example, its maximum feedrate will be about 400 ipm. For the most part, the turning center's maximum feedrate will be sufficient for machining with optimum cutting conditions. About the only exception to this statement might be when machining very coarse threads at high spindle speeds on older machines.

Feedrate (the F word) is also modal. So once a feedrate is selected, it will remain in effect until it is changed. Many operations can be completed with just one feedrate. For example, even though a rough turning operation may require several roughing passes, every pass can be made at the same feedrate as long as feedrate is specified in per revolution fashion (G99). For most tools, you'll need but one feedrate word, regardless of the motion types required for the tool (linear or circular) and how many cutting motions the tool must make.

Remember from Lesson Two that there are *two* feedrate modes, per revolution (G99) and per minute (G98). If in the inch mode (G20), this equates to inches per revolution or inches per minute. If in metric mode (G21), this equates to millimeters per revolution or millimeters per minute. Feed per revolution (G99) is the best feedrate mode for almost all turning center machining operations, especially if you're using constant surface speed spindle speed mode (G96). The bulk of your feedrate specifications will be in per revolution fashion.

As with the rapid motion, there is a way to override the feedrate for all cutting motions, including straight line cutting motions. On most machines, the *feedrate override switch* provides control of feedrates from 0% though about 200% in 10% increments, meaning you can halt the motion – and you can double the programmed feedrate – or you can select any feedrate in between. If programmed feedrates are correct, the feedrate override switch will be left in the 100% position for the entire program.

Again, the G code G01 (or G1) is used to specify the linear interpolation mode. And since all motion types are modal, you need only include a G01 in the first of consecutive linear motion commands. Also included in the G01 command will be the X and/or Z end points for the motion. The first G01 command for each tool must also include a feedrate word (F word).

As with G00, you need only include the moving axes in a G01 command. If the machine will be moving in only the X axis, for example, (possibly facing a workpiece), you need not to include a Z word in the command.

Figure 3.12 shows an example of linear interpolation

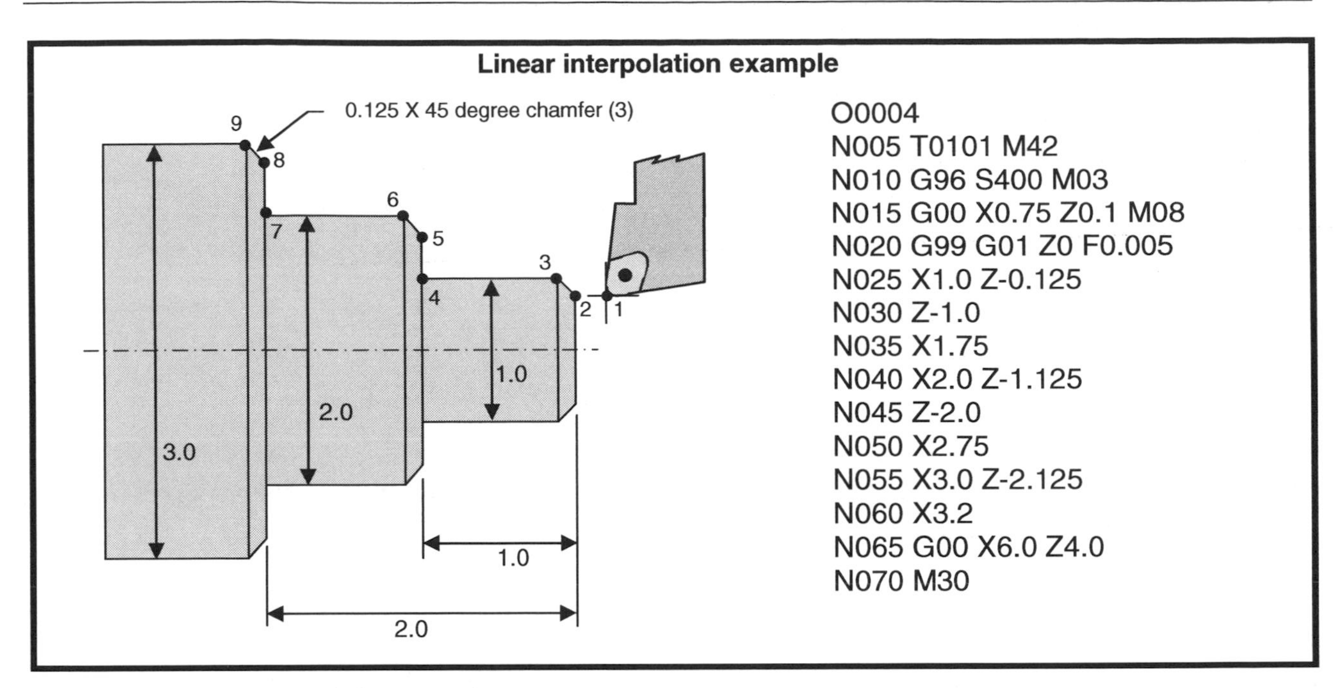

Figure 3.12 – Linear interpolation example

The series of linear motions from point one to point nine finish turns the workpiece. The tool approaches to point one and retracts from point nine with rapid motion commands.

Program with comments:

O0004 (Program number)

N005 T0101 M42 (Index to station number one and invoke wear offset one, select high spindle range)

N010 G96 S400 M03 (Turn the spindle on cw at 400 SFM)

N015 G00 X0.75 Z0.1 M08 (Rapid to point 1, turn on the coolant)

N020 **G99 G01** Z0 **F0.005** (Feed to point 2)

N025 X1.0 Z-0.125 (Feed to point 3)

N030 Z-1.0 (Feed to point 4)

N035 X1.75 (Feed to point 5)

N040 X2.0 Z-1.125 (Feed to point 6)

N045 Z-2.0 (Feed to point 7)

N050 X2.75 (Feed to point 8)

N055 X3.0 Z-2.125 (Feed to point 9)

N060 X3.2 (Feed off diameter)

N065 G00 X6.0 Z4.0 (Rapid away from the workpiece)

N070 M30 (End of program)

After indexing the turret, starting the spindle, and approaching to point one in lines N005, N010, and N015, line N020 begins the series of linear motions. Again, G01 is modal, so it is only required in the first of this series of straight motions (line N020).

The feedrate is also modal, so it is only required in this first cutting command (again, line N020). Every cutting motion will be done at 0.005 ipr, and F0.005 in line N020 specifies this feedrate.

We've also included a G99 in line N020 even though feed per revolution mode is initialized (automatically instated at power-up). This ensures that the feedrate in line N020 will be taken as 0.005 inches per revolution, *not* 0.005 inches *per minute*.

In line N065, the rapid mode is re-selected for the retract motion to the turret index position. Line N070 ends the program. M30 will stop anything that's still running (spindle and coolant in our case) and rewind the program.

Using G01 for a fast feed approach

Some programmers (even experienced programmers) simply don't feel comfortable letting a cutting tool rapid to such a close approach position (0.1 inch in our example). If you want to, you can use G01 to dramatically slow the approach movement once the tool gets near the workpiece. When using this technique, it is best to use the *per minute* feedrate mode for the fast feed approach motion. You must remember to re-select the feed per revolution mode and specify the correct feedrate prior to actually machining the workpiece (otherwise the tool will move at a very fast feedrate, damaging the tool and/or workpiece). Here is the same program just shown, modified to include a fast approach movement using a feedrate of 30.0 ipm.

O0004 (Program number)

N005 T0101 M25 (Index to station one, select high spindle range)

N010 G96 S400 M03 (Turn the spindle on cw at 400 SFM)

N013 G00 **X1.75 Z0.6** M08 (Rapid to fast feed approach position, turn on coolant)

N015 **G98 G01 X0.75 Z0.1 F30.0** M08 (Fast feed to point 1)

N020 **G99 G01** Z0 **F0.005** (Feed to point 2)

N025 X1.0 Z-0.125 (Feed to point 3)

N030 Z-1.0 (Feed to point 4)

N035 X1.75 (Feed to point 5)

N040 X2.0 Z-1.125 (Feed to point 6)

N045 Z-2.0 (Feed to point 7)

N050 X2.75 (Feed to point 8)

N055 X3.0 Z-2.125 (Feed to point 9)

N060 X3.2 (Feed off diameter)

N065 G00 X6.0 Z4.0 (Rapid away from the workpiece)

N070 M30 (End of program)

Notice the addition of line N013. It lets the tool rapid to within 0.5 inch of the final approach position in each axis. In line N015, we've temporarily selected the feed per minute mode (G98) so that the motion to the final approach position can be specified with an inches per minute feedrate (30.0 ipm in our example).

In line N020, we re-select the feed per revolution mode (G99) to allow cutting feedrates to be specified in inches per revolution. As this program is run, the tool will rapid to within 0.5 of the approach position, and then feed at 30.0 ipm the rest of the way. While you may feel (as many programmers do) that this is a safer way to approach, consider the negative impact this approach method will have on program execution time.

G02 and G03 Circular motion commands

Circular interpolation will cause the cutting tool to move along a circular path at a specified feedrate. Circular interpolation is used to machine circular surfaces on a workpiece. While not nearly as often required as linear interpolation, almost all turned workpieces have at least one radius that must be machined, meaning you'll use circular motion in almost every program.

Again, the feedrate (the F word) is modal. So once a feedrate is specified, it will remain in effect until it is changed. This is true even as you switch from linear to circular motions. As stated previously, many machining operations can be completed with but one feedrate word. And as stated during the presentation of G01, all points made during Lesson Two about feed per revolution (G99) and feed per minute (G98) still apply – as do all points made the feedrate override function.

There are two G codes used to command circular motion. G02 (or G2) is used to specify *clockwise* motion and G03 (or G3) is used to specify *counter clockwise* motion. You evaluate which kind of motion is needed by looking at the way the tool will be moving through the motion. In almost all cases, this simply means looking at the print from above. Figure 3.13 shows the difference between clockwise and counter clockwise motion.

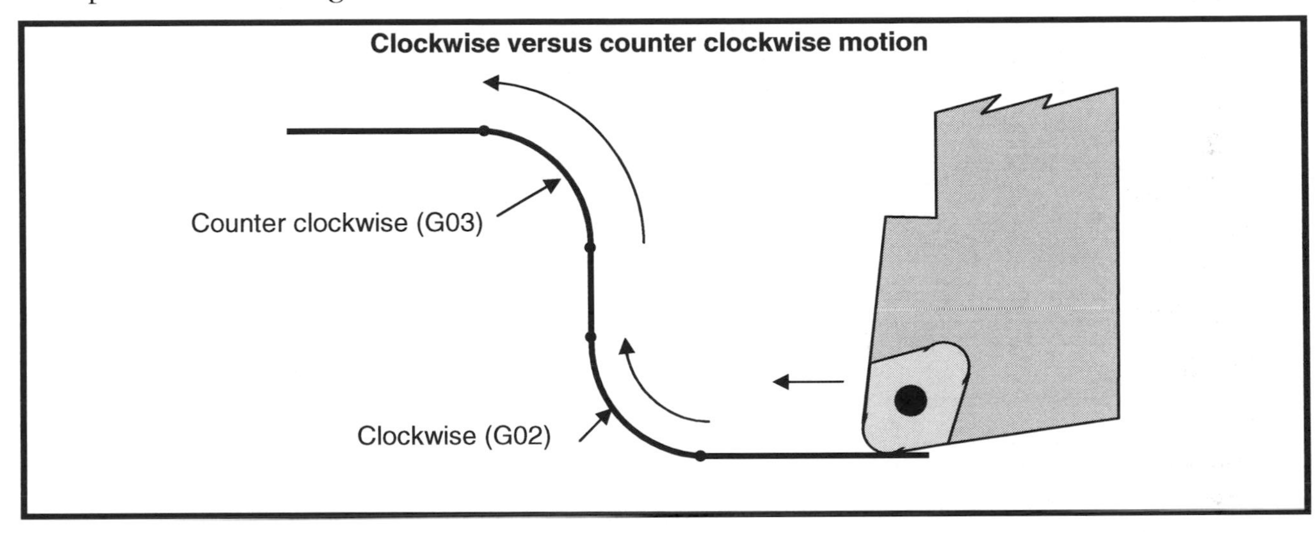

Figure 3.13 – Clockwise versus counter clockwise motion

By the nature of circular motion, both axes will be moving during a circular motion. You'll almost always specify an X and Z coordinate in circular motion commands. As with all other motion types, these coordinates specify the *end point* for the motion.

There is one more specification that must be made within every circular motion command, and it has to do with the *size* of the arc being machined. There are actually two ways to specify arc size. The easiest is to use an R word. We recommend that beginning programmers use this method. The older, more difficult way to specify arc size is to use *directional vectors*. Letter addresses I and K are used for this purpose.

Specifying a circular motion with the radius word

In Figure 3.14, the series of motions from point one to point eleven will finish turn the workpiece. We approach to point one and retract from point eleven with rapid motion commands. Figure 3.5 includes a complete program that will finish turn the workpiece. Notice that this program requires the use of both linear and circular motion. We're using the R word to specify the size of each circular arc.

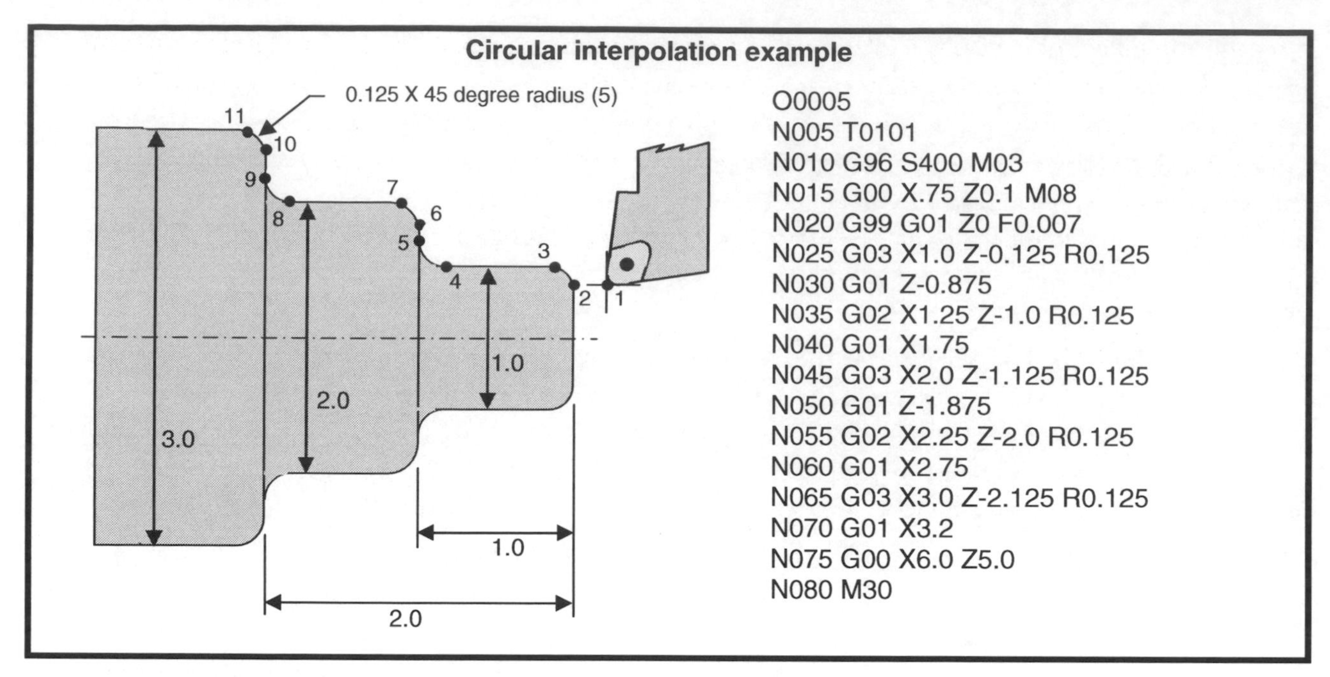

Figure 3.14 - Circular interpolation example

Program with comments:

O0005 (Program number)
N005 T0101 M42 (Index to station number one, select high spindle range)
N010 G96 S400 M03 (Start spindle fwd at 400 SFM)
N015 G00 X.75 Z0.1 M08 (Rapid to point 1, turn coolant on)
N020 G99 G01 Z0 F0.007 (Feed to point 2)
N025 **G03 X1.0 Z-0.125 R0.125** (CCW circular motion to point 3)
N030 **G01** Z-0.875 (Straight move to point 4)
N035 **G02 X1.25 Z-1.0 R0.125** (CW circular motion to point 5)
N040 **G01** X1.75 (Straight move to point 6)
N045 **G03 X2.0 Z-1.125 R0.125** (CCW circular motion to point 7)
N050 **G01** Z-1.875 (Straight move to point 8)
N055 **G02 X2.25 Z-2.0 R0.125** (CW circular motion to point 9)
N060 **G01** X2.75 (Straight move to point 10)
N065 **G03 X3.0 Z-2.125 R0.125** (CCW circular motion to point 11)
N070 **G01** X3.2 (Straight move off workpiece)
N075 G00 X6.0 Z5.0 (Rapid away from workpiece)
N080 M30 (End of program)

After indexing the turret, starting the spindle, and approaching the workpiece in lines N005 through N015, line N020 uses G01 to contact the workpiece. The motion from point two to three is counter clockwise, so a G03 is used (line N025). This command includes the end point in X and Z, as well as the arc size with the R word. The value of the R word is specified right on the print as 0.125 inch. (As you can see, it is relatively simple to specify circular motion using the R word.)

The motion from point three to four is linear, so we *must* to include a G01 in line N030 (again, all motion types are modal). The motion from point four to point five is clockwise, so a G02 is used in line N035. Again the end point is specified with X and Z and the radius is specified with R. word. The balance of this series of motions alternates between linear and circular motion using the same techniques just shown.

Note the importance of including a G01 or G02/G03 as you switch motion types. Since these words are modal, the control will not interpret the program correctly if one is left out. For example, if in line N040, you leave out the G01 during the motion from point five to point six, the control will retain the clockwise motion (G02) from line N035. In this case, it's likely that an alarm will be generated, since only one axis is commanded and no R word is included in line N040.

Important point: ***The R word is not modal.*** In our example, even though each of the radii has a 0.125 radius, the R0.125 must be included in *every* circular command of this program.

Circular motion with directional vectors (I and K)

Almost all current model CNC controls allow you to specify arc size with the simple R word. And we recommend that novice programmers use it whenever commanding circular motions. However, there is another (older) method of specifying arc size with something called *directional vectors*. With directional vectors, two letter addresses (I and K) are used to specify the distance and direction from the start point of the arc to the center of the arc. I specifies this distance and direction in the X axis. K specifies it in the Z axis. Figure 3.15 illustrates this.

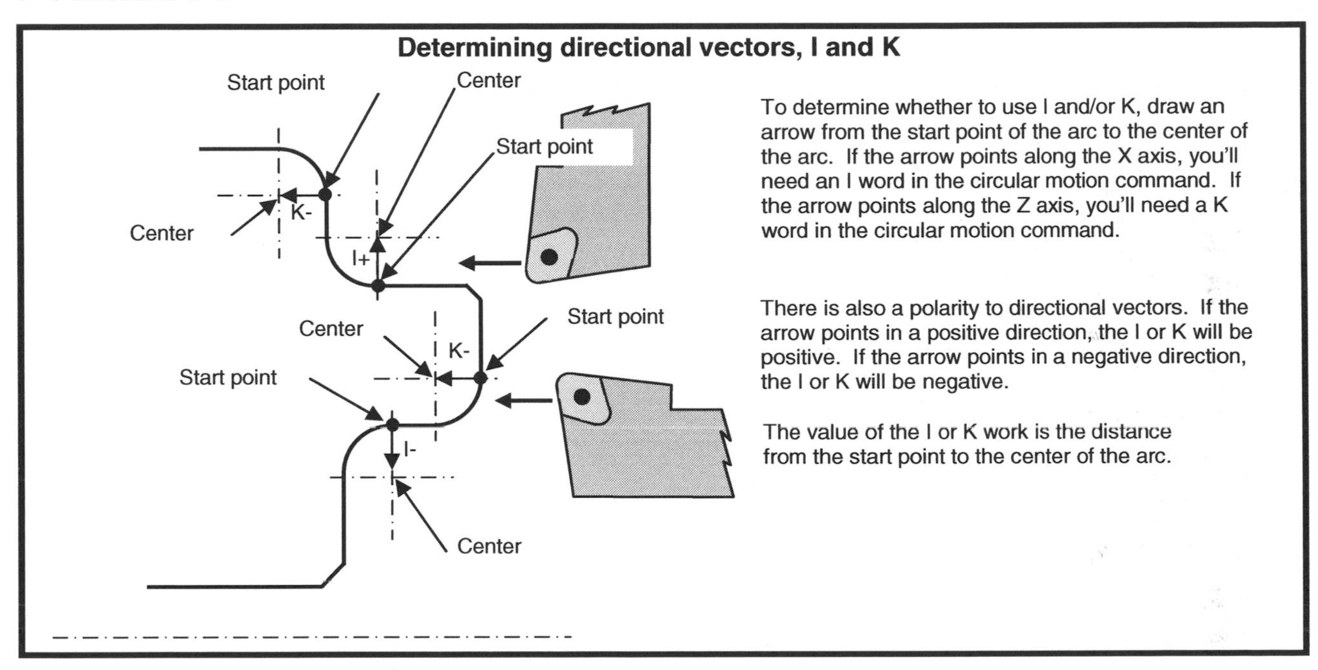

Figure 3.15 – How directional vectors I and K are determined

There is one advantage to using I and K instead of R even with current model CNC turning centers, and it has to do with making mistakes. In a sense, the R word is just a little too forgiving. If a mistake is made with coordinate calculations or in the value of the R word (the correct arc size is not programmed), the machine will still *do something*. No alarm will be sounded. And it's hard to predict exactly what it will do. It will not generate the correct arc. So if you're using the R word to specify arc size, be sure your calculations are correct!

When you use I and K with circular motions (instead of R) most CNC machines are less forgiving. If your calculations are off by as little as 0.0001 inch, most machines will generate an alarm. While this may not sound like much of an advantage, if a mistake is made using R, and if the control simply does its best to generate the motion it thinks you want, the result could be a scrapped workpiece.

Once again, here are the definitions of I and K:

- I is the distance and direction from the starting point of the arc to the center of the arc along the X axis.
- K is the distance and direction from the starting point of the arc to the center of the arc along the Z axis.

As Figure 3.15 shows, I and K have a polarity. If an arrow drawn from the start point of the arc to the center of the arc is pointing in the positive direction, the polarity is plus (no plus sign required). If the arrow is pointing in the negative direction, the polarity is minus (you must only program the minus sign – as you know, the plus sign is assumed).

Here is another program for making the finish pass for the workpiece shown in Figure 3.14. This time we're using directional vectors (I and K) to specify arc center positions instead of R.

Program:

```
O0006 (Program number)
N005 T0101 (Index to station number one)
N010 G96 S400 M03 (Start spindle CW at 400 SFM)
N015 G00 X.75 Z0.1 M08 (Rapid to point 1, turn the coolant on)
N020 G99 G01 Z0 F0.007 (Feed to point 2)
N025 G03 X1.0 Z-.125 K-0.125 (CCW circular motion to point 3)
N030 G01 Z-0.875 (Feed to point 4)
N035 G02 X1.25 Z-1.0 I0.125 (CW circular motion to point 5)
N040 G01 X1.75 (Feed to point 6)
N045 G03 X2.0 Z-1.125 K-0.125 (CCW circular motion to point 7)
N050 G01 Z-1.875 (Feed to point 8)
N055 G02 X2.25 Z-2.0 I0.125 (CW circular motion to point 9)
N060 G01 X2.75 (Feed to point 10)
N065 G03 X3.0 Z-2.125 K-0.125 (CCW circular motion to point 11)
N070 G01 X3.2 (Feed off workpiece)
N075 G00 X6.0 Z5.0 (Rapid away from workpiece)
N080 M30 (End of program)
```

After indexing the turret, starting the spindle and approaching in lines N005 through N015, line N020 feeds the tool up flush with the face and ready to make the first circular movement. It's counter clockwise, so a G03 is being used. X and Z specify the end point. For the directional vector, an arrow drawn from the start point of the arc to the center of the arc will be pointing along the Z axis (the arc center and start point are in the same X position). This means there will be a K word in the circular command (remember, I is related to X, K is related to Z). When it comes to polarity, the arrow is pointing in the Z minus direction, so the polarity of K is minus. The distance from the start point to the arc center is the arc radius (0.125 inch). All other circular commands include I or K following the same rules.

Admittedly, directional vectors I and K are much harder to program than the simple R word to specify the arc size. Again, we recommend that novices use the R word for this purpose. Just be sure you have your coordinates calculated correctly.

What's wrong with this picture?

Frankly speaking, there is a problem with the programs shown in Figures 3.12 and 3.14. It's a minor one, and at this point in the text, it may not be very important. But we better mention it – maybe you've already caught it. For the workpiece machined in Figure 3.12, the chamfers will come out slightly smaller than 0.125 (the specified chamfer size). For the workpiece machined in Figure 3.14, the outside radii will be machined a little smaller than 0.125 – and inside (fillet) radii will be machined a little larger than 0.125. Do you know why?

Hint: *the problem is related to the cutting tool's tool nose radius.*

Since we're programming the extreme tip of the cutting tool in each axis, the radius of the cutting tool will not remain in contact with the work surface during any angular or circular motions. If you're just breaking corners

– as we're doing in Figures 3.12 and 3.14, this may not present a problem. But if you must machine an accurate contour (consider machining a Morse taper, for example), this small deviation will cause a scrap workpiece. The feature *tool nose radius compensation*, which is presented in Lesson Fourteen, will automatically compensate for the tool nose radius as angular and circular surfaces are machined.

Admittedly, we're ignoring the problems caused by the tool nose radius at this point. You'll find (in Lesson Fourteen) that tool nose radius compensation is very easy to program – and with the addition of just two programming words (each), the programs in Figures 3.12 and 3.14 will perfectly machine all angular and circular surfaces.

Working on your first program

Instructions: Using the coordinates you calculated for this practice exercise in Lesson Nine, fill in the blanks for the program that follows. Note that point 1 is now up close to the 1.0 inch diameter (X of point 1 will be X1.2).

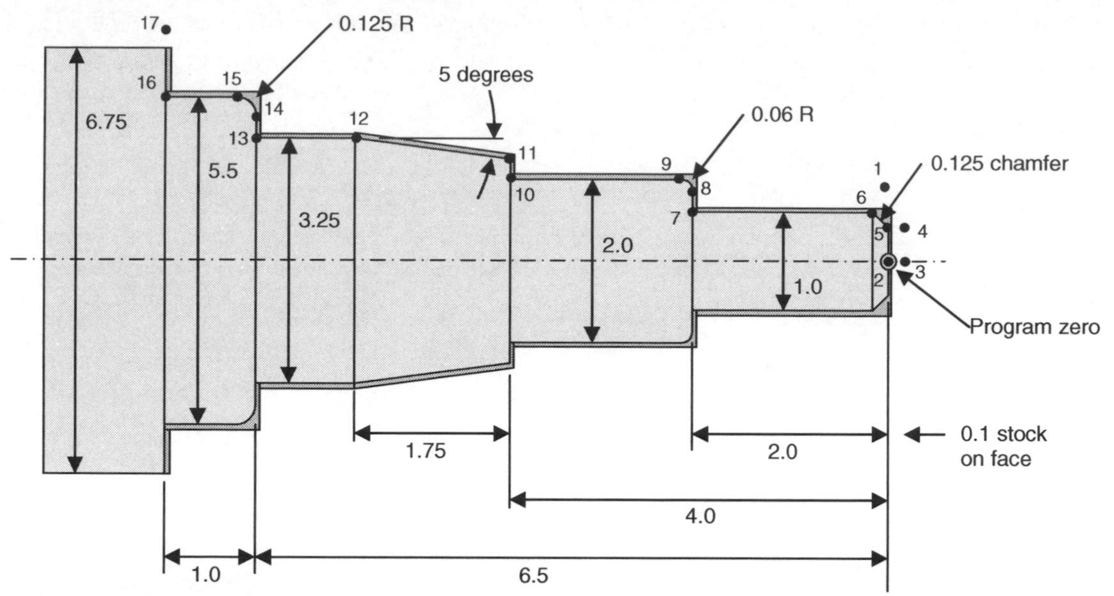

This program will finish face and finish turn the workpiece (it will not do any roughing). Use a feedrate of 0.007 ipr:

O0001

N005 T0202 M42 (Index turret, select high spindle range)

N010 G96 S500 M03 (Start spindle fwd at 500 sfm)

N015 ____ _______ _______ M08 (Rapid to point 1, turn on coolant)

N020 ____ _______ _______ (Face to point 2)

N025 ____ _______ (Rapid away to point 3)

N030 _______ (Rapid up to point 4)

N035 ____ _______ (Feed to point 5)

N040 _______ _______ (Chamfer to point 6)

N045 ________ (Turn 1.0 diameter to point 7)

N050 ________ (Feed up face to point 8)

N055 ____ _______ _______ _______ (Turn radius to point 9)

N060 ____ _______ (Turn 2.0 diameter to point 10)

N065 _______ (Feed up face to point 11)

N070 _______ _______ (Turn taper to point 12)

N075 _______ (Turn 3.25 diameter to point 13)

N080 _______ (Feed up face to point 14)

N085 ____ _______ _______ _______ (Turn radius to point 15)

N090 ____ _______ (Turn 5.5 diameter to point 16)

N095 _______ (Feed up face to point 17)

N100 G00 X8.0 Z6.0 (Rapid to tool change position)

N105 M30 (End of program)

Answer program for exercise on previous page

```
O0001
N005 T0202 M42 (Index turret, select high spindle range)
N010 G96 S500 M03 (Start spindle fwd at 500 sfm)
N015 G00 X1.2 Z0 M08 (Rapid to point 1, turn on coolant)
N020 G01 X-0.062 F0.007 (Face to point 2)
N025 G00 Z0.1 (Rapid away to point 3)
N030 X0.75 (Rapid up to point 4)
N035 G01 Z0 (Feed to point 5)
N040 X1.0 Z-0.125 (Chamfer to point 6)
N045 Z-2.0 (Turn 1.0 diameter to point 7)
N050 X1.875 (Feed up face to point 8)
N055 G03 X2.0 Z-2.0625 R0.0625 (Turn radius to point 9)
N060 G01 Z-4.0 (Turn 2.0 diameter to point 10)
N065 X2.9438 (Feed up face to point 11)
N070 X3.25 Z-5.75 (Turn taper to point 12)
N075 Z-6.5 (Turn 3.25 diameter to point 13)
N080 X5.25 (Feed up face to point 14)
N085 G03 X5.5 Z-6.625 R0.125 (Turn radius to point 15)
N090 G01 Z-7.5 (Turn 5.5 diameter to point 16)
N095 X6.95(Feed up face to point 17)
N100 G00 X8.0 Z6.0 (Rapid to tool change position)
N105 M30 (End of program)
```

Key points for Lesson Ten:

- CNC control manufacturers provide interpolation types for the motions that their machines must make.
- You must know the point on the cutting tool that is being programmed.
- The three most basic motion types are rapid motion – G00, linear motion – G01 (also called straight-line motion), and circular motion – G02 (clockwise) and G03 (counter-clockwise).
- All motion types are modal.
- All motion types require that you specify the end point for the motion in the motion command.
- Use rapid motion whenever the tool is not cutting to minimize cycle time.
- Linear and circular motions additionally require that you specify a feedrate for the motion – and feedrate is also modal.
- Circular motion additionally requires that you specify the arc size for the motion – and the R word is the simplest way to do so.

STOP! Do practice exercise number ten in the workbook.

Know The Compensation Types

CNC turning centers provide three kinds of compensation to help you deal with tooling related problems. In essence, each compensation type allows you to create your CNC program without having to know every detail about your tooling. The setup person will enter certain tooling information into the machine separately from the program.

Key Concept Four is made up of four lessons:

11: Introduction to compensation
12: Geometry offsets
13: Wear offsets
14: Tool nose radius compensation

The fourth Key Concept is that you must know the three kinds of compensation designed to let you ignore certain tooling problems as you develop CNC programs. In Lesson Eleven, we'll introduce you to compensation, showing the reasons why compensation is needed on CNC turning centers.

Two of the compensation types are related to cutting tools: wear offsets and tool nose radius compensation. We discuss them in Lessons Thirteen and Fourteen. The third compensation type, geometry offsets, is more related to work holding devices – and we'll discuss geometry offsets in Lesson Twelve. While we do introduce geometry offsets in Key Concept Number One, we present more about them in Lesson Twelve.

Frankly speaking, not all of the material presented in Key Concept Four may be of immediate importance to you. For example, you may not need to know any more about geometry offsets than what is shown in Lessons Six and Seven. Depending upon your needs, you may want to simply skim this material to gain an understanding of what's included – then move on. You can always come back and dig in should your needs change.

On the other hand, wear offsets and tool nose radius compensation are very important features - features used in *every tool of every program you write.* You'll need to completely master these features.

More on interpreting tolerances

We cannot overstress the importance of being able to interpret tolerances. Every CNC person must possess this ability. Many of the discussions made during Key Concept Four assume you have this ability. Let's begin by reviewing some important points that are related to interpreting tolerances. If you've had machine shop experience, consider this to be a review of what you already know. But if you haven't had shop experience, be sure you understand the material presented here before moving on to Lesson Eleven.

As mentioned previously, *all* dimensions have tolerances. And from Lesson Five, you know that you must *program the mean value of the tolerance band* for every coordinate you include in your programs. The mean value, of course, is right in the middle of the tolerance band. One way to calculate the mean value for any dimension is to divide the overall tolerance for the dimension by two, and add the result to the low limit for the tolerance.

Companies vary when it comes to how tight (small) the tolerances are that they machine on their CNC turning centers. Generally speaking, overall tolerances over about 0.010 inch (about 0.25 mm) are considered pretty open (easy to hold with today's CNC turning centers). Tolerances between 0.002 and 0.010 inch (0.050 – 0.25 mm) are common, and still not considered to be very tight. But tolerances under 0.002 inches can be more difficult to hold. And under 0.0005 inch (0.0013 mm) – which many companies do regularly hold on their CNC turning centers – can be quite challenging – especially when a large number of workpieces must be produced. While there are exceptions, about the smallest overall tolerance most companies regularly hold on their CNC turning centers is 0.0002 inch (0.0051 mm).

Here is a quick summary of the terms related to tolerances:

- **Tolerance band:** total amount of acceptable deviation
- **Mean value:** the value that is right in the middle of the tolerance band
- **High limit:** the largest acceptable size for the dimension
- **Low limit:** the smallest acceptable size for the dimension
- **Measured value:** the value measured on an actual workpiece
- **Target value:** sometimes the mean value, this is the dimension you're shooting for when an adjustment must be made
- **Deviation:** the difference between the measured value and the target dimension
- **Oversize:** a measured value that is too large
- **On-size**: a measured value that is within the tolerance band
- **Undersize:** a measured value that is too small

How design engineers specify tolerances

Design engineers specify tolerances in different ways – and each way has its pros and cons.

High and low limit specified

With this common type of tolerance specification, it's easy to determine the high and low limits – they are directly specified. But you must calculate the mean value.

Consider this example:

4.502/4.498

In this case, the high limit is 4.502. The low limit is 4.498.

To calculate the mean value, the overall tolerance band must first be determined by subtracting the low limit from the high limit. For our example, this results in a tolerance band of 0.004. Next, we must divide the tolerance band by two and add it to the low limit (or subtract it from the high limit). Half of 0.004 is 0.002. When we add 0.002 to 4.498, the result is 4.5. So the mean value of this tolerance is 4.5. You can double-check yourself by *subtracting half the tolerance band from the high limit.* In our example, this means subtracting 0.002 from the 4.502. You should come up with the same mean value. If you do not, you've made a mistake somewhere.

Plus or minus tolerance

With this kind of tolerance specification, it's easy to determine the mean value – it is directly specified. But you must calculate the high and low limits.

Consider this example:

4.5 +/-0.002

In this case, the mean value is 4.5.

To calculate the high limit, you must add the +/- tolerance specification (0.002 in our case) to the mean value. This results in a high limit of 4.502 for our example. To calculate the low limit, you must subtract the +/- tolerance specification (again, 0.002) from the mean value. This results in a low limit of 4.998.

Plus one value, minus another

With this kind of tolerance specification, nothing is easy. You must calculate the mean value as well as the high and low limits.

Consider this example:

4.501 +0.001, -0.003

To calculate the high limit, the plus (+) tolerance specification must be added to the specified dimension. When we add 0.001 to 4.501, the resulting high limit is 4.502. To calculate the low limit, the minus (-) tolerance specification must be subtracted from the specified dimension. When we subtract 0.003 from 4.501, the resulting low limit is 4.498.

To calculate the mean value, half the tolerance band must be added to the low limit. In our case, this means 0.002 (half the tolerance band) must be added to 4.498 (the low limit). The mean value is 4.5.

Is a measured workpiece attribute on-size (acceptable)?

CNC setup people and operators must, of course, be able to tell if a measured workpiece attribute is within its tolerance band (is the measured dimension acceptable?). Again, consider this dimension and tolerance specification:

4.502/4.498

When the high and low limits are specified, it's pretty easy to determine if the measured workpiece attribute is acceptable. Say, for example, this dimension is given for a turned (external) diameter. After machining a workpiece, you measure it and find this diameter to be 4.501. Is this measured dimension acceptable?

Since 4.501 is larger than the low limit and smaller than the high limit, the measured dimension is within the tolerance band. The measured dimension is acceptable.

Now consider this dimension and tolerance:

4.5 +/- 0.002

The high and low limits must be determined before you can determine whether a measured dimension is acceptable. While most CNC people (eventually) get pretty good at doing the related calculations in their heads, beginners might need an electronic calculator.

Consider this dimension and tolerance specification:

4.501 +0.001, -0.003

Since the high and low limits are even more difficult to calculate, determining whether a measured dimension is acceptable also becomes harder.

Again, it is imperative that you possess the ability to interpret tolerances. Without it, you will not be able to correctly determine whether a measured workpiece attribute is acceptable. If you've been in the shop environment for any length of time, much of what has just been presented is probably pretty basic to you – maybe nothing more than a review of what you already know. But if you're new to the shop environment, this presentation may be a bit difficult to understand. You may want to review this presentation several times. Again, you *must* have the ability to interpret dimensions and tolerances. Here is an exercise to test your abilities:

Some practice with tolerances

Instructions: The "Dimension" is the dimension and tolerance that is specified on a blueprint. The "Measured value" is the dimension that you have just measured on an actual workpiece. Based upon these two dimensions, fill in the blanks that follow.

1) Dimension: 3.003/2.997 Measured value: 2.994

Tolerance band: _______

Mean value: _______

Deviation: _______ Acceptable? (y/n): ______

- -

2) Dimension: 1.25 +/-0.002 Measured value: 1.2513

Tolerance band: _______ High limit: _______

Mean value: _______ Low limit: _______

Deviation: _______ Acceptable? (y/n): ______

3) Dimension: 2.501/2.498 Measured value: 2.5003

Tolerance band: _______

Mean value: _______

Deviation: _______ Acceptable? (y/n): ______

- -

4) Dimension: 4.875 +/-0.004 Measured value: 4.8721

Tolerance band: _______ High limit: _______

Mean value: _______ Low limit: _______

Deviation: _______ Acceptable? (y/n): ______

Answers:

1) tb: 0.006, mv: 3.0, dev: 0.006, a: N

2) tb: 0.004, mv: 1.25, hl: 1.252, ll: 1.248, dev: 0.0013, a: Y

3) tb: 0.003, mv: 2.4995, dev: 0.0008, a: Y

4) tb: 0.008, mv: 4.875, hl: 4.879, ll: 4.871, dev: 0.0029, a: Y

What if a measured dimension is not on-size (not acceptable)?

Once again, say this dimension and tolerance is specified for an external diameter being machined by the CNC turning center:

4.502/4.498

After machining this diameter on an actual workpiece, you measure it and find it to be 4.508. This diameter is well above the high limit. It is *oversize*. At the present time, you have an unacceptable workpiece. But because this external diameter is too big (it still has some excess stock), it might be possible to *save* this workpiece. After making an adjustment, you can rerun the tool that machines this diameter – removing more material – and machining it within the tolerance band. (This assumes, of course, that you have not yet removed the workpiece from the turning center.)

On the other hand, after machining, say you measure this diameter and find it to be 4.493. Now the diameter is *undersize*. Since it is an external diameter and too much material has already been removed from it, there will be no (feasible) way to save this workpiece – it is *scrap*.

In either of these cases, an *adjustment* must be made before another workpiece can be machined. Otherwise, of course, you'll just continue making unacceptable workpieces. But how do you determine the amount of needed adjustment?

More on the target value

The target value is the dimension you *shoot for* when you make an adjustment. The target dimension must, of course, be within the tolerance band. And many manufacturing people make the target value equal to the mean value for the tolerance band. That is, when an adjustment must be made, they target the mean value.

Once again, consider this dimension and tolerance:

4.502/4.498

The mean value for this dimension is, of course, 4.5. And again, say this dimension is specifying the size of an external diameter.

After machining, you measure and find the actual diameter on a workpiece is 4.508. Again, the workpiece is oversize – and not acceptable (at least not at the present time). An adjustment must be made. If you are targeting the mean value (4.5), the *amount* of needed adjustment will be 0.008. That is, if you reduce the current diameter on the workpiece by 0.008, it will be right at the mean value.

Every adjustment will have a *polarity* (plus or minus direction). With most turning centers, when a diameter must be made smaller, the adjustment will be negative. When a diameter must be made bigger, the adjustment will be positive. Another way to determine the polarity for a diameter adjustment is to subtract the measured value from the target value:

4.5 minus 4.508 equals -0.008

We've been talking about an external diameter dimension. But the same is true for internal (bored) diameters. If this 4.502/4.498 dimension is specifying the diameter of a bored hole – and if the hole is currently coming out to 4.508, the needed adjustment will still be -0.008 (note that this workpiece is scrap since the hole is oversize).

For length (Z axis dimension), a negative adjustment will cause the cutting tool to machine closer to the chuck face. A positive adjustment will cause it to machine further from the chuck face. If, for example, the overall length of a workpiece is coming out too short, the needed adjustment will be positive.

One more consideration – tool wear

Single point cutting tools (like turning tools and boring bars) will wear as they continue to machine workpieces. Tool wear causes a small amount of material to be removed from the extreme edges of the cutting tool. As tools wear, they leave more and more material on the workpiece.

When tolerances are very large (say over plus or minus 0.003 inch or so), tool wear should not affect workpiece sizing adjustments. If the first workpiece machined by a new cutting tool has come out at the mean value of the tolerance band, the cutting tool can completely wear out without causing the surfaces it machines to go out of their tolerance bands.

But with tighter tolerances, tool wear may eventually cause a machined surface to approach and exceed a tolerance limit. Consider this scenario.

You have a dimension and tolerance specified for an external diameter as 4.5004/4.4996. The high limit is 4.5004, the low limit is 4.4996, and the mean value is 4.5. You decide to target the mean value (again, 4.5). When the first workpiece is machined during setup, the setup person adjusts the machine so that the cutting tool machining is machining the diameter precisely to 4.5 inches – again, right in the middle of the tolerance band.

The first few workpieces will come out to exactly 4.5 inches. But as more workpieces are machined and this finish turning tool wears, the 4.5 inch external diameter will *grow*. Since material is being removed from the cutting edge (due to tool wear), the tool will leave more and more material on the workpiece. After fifty workpieces, you measure the 4.5 inch diameter and find it to be 4.5002. This 0.0002 inch of growth is caused by tool wear.

The diameter is not (yet) out of tolerance. It is still smaller than the high limit (4.5004). So you let the machine continue machining workpieces. After twenty more workpieces are machined, you find that this 4.5 inch dimension is 4.5003. It is still within tolerance, but if this trend continues, the tool will eventually machine the dimension oversize.

Before this occurs, you must make an adjustment. If you're targeting the mean value, the adjustment will be -0.0003. After the adjustment is made, the diameter will come out to 4.5, back at the mean value. And you'll be able to machine another seventy workpieces or so before another adjustment will be needed.

Do you really want to target the mean value?

While targeting the mean value may sound good – and it will be acceptable for large tolerances – when you do so, *you're only working with half the tolerance band.* In the example just shown (targeting the mean value of 4.5 inches), an adjustment is needed after only about seventy workpieces. Knowing the tendency of an external turning tool – causing external diameters to grow as more workpieces are machined – it will be better to target something closer to the low limit.

If, for example, the setup person targets a value of 4.4997 (close to the low limit) when the first workpiece is machined, the number of workpieces machined before an adjustment is necessary can be doubled. And if each subsequent adjustment targets 4.4997, you can reduce the number of adjustments needed during the production run by half.

Again, this strategy only applies when tolerances are quite small and when lot sizes are quite large. With wide open tolerances, the cutting tool will wear out completely before an adjustment is needed. When only machining a few workpieces, you'll complete the production run before an adjustment is required.

Admittedly, this discussion of tolerance interpretation has probably given you a lot to think about. If you're new to CNC turning centers, at the very least, be sure review this presentation a few times. You may want to let a little time pass before you do (giving yourself a chance to absorb this information). Mark this section so you can come back to it later and review.

Lesson 11

Introduction To Compensation

An airplane pilot must compensate for wind direction and velocity when setting a heading. A race-car driver must compensate for track conditions as they negotiate a turn. A marksman must compensate for the distance to the target when firing a rifle. And a CNC programmer must compensate for certain tooling-related problems as programs are written.

What is compensation and why is it needed?

When you compensate for something, you are allowing for some unpredictable (or nearly unpredictable) variation. A *race car driver* must compensate for the condition of the race track before a curve can be negotiated. In this case, the unpredictable variation is the condition of the track. An *airplane pilot* must compensate for the wind direction and velocity before a heading can be set. For them, wind direction and velocity are the unpredictable variations. A *marksman* must compensate for the distance to the target before a shot can be fired – and the distance to the target is the unpredictable variation. The marksman analogy is remarkably similar to what happens with most forms of CNC compensation. Let's take it further…

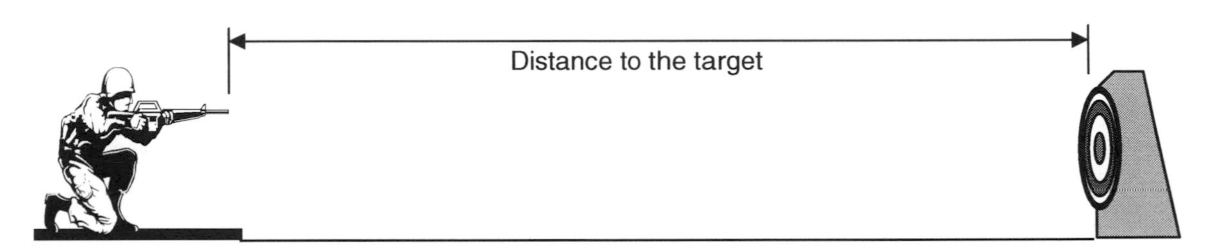

Before a marksman can fire a rifle, they must judge the distance to the target. If the target is judged to be fifty yards away, the sight on the rifle will be adjusted accordingly. When the marksman adjusts the sight, they are *compensating for the distance to the target*. But even after this preliminary adjustment and before the first shot is fired, the marksman cannot be *absolutely* sure that the sight is adjusted perfectly. If they've incorrectly judged the distance – or if some other variation (like wind) affects the sight adjustment – the first shot will not be perfectly in the center of the target.

After the first shot is fired, the marksman will know more. If the shot is not perfectly centered, another adjustment will be needed. And the second shot will be closer to the center of the target than the first. Depending upon the skill of the marksman, it might be necessary to repeat this process until the sight is perfectly adjusted.

With *all* forms of CNC compensation, the setup person will do their best to determine the compensation values needed to perfectly machine the workpiece (just as the marksman does their best in judging the distance to the target and adjusting the sight). But until machining actually occurs, the setup person cannot be sure that their initial compensation values are correct. After machining, they may find that another variation (like tool pressure) is causing the initial adjustment to be incorrect. Depending upon the tolerances for the surfaces being machined, a second adjustment may be required. After this adjustment, machining will be more precise.

There is even a way to make an initial adjustment (prior to machining) that ensures excess material will remain on the machined surface after the first machining attempt (this technique is called *trial machining*). This guarantees that the workpiece will not be scrapped when the cutting tool machines for the first time – and is especially important when very tight (small) tolerances must be held. With tight tolerances, even a small machining imperfection will cause a scrap workpiece.

Once the cutting tool has machined for the first time (under the influence of the trial machining adjustment), the setup person will stop the cycle and measure the surface. If they have used the trial machining technique, there will be more material yet to remove. They will make the appropriate adjustment and re-run the cutting tool. The second time the cutting tool machines, the surface will be within its tolerance band, probably right at the *target* value (the dimension they're shooting for).

The initial setting for compensation

Again, the setup person will do their best to assemble and mount cutting tools. They will also do their best to measure program zero assignment values. They will then enter their measured values into the machine (into something called *offsets*). But even if they *perfectly* measure and enter tooling values, and even if the programmer has specified the mean value for every coordinate in the program, there is still no guarantee that every cutting tool will perfectly machine each dimension to the mean value of its tolerance band.

Tool pressure, which is the tendency for a cutting tool to deflect from the workpiece during machining, will always affect the way a cutting tool machines. It usually has the tendency of pushing the workpiece away from the cutting tool, making external diameters bigger than they should be and internal diameters smaller than they should be. While the surface being machined should be *close* to its mean value (again, assuming the setup person perfectly measures and enters tooling information), it may not be perfectly at the mean value.

How tight is the tolerance? The tighter the tolerance, the more likely it is that the deflection caused by tool pressure will cause the machined surface to be outside the tolerance band when the cutting tool machines for the first time. This will cause a scrap workpiece. And again, this is the reason why trial machining is required – to ensure that the first workpiece gets machined correctly.

And by the way, we've assumed that the setup person has *perfectly* measured and entered tooling information. Any mistakes will, of course, increase the potential for problems holding size on the first workpiece being machined.

When is trial machining required?

First let's talk about when trial machining is *not* required. Trial machining is not required for most roughing operations – like rough turning, rough facing, and rough boring. While a measurement should still be taken right after the roughing operation to confirm that the appropriate amount of finishing stock is being left for the finishing tool (and an adjustment must be made if not), trial machining is not necessary.

There are certain cutting tools that do not allow adjustments, and trial machining cannot be done for these tools. With drills, for example, you cannot control the diameter that the drill machines. Hole-diameter is based upon the diameter of the drill. The same goes for taps and reamers. So for machining operations that cannot be adjusted, you will not be trial machining.

Also, whenever tolerances are greater than about 0.005 inch (0.125 mm) or so, it is not unreasonable to expect the setup person to determine program zero assignment values accurately enough so that the cutting tool will machine within the tolerance band (even considering tool pressure) on the first try. While an adjustment may be necessary *after machining* to bring the dimension precisely to the target value, trial machining should not be necessary.

One last time we'll mention that trial machining is *not* required is when your turning center has a tool touch off probe. As discussed in Lessons Six and Seven, a properly calibrated tool touch off probe eliminates the need for trial machining.

Trial machining is only required when machining tight tolerances (and again, when your machine is not equipped with a tool touch off probe). Though the actual cut-off point for when trial machining is required varies based upon the skill of the setup person and the operation being performed, it is not uncommon to trial machine for dimensions that have tolerances under about 0.004 inch (0.10 mm). This includes most finish turning, finish facing, finish boring, and grooving operations.

What happens as tools begin to wear?

Say the setup person uses trial machining techniques for critical dimensions and the first workpiece passes inspection. Now the production run begins. As cutting tools continue to machine workpieces, of course, they will begin to show signs of wear. Certain tools, like drills, will continue to machine properly for their entire lives without having an impact on the dimensions they machine (hole size and depth).

But as certain tools wear (like finish turning tools and finish boring bars), a small amount of material will wear away from the tool's cutting edge. Depending upon the tolerance band for the dimension being machined by the cutting tool, it is often necessary to make additional sizing adjustments *during the cutting tool's life* to ensure that the cutting tool continues to machine acceptable workpieces.

What do you shoot for?

As stated earlier, the dimension that you are shooting for as you make adjustments is called the *target value.* In some companies, CNC setup people and operators are told to always target the mean value of the tolerance band (the same value the programmers includes in the program). But to prolong the time between needed adjustments, some companies ask their CNC people to target a value that is closer to the high or low limit of the tolerance band, whichever will prolong the time between adjustments. In any event, CNC people must know the target value for each dimension they must machine (this information should be in the production run documentation).

Why do programmers have to know this?

Admittedly, much of this discussion is more related to setup and operation than it is to programming. But again, a programmer must understand enough about machine usage to be able to direct people that run the machine. As a programmer, you must develop the production run documentation – this documentation should include target values for all critical workpiece dimensions. And these discussions go to the heart of one of the most important reason why compensation is required on CNC turning centers: *to allow workpiece sizing adjustments without needing to modify the program.*

Understanding offsets

All three compensation types use *offsets.* Offsets are storage registers for numerical values. They are very much like memories in an electronic calculator. With a calculator, if a value is needed several times during your calculations, you can store the value in one of the memories. When the value is needed, you simply type one or two keys and the value returns. In similar fashion, the setup person or operator can enter important tooling-related values into offsets. When they are needed by the program, a command within the program will *invoke* the value of the offset. And by the way, just as a calculator's memory value has no meaning to the calculator until it is invoked during a calculation, neither does a CNC offset have any meaning to the CNC machine until it is invoked by a CNC program.

Like the memories of most calculators, offsets are designated with *offset numbers.* Geometry offset number one's X register may have a value of -12.5439 (each geometry offset has four registers). Geometry offset number two's X register may have a value of -11.2957. Almost all offsets used on turning centers are related to cutting tools (the only exception being the work shift offset discussed in Lesson Seven), and offset numbers are made to correspond to *tool station numbers.* For example, program zero assignment values for tool number one are entered in geometry offset number one.

Unlike the memories of an electronic calculator that will be lost when the calculator's power is turned off, CNC offsets are more permanent. They will be retained even after the machine's power is turned off – and until an operator or setup person changes them.

Offsets are used with each compensation type to tell the control important information about tooling. From the marksman analogy, you can think of offset values as being like the amount of sight adjustment a marksman must make prior to firing a shot. Tooling related information entered into offsets includes each cutting tool's program zero assignment values (in geometry offsets), the tool nose radius for single point tools (also in geometry offsets), and the amount of wear a tool experiences during its life (in wear offsets).

Offset organization

First of all, rest assured that you will always have enough offsets to handle your applications. Most machine tool builders supply many more offsets than are required even in the most elaborate applications. Most turning centers have at least two distinct groups of offsets, one for cutting tools and another for program zero assignment.

Offset pages on the display screen

These display screen pages are introduced in Lessons Six and Seven. Figure 4.1 shows the geometry offset page, used for two purposes: to enter program zero assignment values and to enter tool nose radius compensation values.

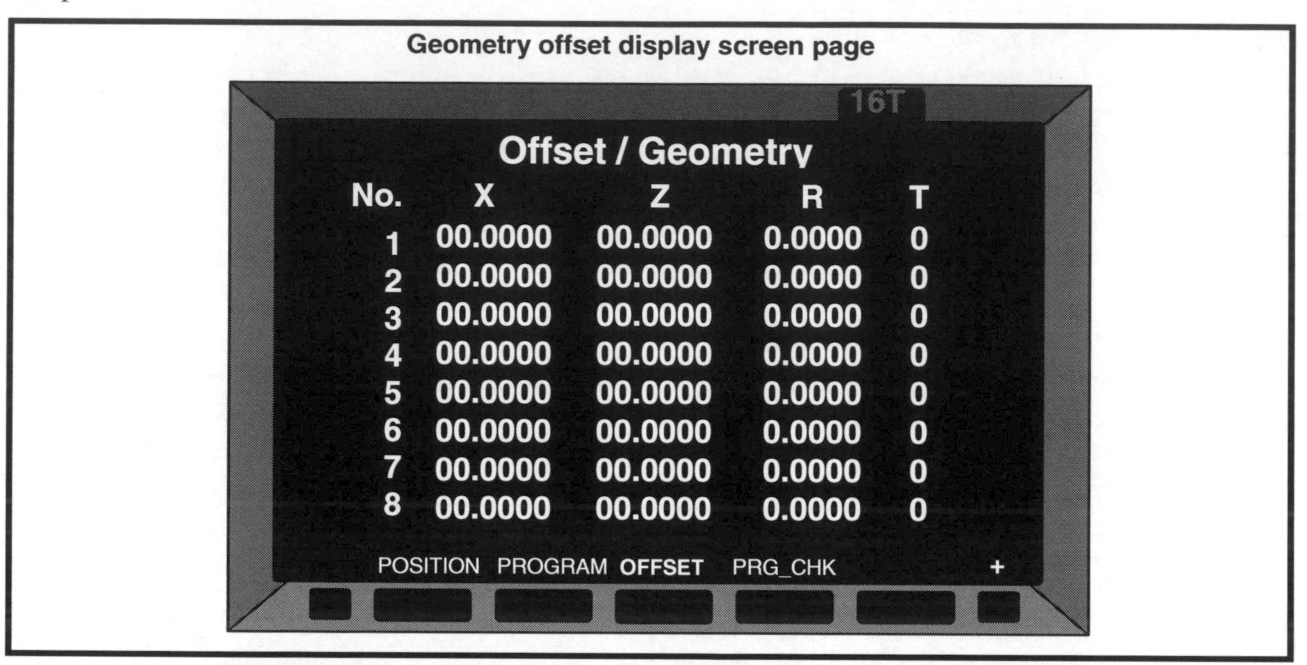

Figure 4.1 – Geometry offset display screen page

Actually, Figure 4.1 shows the *first page* of geometry offsets for a popular Fanuc control model. Only the first eight geometry offsets are shown. All Fanuc controls have at least sixteen sets of geometry offsets (most have thirty-two sets). Since most turning centers have turrets that can hold no more than twelve tools, you'll surely have more than enough geometry offsets.

Notice that each offset number has four registers, X, Z, R, and T. As you know from Lesson Seven, The X and Z registers are use to assign program zero. It is in these registers that you enter program zero assignment values (more will be presented about geometry offsets in Lesson Twelve).

The R and T registers are used to specify tool nose radius compensation values. These registers will be discussed in Lesson Fourteen.

Figure 4.2 shows the wear offset display screen page (again, only the first page is shown).

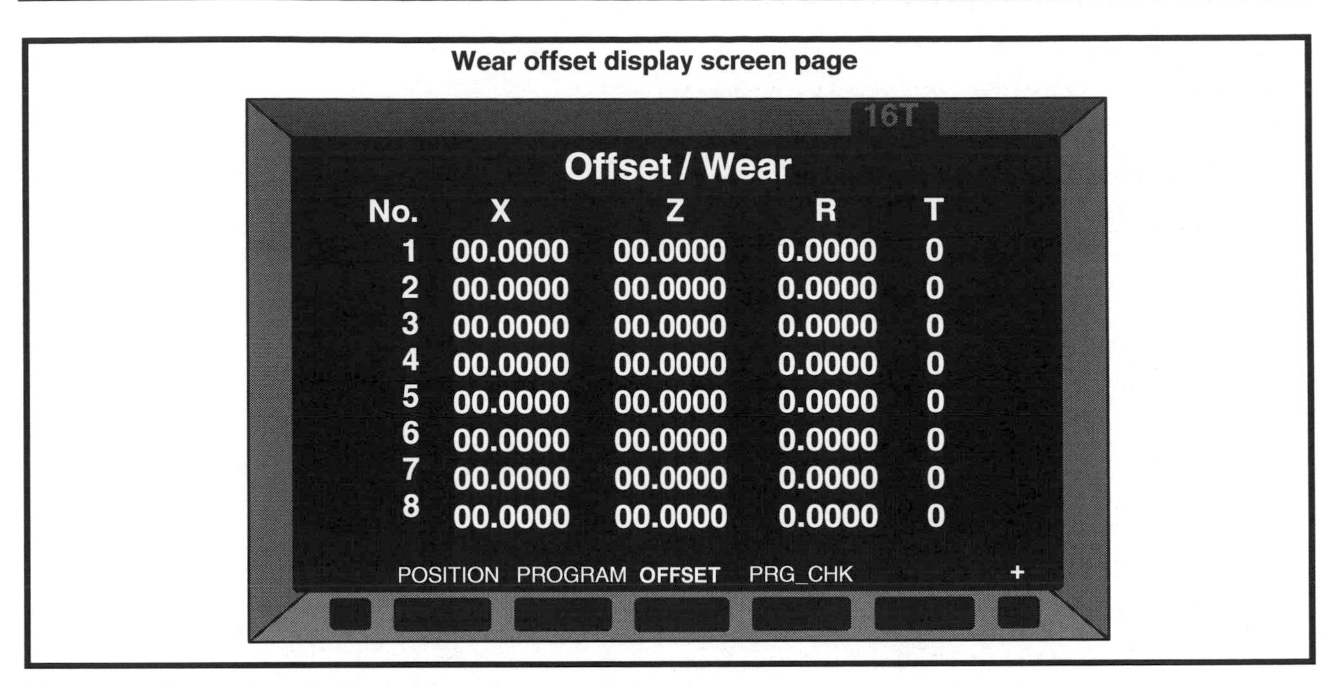

Figure 4.2 – Wear offset display screen page

First of all, notice how similar the wear offset display screen page is to the geometry offset display screen page. While some control models do make the difference a little more distinct, you'll always want to confirm that you're looking at the correct page before you make an entry.

The X and Z registers of the wear offset page are used to enter sizing adjustments that are needed as a cutting tool wears. We discuss wear offsets in Lesson Thirteen.

Do notice that the wear offset page also has a set of R and T registers. Though you can use either the geometry offset page or the wear offset page to enter tool nose radius compensation values, we recommend doing so on the geometry offset page. We explain why in Lesson Fourteen.

Figure 4.3 shows the work shift display screen page.

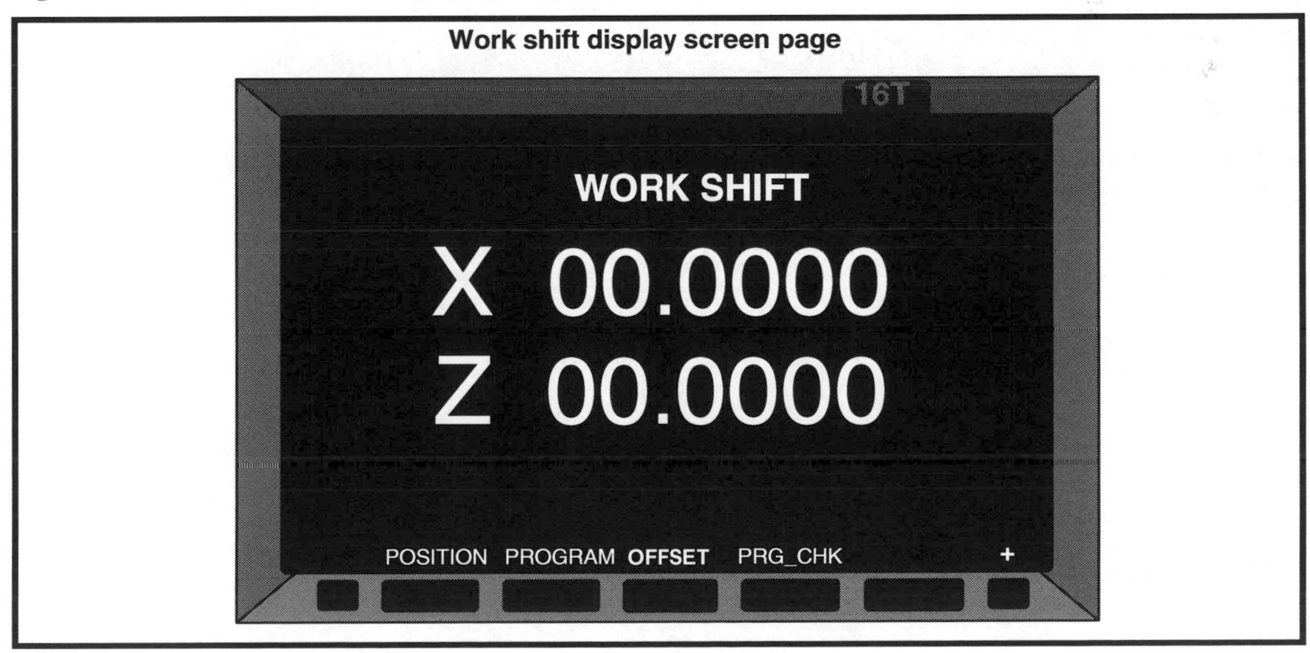

Figure 4.3 – Work shift display screen page of one popular control model

As is presented in Lesson Seven, the work shift page is used to help specify the program zero position in the Z axis. We discuss it in more detail during Lesson Twelve.

How offsets are instated

Again, offsets have no meaning to the control until they are *instated* by a program command. Geometry offsets and wear offsets are instated by the T word alone. Tool nose radius compensation additionally requires a special command in the program to completely instate them. A G41 or G42 is required to specify the relationship between the cutting tool and the surface/s being machined. These G codes are presented in Lesson Fourteen.

The T word for most turning centers has a four digit format. The first two digits cause the machine to index the turret to the specified station and instate the geometry offset. The second two digits of the T word instate the wear offset. Since the R and T registers are included in each offset table, The T word will cause the machine to know the tool nose radius compensation values for the time when a G41 or G42 is executed. For example, consider this command.

```
N160 T0505
```

The first two digits of the T word (05) will cause the machine to index the turret to station number five and instate geometry offset number five. This means the machine will know the program zero assignment values (previously placed in the X and Z registers of geometry offset number five). If the tool requires tool nose radius compensation, and if the setup person has entered the cutting tool's tool nose radius value and tool type in the R and T registers of the geometry offset, the machine will also know the tool nose radius and tool type.

The second two digits of the T word (also 05) will cause the machine to instate the wear offset.

Most programmers will always make the geometry offset number the same as the wear offset number (T0505, T0202, T0404, etc.).

Key points for Lesson Eleven:

- You must know how to interpret tolerances
- Compensation types allow you to ignore certain tooling-related problems as you write programs
- Compensation types allow sizing and trial machining for critical workpiece attributes.
- Setup people and operators must know the *target value* for each dimension they must machine.
- Critical dimensions require trial machining for the first workpiece being machined.
- Tool wear can affect the dimensions machined by a cutting tool. When small tolerances must be held, adjustments may have to be made during the cutting tool's life.
- Offsets are used in which to store compensation values.

Talk to experienced people in your company...

...to learn more about how tight tolerances are held.

1) What are the tightest tolerances held on your CNC turning centers?
2) Are any of the tolerances you hold so tight that trial machining is required? If so, ask to see the trial machining techniques being used.
3) Once as setup is made and production is run, do CNC operators have to make sizing adjustments to deal with tool wear? If so, ask to see it demonstrated.
4) How do you determine the target value for dimensions being machined?
5) Ask to see the offset display screen pages on your CNC turning centers.

STOP! Do practice exercise number eleven in the workbook.

Lesson 12

Geometry Offsets

Most of what you need to know about geometry offsets is presented in Lessons Six and Seven. There are just a few more points we want to make about this important CNC feature.

You know from Lessons Six and Seven that current model Fanuc controls are equipped with geometry offsets. You also know that geometry offsets will be your method of choice for assigning program zero. The only reason to assign program zero in the program with G50 is if the turning center does not have geometry offsets – as will be the case with older turning centers (made before about 1987). And of course, if your machine does not have geometry offsets, you can skip this lesson.

In this lesson, we're assuming that your turning center has geometry offsets, and that you're using them to assign program zero.

Review of reasons for using geometry offsets to assign program zero

Again, there are several advantages to using geometry offsets as compared to assigning program zero within the program. Here is a quick list of the major advantages shown in Lesson Seven.

- Program zero assignment is separated from the program (improved ease of use)
- Only two values must be entered for each tool (improved ease of use)
- If the control is equipped with the measure function, error prone calculations and offset entries are eliminated (safety is improved)
- With the feature *work shift*, redundant Z measurements for cutting tools left in the turret from job to job can be eliminated (improved efficiency)
- The turret can be in any location at the beginning of the program (improved efficiency and ease of use)
- The turret can be in any location when re-running tools (improved efficiency and ease of use)
- For tool changing, the turret can be sent to any safe index point (improved efficiency)

How geometry offsets work

As you know from Lessons Six and Seven, the control must be told the distances in X and Z from the cutting tool tip while the machine is resting at its zero return position to the program zero point. We call these distances the *program zero assignment values*. You know there are several ways to actually determine and enter these values, and we won't repeat those presentations from Lessons Six and Seven here.

While it may not be of the utmost importance, you may find it helpful to understand what actually happens when geometry offsets are used. (Of course, what is of utmost importance is that you know how to use geometry offsets – as is presented in Lessons Six and Seven.)

You know that each cutting tool will have a different set of program zero assignment values. When geometry offsets are used, program zero assignment values (one set per tool) are placed in each tool's geometry offset. The geometry offset number will match the tool station number, so the program zero assignment values for tool number six will be placed in geometry offset number six.

You can get a better understanding of how geometry offsets actually work by looking at the absolute position display page (*not* the relative page) right after power up and before a program is executed. While monitoring this page, send the machine to its zero return position in the X and Z axes.

With the machine is resting at its zero return position right after powering up, most machines will set the absolute position display page to X00.0000 and Z00.0000. Think about this. The absolute position display page always shows the machine's position relative to the program zero point. At this point in time, the control actually thinks the zero return position is the program zero point!

Whenever a geometry offset is instated (with the first two digits of the T word) the distances from the tool tip at the zero return position to the program zero point will be transferred to the absolute position display screen registers. If the machine is not currently at the zero return position in one axis of the other, it will even take this into consideration when it sets the absolute position display values.

For example, say the current X and Z registers of geometry offset number one are set to:

X-12.2437 Z-11.8476

The machine is currently at the zero return position when the T word T0101 is executed. At this point the absolute position display registers will show X12.2437 and Z11.8476 (again *if* the machine is currently at the zero return position in X and Z – and if the work shift offset is currently zero). The machine now knows how far it is from the tool tip to the program zero point, and can correctly make the motions commanded by the program to machine the workpiece.

Again, if the machine is not at the zero return position when a geometry offset is instated, it will consider this when it sets the absolute position display. For example, say the machine is precisely one inch in each axis from the zero return position (closer to the workpiece – on the negative side of the zero return position). In this case, when the T word T0101 is executed, the absolute position displays will show X11.2437 and Z10.8476. The machine has correctly compensated for the turret's position, and still knows the correct distance between the program zero point and the tool tip.

This is why the machines starting position is not critical when geometry offsets are used. The turret can be in any location and the machine will still correctly determine the cutting tool's position – and set the absolute position displays accordingly.

What about work shift? If you're using geometry offsets in conjunction with the work shift value, the machine will also consider this when setting the absolute position displays as geometry offsets are instated. In the scenario just given (machine is one inch from the zero return position), say you have placed a value of 2.5 in the Z register of the work shift offset (this is the distance from the chuck face to the program zero point). The Z value in the geometry offset now represents the distance from the tool tip at the zero return position and the chuck face. When the T word T0101 is executed, now the machine will set the absolute position display to X11.2437 and Z08.3476 (Z value is calculated by taking the 11.8476 geometry offset value minus the 1.0 distance to Z axis zero return position minus the 2.5 work shift value). Again, the machine will always be able to determine the distance between the program zero point and the tool tip.

The total program zero assignment value

As you now know, the absolute position displays will be correctly updated whenever a T word is executed. And the machine will take into consideration the geometry offset value for the tool, the work shift value (for the Z axis), and the machine's current position. But there's one more thing that affects the values being placed in the absolute position displays when a T word is executed – the *wear offset values* for the tool.

Wear offsets are presented in Lesson Thirteen. For now, we simply want to point out that a wear offset will also be instated when a T word is executed. The machine will total the geometry offset, the wear offset, the work shift value, and the machine's current position relative to the zero return position. It will use the result (in each axis) as the total offset for the tool. *These are the true program zero assignment values for the tool* – they are the distances from the program zero point to the tool tip (in each axis) at the very moment the T word is executed. It will place these values on the absolute position display page. Figure 4.4 shows an example.

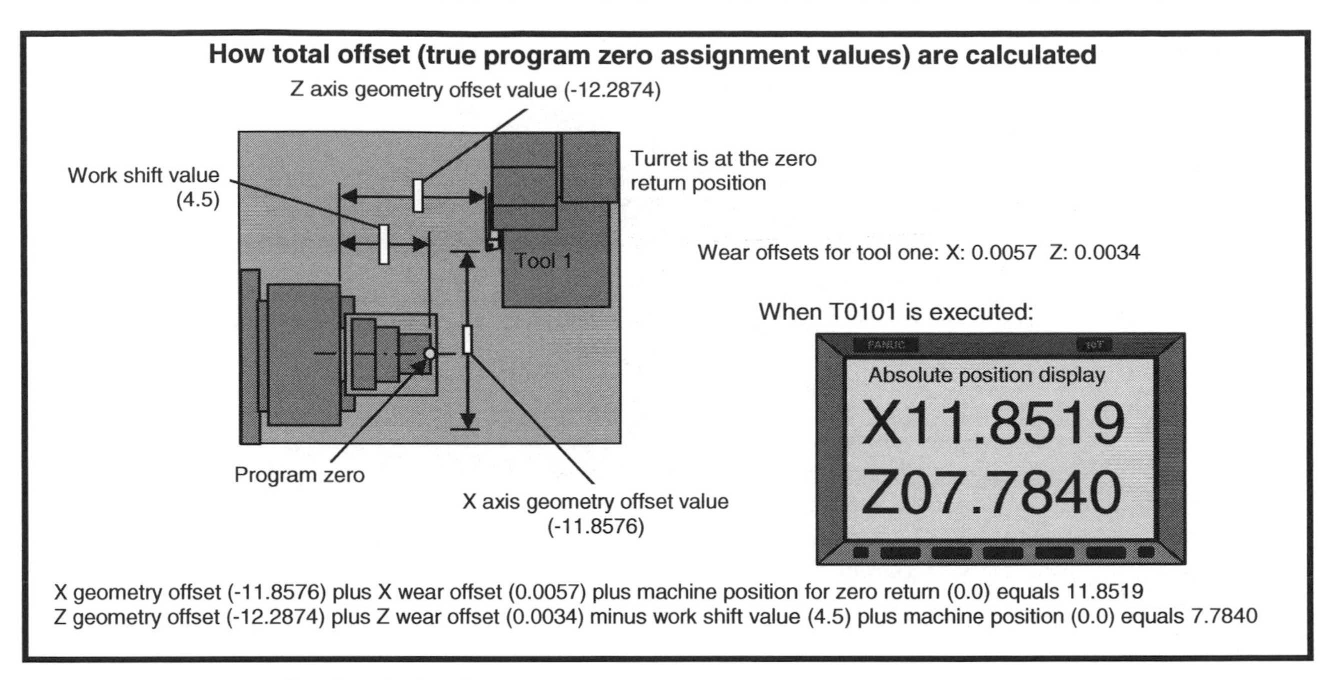

Figure 4.4 – How total offset is calculated

Admittedly, you don't have to know how geometry offsets work to use them. But it's important to know that nothing magical is happening. Additionally, this presentation should help you understand the function of each value involved with program zero assignment.

Warning about the machine lock feature:

Though most current turning centers do not have this function, if you own an older Fanuc-controlled turning center, it might have a feature called *machine lock*. If it does, you'll notice and on/off switch on the control panel named machine lock.

In the on position, machine lock will keep the machine's axes from moving. It can be used during verification to allow the control to scan a questionable (new) program for syntax mistakes. With machine lock turned on, you can run the program and let the control check for syntax mistakes. The control will go through the entire program, *thinking* the machine's axes are moving when in fact, they are not.

While machine lock may be helpful for finding syntax mistakes, it can cause *major* problems with geometry offsets. If the program does not leave the turret in exactly the same location it starts from (most turning center programs do not), the machine lock function will cause the machine to be out of sync. Again, when machine lock is turned on, the control thinks the axes are moving. It continues to track axis position, regardless of how the motion is commanded.

If the control is out of sync with the true axis position, the results can be disastrous. In essence, the machine does not know where the axes are positioned, and after machine lock is turned off, it will not be able to correctly move them when a program is activated. The result is commonly a crash.

If your machine has the machine lock function, we urge you to always leave it off (don't use it). If you do decide to use it for program verification purposes, remember that you must turn the power off in order to clear the confusion it has caused. After power-up, the machine will be in sync again once you do a zero return.

Minimizing program zero assignment effort from job to job

Figure 4.5 shows (once again) our recommended method of program zero assignment when geometry offsets are used. Remember, this method can be used regardless of whether or not your machine has a tool touch off probe – and regardless of whether or not it has the measure function. It does require, of course, that your machine has the work shift offset.

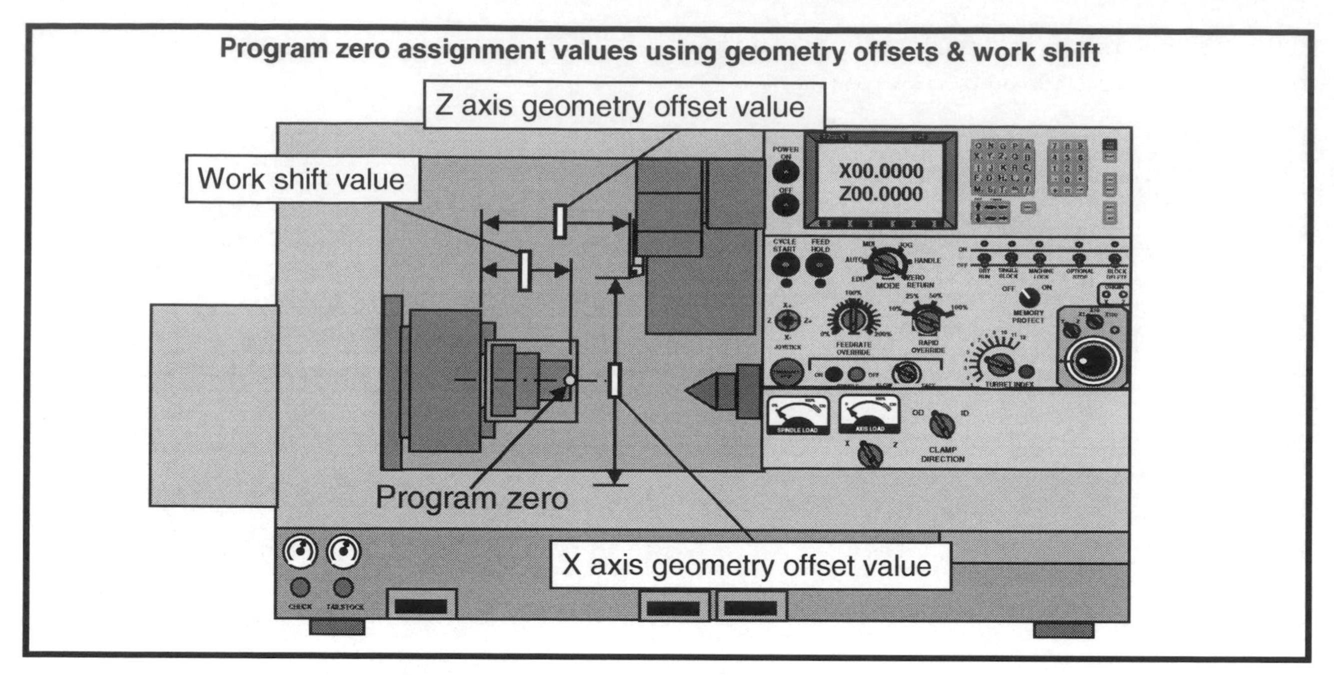

Figure 4.5 – Program zero assignment values when our recommended method is used

Many of the cutting tools used in turning centers are used from job to job. Indeed, some tools remain in the turret on a permanent basis. Consider, for example, a rough turning tool. In many companies, all jobs require a rough turning operation. The same is often true for a finish turning tool. And if your company performs threading operations on a regular basis, a threading tool will be often needed.

For internal work, a drill is required – unless the raw material comes in the form of tubing. While the size of the drill will vary from job to job, most companies will dedicate a turret station to hold the drill. The same goes for rough boring and finish boring tools. While boring bar sizes will often change from job to job, turret stations are dedicated to rough and finish boring bars.

For this reason, programmers often come up with a series of *standard tool stations* in the turret. Here is an example:

- Station one: Rough turning and facing tool (80 degree diamond shaped insert)
- Station two: Finish turning and facing tool (55 degree diamond shaped insert)
- Station three: Threading tool (Insert may change from job to job based upon thread pitch)
- Station four: Grooving tool (Insert will change from job to job based upon groove width)
- Station six: Drill (Drill size will change from job to job based upon hole size)
- Station eight: Rough boring bar (boring bar size will change from job to job)
- Station ten: Finish boring bar (boring bar size will change from job to job)

The goal, of course, is to minimize the work that must be done between jobs during setup. The physical tasks related to placing tools into the turret can be dramatically minimized. But so can many of the tasks related to program zero assignment – as well as the tasks related to trial machining and sizing as the first workpiece in the new job is machined.

As a simple example, consider this scenario. Say two jobs run consecutively that require a rough turning tool, a finish turning tool, and a threading tool (stations 1, 2, and 3 above). When setting up the first job, the three tools must be loaded into the turret (if they're not already in place). Program zero assignment values must be measured and entered. We'll say the work shift value is being used, so it must also be measured and entered.

When the setup person runs the fist workpiece, they'll size in all dimensions – making adjustments as necessary. The rest of workpieces are then run.

When the first job is completed, the second job must be run. Though it may be of a totally different configuration, it requires the same three tools. In this scenario, there won't be much work for the setup person to do – at least related to cutting tools and program zero assignment. Since all three tools are currently in the turret and in the correct turret stations, there will be no need to remove or mount cutting tools. And since chuck face is the point of reference for geometry offsets, and since the chuck face position doesn't change from job to job, the geometry offsets need not be measured. Only one value – the work shift value – must be measured and entered – assuming the workpiece in the second job is not of the same length as the workpiece in the first job.

And our point is: *If a cutting tool is machining properly in one job, it will continue to do so in the next job. As long as you use our recommended method of program zero assignment (working from the chuck face with geometry offsets and using the work shift value), any tool used in the previous job will remain properly set for the next job.*

In our given scenario, even the sizes machined by each tool will come out correctly in the second job. If tools were machining surfaces within tolerance bands in the first job, they will continue to do so in the second job.

There is an exception to this statement having to do with workpiece material. If workpiece material changes dramatically (consider machining tool steel in one job and machining aluminum in the next), there will be a difference in tool pressure. This variation may cause the need to trial machine for finishing operations in the new job. But by the way, this dramatic difference in workpiece material will probably require a different insert grade to be used from one job to the next. In effect, the same tool is *not* being used from job to job.

So when do you clear geometry offsets?

This brings up a good point about clearing (setting to zero) geometry offset values. For as long as a cutting tool remains in the turret, the related program zero assignment values will be correct. Even if a cutting tool is not needed in a given job (and as long as the turret station is not needed), its related program zero assignment will remain intact as the job is run.

Some setup people begin their setups by clearing all offsets. This can be a dreadful (time wasting) mistake – requiring duplicated effort during every setup.

We recommend that you get in the habit of clearing offsets for a tool (geometry and wear) *as you remove the tool from the turret.* With this technique, you can rest assured that the program zero assignment values are correct for any cutting tool that is currently in the turret (even for tools not used in every setup).

Key points for Lesson Twelve:

- Geometry offsets should be your method-of-choice for program zero assignment.
- The machine will total the geometry offset, wear offset, work shift value, and the machine's position relative to the zero return position and use the result as the total offset. It will place these values on the absolute position display page. These are the true program zero assignment values – the distances from the program zero point to the tool tip in each axis.
- With our recommended method of program zero assignment, program zero assignment values will remain correct from job to job.
- Program zero assignment must only be done for new tools placed in the turret during setup.
- The work shift value must be measured and entered for each job.
- Get in the habit of clearing offsets (geometry and wear) for a tool as you remove the tool from the turret.

Work on a multi-tool program

Drawing:

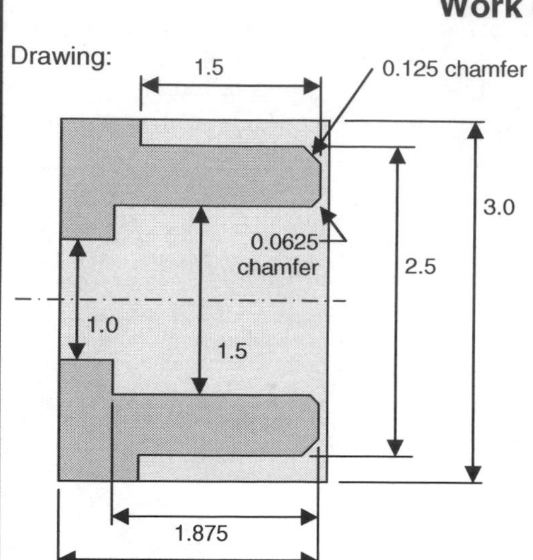

Instructions: This is the exercise from Lesson Nine. In it, you used our recommended process to plan tool paths and calculate coordinates needed in the program. The process, tool paths, and coordinates are shown below. Use them to fill in the blanks for the program that follows.

Process:

Tool one: Rough face leaving 0.005 for finishing. Rough turn 0.125 chamfer and 2.5 diameter (one pass) leaving 0.080 on the 2.5 diameter and 0.005 on the 1.5 face for finishing (**Feed: 0.012 ipr, Speed: 500 sfm**)

Tool two: Drill (1.0 diameter drill) through the workpiece (**Feed: 0.008 ipr, Speed: 1,150 rpm**)

Tool three: Rough bore 0.0625 chamfer and 1.5 diameter (one pass), leaving 0.040 on the diameter and 0.005 on the 1.875 face (**Feed: 0.007 ipr, Speed: 400 sfm**)

Tool four: Finish bore 0.0625 chamfer and 1.5 diameter to size (**Feed: 0.005 ipr, Speed: 500 sfm**)

Tool five: Finish turn 0.125 chamfer and 2.5 diameter to size (**Feed: 0.005 ipr, Speed: 500 sfm**)

Tool paths:

Tool one: Rough face & turn

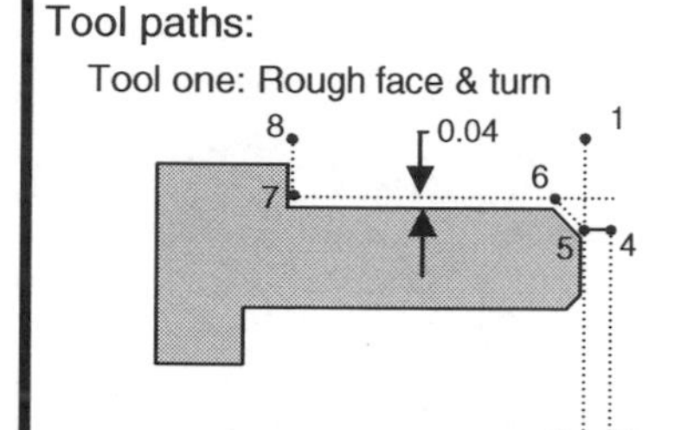

Tool two: Drill 1.0 hole

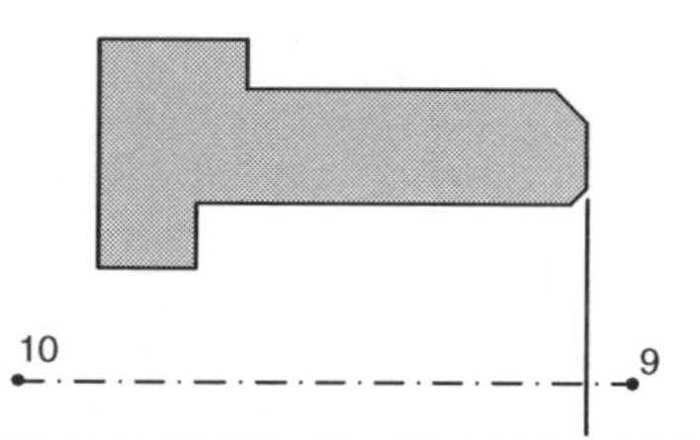

Tool three: Rough bore

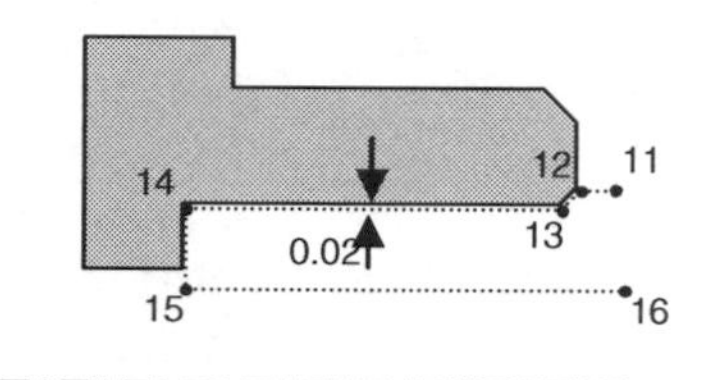

Tool four: Finish bore

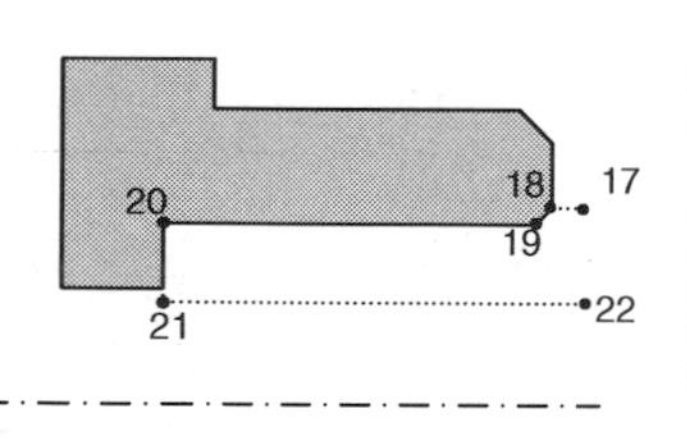

Tool five: Finish face & turn

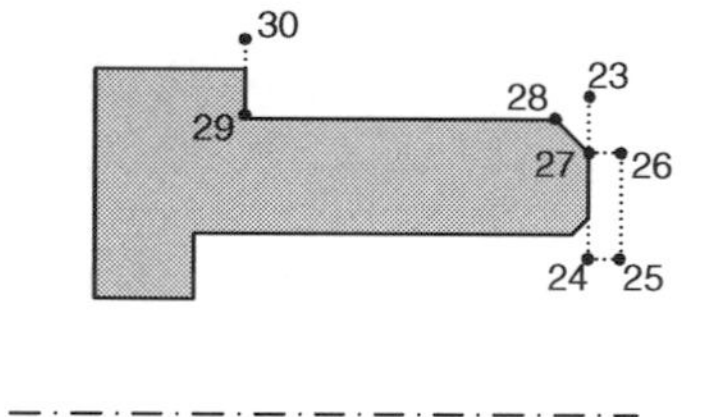

Coordinate sheet:

1: X3.2 Z0.005
2: X-0.062 Z0.005
3: X-0.062 Z0.1
4: X2.33 Z0.1
5: X2.33 Z0.005
6: X2.58 Z-0.12
7: 2.58 Z-1.495
8: 3.2 Z-1.495
9: X0 Z0.1
10: X0 Z-2.705
11: X1.585 Z0.1
12: X1.585 Z0.005
13: X1.46 Z-0.0575
14: X1.46 Z-1.87
15: X0.95 Z-1.87
16: X0.95 Z0.1
17: X1.625 Z0.1
18: X1.625 Z0
19: X1.5 Z-0.0625
20: X1.5 Z-1.875
21: X0.95 Z-1.875
22: X0.95 Z0.1
23: X2.7 Z0
24: X1.3 Z0
25: X1.3 Z0.1
26: X2.25 Z0.1
27: X2.25 Z0
28: X2.5 Z-0.125
29: X2.5 Z-1.5
30: X3.2 Z-1.5

O0002 (Program number)

(ROUGH FACE AND TURN)

N005 T0101 M41 (Index turret, select low spindle range)

N010 ____ ________ ____ (Start spindle fwd at 500 sfm)

N015 ____ ________ ________ M08 (Rapid to point 1, turn on coolant)

N020 G99 ____ ________ ________ (Face to point 2 at 0.012 ipr)

N025 ____ ________ (Rapid away to point 3)

N030 ________ (Rapid up to point 4)

N035 ____ ________ (Feed up to point 5)

N040 ________ ________ (Chamfer to point 6)

N045 _________ (Turn 2.5 diameter to point 7)

N050 _________ (Feed up face and off part to point 8)

Answer program is on the next page.

Work on a multi-tool program (continued)

N055 G00 X8.0 Z7.0 (Rapid to tool change position)

N060 M01 (Optional stop)

(DRILL 1.0" HOLE)

N065 T0202 M41 (Index turret, select low spindle range)

N070 ____ ________ ____ (Start spindle at 1,150 rpm)

N075 ____ ________ ________ M08 (Rapid to point 9, turn on coolant)

N080 ____ ________ ________ (Feed to point 10 at 0.008 ipr)

N085 ____ ________ (Rapid back to point 9)

N090 X8.0 Z7.0 (Rapid to tool change position)

N095 M01 (Optional stop)

(ROUGH BORE)

N100 T0303 M41 (Index turret, select low spindle range)

N103 ____ ________ ____ (Start spindle at 400 sfm)

N105 ____ ________ ________ M08 (Rapid to point 11, start coolant)

N110 ____ ________ ________ (Feed to point 12 at 0.007 ipr)

N115 ________ ________ (Chamfer to point 13)

N120 ________ (Rough bore 1.5 to point 14)

N125 ________ (Feed down face to point 15)

N130 ____ ________ (Rapid to point 16)

N135 X8.0 Z7.0 (Rapid to tool change position)

N140 M01 (Optional stop)

(FINISH BORE)

N145 T0404 M42 (Index turret, select high spindle range)

N150 ____ ________ _____ (Start spindle at 500 sfm)

N155 ____ ________ ________ M08 (Rapid to point 17, turn on coolant)

N160 ____ ________ ________ (Feed to point 18 at 0.005 ipr)

N165 ________ ________ (Chamfer to point 19)

N170 ________ (Bore 1.5 to point 20)

N175 ________ (Feed down face to point 21)

N180 ____ ________ (Rapid to point 22)

N185 X8.0 Z7.0 (Rapid to tool change position)

N190 M01 (Optional stop)

(FINISH FACE AND TURN)

N195 T0505 M42 (Index turret, select high spindle range)

N200 ____ ________ ____ (Start spindle at 500 sfm)

N205 ____ ________ ________ M08 (Rapid to point 23, turn on coolant)

N210 ____ ________ ________ (Face to point 24 at 0.005 ipr)

N215 ____ ________ (Rapid to point 25)

N220 ________ (Rapid to point 26)

N225 ____ ________ (Feed to point 27)

N230 ________ ________ (Chamfer to point 28)

N235 ________ (Turn 2.5 to point 29)

N240 ________ (Feed up face to point 30)

N245 G00 X8.0 Z7.0 (Rapid to tool change position)

N250 M30 (End of program)

Answer program is on the next page.

Answer program

O0002 (Program number)
(ROUGH FACE AND TURN)
N005 T0101 M41 (Index turret, select low spindle range)
N010 **G96 S500** M03 (Start spindle fwd at 500 sfm)
N015 **G00 X3.2 Z0.005** M08 (Rapid to point 1, turn on coolant)
N020 G99 **G01 X-0.062 F0.012** (Face to point 2 at 0.012 ipr)
N025 **G00 Z0.1** (Rapid away to point 3)
N030 **X2.33** (Rapid up to point 4)
N035 **G01 Z0** (Feed up to point 5)
N040 **X2.58 Z-0.12** (Chamfer to point 6)
N045 **Z-1.495** (Turn 2.5 diameter to point 7)
N050 **X3.2** (Feed up face and off part to point 8)
N055 G00 X8.0 Z7.0 (Rapid to tool change position)
N060 M01 (Optional stop)
(DRILL 1.0" HOLE)
N065 T0202 M41 (Index turret, select low spindle range)
N070 **G97 S1150 M03** (Start spindle at 1,150 rpm)
N075 **G00 X0 Z0.1 M08** (Rapid to point 9, turn on coolant)
N080 **G01 Z-2.705 F0.008** (Feed to point 10 at 0.008 ipr)
N085 **G00 Z0.1** (Rapid back to point 9)
N090 X8.0 Z7.0 (Rapid to tool change position)
N095 M01 (Optional stop)
(ROUGH BORE)
N100 T0303 M41 (Index turret, select low spindle range)
N103 **G96 S400 M03** (Start spindle at 400 sfm)
N105 **G00 X1.585 Z0.1** M08 (Rapid to point 11, start coolant)
N110 **G01 Z0 F0.007** (Feed to point 12 at 0.007 ipr)
N115 **X1.46 Z-0.0575** (Chamfer to point 13)
N120 **Z-1.87** (Rough bore 1.5 to point 14)
N125 **X0.95** (Feed down face to point 15)
N130 **G00 Z0.1** (Rapid to point 16)
N135 X8.0 Z7.0 (Rapid to tool change position)
N140 M01 (Optional stop)
(FINISH BORE)
N145 T0404 M42 (Index turret, select high spindle range)
N150 **G96 S500 M03** (Start spindle at 500 sfm)
N155 **G00 X1.625 Z0.1** M08 (Rapid to point 17, turn on coolant)
N160 **G01 Z0 F0.005** (Feed to point 18 at 0.005 ipr)
N165 **X1.5 Z-0.0625** (Chamfer to point 19)
N170 **Z-1.875** (Bore 1.5 to point 20)
N175 **X0.95** (Feed down face to point 21)
N180 **G00 Z0.1** (Rapid to point 22)
N185 X8.0 Z7.0 (Rapid to tool change position)
N190 M01 (Optional stop)
(FINISH FACE AND TURN)
N195 T0505 M42 (Index turret, select high spindle range)
N200 **G96 S500 M03** (Start spindle at 500 sfm)
N205 **G00 X2.7 Z0** M08 (Rapid to point 23, turn on coolant)
N210 **G01 X1.3 F0.005**(Face to point 24 at 0.005 ipr)
N215 **G00 Z0.1** (Rapid to point 25)
N220 **X2.25** (Rapid to point 26)
N225 **G01 Z0** (Feed to point 27)
N230 **X2.5 Z-0.125** (Chamfer to point 28)
N235 **Z-1.5** (Turn 2.5 to point 29)
N240 **X3.2** (Feed up face to point 30)
N245 G00 X8.0 Z7.0 (Rapid to tool change position)
N250 M30 (End of program)

STOP! Do practice exercise number twelve in the workbook.

Lesson 13

Wear offsets

This compensation type allows the setup person or operator to deal with minor size variations as workpieces are machined. While many CNC people (somewhat inappropriately) use wear offsets to compensate for minor setup imperfections and tool pressure, the major application for wear offsets is to compensate for tool wear during a tool's life.

You know that the tolerances commonly held on CNC turning centers are quite small. It is not unusual to see at least one overall tolerance of under 0.001 inch (0.254 mm) on turned workpieces.

You also know that each cutting tool has its own program zero assignment – and that there are several ways to assign program zero. And you know that unless you are using a properly calibrated tool touch off probe, mistakes with program zero assignment – even minor ones – as well as the effects of tool pressure, make it difficult to *perfectly* assign program zero. That is, even after program zero is assigned, there is no guarantee that every tool will machine the workpiece perfectly – or even within specified tolerances. The tighter the tolerances you must hold, the greater the potential there will be for sizing problems.

Wear offsets provide a way to make minor adjustments when machined surfaces are not within their tolerance bands – or when they're close to a tolerance limit.

There are at least four times when a typical CNC setup person or operator will use wear offsets:

- **During setup and after mounting a cutting tool in the turret** - After machining with the new tool, if the setup person discovers that the cutting tool has not machined a surface within the tolerance band, or if the surface is close to a tolerance limit, they can change a wear offset to make the needed adjustment.
- **When trial machining** – Trial machining is done when the setup person or operator is worried that a cutting tool (that has just been placed in the turret) will not machine within the tolerance band. Wear offsets are commonly used with which to make trial machining adjustments (Remember, if you are using a properly calibrated tool touch off probe, you shouldn't need to trial machine.)
- **When compensating for tool wear** – As a cutting tool wears, it will cause the surfaces it machines to grow or shrink in size. Wear offsets are used to keep cutting tools machining on-size for their entire lives.
- **After a dull tool is replaced** – Again, during a cutting tool's life, tool wear commonly causes the need for sizing adjustments in wear offsets. When a dull tool is replaced with a new one, the wear offset must be set back to its initial value – otherwise the new tool will machine too much material from the workpiece.

In the introduction to Key Concept Four, there is a lengthy presentation about *tolerance interpretation*. From this presentation, you know that *every* dimension has a tolerance. You know that each tolerance will have a high limit (largest acceptable dimension), a low limit (smallest acceptable dimension), and a mean value (the dimension right in the middle of the tolerance band.

You also know that each dimension to be machined on a workpiece will have a *target value* – this is the dimension you shoot for when an adjustment must be made. Many CNC people use the mean value of the tolerance band as the target value. That is, when an adjustment must be made, they target the mean value.

While there are times when this may not be appropriate (large lot sizes with small tolerances), we'll use this technique throughout Lesson Thirteen. That is, for each surface needing an adjustment, we'll be targeting the mean value.

You know that the *deviation* is the amount of needed adjustment. It is the difference between the measured value and the target value. In all cases, there will be a polarity to the deviation (plus or minus). The current wear offset value must be either increased or decreased by the amount of the deviation. The polarity is determined by judging which way the cutting tool must move (plus or minus) in order to bring the dimension back to the target value. We'll discuss more abut the deviation and its polarity later in this lesson.

Which dimension do you choose for sizing?

It is very common for a finishing tool to machine several surfaces. One finish turning tool may, for example, finish three external diameters and faces. Each surface may have its own tolerance specification. And of course, you program the mean value for each tolerance. In almost all cases, only one wear offset will be used to control *all* surfaces machined by each tool, meaning one adjustment will handle all surfaces machined by the tool.

For example, say a finish turning tool machines three external diameters: a 1.5 inch diameter, a 2.0 inch diameter, and a 3.0 inch diameter. Unless there is a substantial tool pressure problem, when the 1.5 inch diameter is coming out correctly (based on a wear offset adjustment in X), so will the 2.0 and 3.0 inch diameters.

When it comes to making sizing adjustments, you should use the two surfaces (one for diameters and one for lengths) that have the smallest tolerances. Use them to make sizing adjustment decisions.

How wear offsets are programmed

As you know, wear offsets are invoked by the second two digits of the T word. The command

```
N140 T0303
```

for example, indexes the turret to station number three, instates geometry offset number three (the first two digits) and instates wear offset number three (the second two digits). In almost all applications, you'll be making the wear offset number the same as the tool station number (tool one: wear offset one, tool two: wear offset two, and so on).

What actually happens when a wear offset is instated will vary based upon the CNC control (model and age). With newer controls (and especially if geometry offsets are used to assign program zero), the T word appears to simply index the turret. The geometry and wear offsets will not actually be instated (made active) *until the next motion command.* But with some older controls – and especially when program zero must be assigned in the program – the T word indexes the turret and the wear offset will be immediately instated. This will cause the turret to actually move by the amount of the wear offset.

What if my machine doesn't have geometry offsets?

It's much better if the turret remains in position during a turret index, and you can rest assured it will if you're assigning program zero with geometry offsets.

If you're assigning program zero in the program with G50 commands, you may notice that the turret *jumps* at each turret index. This motion can be troublesome for two reasons. First, the additional motions will add to cycle time. The turret may have to move in the plus direction when instating the wear offset. Then it will have to move in the negative direction when approaching the workpiece. Second, and more importantly, if the machine's starting position is close to a plus over travel limit in either axis (and the zero return position usually is), any large plus wear offset will cause an axis over-travel as soon as the turret index command is executed.

If you find that your machine's turret moves by the amount of the wear offset during turret indexes, here's a tip that will eliminate the associated motion problems. As you index the turret in your program, do not (yet) instate the wear offset. The command

```
T0300
```


for example, indexes the turret to station number three but will not instate a wear offset. In the first approach motion for the tool, instate the offset. Here is a program segment that eliminates unwanted motion for a machine that will move by the amount of the wear offset as soon as the offset is instated.

```
O0003 (Program number)
N005 G28 U0 W0
N010 G50 X8.3432 Z10.2383 (Assign program zero)
N015 T0100 M41 (Index to station one, no wear offset, and select the low range)
N020 G96 S600 M03 (Start the spindle CW at 600 sfm)
N025 G00 X2.2 Z0.005 T0101 M08 (Instate wear offset during the approach movement)
.
.
.
```

In line **N015**, the turret is indexed to station one but no wear offset is being instated (yet). During the first approach movement in line **N025**, notice the **T0101**. Since the turret is already indexed to station one, it will not index – but the wear offset will be instated. The machine will actually consider the current values in wear offset number one as it makes its approach movement. In essence, the machine will modify the commanded end point in X and Z by the amounts stored in wear offset number one, and then rapid to the modified position. Note that this technique will only be required older machines when program zero assignment is done in the program – and only if the machine moves as soon as a wear offset is instated.

What about wear offset cancellation?

Canceling a wear offset has the inverse effect of instating it. Programmed positions will not be affected by the values stored in wear offsets once the wear offset is cancelled.

If you're using geometry offsets to assign program zero, *you don't need to cancel wear offsets* (but doing so will have no adverse affect). If you are assigning program zero in the program with **G50** commands, you must cancel the active wear offset at the end of every tool. The wear offset is canceled during the tool's return to the tool change position.

One way to cancel a tool's offset is to repeat the tool station number in the T command but make the last two digits zero. The command

```
N240 T0500
```

for example, will cancel tool station number five's wear offset. However, there's an easier way that will cancel *any* tool's wear offset. The command

```
T0
```

will do so.

Here is a program segment that simply faces a workpiece. The technique shown in this program is not required if you are using geometry offsets to assign program zero. We're using **G50** to assign program zero (it's an older machine), so the tool offset must be canceled at the end of the tool.

Sample program that uses **G50** to assign program zero:

```
O0003 (Program number, example when G50 is used to assign program zero)
N003 G28 U0 W0
N005 G50 X8.3432 Z10.2383 (Assign program zero)
N010 T0100 M41 (Index to station one, no wear offset, and select the low range)
N015 G96 S600 M03 (Start the spindle fwd at 600 sfm)
N020 G00 X2.2 Z0 T0101 M08 (Invoke offset during the approach movement)
N025 G01 X-0.062 F0.010 (Face workpiece)
```

```
N030 G00 Z0.1 (Move away in Z)
N035 G00 X8.3432 Z10.2383 T0 (Move back to tool change position, cancel offset)
N040 M01 (Optional stop)
N045
.
. (Program continues)
.
```

In line N035, we're sending the tool back to where it started and canceling the offset with T0.

Once again, remember that programming wear offsets is simpler when geometry offsets are used to assign program zero. Here is the same program segment modified to use *geometry offsets* to assign program zero.

```
O0003 (Program number, example when geometry offsets are used to assign program zero)
N005 T0101 M41 (Index to station one, select wear offset one, and select the low range)
N010 G96 S600 M03 (Start the spindle fwd at 600 sfm)
N015 G00 X2.2 Z0 M08 (Move into approach position, turn on the coolant)
N020 G01 X-0.062 F0.010 (Face workpiece)
N025 G00 Z0.1 (Move away in Z)
N030 G00 X7.0 Z5.0 (Move back to tool change position)
N035 M01 (Optional stop)
N040
.
. (Program continues)
.
```

Again, geometry offsets make things easier. Notice that program zero assignment values are not in the program (they're in the geometry offsets). In line N005, the wear offset can be easily instated right in the turret index command. And at the end of the tool in line N030, there's no need to cancel the wear offset (though doing so won't hurt anything).

How wear offsets are entered

Figure 4.6 shows the wear offset display screen page.

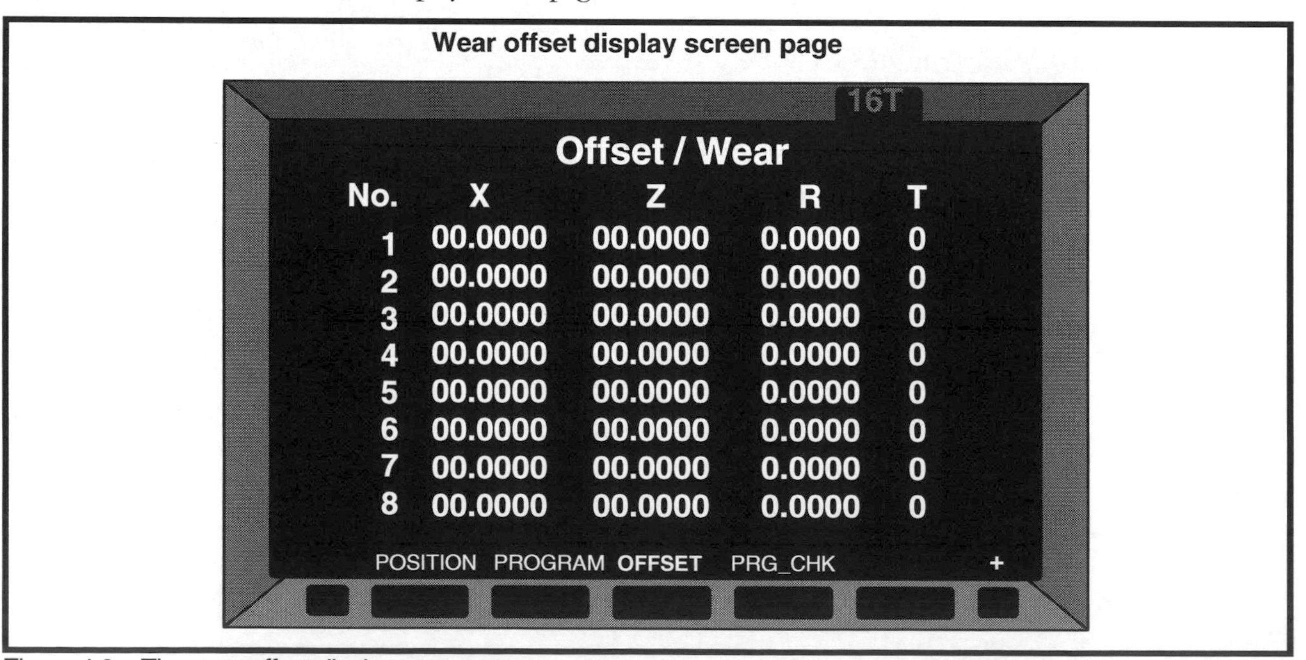

Figure 4.6 – The wear offset display screen page

Wear offsets use only the X and Z registers on this page. The X registers allow adjustments to diameters (adjustments related to the X axis). The Z registers allow adjustments to lengths (adjustments related to the Z axis). The R and T registers are related to tool nose radius compensation, and are discussed in Lesson Fourteen. Notice once again that wear offsets are organized by number (wear offsets one through eight are shown in Figure 4.6). As you know, you'll be making the primary wear offset number for a cutting tool the same as the tool station number. That is, tool number five will use wear offset number five. Tool number ten will use wear offset number ten. And so on.

The wear offset page shown in Figure 4.6 is blank. That is, none of the wear offsets have been used – they're all set to zero. Let's look at how wear offsets can be entered.

There are two common ways to actually enter a wear offset, with **INPUT** and with **+INPUT**. As you begin typing the value of a wear offset, the soft keys at the bottom of the screen will change. Look at Figure 4.7.

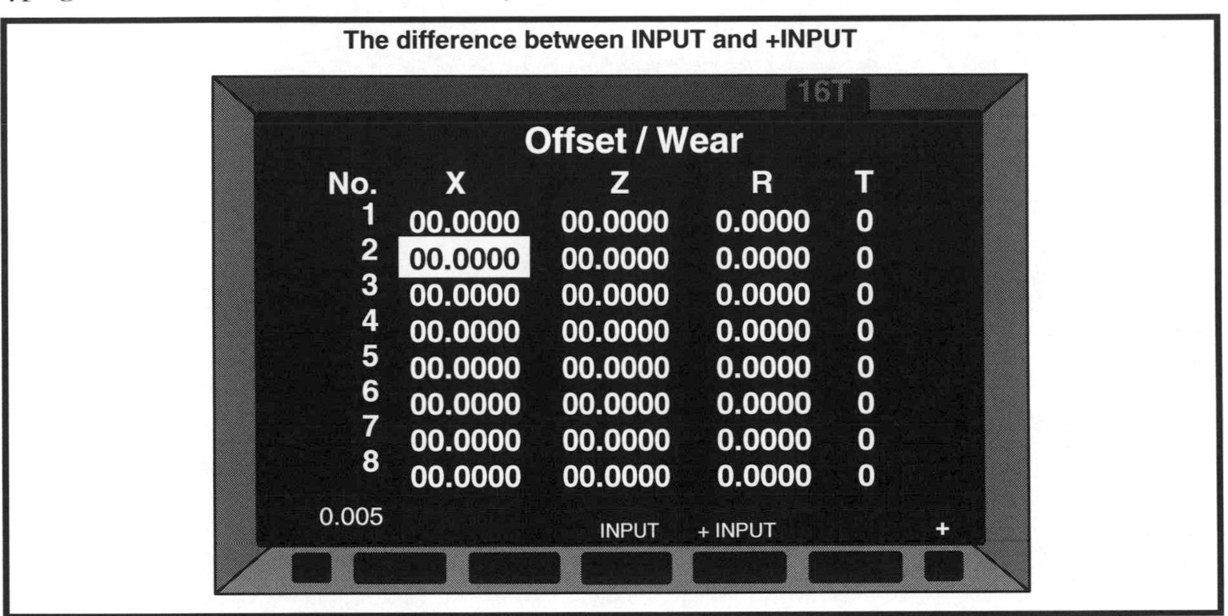

Figure 4.7 – INPUT versus +INPUT

Notice that the cursor is currently on the X register of wear offset number two. The operator has just typed 0.005 from the keyboard on the control panel (the entry shows up in the lower-left corner of the display screen. As the operator types the first digit (zero in our case), the soft keys change. While Fanuc control models vary with regard to what the soft keys show, two of the soft keys will be **INPUT** and **+INPUT**. These two keys are used to enter offset adjustments.

INPUT – This soft key (or the **INPUT** key on the control panel) will cause the machine to *replace* the value that is currently in the offset register with the entered value. For the example in Figure 4.7, if the operator presses the **INPUT** soft key, the X register of offset number two will be set to 0.005.

+INPUT – This soft key will cause the machine to *modify* the value that is currently in the offset register by the entered value. For the example in Figure 4.7, if the operator presses the +**INPUT** soft key, the X register of offset number two will be set to 0.005.

Since the initial value of the offset register is zero, **INPUT** and **+INPUT** happen to have the same result in this example (0.005 being placed in the register). But consider what happens after an initial entry. Figure 4.8 illustrates.

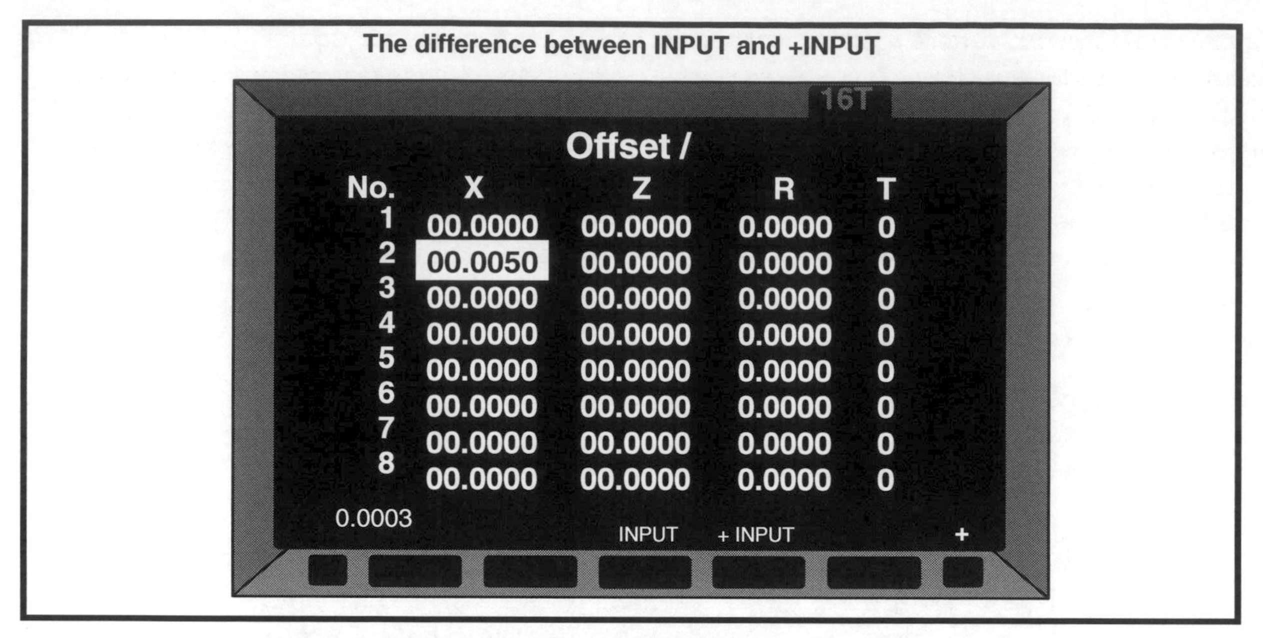

Figure 4.8 – Better example to stress the difference between INPUT and +INPUT

This time, the current value in the offset register is 0.005. The operator has typed the value 0.0003. If they press the INPUT soft key, the register's value will change to 0.0003. If they press the +INPUT soft key, the register's value will change to 0.0053.

Which is better, INPUT or +INPUT?

In almost all cases, a CNC setup person or operator will need to *modify* the current value in the offset register by the amount of the offset entry, meaning the +INPUT soft key will be the offset entry soft key. Think about it. Whenever a wear offset adjustment is required, you will have calculated the deviation – which is the difference between the target value and the dimension you have measured on the workpiece. You must modify the offset's current setting by the amount of this deviation. The +INPUT soft key will keep you from having to calculate the resulting offset value – the machine will make this calculation for you.

Remember that the deviation has a polarity. Sometimes it will be negative. The +INPUT soft key is still used. Say for example, the current value in the offset register is 0.005 (like it is in Figure 4.8). You've determined the deviation – and it is -0.0007. To make the adjustment, you will type -0.0007, and then press the +INPUT key. After you do, the register will show 0.0043 (when you add 0.005 to -0.0007, the result is 0.0043). Again, the machine will do the calculation for you when you use the +INPUT soft key.

What if my machine doesn't have a +INPUT soft key?

Older Fanuc controls don't have a soft key labeled +INPUT. Indeed, they don't have soft keys. If you have one of these controls, rest assured that there is still a way to enter the amount of deviation and let the control do the needed calculation. If your machine doesn't have a +INPUT soft key, you will use letter addresses X and Z to cause the control to overwrite the offset (like the INPUT soft key) and U and W to enter modify the offset (like the +INPUT soft key). In either case, the INPUT key will be used to enter the value.

For example, say the current value of the X offset register is 0.005. If you type X0.0007 and then press the INPUT key (this INPUT key is on the keyboard), its new value will be 0.0007. But if you type U-0.0007 and then press the INPUT key, its new value will be 0.0043.

While operation techniques are different, the result is the same. You can enter the amount of deviation and cause the machine to do the calculation if you use U (for X axis registers) and W (for Z axis registers) when making offset entries.

Sizing in a tool after it has just been placed in the turret

This example shows the first of four times when wear offsets can be used. While this example doesn't show the method we recommend, it should help you understand how wear offsets work. Look at Figure 4.8.

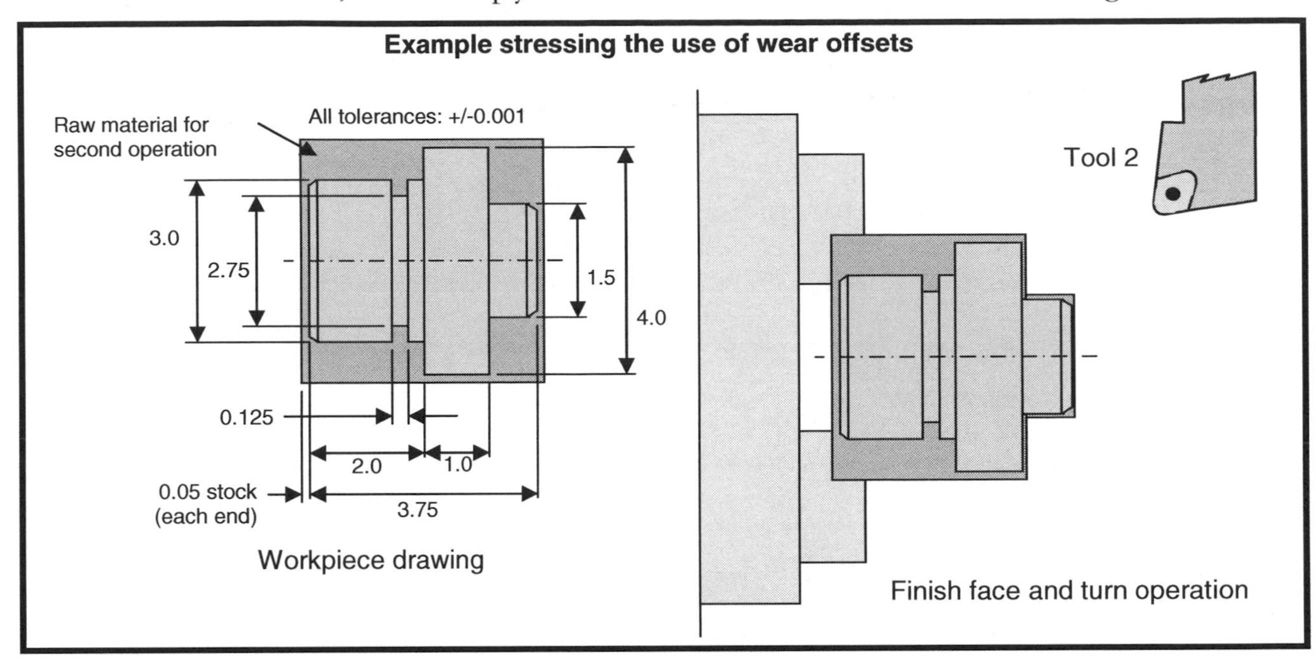

Figure 4.8 – Example to help stress the use of wear offsets

From the drawing on the left side of Figure 4.8, notice that all dimensions have a tolerance specified as +/- 0.001 inch (0.002 overall). On the right side, the finish face and turning operation is about to take place and is done by tool number two. The roughing operation has already been done.

We'll say that this is the first workpiece being machined in the job, and during setup, tool number two (the finish face and turn tool) has just been placed in the turret. The setup person has done their best to measure program zero assignment values and enter them into geometry offsets. The X and Z wear offset registers for this tool are currently set to zero.

Notice that tool number two will be machining the 1.5 inch diameter – and it is this diameter we'll concentrate on for this example.

After tool number two runs – machining the 1.5 diameter – the setup person stops the cycle and measures the 1.5 inch diameter. Say they find that this diameter is currently 1.5024. It is, of course, oversize – larger than the high limit for the tolerance band (1.501). An adjustment must be made.

Since we're targeting the mean value, the adjustment must bring the diameter to 1.5 inches. The amount of the adjustment – the deviation – will be the difference between the target dimension and our measured dimension of 1.5024 (this comes out to 0.0024). Since the measured diameter is *larger* than the target diameter, the deviation will be *negative*. And since this is a diameter deviation, we must modify the X register of the offset (not the Z register).

So we must modify the current value of wear offset number two's X axis register by -0.0024. After bringing the cursor to the X register of offset number two, we type -0.0024 and press the soft key under **+INPUT**. After doing so, the value in this register will be -0.0024 (remember that it started at zero). The next time tool number two machines a workpiece, this diameter will come out 0.0024 smaller – and right at our target value of 1.5 inches.

In the scenario just given, we got lucky. Since the 1.5 inch external diameter is oversize, there is still a little stock to be removed. After making the offset adjustment, the setup person can re-run tool number two on the same workpiece. After tool number two runs the second time, the 1.5 diameter will be within its tolerance band (though it may not be exactly 1.5 since the amount of material being removed now is only 0.0024).

Consider the other possibility. After machining with tool number two for first time, maybe the setup person measures the 1.5 inch diameter and finds it to be 1.4975. Now the diameter is *undersize* – smaller than the low limit. Since there is no (feasible) way t put material back on this diameter, the workpiece is scrap. The deviation in this case is (positive) 0.0025, and the X register of wear offset number two must be modified by this amount before the *next* workpiece can be machined.

We call the technique just shown *sizing after the fact.* While it may be acceptable for some cutting tools (like roughing tools), it is not recommended when the cutting tool must machine surfaces having tight tolerances.

Sizing in a new tool with trial machining

This example stresses the second of four times when wear offsets can be held. As you have just seen, when tight tolerances must be held, there is a fifty-fifty chance that the workpiece will be scrapped when a cutting tool machines for the first time. You should use trial machining techniques whenever there is *any* doubt about whether a cutting tool will machine all surfaces within their tolerance bands.

Trial machining involves five steps:

- Recognizing a workpiece attribute with a tight tolerance
- Making an initial adjustment to machine the surface with some excess stock
- Letting the cutting tool machine under the influence of the trial machining offset
- Measuring the surface and adjusting accordingly
- Making the cutting tool run a second time (this time the surface will be within its tolerance band)

Consider these five steps to trial machining for the example just shown. In the first step, we know there could be a problem holding the +/-0.001 tolerance on the 1.5 inch diameter. We doubt whether the initial geometry offset setting for tool number two is accurate enough to make the tool machine this diameter within its tolerance band.

In the second step, the trial machining adjustment will be made in the X register of offset number two. To make this tool leave more material on the diameter, the adjustment must be positive. And usually an amount of about 0.010 inch (0.25 mm) is an appropriate amount of excess material to leave. So the X register of wear offset number two must be increased by 0.01 inch (the setup person types 0.01 and presses the soft key under +INPUT).

In the third step, the setup person lets tool number two run. The 1.5 inch diameter is machined with excess material.

In the fourth step, the setup person measures the 1.5 inch diameter. It will, of course, be oversize. Say they find it to be 1.5082. The needed adjustment is -0.0082 to bring the diameter back to the target value of 1.5 inch. So the X register of wear offset number two is modified by -0.0082 (setup person types -0.0082 and presses the soft key under +INPUT). While it may not be important at this time, the X value of wear offset number two will end up as (positive) 0.0018.

In the fifth step, the setup person reruns tool number two. This time it will machine the 1.5 inch diameter within its tolerance band – and for the next workpiece to be machined, the 1.5 inch diameter should come out to precisely to its target value, 1.5 inches.

By the way, in the trial machining example just given, trial machining saved us from machining a scrap workpiece. After trial machining, the diameter came out to only 0.0082"oversize. Yet we had initially increased the offset by 0.01". If we had not used trial machining techniques, the 1.5 inch diameter would have come out to 1.4982, which is smaller than the low limit – again, this diameter would have been undersize.

What causes the initial deviation?

Before we show an example of the third time when wear offsets can be used, we want to make an important point. This point will only apply if geometry offsets are being used to assign program zero.

In each example just given, when the (new) finish turning tool machines for the first time, there is a substantial deviation. That is, the difference between the target value and the measured value is quite large – large enough to cause the 1.5 inch diameter to be out of its tolerance band. And by the way, unless you have a properly calibrated tool touch off probe, these are realistic examples. It is not unusual for the initial deviation to be large enough to cause the surface being machined to be out of its tolerance band.

What is causing such a large deviation? As you know from Lesson Eleven, it could be one of two things, or a combination of both. First, it is possible that the setup person has made a mistake with the measurement and/or entry of the program zero assignment values for the finish turning tool. And second, even if these values are perfectly measured and entered, the difference in tool pressure from when the program zero assignment values are measured to the actual machining of the workpiece may be causing the initial deviation.

Either way, *this initial deviation is more related to incorrect program zero assignment than it is to tool wear.* Keep in mind that you can just as easily make the initial adjustment (including trial machining adjustments) in *geometry offsets* as you can in wear offsets. The machine will behave in *exactly* the same way. The **+INPUT** key can even be used to modify the current value of the (geometry) offset.

If you make the initial adjustment in the geometry offset, the wear offset values for the (new) cutting tool will start out at zero. The benefit of having the wear offset begin at zero with a new cutting edge will become clear as this presentation continues.

Dealing with deviations caused by tool wear

This is the third time when wear offsets can be used. As you know, when certain cutting tools wear, the surfaces they machine will change. Single point turning tools and boring bars are prime examples. When a finish turning tool wears, a small amount of material will be removed from its cutting edge. When this happens, any external diameter machined by the tool will *grow.* That is, external diameters will get larger as turning tools wear. At this point, the cutting tool is not worn out – it is just showing signs of wear.

It is not unusual for as much as 0.002 inch (0.050 mm) or more material to be removed from an insert before it must be replaced. This means a diameter machined by the finish turning tool will grow by 0.004 inch during the cutting tool's life.

The same occurs with boring bars. But an internal diameter will *shrink* as the boring bar insert wears.

Whether or not this will present a problem holding size depends upon two factors: lot size and the smallest tolerance machined by the tool.

With small lots, it is likely that the production run will be completed before the cutting tools wear enough to require an adjustment. And with wide open tolerances, it is likely that a new cutting tool can completely wear out without causing the surfaces being machined to approach tolerance limits.

But with large lots, cutting tools will eventually wear out and will need to be replaced. And if tolerances are small, surfaces will eventually grow (or shrink) out of their tolerance bands long before the tool is completely dull. This means adjustments must be made during the tool's life in order to keep the cutting tool machining all surfaces within tolerance bands.

In the drawing shown on the left side of Figure 4.8, notice once again that the 1.5 inch diameter has a tolerance of +/-0.001. The high limit is, of course, 1.501 and the low limit is 1.499. During setup, we'll say the setup person uses the trial machining techniques shown in the second example to size in the 1.5 inch diameter – they use wear offsets to do so. When the production run begins, a value of 0.0018 is in the X register of wear offset number two and the 1.5 inch diameter is coming out to *precisely* 1.5 inches.

For this discussion, we'll say there are thousands of workpieces to produce in this production run. So maybe the job is turned over to a CNC operator at this point to run out the job. As they begin running workpieces, the 1.5 inch diameter will continue coming out very close to 1.5 inches. But after fifty workpieces are machined, the operator notices that the 1.5 inch diameter is 1.5002. The finish turning tool has worn by 0.0001 inch. The diameter is still well within the tolerance band, so the operator simply continues running workpieces.

After fifty more workpieces, the 1.5 inch diameter comes out to 1.5005. The growth trend is continuing. But the diameter is still within the tolerance band – and nothing needs to be done (yet).

After fifty more workpieces (one-hundred-fifty total), the 1.5 inch diameter comes out to 1.5008. While this diameter is still within its tolerance band, it is getting dangerously close to the high limit (1.501). If this trend continues much longer, the diameter will be larger than the high limit (oversize). So at this point, the operator decides to make a sizing adjustment.

Since the target value is the mean value of the tolerance band, the operator will target 1.5. Currently the workpiece is 1.5008. So the operator must modify the current X value of wear offset number two by -0.0008.

To do so, they position the cursor to the X register of wear offset number two (which currently happens to have a value of 0.0018). Then they type -0.0008 and press the soft key under **+INPUT**. The machine will subtract 0.0008 from the current value of the offset.

With the next workpiece machined, the 1.5 inch diameter will come out to precisely 1.5 inches – which is back at the target value. The operator will be able to run another one-hundred-fifty workpieces or so before another offset adjustment will be required.

After running one-hundred-forty more workpieces, say they find the 1.5 inch diameter has grown to 1.5008 again. So they must make another adjustment. To do so, they position the cursor to the X register of wear offset number two (which currently happens to have a value of 0.0010). Then they type -0.0008 and press the soft key under **+INPUT**. The machine will subtract 0.0008 from the current value of the offset.

The operator may have to make several such adjustments during a given cutting tool's life – the smaller the tolerance, the more adjustments that will have to be made. Eventually, the cutting tool will be completely dull and in need of replacement.

After a dull tool is replaced

This is the fourth time when wear offsets can be used. While different types of cutting tools vary when it comes to what must be done to replace them, most single point turning tools used on turning centers (like turning tools and boring bars) incorporate *inserts*. With these cutting tools, the task of replacing a dull cutting tool simply means indexing or replacing the insert.

One advantage of most inserts is that they have more than one cutting edge (some have as many as eight cutting edges). When one edge of the insert is dull, the operator will remove the insert from the cutting tool holder, rotate it to an unused cutting edge, and replace it in the cutting tool holder. This task is known as *indexing* the insert. When all cutting edges on the insert have been used, of course, the insert must be replaced.

The key to insert indexing and replacement is *consistency*. If you can index or replace the insert in such a way that the new cutting edge is in exactly the same location as the last cutting edge, the new cutting edge will machine in exactly the same way as the last cutting edge *when it was new*. The only problem you'll have to deal with is related to the amount of offset change you made during the last cutting edge's life.

In the previous example, after sizing the first workpiece in during setup with a new cutting edge, the X value of wear offset number two started out at a value of 0.0018. During this tool's life, several adjustments are made. Eventually this cutting edge will be dull and the insert must be indexed or replaced. As long as it can be perfectly replaced (the new cutting edge is in the same location as the old cutting edge), the operator will set the X axis register for wear offset number two back to the initial value of 0.0018. This will cause the new cutting edge to machine the 1.5 diameter precisely to 1.5, just as the last cutting edge did when it was new.

This is one time when the **INPUT** soft key will be used. Since you know the value you want in the offset (0.0018 in our case), it will be easier to use than the **+INPUT** soft key – which would first require a calculation to be made.

By the way, this is the reason why we recommend making the initial adjustment (including trial machining adjustments) in geometry offsets. If you do, the wear offset will *begin at zero*. When cutting tools are replaced, you won't have to remember the initial wear offset value (0.0018 in our example). Instead, you can simply set the wear offset value back to zero.

What if my machine has a tool touch off probe?

One of the advantages of a properly calibrated tool touch off probe is that it can *perfectly* set the geometry offset – even allowing for tool pressure. It will also set the wear offset to zero. If, during dull tool replacement, you use the tool touch off probe (just as you do during setup), you need not be concerned about resetting the wear offset for the cutting tool. And as previously mentioned, the tool touch off probe will also eliminate the need for trial machining – even when very small tolerances must be held. While an adjustment may be required after the next workpiece is run to make the cutting tool machine right at its target value, the surfaces machined by the new tool will be within their tolerance bands.

Consistently replacing inserts

Unfortunately, inserts used for turning center applications may not be very consistent. That is, the insert itself may have a rather large tolerance. If inserts vary in size – even by a little bit – it will be impossible to *perfectly* replace them.

Consider this specification for a very common 80 degree diamond-shaped insert:

CNMG-432

Each letter and number of an insert's specification has a special (and universal) meaning. It just so happens that the third letter in the specification (M in our case), specifies the tolerance for the insert.

- A- Included circle = plus or minus .0002, thickness = plus or minus .001
- B- Included circle = plus or minus .0002, thickness = plus or minus .005
- C- Included circle = plus or minus .0005, thickness = plus or minus .001
- D- Included circle = plus or minus .0005, thickness = plus or minus .005
- E- Included circle = plus or minus .001, thickness = plus or minus .001
- G- Included circle = plus or minus .001, thickness = plus or minus .005
- **M Included circle = plus or minus .002, thickness = plus or minus .005**
- U- Included circle = plus or minus .005, thickness = plus or minus .005

The M specification in CNMG-432 specifies a tolerance for the included circle of 0.004 (+/-0.002). The tolerance for the insert's thickness is +/-0.005. While the thickness of an insert isn't very critical, the size of its included circle controls the position of the cutting edge, as Figure 4.9 shows.

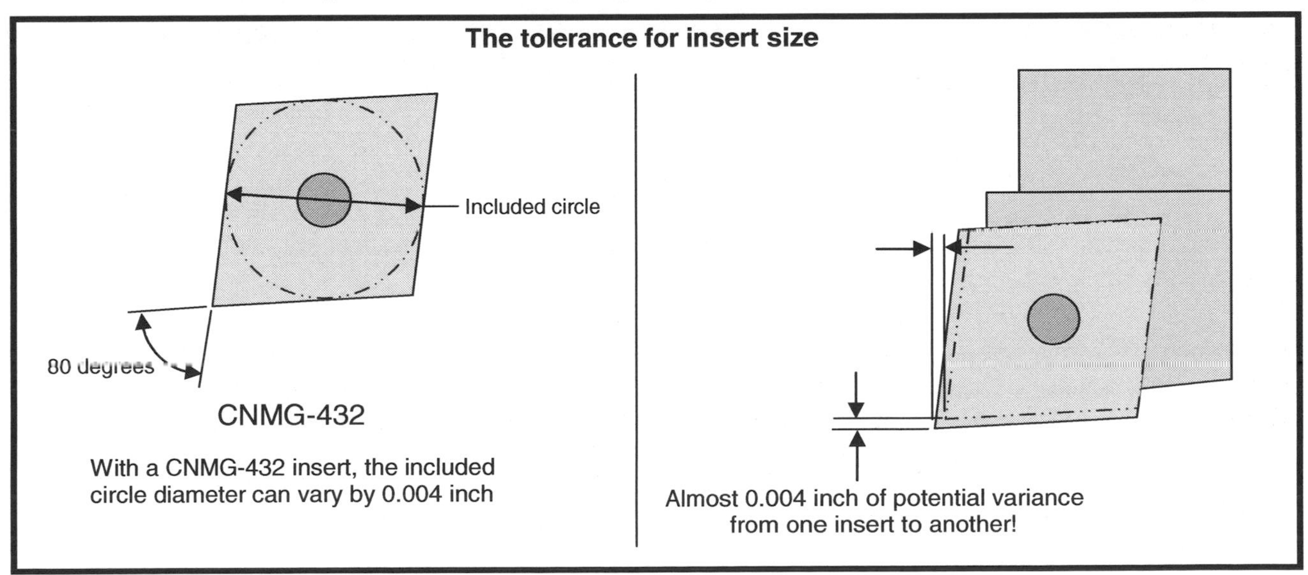

Figure 4.9 – Included circle of an 80 degree insert

A position variance of almost 0.004 inch is possible with a **CNMG-432** insert. This can result in as much as 0.008 inch variance when machining a diameter. With this much variation, of course, you will not be able to change inserts precisely enough to eliminate trial machining or using a tool touch off probe when inserts are replaced.

Note that the tolerances specified above are *maximum* variations. In practice, you'll probably find them to be much less from insert to insert – especially with inserts coming out of the same package. You can easily measure the amount of variation from insert to insert with a micrometer. If you find the variance to be under about 0.0005 inch from insert to insert, you may be able to replace inserts precisely enough to eliminate the need for trial machining or the need to use the tool touch off probe after insert replacement.

Consistently indexing inserts

As stated, almost all inserts have more than one cutting edge. The **CNMG-432**, for example, has *four* cutting edges. This means after a new insert is placed in the tool holder, it can be *indexed* three times.

The variation among new inserts will not apply to an insert when it is indexed. If an insert is consistently placed in a tool holder, there will be no variation from one cutting edge of an insert to another. But the key word is *consistently*. In order to eliminate the need for trial machining or the tool touch off probe, you must be able to index inserts perfectly – the next cutting edge must be in precisely the same location as the last cutting edge. While it may take a little practice, you can save much time and effort if you can master the ability to consistently index inserts.

Many inserts used on turning centers incorporate an eccentric pin for clamping and location. This pin can be turned in either direction to clamp the insert. When it is turned in one direction, the pin will press the insert against one of the location surfaces of the tool holder. When it is turned in the other direction, the eccentric pin will press the insert against the other location surface of the tool holder. Figure 4.10 shows this.

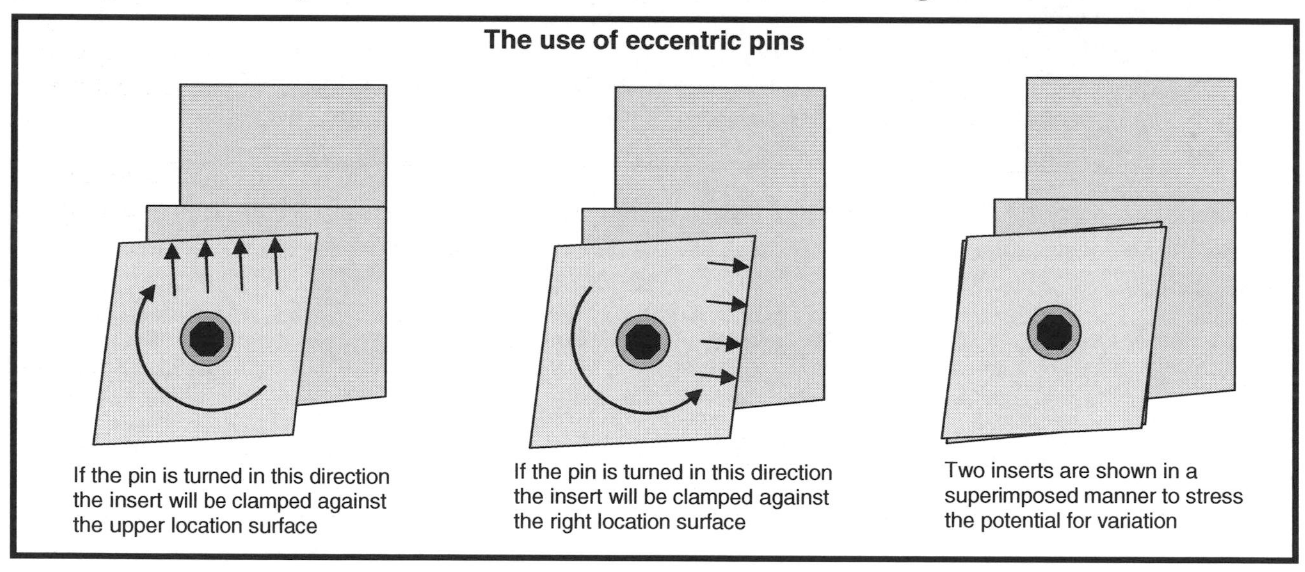

Figure 4.10 – How an eccentric pin locates an insert in a tool holder

When you turn the pin in the direction shown in the left-most illustration of Figure 4.10, the insert will be pressed against the upper location surface of the tool holder. When you turn the pin in the direction shown in the middle illustration, the insert will be pressed against the right location surface of the tool holder. The right-most illustration super-imposes two inserts that are clamped in opposite directions. If the seat in the tool holder is not perfect, you can see how inconsistently clamped inserts will vary in position.

To determine which way the pin should be turned, consider what the cutting tool is doing. For the tool shown in Figure 4.10, if the tool will be predominantly facing, turn the eccentric pin in the direction shown in the left-most illustration. If it will be predominantly turning, turn the eccentric pin in the direction shown in the middle drawing. Once you determine which way the eccentric pin should be turned, *use the same method every time an insert must be indexed or changed.*

The method by which inserts are located in the tool holder affects more than just the position of the cutting edge. It also affects basic machining practice – especially for tools that perform powerful machining operations. If an insert is not properly located in the tool holder, it will prematurely fail.

Minimizing the need for trial machining

If inserts vary from one to the next, you will have to trial machine whenever you replace an insert (or you must use the tool touch off probe). Keep track of the initial wear offset value right after sizing in the new insert (again, this value will be zero if you use the tool touch off probe – or if you use geometry offsets to trial machine). When it comes time to index the insert (which you may have to do several times) be sure to do so consistently. Now you can simply set the wear offset back to its initial value. You will not have to trial machine of use the tool touch off probe when inserts are indexed.

Going from job to job

You know from Lesson Twelve that if a cutting tool is machining properly at the completion of one job, it will continue to machine properly in the next job. And you know that many cutting tools are often used from job to job. Let us expand an example scenario from Lesson Twelve, adding some important points related to wear offsets.

Say two jobs run consecutively that require a rough turning tool, a finish turning tool, and a threading tool (stations 1, 2, and 3). When setting up the first job, the three tools must be loaded into the turret (if they're not already in place). Program zero assignment values must be measured and entered. We'll say the work shift value is being used, so it must also be measured and entered.

When the setup person runs the fist workpiece in the first job, they'll size in all dimensions – making adjustments as necessary. For the rough turning tool, they'll ensure that the appropriate amount of finishing stock is being left for the finish turning tool. For the finish turning tool, they use trial machining techniques to keep from scrapping the first workpiece. For the threading tool, they'll use trial machining to ensure that the thread will be properly machined on the first workpiece.

We'll say one-hundred-fifty workpieces must now be run. During this time, the CNC operator must make adjustments to the wear offset for the finish turning tool (wear offset number two). At the completion of the first job, tool number two is not worn out, but two wear offset adjustments have been made to keep it machining on size.

When the first job is completed, the second job must be run. Though it may be of a totally different configuration, it requires the same three tools. In this scenario, there won't be much work for the setup person to do – at least related to cutting tools and program zero assignment. Since all three tools are currently in the turret and in the correct turret stations, there will be no need to remove or mount cutting tools. Since chuck face is the point of reference for geometry offsets, and since the chuck face position doesn't change from job to job, the geometry offsets need not be measured. Only one value – the work shift value – must be measured and entered – assuming the workpiece in the second job is not of the same length as the workpiece in the first job.

Additionally, there will be no need for trial machining in the second job. Again, all three cutting tools were machining properly at the completion of the first job. They will continue to machine properly as the second job begins.

When do you clear wear offsets?

Clearing a wear offset is the act of setting its registers to zero. In some cases, wear offsets are automatically cleared. When you use the tool touch off probe, for example, it will perfectly set the tool's geometry offset *and clear the tool's wear offset.* The same happens when you use the measure function to set geometry offsets.

We make the same recommendation for clearing wear offsets as we do for clearing geometry offsets. Get in the habit of clearing wear offsets for tools as you remove them from the turret. This way, the offsets (wear and geometry) will be correct for any tool that is currently in the turret.

A more complex example

At this point, you should have a clear understanding of how wear offsets work. But we've only scratched the surface of how they're used. We've shown the four times when wear offsets can be used for one tool and only for controlling a diameter (X axis adjustment). Now we'll show a little more complex example that stresses the use of wear offsets for three tools – as well as how they can be used to control length-related dimensions (Z surfaces) as well as diameters. Figure 4.11 shows the application.

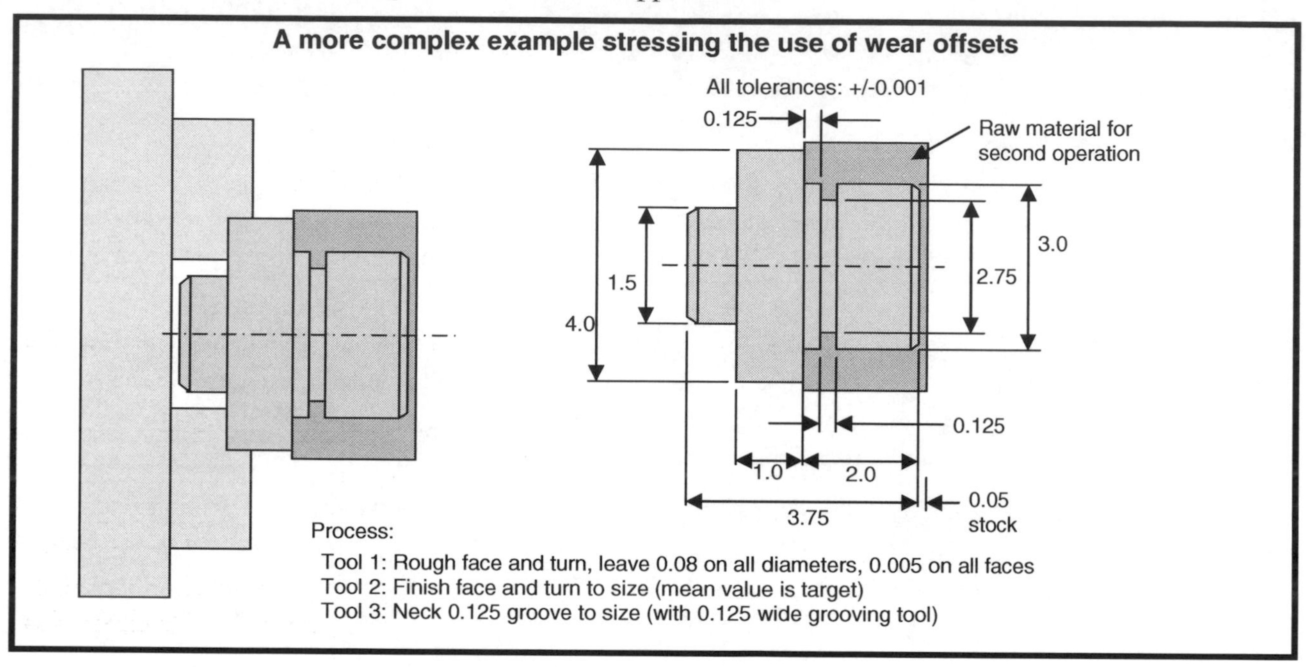

Figure 4.11 – Three cutting tools are used to machine this workpiece

This is the same workpiece shown in the previous example, but this time we're working on the other side. Notice from the illustration on the left that the workpiece is being held in a three-jaw chuck (in soft jaws) on the 4.0 inch outside diameter. It is being located in Z against the jaw on the largest face. This exposes the other end of the workpiece to the turret for machining.

We'll say that we're machining one thousand of these workpieces – and that all of the tools have been newly placed in the turret (though in reality, the rough face and turn tool [tool number one] and the finish face and turn tool [tool number two] were probably used to machine the other side of this workpiece).

The setup person has measured the program zero assignment values for each tool (*without* the help of a tool touch off probe) and entered the related values into geometry offsets. They have also measured and entered the work shift value.

Running the first workpiece – one tool at a time

Assuming that it is important to make the first workpiece being machined a good one, and assuming that the program is correct, there is really no excuse for scrapping the first workpiece being machined due to wear offset settings. That is, there will always be a way to ensure that each tool machines properly when machining the first workpiece.

The key to making the first workpiece a good one is dealing with each tool as you come to it. Consider what the tool will be doing. If necessary, use trial machining to ensure that the tool machines the workpiece properly. Make the tool machine correctly before moving on to the next tool in the program. Do the same for each tool. With this approach, each tool will machine as it should. When you finish with the last tool in the program, all tools will have machined properly – and you'll have a good workpiece.

Tool number one: Rough face and turn tool

Notice that this tool will be leaving 0.080 inch of finishing stock on all diameters (0.040 inch on the side) and 0.005 inch of finishing stock on all faces. While the roughing operation itself is not all that critical (trial machining is not required for most roughing operations), it is quite important to leave the specified amount of finishing stock. The amount of finishing stock to be left on all diameters is substantial (again, 0.080 inches on all diameters). But the amount of finishing stock to be left on all faces is minimal (again, only 0.005 inch).

For the diameters machined by this tool, there is ample material being left for finishing. And if we're off by a few thousandths, it won't affect the way the finishing tool machines. So we will not trial machine for diameters. But to be on the safe side, we'll trial machine to ensure that this tool truly leaves precisely 0.005 on all faces of the workpiece for finishing. To do so, we'll add 0.01 to the Z register of wear offset number one. (The Z adjustment must be positive in order to leave more material on all faces.)

After running the tool, say we measure the 3.0 inch diameter (while the workpiece is still in the chuck) and find it to be 3.082. We then measure the 1.0 inch thickness of the flange and find it to be 1.0139.

We're currently leaving 0.082 inch on all diameters, so we'll reduce the X register of wear offset number one by 0.002 inch (type -0.002 and press the soft key under **+INPUT**). To make this tool leave precisely 0.005 inch on all faces for finishing, we'll reduce the Z register of wear offset number one by 0.0089 inch (type -0.0089 and press the soft key under **+INPUT**).

Now we re-run tool number one. After we do, we measure the 3.0 inch diameter and find it to be 3.0796. We measure the 1.0 inch flange thickness and find it to be 1.0049 inch.

We've made the rough turning tool machine properly. It is now leaving the appropriate amount of finishing stock for the next tool.

Tool number Two: Finish face and turn tool

We're now ready to run the second tool. Again, this tool will be machining some rather small tolerances (+/- 0.001), and we're worried that our program zero assignment may not be precise enough. So we decide to trial machine in both axes – in X to ensure that the three inch diameter comes out correctly, and in Z to ensure that the faces are machined correctly.

We increase the X register of wear offset number two by 0.01 inches. In Z, we cannot increase it by such a large amount (there is only 0.005 left on all faces at this point). So we'll increase it by a smaller amount – say 0.003. When this tool is actually machining the faces, we must visually confirm that it is truly machining material from the workpiece (you must see chips being formed). If this tool happens to be perfectly set, it will only be machining 0.002 inch of material from the faces. If it is off at all in the positive direction, it won' machine any material at all from the faces.

Now we let the finish turning tool machine the workpiece. And sure enough, during facing we notice some chips are being formed.

When we measure the 3.0 inch diameter, we find it to be 3.0087. We must reduce the X register of wear offset number two by 0.0087 (type -0.0087 and press the soft key under **+INPUT**). And note that if we had not trial machined, we would have scrapped the workpiece.

When we measure the 1.0 inch flange thickness, we find it to be 1.0023. We must reduce the Z register of wear offset number two by 0.0023 (type -0.0023 and press the soft key under **+INPUT**).

At this point, tool number two must be run again. After re-running this tool, we find the 3.0 inch diameter to be 2.9998. We find the 1.0 inch flange thickness to be 0.9999.

(You might think these dimensions should come out to precisely 3.0 and 1.0 – and in some cases they will. But there is a tool pressure difference between the first time the tool machines (when trial machining) and the second time it machines. The first time – it will be removing about 0.07 inch of material from all diameters. The second time it machines, it will be removing only 0.0087 inches. This difference in material removal will also cause a difference in tool pressure. When you go on to the *next* workpiece, you should expect these dimensions to come out to precisely 3.0 and 1.0.)

We happened to use the 1.0 inch flange thickness to determine whether this tool is machining properly. We did so only because this workpiece attribute is very easy to measure while the workpiece is still in the chuck. We could also have chosen the 3.75 inch overall length, but it may not be possible to measure this value while the workpiece is in the chuck. We could also have used the distance from the right end of the workpiece to the left side of the flange (a total dimension or 3.0 inches). But again, this might be tougher to measure.

Tool number two is now machining properly. And at this point, we still have a good workpiece. Now it's time to move on to tool number three.

Tool number Three: 0.125 grooving tool

We'll trial machine to ensure this tool correctly machines the 2.75 inch diameter. We'll add 0.01 to the X register of wear offset number three.

After letting this tool machine, we measure the 2.75 inch diameter and find it to be 2.762. The X register of wear offset number three must be reduced by 0.012 (type -0.012 and press the soft key under **+INPUT**).

After this tool is re-run, we measure the 2.75 inch diameter and find it to be precisely 2.75 inches.

We've carefully sized in each tool, making it cut properly before going on to the next tool. And now that the last tool has machined, we still have a good workpiece that will pass inspection. While this is still a rather simple process, this method can be applied to the most complex processes. Again, there is really no excuse for scrapping a workpiece for reasons related to wear offset settings.

What about the Z position of the groove?

Notice from the drawing in Figure 4.11 that this groove is supposed to be 0.125 inches from the large flange face. This may be a difficult (though not necessarily impossible) situation for trial machining. So we just went for it – hoping the groove would be in the right position. When we can measure this dimension (it may require removing the workpiece from the chuck), we find the 0.125 dimension to be 0.01258. We're within tolerance, but not by much. To put this groove in precisely the correct position for the next workpiece, we'll make an adjustment to the Z register of wear offset number two. Since the groove is coming out to the right (positive side) of where it should be, the adjustment will be negative 0.0008 (type -0.0008 and press the soft key under **+INPUT**).

If it is absolutely mandatory that this groove be machined correctly in the first workpiece, you can use a modified form of trial machining. Once the grooving tool is in position and ready to machine the groove, you can stop the cycle and measure the distance from the large flange face to the leading edge of the grooving tool (possibly with feeler gauges). Since you're doing this before any machining takes place, you can confirm that the grooving tool is in the correct position before a workpiece is scrapped. If the measurement isn't precisely 0.125 inches, an adjustment can be made to the Z register of wear offset number three.

Completing the production run

With the first workpiece successfully machined and having passed inspection, the job is turned over to a CNC operator. Again, 1,000 workpieces must be machined. Let's consider what will happen to the three tools during the production run.

Tool number one: Rough face and turn tool

For the rough face and turn tool, say we find that it lasts for about 600 workpieces before the cutting edge is dull. So it will have to be replaced once during the production run. Throughout its life, tool wear may cause the diameters machined by this tool to grow (maybe by as much as 0.006 inch or so), but there will be no need for any kind of sizing adjustments.

This small deviation in finishing stock will not affect the finish turning tool (in most jobs). When the insert for this roughing tool is indexed or replaced, care must be given to ensure that the new cutting edge is in the same location as the last cutting edge. If inserts are varying dramatically, it may be necessary to repeat the trial machining done during the machining of the first workpiece when the insert is changed (not just indexed).

Tool number Two: Finish face and turn tool

We'll say this tool lasts for about 500 workpieces before it gets dull. But during this time, tool wear will cause the diameters machined by this tool to vary by as much as 0.005 inch. This means several sizing adjustments must be done during the tool's life to ensure that the high limits of tolerances are not exceeded.

When the insert for this tool is indexed, the operator must be careful to get the new cutting edge in the same location as the previous cutting edge. They must also reset the X register of wear offset number two to its initial value to keep the tool from machining too much material from the next workpiece.

If it is not possible to perfectly index inserts, trial machining must be done every time the insert is indexed. The same goes for insert replacement. If inserts are varying, trial machining must be done whenever they are replaced.

Tool number Three: 0.125 grooving tool

We'll say this tool lasts for 700 workpieces before it gets dull. And during its life, the diameter it machines doesn't vary by more than about 0.0005 inch, meaning there will be no need for sizing adjustments. But as with all insert indexing and replacement, it is critical that the new cutting edge position is in precisely the same location as the previous cutting edge. This is true whether the insert is indexed of replaced. If there is any doubt about insert placement, trial machining must be done when indexing or replacing inserts.

A reminder about up-coming jobs

These three tools have now been *tweaked-in*. They are all machining properly in the current job. When this job is finished, it is likely that the rough face and turn tool and the finish face and turn tool will be used in the next job. And remember, if a cutting tool is machining properly in one job, it will continue to do so in the next.

This requires, of course, that the mean value of each dimension is programmed for every coordinate in every job. It also requires you to use the work shift function with geometry offsets as described in Lessons Seven and Twelve.

As you go on to the next job you'll know, for example, that the rough face and turn tool will be leaving the programmed amount of finishing stock on all diameters and faces. And you'll know that the finishing tool will machine the first workpiece in the next job within its tolerances (assuming the next job has similar tolerance requirements). If the grooving tool is used in the next job, you'll know it will machine to the programmed diameter and at the programmed Z position.

A reminder about target values

When adjustments are required, we have been targeting the mean value of each tolerance. And many CNC people are told to do so. However, targeting the mean value actually *doubles* the number of sizing adjustments that must be done during a cutting tool's life. This means the CNC operator will have to more closely monitor the CNC cycle.

For the 3.0 inch diameter that is machined by the rough face and turn tool in our example, it might be wiser to target a value closer to the low limit – say a dimension of 2.9992. This will allow the cutting tool to machine for *twice the amount of time* before an adjustment must be made.

Secondary wear offset applications

As stated, most operations require but one wear offset per tool. And we've called this offset the tool's *primary offset*. But we do want to prepare you for a few times when you may need two or more wear offsets for a cutting tool. We call these additional wear offsets *secondary offsets*.

Whenever you use a secondary offset, be sure to come up with a logical method of selecting secondary offset numbers. The people setting up and running your programs must be told which offsets are related to each tool (include this information on the setup sheet). If your machine has a twelve station turret and (at least) thirty-two offsets, we recommend that you *add twenty to the tool's station number to come up with its secondary offset number*. For example, tool station number five will use wear offset number five as its primary offset and wear offset number twenty-five as its secondary offset.

Here are some applications for secondary offsets. All but the first should only be necessary if you're experiencing problems caused by tool pressure deviations as finishing tools machine workpiece surfaces. This should be taken as a signal that the rigidity and strength of your setup is marginal.

Flip jobs

Many companies like to complete the workpiece (both ends) in one operation. The operator will load the workpiece and activate the cycle to machine one end. In the middle of the program the machine will stop and the operator will turn the workpiece around. The cycle will be reactivated to run the second end.

In this application, it's likely that at least some of the cutting tools used to machine one end will be used to machine the other. If this is the case, it is wise to use different wear offsets on each end. We recommend using wear offsets one through twelve for tool one through twelve for the first end and wear offsets twenty-one through thirty-two for tool one through twelve for the second end.

Two or more critical diameters

Unsupported sections of a workpiece are prone to *deflection* (the tool pushes the workpiece away). If a workpiece deflects, the diameter machined by the cutting tool will vary. Say for example, you're holding a rather long workpiece and using a tailstock to support the right end. In this case, the workpiece will be well supported at both ends, but not well supported in the middle. It will be less prone to deflection as tools machine at the ends, but more prone to deflection in the middle. This kind of problem can be difficult (if not impossible) to handle with only one offset. Given this problem, the primary offset can be used to help the operator hold size on diameters close to the tailstock or chuck ends (less deflection). A secondary wear offset can be used to help them hold size on diameters close to the middle of the workpiece (more deflection).

Unwanted taper

In similar fashion, your tools may experience changes in deflection even as they machine one diameter. This commonly appears as taper in the diameter (one end is larger than the other). A secondary offset can be instated during the machining of the diameter to induce a reverse tapered movement to counteract the taper being machined.

Grooving into different areas of the workpiece

Yet another tool pressure problem has to do with grooving. You may be necking two grooves with a grooving tool. One of the grooves may be close to the tailstock support (good support) and another groove may be right in the middle of the workpiece (poor support). The grooving tool will cause more deflection in the middle than at the tailstock end. Two wear offsets (one for the end groove and one for the middle) can be used to give the setup person and operator individual sizing control for the grooves.

Key points for Lesson Thirteen:

- Wear offsets can be used four times: after tools are first mounted in the turret, when trial machining is done, to compensate for tool wear, and after dull tools are replaced.
- Wear offsets are specified in the program with the second two digits of the T word.
- The wear offset number for a cutting tool should match the tool station number for the tool.
- There are two ways to enter wear offsets, with the INPUT soft key and the +INPUT soft key. The +INPUT soft key allows you to modify the value that is currently in the offset register.
- A properly calibrated tool touch off probe eliminates the need for trial machining.
- Trial machining should be done whenever you're in doubt about whether or not a cutting tool will machine the first time within tolerance bands.

STOP! Do practice exercise number thirteen in the workbook.

Lesson 14

Tool Nose Radius Compensation

This compensation type allows you to deal with problems caused by the small radius that is on the cutting edge of all single point turning tools and boring bars. With the addition of two G codes per tool in your program – and with the entry of two simple values in the tool's offset – the machine will automatically keep the cutting edge of cutting tools flush with the surfaces they machine.

You know from Lesson Ten that you must sometimes compensate for tooling attributes when you calculate the coordinates for your programs. With a twist drill, for example, you must compensate for the lead of the drill when calculating hole depth. In similar fashion, you must compensate for the lead of a reamer – or the number of imperfect threads on a tap.

In Lesson Ten, we mention one other important time when you must compensate for attributes of cutting tools. It has to do with the small radius that is on the cutting edge of any single point cutting tool – like a turning tool or a boring bar. Figure 4.12 shows this radius.

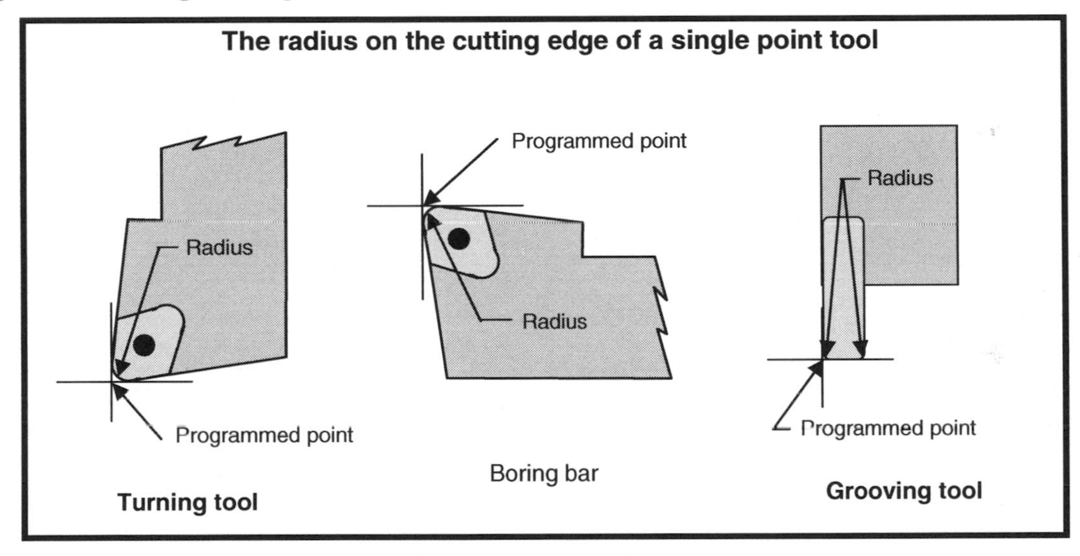

Figure 4.12 – All single point tools have a small radius on the cutting edge

For cutting tools used in the United States, the actual size of the radius will be specified in inches – and there are four standard tool nose radius sizes for turning and boring inserts:

- 1/64 inch (0.0156)
- 1/32 inch (0.0316)
- 3/64 inch (0.0468)
- 1/16 inch (0.0625)

Though you may consider these radii to be quite small, this small nose radius on the edge of the cutting tool will be sufficient to cause a small deviation from the programmed shape of your workpiece – at least when angular and circular surfaces must be machined.

Remember that you are programming the extreme edges of the cutting tool in each axis. This is illustrated in Figure 4.12 (specified as *programmed point*). Notice the small gap between the programmed point and the actual cutting edge.

This small gap will not affect the turning of diameters (parallel to the Z axis) and the machining of faces (parallel to the X axis). Figure 4.13 shows this.

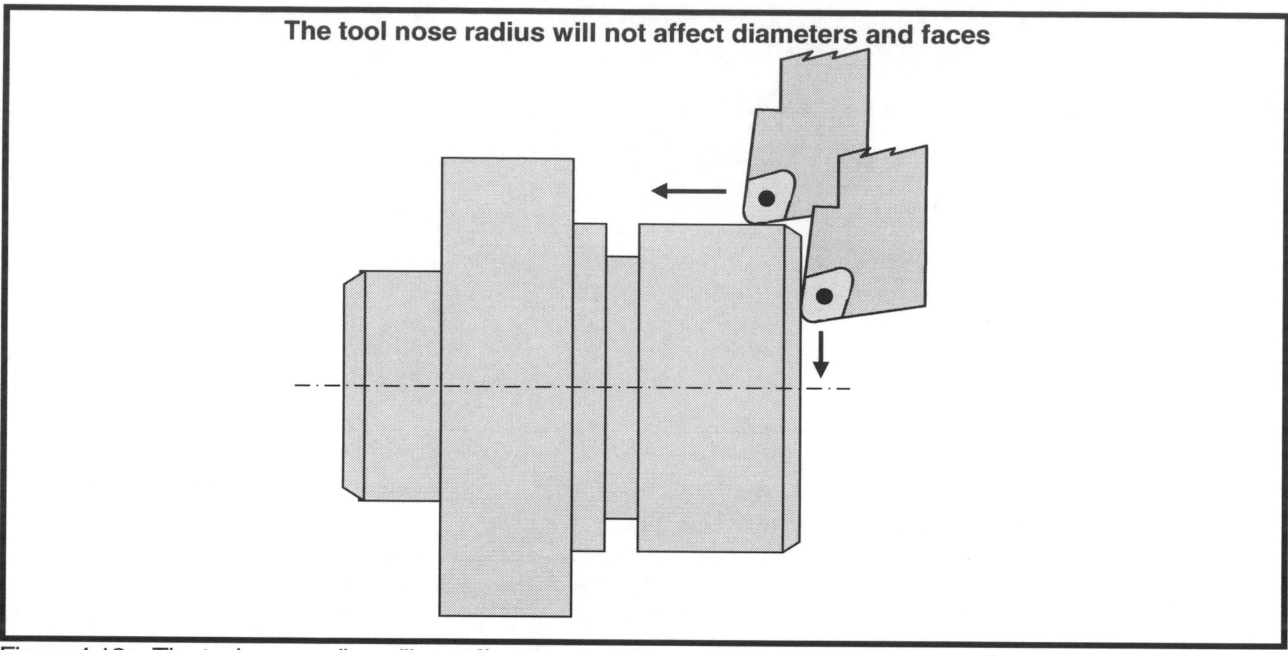

Figure 4.13 – The tool nose radius will not affect the machining of straight turns and faces

When a cutting tool is turning diameters and machining faces, the cutting edge is in direct contact with the workpiece surface being machined.

But when angular (tapered) surfaces and circular surfaces must be machined, the gap between the programmed point and the cutting edge will affect machining – as Figure 4.14 shows.

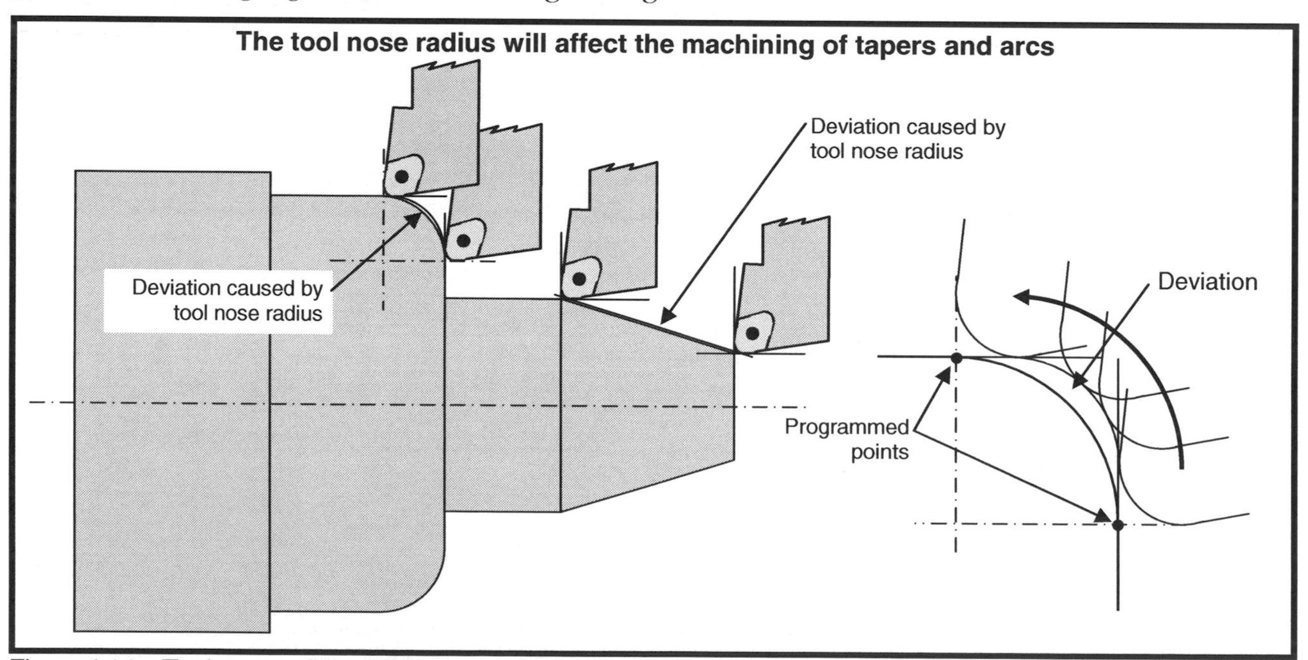

Figure 4.14 – Tool nose radius affects the machining of angular and circular surfaces

How much deviation are we talking about?

The deviation is at its worst at a forty-five degree angle – when the cutting tool is half-way through a ninety-degree arc – or when it is machining a forty-five degree chamfer. Figure 4.15 shows this.

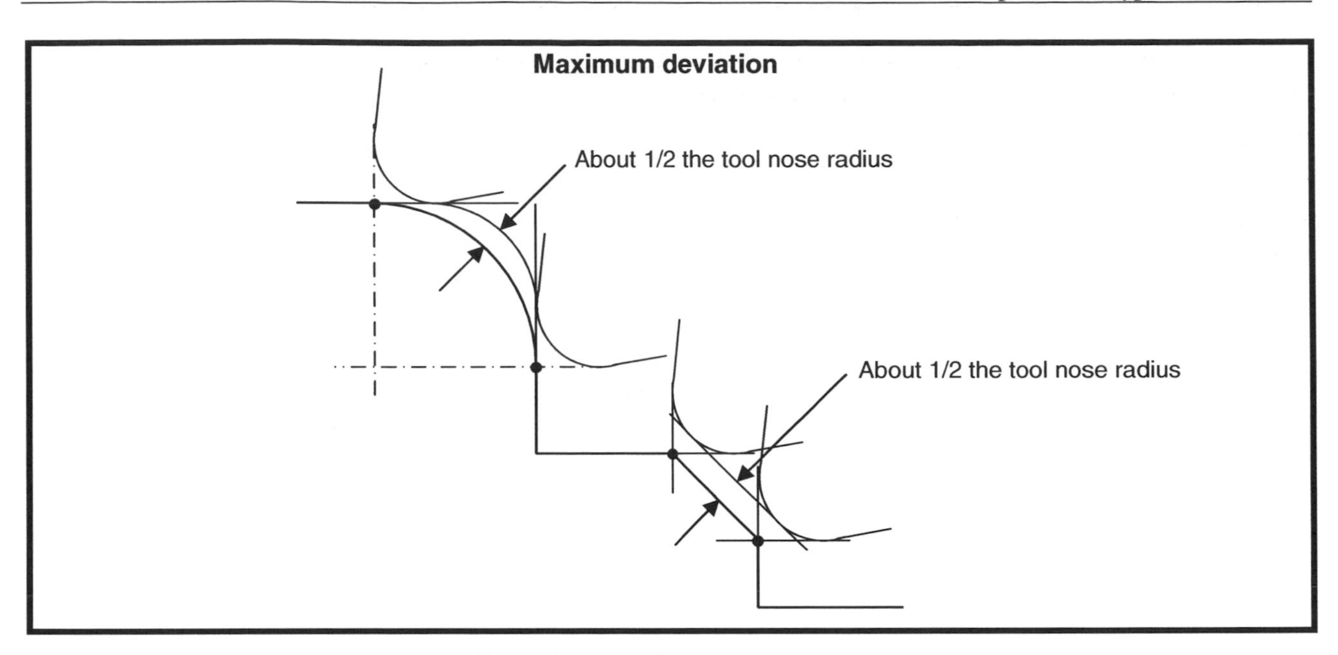

Figure 4.15 – Maximum deviation caused by tool nose radius

If you are using a cutting tool with a 1/32 inch (0.0312) nose radius, the deviation at forty-five degrees will be about 0.0155 inch.

In some cases, this may not be enough to cause problems. Maybe you are machining a chamfer or radius for the purpose of breaking sharp corners. On the other hand, there are many applications when even a tiny deviation will cause a scrap workpiece. Consider, for example, machining a Morse taper. The taper angle and position will be critical to the function of the workpiece.

Keeping the cutting edge flush with the work surface at all times

One way to handle the deviations is to manually compensate for the tool nose radius with your programmed coordinates. Note that this is *not* the method we recommend. We're only showing it to help you gain an understanding of how tool nose radius compensation works – and what it's doing for you. Look at figure 4.16.

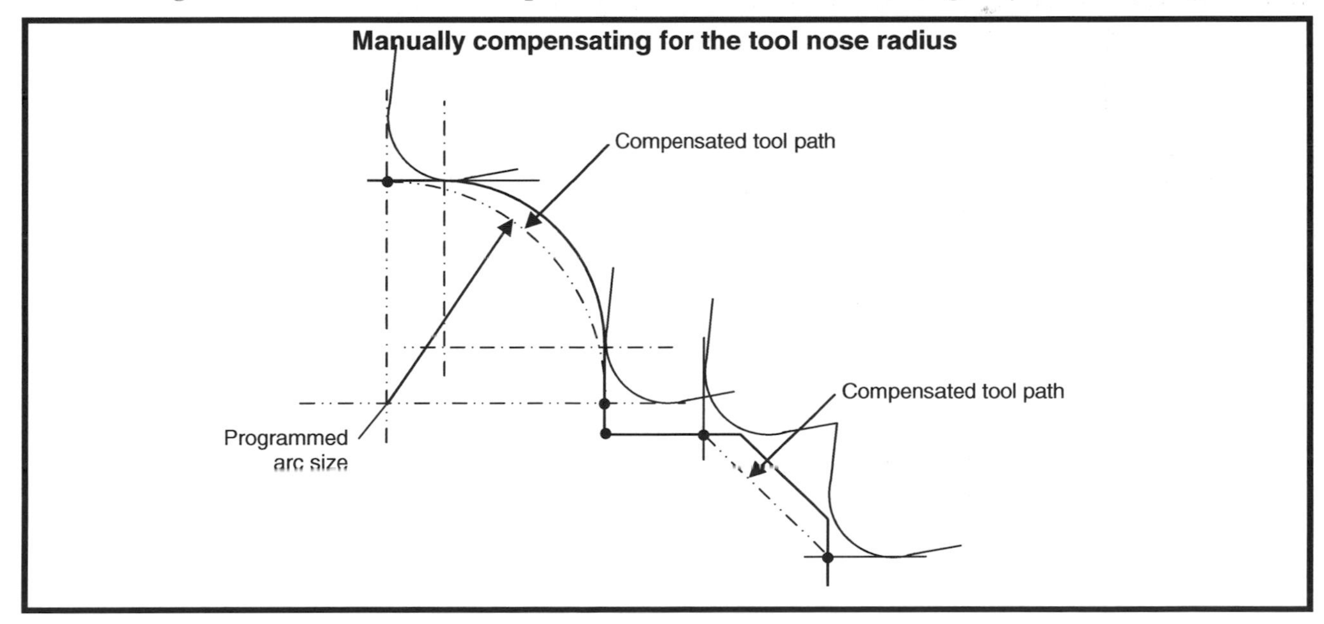

Figure 4.16 – Coordinates that compensate for the tool nose radius

Notice that the programmed points are no longer related to workpiece coordinates. For the chamfer, the start point is at a smaller diameter than the chamfer's beginning position. The end point is more negative in Z.

With these programmed points (which we're calling the compensated tool path), the cutting edge of the tool will remain in contact with the workpiece chamfer during the motion.

We've done the same for the radius motion. The start point is smaller in X and the end point is more negative in Z. And the size of the radius must be increased by the tool nose radius. With this compensated tool path, the cutting edge will remain flush with the workpiece radius during the motion – and the radius on the workpiece will be machined properly.

Again, *this is not the method we recommend* for dealing with the deviations caused by the tool nose radius. We're showing it for two reasons. First, this is exactly what the feature *tool nose radius compensation* will (automatically) do for you. Tool nose radius compensation allows you to program the work surface path (all coordinates right on the workpiece). It will automatically create the compensated path – and keep the cutting edge flush with all workpiece surfaces.

Second, some computer aided manufacturing (CAM) systems will create a CNC program with compensated tool paths. During programming, the programmer tells the CAM system the size of the tool nose radius and the CAM system outputs a CNC program with compensated tool paths. This eliminates the need for CNC-based tool nose radius compensation, which in turn eliminates some of the work a setup person must do during setup. If your company uses a CAM system, you'll want to determine whether it is generating compensated tool paths – or whether it is generating programs with CNC-based tool nose radius compensation commands (G41 and G42).

When to use tool nose radius compensation

As stated, tool nose radius compensation is used with single point cutting tools, like turning tools and boring bars. And it should only be used for *finishing* operations. We don't recommend using tool nose radius compensation for roughing operations. And it is never required for center-cutting operations like drilling, reaming, and tapping.

Steps to programming tool nose radius compensation

Programming tool nose radius compensation is relatively easy. Here are the three programming steps:

- Instate tool nose radius compensation
- Program the motions to machine the workpiece
- Cancel tool nose radius compensation

Instating tool nose radius compensation

To instate tool nose radius compensation, you must first be able to determine how the tool will be related to the work surface during the machining operation. Look in the direction the tool will be moving during the operation (rotate the print if it's necessary). Looking in this direction, ask the question, *"What side of the work surface is the cutting tool on, the left side or the right side?"* If the cutting tool is on the left side of the surface to be machined, you will use a G41 to instate tool nose radius compensation. If the cutting tool is on the right side of the surface to be machined, you will use a G42 to instate tool nose radius compensation. Look at Figure 4.17 to see some examples of how this is done.

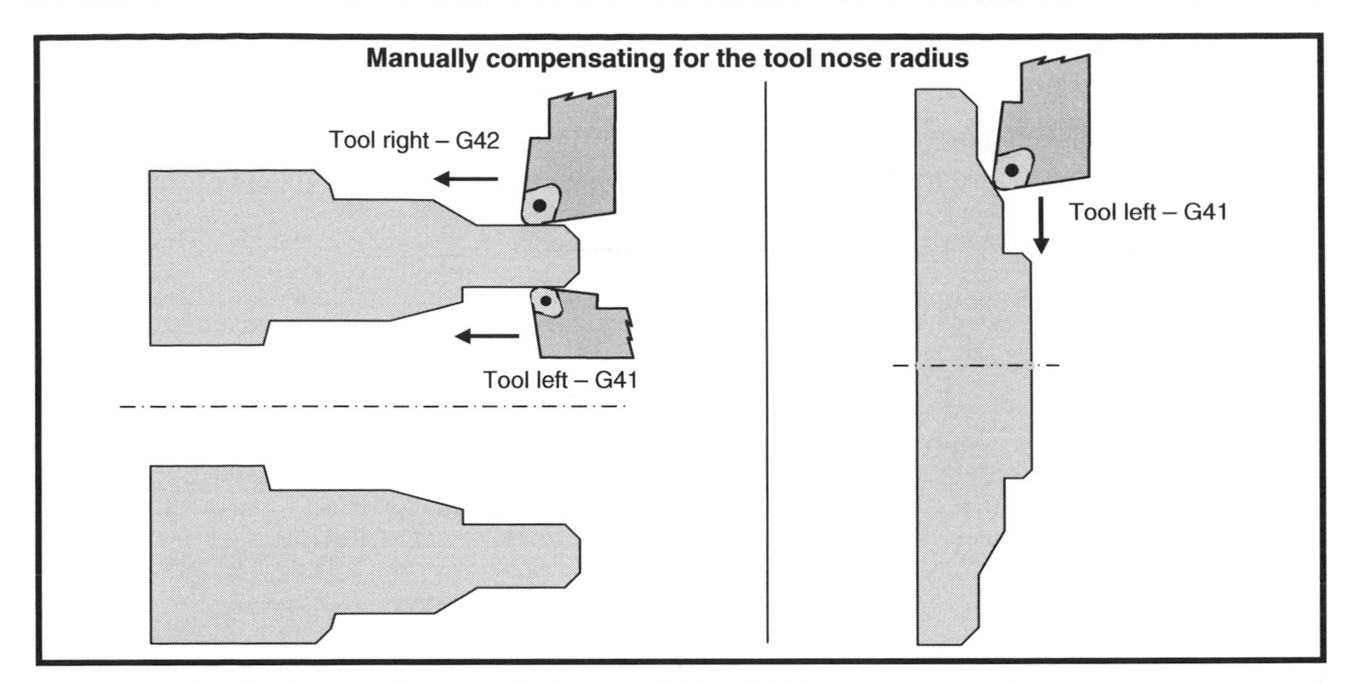

Figure 4.17 – Drawing illustrates how to decide between G41 and G42

Notice that G42 is always used for turning toward the chuck and G41 is always used for facing (toward the spindle center) and boring operations.

Once you know which side of the surface to be machined the tool is on (left or right), you simply include the appropriate G code (G41 or G42) in the cutting tool's approach movement.

Programming motion commands to machine the workpiece

Once you've instated tool nose radius compensation, you simply program the movements to finish turn or bore the workpiece using work surface coordinates. Again, these programmed coordinates are right on the work surface. The control will automatically keep the cutting edge radius on the specified side of the workpiece (left or right), and tangent to surfaces being machined.

You must be sure that the machine has the ability to keep the tool radius tangent to one surface without violating an upcoming surface. The most common problem in this regard has to do with machining narrow recesses. The recess must be wide enough to allow the tool nose radius into the recess. If it's not, the tool will actually violate one side of the recess before the depth of the recess is reached. Before this will be allowed to happen, an alarm will be sounded. Figure 4.18 demonstrates this possible problem.

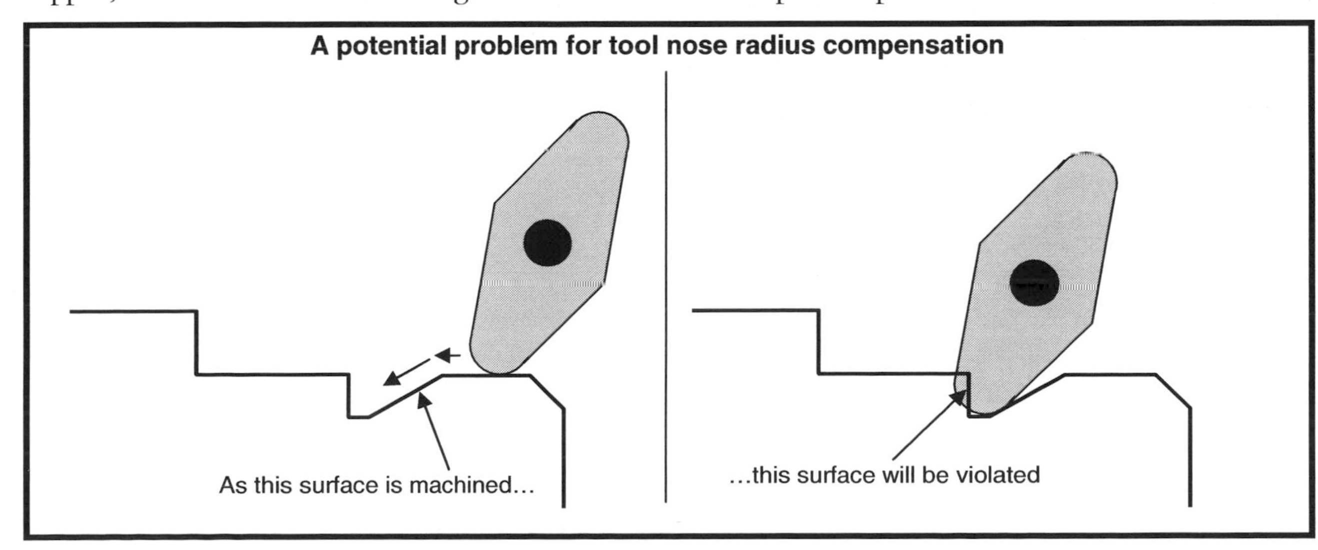

Figure 4.18 – Narrow recesses can present problems for tool nose radius compensation

Once tool nose radius compensation is instated, the machine will simply keep the tool on the left or right side of all programmed surfaces until tool nose radius compensation is canceled.

Canceling tool nose radius compensation

You must remember to cancel tool nose radius compensation. If it is not canceled, the machine will remain under its influence even with subsequent tools.

Canceling tool nose radius compensation is easy. Just include a G40 word in the command that sends the tool back to the tool change position.

An example program

Figure 4.19 shows the workpiece to be used for our first example program. It is the drawing used for the exercise you worked on in Lesson Ten. Again, we're only finishing this workpiece – the roughing has already been done by another tool.

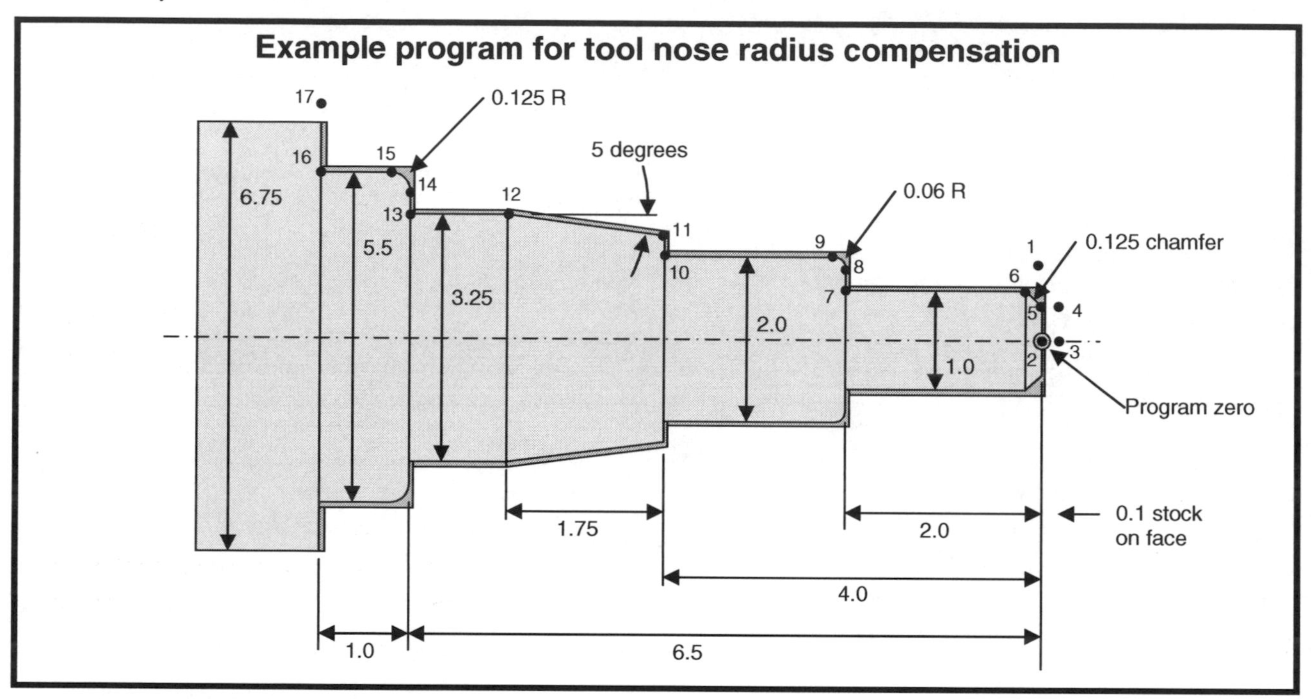

Figure 4.19 – Drawing for tool nose radius compensation example program

Here is the program. Notice that it is identical to the one shown in Lesson Ten except for the addition of two G codes.

```
O0001
N005 T0202 M42 (Index turret, select high spindle range)
N010 G96 S500 M03 (Start spindle fwd at 500 sfm)
N015 G00 X1.2 Z0 M08 (Rapid to point 1, turn on coolant)
N020 G01 X-0.062 F0.007 (Face to point 2)
N025 G00 Z0.1 (Rapid away to point 3)
N030 G42 X0.75 (Instate tool nose radius compensation, rapid up to point 4)
N035 G01 Z0 (Feed to point 5)
N040 X1.0 Z-0.125 (Chamfer to point 6)
N045 Z-2.0 (Turn 1.0 diameter to point 7)
N050 X1.875 (Feed up face to point 8)
N055 G03 X2.0 Z-2.0625 R0.0625 (Turn radius to point 9)
N060 G01 Z-4.0 (Turn 2.0 diameter to point 10)
```



```
N065 X2.9438 (Feed up face to point 11)
N070 X3.25 Z-5.75 (Turn taper to point 12)
N075 Z-6.5 (Turn 3.25 diameter to point 13)
N080 X5.25 (Feed up face to point 14)
N085 G03 X5.5 Z-6.625 R0.125 (Turn radius to point 15)
N090 G01 Z-7.5 (Turn 5.5 diameter to point 16)
N095 X6.95 (Feed up face to point 17)
N100 G00 G40 X8.0 Z6.0 (Rapid to tool change position, cancel tool nose radius compensation)
N105 M30 (End of program)
```

In line N030, notice the addition of the G42 word. Since this is a finish turning tool – and it will be on the right side of the work surfaces being machined, a G42 is being used to instate tool nose radius compensation.

Do notice that we waited until after the finish facing has been done before instating tool nose radius compensation. Since the face being machined is parallel to the X axis, there is no need for tool nose radius compensation. (If you must face surfaces that include angular and circular movements, tool nose radius compensation must be used – and a G41 will be used to instate it.)

In line N020, we do manually compensate for the tool nose radius. We send the facing tool past the workpiece centerline to a diameter that is twice the tool nose radius. This ensures that the face will be completely machined.

From lines N035 through N095, the programmed coordinates are directly on the work surface. The G42 in line N030 will ensure that the cutting tool will remain on the right side of all programmed surfaces. That is, the cutting edge radius will remain tangent to (flush with) all surfaces being machined.

In line N100, the G40 cancels tool nose radius compensation on the tool's motion to the tool change position.

Tool nose radius compensation from a setup person's point of view

As you've seen, programming tool nose radius compensation is relatively easy. You simply instate tool nose radius compensation during the tool's approach movement and cancel it during its return to the tool change position. But there is one more thing you must know – and it is related to the cutting tool's offset.

By one means or another, two offset registers must be entered for each tool that is using tool nose radius compensation. In many companies, the setup person manually enters this information during setup. Figure 4.20 shows the geometry offset display screen page.

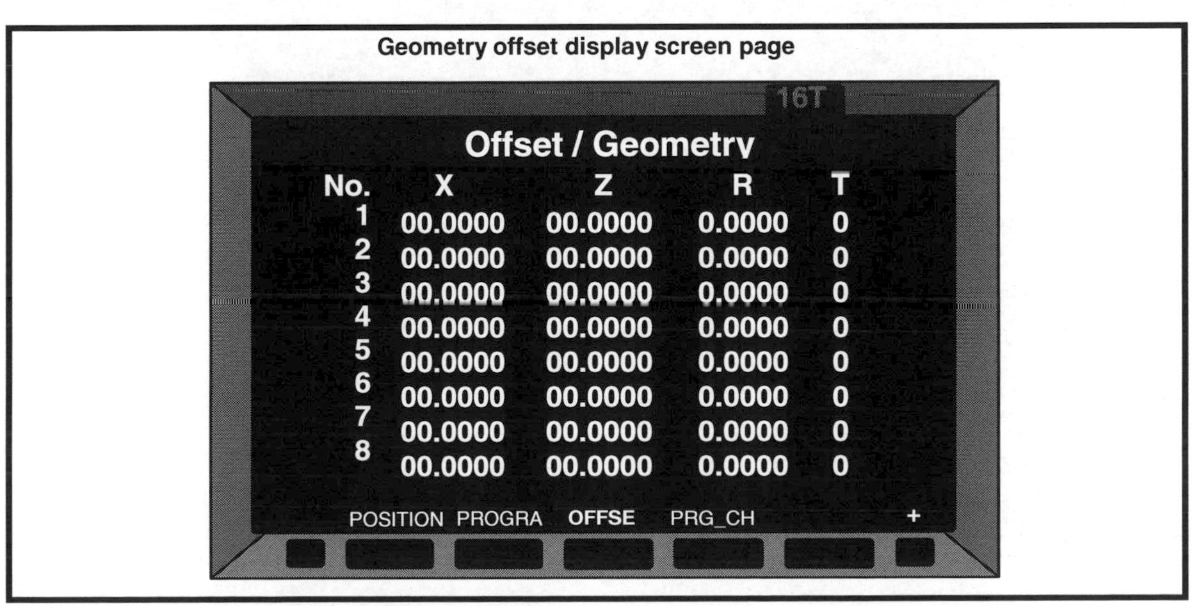

Figure 4.20 – Geometry offset display screen page

Notice the R and T registers. These registers are used with tool nose radius compensation. The R register is used to specify the tool nose radius. As you can see, this register requires a decimal value. For a 1/32 inch tool nose radius, a value of 0.0312 must be entered.

The T register is used to specify the cutting tool *type*. It is a code number. Figure 4.21 provides a chart, showing the various T register values.

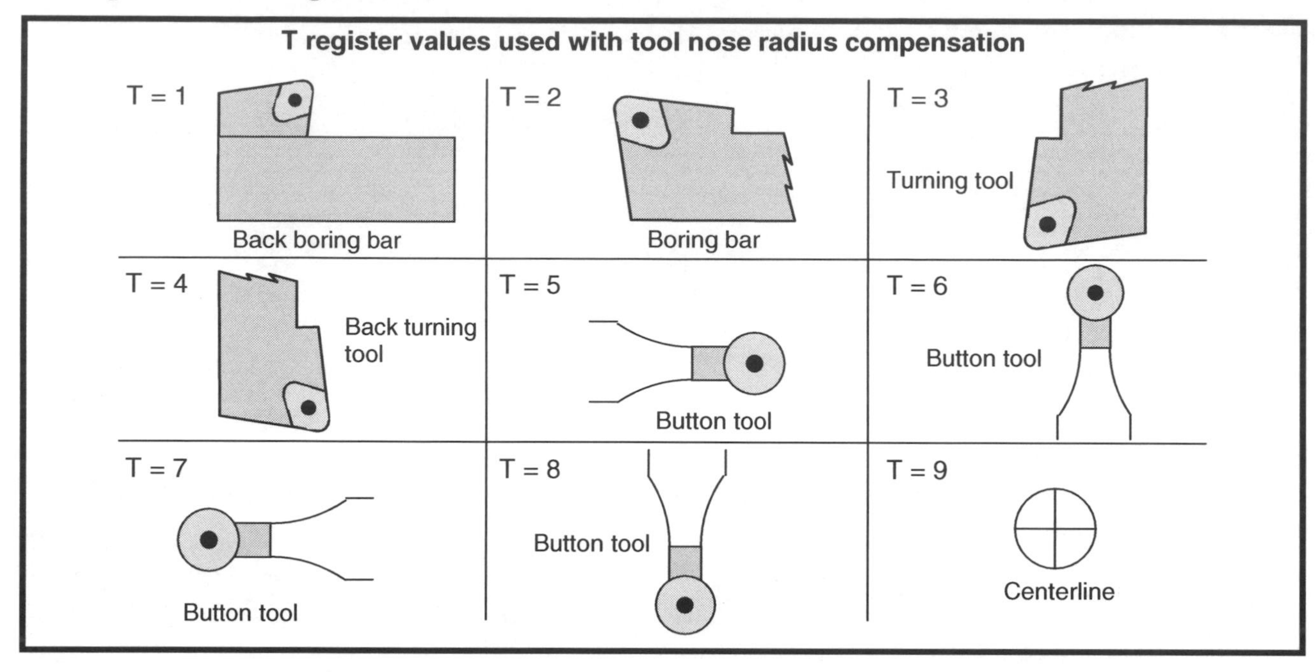

Figure 4.21 – Types of cutting tools used with tool nose radius compensation and the related T values

Of the tools shown on the chart in Figure 4.21, by far the two most popular types are the boring bar (type number two) and the turning tool (type number three). You should try to remember these two types.

In order for the example program for the workpiece shown in Figure 4.19 to work, geometry offset number two's R and T registers must be entered. Figure 4.22 shows this. Note that tool number two is a finish turning tool and we'll say it has a 0.0312 nose radius.

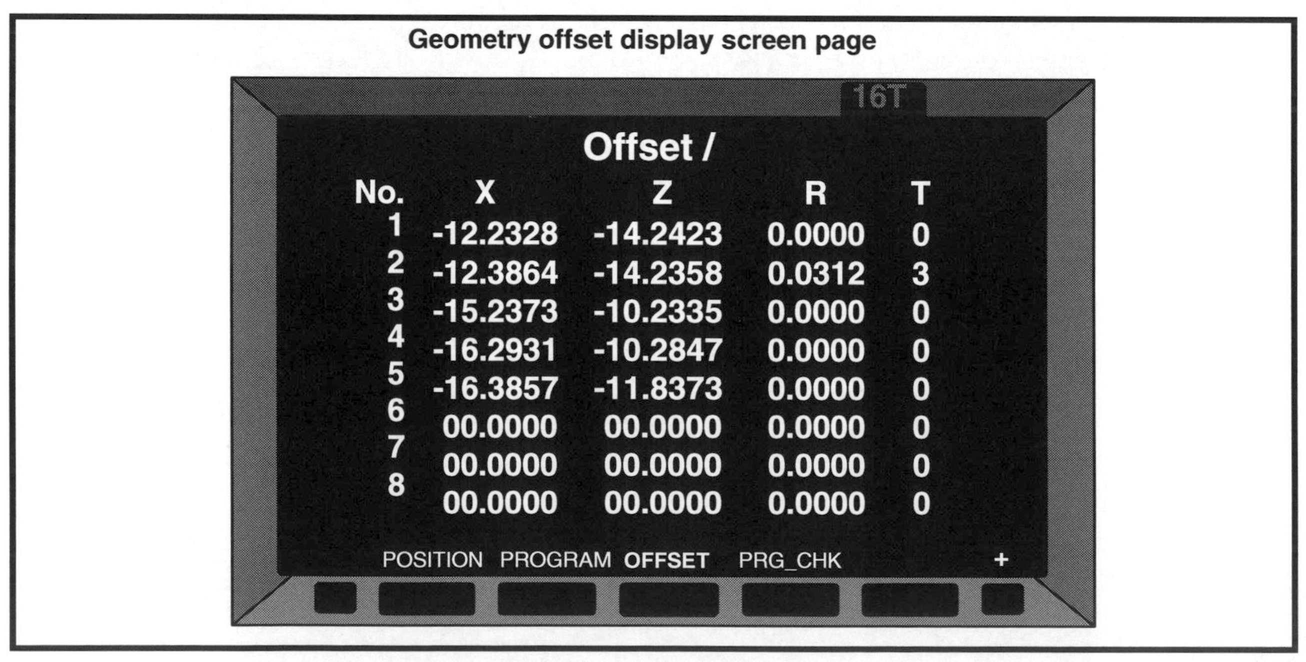

No.	X	Z	R	T
1	-12.2328	-14.2423	0.0000	0
2	-12.3864	-14.2358	0.0312	3
3	-15.2373	-10.2335	0.0000	0
4	-16.2931	-10.2847	0.0000	0
5	-16.3857	-11.8373	0.0000	0
6	00.0000	00.0000	0.0000	0
7	00.0000	00.0000	0.0000	0
8	00.0000	00.0000	0.0000	0

Figure 4.22 – How the R and T registers must be set for the example program

While the X and Z program-zero-assignment registers are set for several tools (there must be other tools currently in the turret), we're only concerned with offset number two. Notice that the R register is set to the cutting tool's nose radius (0.0312) and the T register is set to the tool type (3, which specifies a turning tool).

What if my machine does not have geometry offsets?
Older machines that do not have geometry offsets have only one offset page. For these machines, you must use the wear offset page to enter tool nose radius compensation values.

What if I forget to enter tool nose radius compensation values?
If you forget to enter tool nose radius compensation values into offsets, and if the current values in the offset are set to zero, the machine will think the cutting tool has a zero radius. The program will run, but the machine will not modify the programmed path in any way. In essence, it will be as if there is no G41 or G42 in the program.

What if I enter tool nose radius compensation values into wear offsets?
If your machine has geometry offsets, we recommend that you use them to enter tool nose radius compensation values. Many machines will clear (set to zero) wear offset values for an offset whenever geometry offsets are automatically entered by a tool touch off probe or by the measure function. This means that if you use the tool touch off probe during a production run during dull tool replacement, you could lose the tool nose radius values if they are placed in wear offsets.

What if I enter tool nose radius compensation values into both the geometry and wear offsets?
Remember that the machine will total the values that are in the wear and geometry offset registers. If for example, you have placed a value of 0.0312 in both the geometry and wear offset register for a tool, most machines will use a value of 0.0624 when creating the compensated tool path for the tool.

Programming tool nose radius compensation value entries
While offset entries for tool nose radius compensation are easy to make, you know that the control will not machine the workpiece correctly without them. Again, if the setup person forgets to enter the tool nose radius compensation offsets for a given tool (and the current offset values are zero), it will be just as if tool nose radius compensation is not being used. The workpiece, of course, will not be correctly machined.

For this reason, you should *program* the tool nose radius compensation offset entries as long as you're sure that the operator will not be changing tool nose radius compensation values during the production run (they normally don't). A G10 command is used to program offset entries. Here is an example command that enters geometry offset number two for our example program:

```
N005 G10 L2 P2 R0.0312 T3
```

G10 tells the control that you wish to set data. The L word specifies what kind of data. For one popular Fanuc control model, L1 specifies wear offsets and L2 specifies geometry offsets. Unfortunately, L word values vary from one control model to another. You'll have to look up the L word value related to geometry offsets in the Fanuc programming manual. The P word specifies the offset number being set. The R word specifies the value going into the R register (tool nose radius). And the T word specifies the value going into the T register (tool type). In our example command, were setting geometry offset number two's R register to 0.0312 and its T register to 3. This command can be included at the very beginning of the program to ensure that the tool nose radius compensation registers are properly set each time the program is run.

Another example program showing tool nose radius compensation
Figure 4.23 shows the workpiece to be machined. Notice that the part has already been rough machined. All that is left is to finish turn and finish bore the workpiece. Note that the format we follow for this program assumes geometry offsets are used to assign program zero.

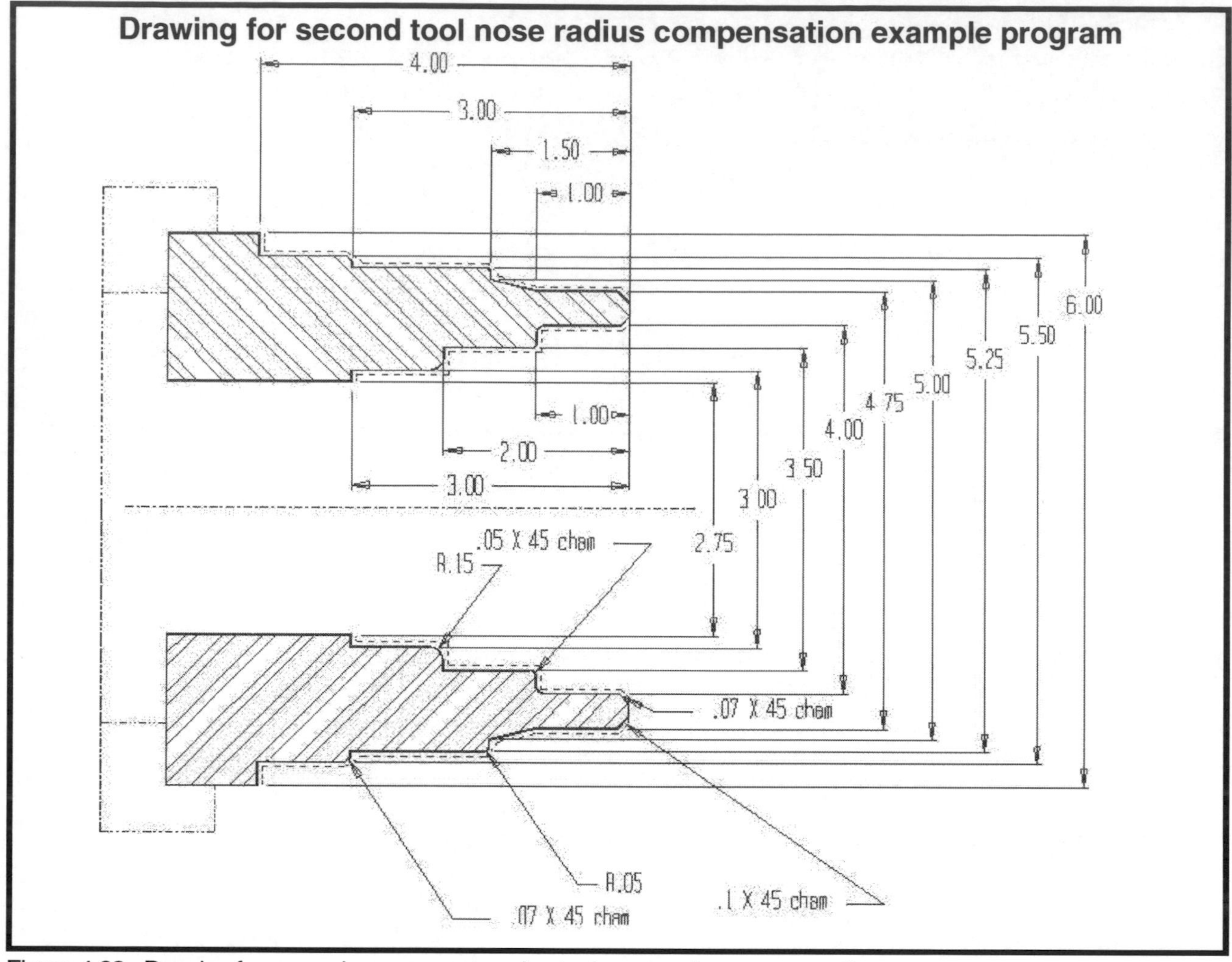

Figure 4.23 Drawing for example program stressing tool nose radius compensation

Geometry offset settings:

Both tools have a 0.0312 inch tool nose radius.

#	X	Z	R	T
1	-12.3324	-8.2342	0.0316	3
2	-11.2343	-3.9576	0.0316	2

Program:

Once again, we're only showing the finishing tools.

O0008 (Program number)

N002 **G10 L2 P1 R0.0312 T3**

N003 **G10 L2 P2 R0.0312 T2**

N005 T0101 M42 (Index to tool turning tool, select high range)

N010 G96 S300 M03 (Turn spindle on fwd at 300 sfm)

N015 G00 **G42** X4.55 Z.1 M08 (Instate tool nose radius compensation, rapid to first position, turn coolant on)

N020 G99 G01 Z0 F0.007

N025 X4.75 Z-0.1

N030 Z-1.0

N035 X5.0 Z-1.5

N040 X5.15

N045 G03 X5.25 Z-1.55 R0.05
N050 G01 Z-3.0
N055 X5.36
N060 X5.5 Z-3.07
N065 Z-4.0
N070 X6.1
N075 G00 **G40** X8.5 Z7.0 (Rapid to safe index point, cancel tool nose radius compensation)
N080 M01 (Optional stop)
N085 T0202 M42 (Index to boring bar, select high range)
N090 G96 S300 M03 (Turn spindle on fwd at 300 sfm)
N095 G00 **G41** X4.14 Z.1 M08 (Instate tool nose radius compensation, rapid to first position, turn coolant on)
N100 G01 Z0 F0.006
N105 X4.0 Z-0.07
N110 Z-.85
N115 G03 X3.7 Z-1.0 R0.15
N120 G01 X3.6
N125 X3.5 Z-1.05
N130 Z-2.0
N135 X3.3
N140 G02 X3.0 Z-2.15 R0.15
N145 G01 Z-3.0
N150 X2.65
N155 G00 Z0.1 (Rapid out of hole)
N160 **G40** X8.0 Z5.5 (Cancel tool nose radius compensation, rapid to safe index position)
N165 M30 (End of program)

In lines N002 and N003, we've included G10 commands to keep the setup person from having to enter tool nose radius compensation values into the geometry offsets. Since we know that both tools will have a 0.0312 inch tool nose radius – and that the setup person will *always* use inserts having this radius when inserts are indexed or replaced, we can include these offset entries right in the program.

Notice once again that tool nose radius compensation is instated on each tool's first approach to the workpiece (in lines N015 and N095). The motions for the cutting tool are then made. Finally, tool nose radius compensation is canceled (in lines N075 and N160).

Key points for Lesson Fourteen:

- All single point tools have a small radius on their cutting edges. This radius will affect the shape of the workpiece being machined.
- Tool nose radius compensation keeps the radius on the cutting edge flush with all workpiece surfaces during machining.
- There are three steps to using tool nose radius compensation: instate, make machining motions, and cancel. To instate too nose radius compensation you must include a G41 (tool left) or G42 (tool right) in the cutting tool's approach movement to the workpiece. All movements under the influence of tool nose radius compensation will be specified right on the workpiece. To cancel tool nose radius compensation, you include a G40 in the cutting tool's return motion to its tool change position.
- By one means or another, the R and T registers of the cutting tool's offset must be set. The setup person can enter them manually, or they can be programmed with a G10 command.

Add tool nose radius compensation to this program

Drawing:

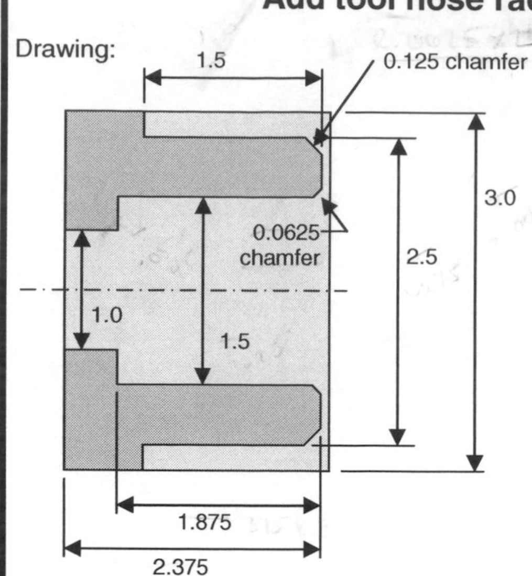

Instructions: This is the exercise from Lessons Nine and Twelve. After Lesson Twelve, you created a program for this workpiece. In this new exercise, we're asking you to add tool nose radius compensation. You will be doing so in tool's four and five (the finish boring bar and the finish turning tool). Find the appropriate commands in the program below and add the appropriate G codes. Also specify how offsets must be set.

Process:

Tool one: Rough face leaving 0.005 for finishing. Rough turn 0.125 chamfer and 2.5 diameter (one pass) leaving 0.080 on the 2.5 diameter and 0.005 on the 1.5 face for finishing (**Feed: 0.012 ipr, Speed: 500 sfm**)

Tool two: Drill (1.0 diameter drill) through the workpiece (Feed: 0.008 ipr, Speed: 1,150 rpm)

Tool three: Rough bore 0.0625 chamfer and 1.5 diameter (one pass), leaving 0.040 on the diameter and 0.005 on the 1.875 face (Feed: 0.007 ipr, Speed: 400 sfm)

Tool four: Finish bore 0.0625 chamfer and 1.5 diameter to size (Feed: 0.005 ipr, Speed: 500 sfm). **This tool has a 0.0156 nose radius.**

Tool five: Finish turn 0.125 chamfer and 2.5 diameter to size (Feed: 0.005 ipr, Speed: 500 sfm). **This tool has a 0.0312 nose radius.**

```
O0002 (Program number)
(ROUGH FACE AND TURN)
N005 T0101 M41 (Index turret, select low spindle range)
N010 G96 S500 M03 (Start spindle fwd at 500 sfm)
N015 G00 X3.2 Z0.005 M08 (Rapid to point 1, turn on coolant)
N020 G99 G01 X-0.062 F0.012 (Face to point 2 at 0.012 ipr)
N025 G00 Z0.1 (Rapid away to point 3)
N030 X2.33 (Rapid up to point 4)
N035 G01 Z0 (Feed up to point 5)
N040 X2.58 Z-0.12 (Chamfer to point 6)
N045 Z-1.495 (Turn 2.5 diameter to point 7)
N050 X3.2 (Feed up face and off part to point 8)
N055 G00 X8.0 Z7.0 (Rapid to tool change position)
N060 M01 (Optional stop)
(DRILL 1.0" HOLE)
N065 T0202 M41 (Index turret, select low spindle range)
N070 G97 S1150 M03 (Start spindle at 1,150 rpm)
N075 G00 X0 Z0.1 M08 (Rapid to point 9, turn on coolant)
N080 G01 Z-2.705 F0.008 (Feed to point 10 at 0.008 ipr)
N085 G00 Z0.1 (Rapid back to point 9)
N090 X8.0 Z7.0 (Rapid to tool change position)
N095 M01 (Optional stop)
(ROUGH BORE)
N100 T0303 M41 (Index turret, select low spindle range)
N103 G96 S400 M03 (Start spindle at 400 sfm)
N105 G00 X1.585 Z0.1 M08 (Rapid to point 11, start coolant)
N110 G01 Z0 F0.007 (Feed to point 12 at 0.007 ipr)
N115 X1.46 Z-0.0575 (Chamfer to point 13)
N120 Z-1.87 (Rough bore 1.5 to point 14)
N125 X0.95 (Feed down face to point 15)
N130 G00 Z0.1 (Rapid to point 16)
N135 X8.0 Z7.0 (Rapid to tool change position)
N140 M01 (Optional stop)
```

Specify the tool nose radius compensation offset entries for this job:

#	R	T
4	______	____
5	______	____

```
(FINISH BORE)
N145 T0404 M42 (Index turret, select high spindle range)
N150 G96 S500 M03 (Start spindle at 500 sfm)
N155 G00 X1.625 Z0.1 M08 (Rapid to point 17, turn on coolant)
N160 G01 Z0 F0.005 (Feed to point 18 at 0.005 ipr)
N165 X1.5 Z-0.0625 (Chamfer to point 19)
N170 Z-1.875 (Bore 1.5 to point 20)
N175 X0.95 (Feed down face to point 21)
N180 G00 Z0.1 (Rapid to point 22)
N185 X8.0 Z7.0 (Rapid to tool change position)
N190 M01 (Optional stop)
(FINISH FACE AND TURN)
N195 T0505 M42 (Index turret, select high spindle range)
N200 G96 S500 M03 (Start spindle at 500 sfm)
N205 G00 X2.7 Z0 M08 (Rapid to point 23, turn on coolant)
N210 G01 X1.3 F0.005(Face to point 24 at 0.005 ipr)
N215 G00 Z0.1 (Rapid to point 25)
N220 X2.25 (Rapid to point 26)
N225 G01 Z0 (Feed to point 27)
N230 X2.5 Z-0.125 (Chamfer to point 28)
N235 Z-1.5 (Turn 2.5 to point 29)
N240 X3.2 (Feed up face to point 30)
N245 G00 X8.0 Z7.0 (Rapid to tool change position)
N250 M30 (End of program)
```

Offset entry answers:

#	R	T
4	0.0156	2
5	0.0312	3

Program answers:
In line N155, add G41
In line N185, add G40
In line N220, add G42
In line N245, add G40

STOP! Do practice exercise number Fourteen in the workbook.

Key Concept

5

You Must Provide Structure To Your CNC Programs

While there are many ways to write programs, you must ensure that your programs are safe and easy to use, yet as efficient as possible. This can be a real challenge since safety and efficiency usually conflict with one another.

Key Concept Number Five contains two lessons:

15: Introduction to program structure

16: Four types of program format

The fifth Key Concept is you must structure your CNC programs using a strict format – while incorporating a design that accomplishes the objectives you intend. Though it is important to create *efficient* programs, *safety and ease-of-use* must take priority – at least until you gain proficiency. In Key Concept Number Five, we'll be showing techniques that stress *safety as the top priority.*

To this point in the text, we have been presenting the building blocks of CNC programming – providing the needed individual tools. Machine components, axes of motion, program zero, and programmable functions are presented in Key Concept One. Preparation steps in Key Concept Two. Motion types in Key Concept Three. Compensation types in Key Concept Four. In Key Concept Five, we're going to draw all of these topics together, showing you what it takes to write CNC turning center programs completely on your own – making you a *self-sufficient programmer.*

Lesson Fifteen will introduce you to a CNC program's structure, showing you the reasons *why* programs must be strictly formatted. We'll also review some the program-structure-related points we've made to this point – and present a few new ones. And we'll address some variations related to how certain machine functions are handled.

In Lesson Sixteen, we'll show the four types of program format as they are applied to CNC turning centers. These formats can be used as a crutch to help you write your first few programs.

You may be surprised at how much you already know about a program's structure, especially if you have been doing the exercises included in this text and/or those in the workbook. We have introduced many of the CNC words used in programming, and we have been following the structure-related suggestions that we will be recommending here in Key Concept Five.

For example, do you recognize any of these words and commands? If so, write their meanings in the space provided.

O0001: ____________________________________

T0101 M41: __________________________________

G96 S500 M03: ____

M01: ____

G00 X4.2 Z0.005 M08: ____

M30: ____

If you can't remember one or more of these CNC words or commands, don't worry. We'll be explaining them in detail during Key Concept Number Five. But if you do recognize most of them, you're already well on your way to understanding program structure.

Lesson 15

Introduction To Program Structure

Structuring a CNC program is the act of writing a program in a way that the CNC machine can recognize and execute safely, efficiently, and with a high degree of operator-friendliness.

You know that CNC programs are made up of commands, that each command is made up of words, and that each word is made up of a letter address and a numerical value.

You also know that programs are executed sequentially – command by command. The machine will read, interpret, and execute the first command in the program. Then it will move on to the next command. Read – interpret – execute. It will continue doing so until the entire program has been executed.

You have seen several complete programs so far in this text – you have even worked on a few if you have done the exercises in this text and in the workbook. You have probably noticed that there is quite bit of consistency and *structure* in the CNC programs we have shown.

Our focus in Lesson Fifteen will be to help you understand more about the structure that is used in CNC programming.

Objectives of your program structure

CNC machines have come a long way. In the early days of NC (before computers), a program had to be written *just so*. If *anything* was out-of-place, the machine will go into an alarm state – failing to execute the program. While today's CNC machines are *much* more forgiving, you must still write CNC programs in a rather strict manner.

There are many ways to write a workable program – and the methods you use in structuring your programs will have an impact on the three most important objectives:

- Safety
- Efficiency
- Ease-of-use (operator friendliness)

It may be impossible to come up with a perfect balance among these objectives. Generally speaking, what you do to improve one objective will negatively affect the other two. When faced with a choice, a beginning programmer's priorities should always lean toward *safety* and *ease-of-use*. Our recommended programming structure stresses these two objectives. We will, however, show some of the efficiency-related short-comings of our recommended methods – so you can improve efficiency as you gain proficiency.

We're going to be assuming that *you have control* of the structure you use to write programs. Your company may, however, already have a programmer that is writing programs with a different structure. As long as these programs are working – and satisfying the company's objectives – you're going to have to adapt to the established structure. If you understand the reasons for formatting, and if you understand one successful method for structuring programs, it shouldn't be too difficult to adapt.

Reasons for structuring programs with a strict and consistent format

Let's begin by discussing the reasons *why* you must write your programs using a strict structure.

Familiarization

You must have some way to get familiar with CNC programming. You'll need some help writing your first few programs. The formats we show in Lesson Sixteen will provide you with this help. You'll be able to use our given formats as a crutch until you (eventually) have them memorized.

Actually, the formats we show in Lesson Sixteen will keep you from having to memorize anything. Instead, you will look at a word or command in the format and then you must remember its use. Compare this to recognizing the road signs you see as you drive an automobile. It is unlikely that you could recite every road sign from memory – but when you see one – you immediately know its meaning. Think of our given formats as like a set of road signs designed to help you write CNC programs.

As you'll see in Lesson Sixteen, you'll have road signs to help you *when you begin writing a program* (program start-up format), *when you're finished with a tool* (tool ending format), *when you begin a new tool* (tool start-up format), and *when you end a program* (program ending format).

Consistency

If you have been doing the exercises in this text and in the workbook, you've already worked on a few actual programs, filling in the blanks with needed CNC words. You have also seen several complete example programs in this text. You have probably noticed that these programs are written in a very consistent manner. The commands within each tool of each program are consistent with the other tools in the program.

Consistency within programs is important for three reasons. First, consistency helps you to become familiar with programming. Repeated commands soon become memorized.

Second, and more importantly, consistency among programs will help *everyone* that must work with your programs, including other programmers, setup people, and operators. Your programs will be easier to work with if you always structure them in the same manner.

Third, it is important to be able to repeat your past successes. If a program is running properly – achieving all of the objectives you intend – using its structure in your *next* program will ensure continued success.

Re-running tools in the program

This is the most important reason for structuring your programs using a strict format. As you know, *trial machining* involves five steps: recognizing a close tolerance, adjusting an offset to leave additional stock, machining under the influence of the trial machining offset, measuring the machined dimension and adjusting the offset accordingly, *and re-running the tool.*

Trial machining is but one time when tools must be re-run. Say you're verifying a lengthy program. You are five tools into a nine-tool program when you find a mistake. You must stop the cycle to correct the mistake. With the mistake corrected, you'll want to pick up where you left off – at the beginning of the fifth tool. You won't want to re-run the entire program (from the beginning) just to get to tool number five (doing so would be a waste of time).

Your ability to re-run tools is directly related to the structure you use to program. *If you don't structure your programs properly, it will not be possible to re-run tools.*

Here is a specific example that stresses why program formatting is so very important for re-running tools. Say you are programming two tools that run in sequence (say, tools one and two). Both tools happen to run at 400 sfm. Close to the beginning of the first tool, you have an S400 word in the command that starts the spindle. As you continue writing the commands for the second tool, you decide to leave out the S400 word, since the spindle speed is *modal.*

Everything will work just fine as long as the program runs in sequence – from beginning to end (the second tool immediately follows the first tool). But say the operator runs the entire program before they discover that the second tool has done something wrong (maybe it's a grooving tool that has not gone deep enough). After correcting the problem, if the operator attempts to re-run the second tool by itself (as they should), it will run at the same speed as the last tool in the program —*probably not 400 sfm!* This is but one of several times when you must include redundant words (words that are already instated) in each tool – just to gain the ability to re-run tools.

In essence, you must make each tool in the program *independent from the rest of the program.* Think of the programming commands for each tool in a program as making a *mini-program*, self sufficient and capable of activating all necessary machine functions.

You will often be tempted to leave out (necessary) redundant words because they are modal and remain in effect from tool to tool. To fight this temptation, think of each tool in the program as if it is the first tool. Everything you include in the program for the first tool (except the program number) will be needed in the program for each successive tool. This includes the turret index word (like T0101), the spindle range selecting word (like M41), the spindle starting words (like G96 S500 M03), movement to the first XZ position (like G00 X2.0 Z0.1), the coolant activation word (M08), and the feedrate word in the tool's first cutting command (like F0.012).

Here is another example that should help you understand why certain redundant words must be repeated at the beginning of each tool. This is the program from the exercise you worked on at the end of Lesson Fourteen:

```
O0002 (Program number)
(ROUGH FACE AND TURN)
N005 T0101 M41 (Index turret, select low spindle range)
N010 G96 S500 M03 (Start spindle fwd at 500 sfm)
N015 G00 X3.2 Z0.005 M08 (Rapid to point 1, turn on coolant)
N020 G99 G01 X-0.062 F0.012 (Face to point 2 at 0.012 ipr)
N025 G00 Z0.1 (Rapid away to point 3)
N030 X2.33 (Rapid up to point 4)
N035 G01 Z0 (Feed up to point 5)
N040 X2.58 Z-0.12 (Chamfer to point 6)
N045 Z-1.495 (Turn 2.5 diameter to point 7)
N050 X3.2 (Feed up face and off part to point 8)
N055 G00 X8.0 Z7.0 (Rapid to tool change position)
N060 M01 (Optional stop)
(DRILL 1.0" HOLE)
N065 T0202 M41 (Index turret, select low spindle range)
N070 G97 S1150 M03 (Start spindle at 1,150 rpm)
N075 G00 X0 Z0.1 M08 (Rapid to point 9, turn on coolant)
N080 G01 Z-2.705 F0.008 (Feed to point 10 at 0.008 ipr)
N085 G00 Z0.1 (Rapid back to point 9)
N090 X8.0 Z7.0 (Rapid to tool change position)
N095 M01 (Optional stop)
(ROUGH BORE)
N100 T0303 M41 (Index turret, select low spindle range)
N103 G96 S400 M03 (Start spindle at 400 sfm)
N105 G00 X1.585 Z0.1 M08 (Rapid to point 11, start coolant)
N110 G01 Z0 F0.007 (Feed to point 12 at 0.007 ipr)
N115 X1.46 Z-0.0575 (Chamfer to point 13)
N120 Z-1.87 (Rough bore 1.5 to point 14)
N125 X0.95 (Feed down face to point 15)
N130 G00 Z0.1 (Rapid to point 16)
N135 X8.0 Z7.0 (Rapid to tool change position)
N140 M01 (Optional stop)
```

```
(FINISH BORE)
N145 T0404 M42 (Index turret, select high spindle range)
N150 G96 S500 M03 (Start spindle at 500 sfm)
N155 G00 G41 X1.625 Z0.1 M08 (Rapid to point 17, turn on coolant)
N160 G01 Z0 F0.005 (Feed to point 18 at 0.005 ipr)
N165 X1.5 Z-0.0625 (Chamfer to point 19)
N170 Z-1.875 (Bore 1.5 to point 20)
N175 X0.95 (Feed down face to point 21)
N180 G00 Z0.1 (Rapid to point 22)
N185 G40 X8.0 Z7.0 (Rapid to tool change position)
N190 M01 (Optional stop)
(FINISH FACE AND TURN)
N195 T0505 M42 (Index turret, select high spindle range)
N200 G96 S500 M03 (Start spindle at 500 sfm)
N205 G00 X2.7 Z0 M08 (Rapid to point 23, turn on coolant)
N210 G01 X1.3 F0.005 (Face to point 24 at 0.005 ipr)
N215 G00 Z0.1 (Rapid to point 25)
N220 G42 X2.25 (Rapid to point 26)
N225 G01 Z0 (Feed to point 27)
N230 X2.25 Z-0.125 (Chamfer to point 28)
N235 Z-1.5 (Turn 2.5 to point 29)
N240 X3.2 (Feed up face to point 30)
N245 G00 G40 X8.0 Z7.0 (Rapid to tool change position)
N250 M30 (End of program)
```

Though there are no words in this program to turn the spindle and coolant off (M05 and M09), notice that an M03 and an M08 are specified at the beginning of every tool. As you worked on this exercise, you may have been wondering why. Again, there will be no need for these words (except in the first tool) as long as the program is executed in its entirety – from tool one to tool five. But if a tool must be rerun for any reason, these redundant words must be in the program.

Again, consider the need for trial machining. If the finish boring bar (tool number four) requires trial machining, the setup person will adjust the wear offset number four to allow excess material, run the tool, measure what tool four has done, re-adjust the offset, and re-run tool four. To re-run tool number four, the restart command will be line N145. Without an M03 in line N150 or an M08 in line N155, the spindle and coolant will not come on when this tool is restarted.

In Lesson Sixteen, we provide two complete sets of program format – one for use when geometry offsets are used to assign program zero and the other for use when program zero is assigned in the program. These formats include all of the words needed in each tool – allowing every tool in the program to be re-run.

Efficiency limitations

Again, the highest priority for our given formats is safety. Even so, beginners must exercise extreme caution when verifying their first few programs. Next we emphasize *ease-of-use*. Programs written with our recommended formats will be relatively easy to work with.

Our given formats do *not* place a large emphasis on efficiency. While programs written with these formats will not be terribly wasteful, they will not be as efficient as they could be. But remember, when you start emphasizing efficiency, safety and ease-of use will probably suffer. Here are some of the efficiency-related limitations of our given formats:

One activity per command – For the most part, our given formats will cause the machine to do *one thing at a time.* In reality, the machine can sometimes be doing more. For example, our recommended format for tool start-up will have the spindle start in one command and the movement to the first XZ position in the next. While these two functions can be done together, it is helpful for beginners to concentrate on but one thing at a time.

Use of constant surface speed – Our given formats pay no special attention to spindle acceleration and deceleration time. As a cutting tool retracts to the tool change position, the spindle will commonly slow down if constant surface speed is used. When the next tool approaches (assuming constant surface speed is still used), the spindle will speed up. Spindle acceleration and deceleration takes time – but our recommended formats do nothing to minimize this time.

Rapid approach distance – In our given formats, we recommend using an approach distance of 0.100 inch (about 2.5 mm) when approaching qualified (known, consistent) surfaces. While it is more efficient to reduce rapid approach distance, it is not as safe. As you gain confidence and proficiency, you should consider reducing rapid approach distance to improve efficiency.

Machine variations that affect program structure

As mentioned, we provide two sets of format in Lesson Sixteen – one for use when geometry offsets are used to assign program zero and the other for use when program zero is assigned in the program. These formats can be applied to just about any of the machine configurations shown in Lesson One. But even within a given machine type, there are lots variations among machine tool builders.

Our given formats are aimed at basic turning centers – those without a lot of bells and whistles. About the only programmable functions we address are turret indexing, spindle, coolant, and feedrate. If your machine has other accessories (like a tailstock, steady rest, or bar feeder), you must consult your machine tool builder's programming manual to learn how they are programmed. As is discussed in Lesson One, most accessories are handled with M codes – so look first for the machine's list of M codes. (Note that we also provide information about certain accessories in the Appendix that follows Lesson Twenty-Three.)

Even within our limited coverage of accessories, there are variations. Here we list them.

M code differences

M code number selection is left completely to the discretion of your machine tool builder/s, and no two builders seem to be able to agree on how all M codes should be numbered. For *very* common machine functions like spindle on/off (M03, M04, M05) and flood coolant on/off (M08, M09), machine tool builders have standardized. But for less common functions like tailstock activation, chuck open/close, bar feeder activation, chip conveyer activation, door open/close, and high pressure coolant systems, you must find the list of M codes in your machine tool builder's programming manual.

Here is a list of very common M codes you will need for our given program formats. Fortunately, most machine tool builders do utilize these M code numbers just as we show.

M00 - Program stop (halts the program's execution until the operator reactivates the cycle)

M01 - Optional stop (used to cause the machine to stop between tools during program verification)

M03 - Spindle on forward (for right hand tools)

M04 - Spindle on reverse (for left hand tools)

M05 - Spindle off (not normally needed with our formats since the spindle need not stop when the turret is indexed)

M08 - Flood coolant on (cools and lubricates the machining operation)

M09 - Coolant off (not normally needed with our formats since the coolant can continue flowing when the turret is indexed)

M41 - Low spindle range selection

M42 - High spindle range selection

M30 - End of program (on many machines, M02 can also be used)

Here are some other functions commonly handled with M codes, but that are not consistently numbered by machine tool builders. We've left a blank next to each M code number. After checking your machine tool builder's programming manual, fill in the blanks for those machine functions that are equipped on your CNC turning center/s.

________ - Coolant on and spindle on forward (together)
________ - Coolant on and spindle on reverse (together)
________ - Open chuck jaws
________ - Close chuck jaws
________ - Open machine door
________ - Open machine door
________ - Tailstock body forward
________ - Tailstock body back
________ - Tailstock quill forward
________ - Tailstock quill back
________ - Activate bar feeder
________ - Deactivate bar feeder
________ - Advance part catcher
________ - Retract part catcher
________ - Activate tool touch-off probe
________ - Deactivate tool touch-off probe
________ - Select normal turning mode
________ - Select live tooling mode
________ - Turn on chip conveyor
________ - Turn off chip conveyor

G code numbering differences

Machine tool builders even vary when it comes to certain G code numbers. Even if you have a Fanuc or Fanuc-compatible control, you must be aware of this potential for variation. Generally speaking, imported machines (Japanese, Korean, Taiwanese, etc.) use what Fanuc calls the *standard G code specifications*. We use this set of G code specifications throughout this text. On the other hand, American machine tool builders tend to use the *special G code specifications*. Rest assured that the usage of these G codes remains exactly the same, only the invoking G code *number* will vary. Again, there are just a few differences. Here is a list of the G codes that vary based upon being standard or special.

Standard	Special	Description
G50	G92	Program zero assignment and maximum rpm designator
G90	G77	One-pass turning cycle
G92	G78	One pass threading cycle
G94	G79	One pass facing cycle
G98	G94	Feed per minute
G99	G95	Feed per revolution
X/Z	G90	Absolute positioning
U/W	G91	Incremental positioning

Again, most of these differences simply require that you translate a G code number (G92 instead of G50, for example) if your machine/s require the special G code specifications. However, notice how incremental and absolute motions are commanded. With the standard G code specification (used in this text), X and Z specify absolute positions. U and W specify incremental motions. With the special G code specification, G90 specifies incremental mode and G91 specifies incremental mode. With both special G codes, X and Z are used to specify motion. The command

```
N040 G90 G00 X1.0 Z1.0
```

tells the control to rapid to a position (relative to program zero) of one inch in X and one inch in Z. On the other hand, the command

```
N040 G91 G00 X1.0 Z1.0
```

tells the control to rapid from its current position one inch larger (positive) in diameter, and one inch in the Z plus direction.

The only other functions affected by the G code numbering are feedrate modes and simple canned cycles (discussed in Key Concept Six).

Turret variations

The design of a turning center's turret is left completely to the machine tool builder. Some incorporate external tool holders that are mounted on the outside of the turret. Others hold cutting tools directly in the turret itself. Some make it very easy to direct coolant lines to the cutting tool tip. Others make it more difficult. Some are large enough to eliminate interference when a long boring bar or drill is placed in a station that is adjacent to a face and turn tool. With others, you must be very careful with cutting tool placement in order to avoid interference problems.

As you know, a T word is used to specify turret index. This T word also instates the geometry and wear offsets. The command

```
N240 T0404
```

for example, indexes the turret to station number four, instates geometry offset number four, and instates wear offset number four.

Most machines have a *bi-directional* turret, meaning the turret can index in either direction. To select *which* direction, most turning centers are designed to automatically select the direction that provides the shortest rotational distance to the specified tool.

There are some turning centers, however, that let you specify the direction of turret rotation. Three M codes are usually involved. One M code specifies that the machine will select the shortest rotational distance (this is commonly the initialized M code). A second M code will force the turret to index in a clockwise direction. And a third M code will force the turret to index in the counter clockwise direction. You'll have to reference the machine tool builder's programming manual to find out whether your machine has this ability – and if so, which M codes are involved.

How do you determine a safe yet efficient index position?

When geometry offsets are used to assign program zero, the machine does not have to be sent to its zero return position in order to index the turret. The turret index position can be in just about any location, as long as the turret can index without a cutting tool crashing into the workpiece or chuck. For our example programs, we've been selecting some rather arbitrary turret indexing positions. Here is a more logical method that may not be as efficient as possible, but it is safe – and easy to determine. And it will work for *all* of your programs.

Like any other motion command, the rapid motion to the turret index position is specified from the program zero point. And with most programs, the Z axis program zero point is the right end of the finished workpiece. So you must come up with a turret index position (relative to program zero) that will allow the turret to rotate without interference.

The trick is to determine the Z axis distance between your longest cutting tool (usually an internal tool like a drill, boring bar, reamer, etc.) and your shortest cutting tool (commonly a face and turn tool). Figure 5.1 shows an example.

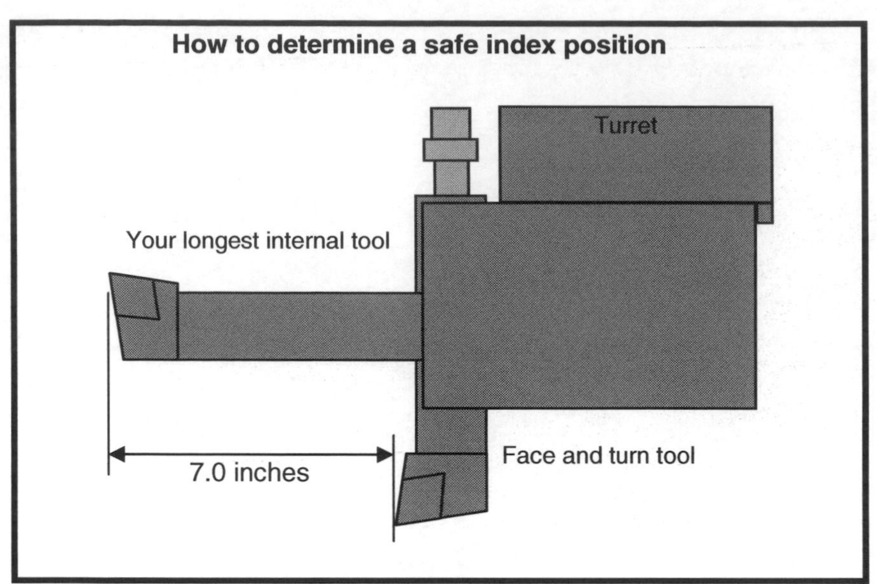

Figure 5.1 – The distance between your longest tool and your shortest tool

In Figure 5.1, notice that our longest boring bar extends seven inches from the shortest tool. In order for the turret to index safely, a turning tool must be more than seven inches in the Z axis from the end of the workpiece (the program zero point in most programs). Most turrets will extend about 0.5 inch during the turret index, but we recommend adding at least two inches to the distance from the longest tool to the shortest tool. In our example, we'd program all turning tools to move to a position of Z9.0 before making a tool change.

In X, we recommend making the safe index position something a little larger than the diameter of the workpiece that is being machined. For a 5.0 inch diameter workpiece, we recommend an X axis safe index position of X6.0.

While you can come up with a safe index position for each job (which might be advisable with very large lots), we're recommending that you come up with one safe index position for *all of your jobs*. This means, of course, that you must determine the distance between the face/turn tool and the longest internal tool you will ever use.

Though you may consider it a little wasteful, we recommend that you use the same safe index position (Z9.0 in our example) for every tool in the program. You can always optimize once the job is up and running. But if you make a mistake, the results could be disastrous.

This is yet another time when our recommended format will not be very efficient. The larger the lot sizes you typically run, the more important that it will be to make your programs as efficient as possible. But until you gain some experience on the machine, we urge you to make your programs as safe and easy to run as possible.

A warning: Even when you do decide to start optimizing turret indexing, proceed with extreme caution. Remember that cutting tools often remain in the turret from job to job, even when they are not needed in the current program. It is possible that the turret's makeup will be different from one time you run a job to the next time you run it. For example, today (when you optimize turret indexing), maybe there are only turning tools are in the turret (no boring bars). But the next time you run the job, maybe there are two long boring bars in the turret, left over from the previous job. If the turret index position in your program does not consider the long boring bars, the boring bars will crash during turret indexing.

What if my machine doesn't have geometry offsets?

If you are assigning program zero in the program, we recommend that you make the zero return position the starting point for the program as well as the turret index position. While this isn't very efficient, it is quite safe.

A reminder about spindle limiting

We discuss spindle limiting in Lesson Two, but none of the example programs shown in this text or in the exercises have required it. Since it is part of program formatting, we provide a review.

A very important format-related feature of turning centers is spindle limiting. G50 is the command word to specify spindle speed limiting. The command

N005 G50 S2000

for example, will limit the main spindle to 2,000 rpm, regardless of how fast the program (or constant surface speed) commands the spindle to run. G50 in this application has nothing to do with program zero assignment. The S word in the G50 command specifies the maximum rpm.

This application for G50 is extremely important when you must run unbalanced (out-of-round) workpieces. Say, for example, you must machine a large, unbalanced casting. Maybe your turning center's maximum spindle speed (the fastest it can run) is 5,000 rpm. But during setup you find (by testing) that if this unbalanced workpiece is run at anything over 1,500 rpm, the machine starts to vibrate. It's likely that if the machine is allowed to run the spindle up to its maximum (5,000 rpm), the chuck jaws will not hold the workpiece and the workpiece will be thrown from the chuck. This will cause damage to the machine and possibly injure the operator. Knowing the fastest safe rpm for this workpiece is 1,500 rpm, the command

N005 G50 S1500

can be specified at the beginning of the program. Since the G50 spindle limiting command is modal, you can rest assured that the machine will not attempt to rotate the workpiece faster than 1,500 rpm, even if it is commanded to do so.

This spindle limiting command is most important when working in the constant surface speed (G96) mode, since you will not have direct control of the spindle rpm (the control will automatically adjust rpm based on the specified speed in sfm [meters per minute in metric mode] and the tool's current diameter). If the machine is under the influence of a spindle limiting command, it will not be allowed to rotate the spindle faster than the speed specified in the last spindle limiting command, even when constant surface speed mode is being used.

For example, consider what happens when you face a workpiece to center under the influence of constant surface speed. As the tool approaches center, the spindle will accelerate. When the tool reaches center, the spindle will be running at its maximum speed in the selected spindle range. Again, if the workpiece is unbalanced, this will cause real problems if you haven't specified a spindle limiting command. In the example command given above (N005 G50 S1500), as the facing tool machines to smaller and smaller diameters, the spindle will stop accelerating when it reaches 1500 rpm.

Keep in mind that the majority of workpieces to be machined on turning centers are symmetrical (not unbalanced). If the workpiece is nicely balanced, there will be no need to limit the spindle. The workpieces shown in every example and programming activity to this point, for example, will require spindle limitation. However, our recommendation is to include a spindle limiting command as part of the program startup format for every program you write. If you don't need to limit the spindle for the workpiece, simply specify the machine's maximum rpm as the spindle limitation. If you do need to limit the spindle rpm (the workpiece is unbalanced), set the S word to an rpm that you're sure will work (test during setup and start slow if you're in doubt). We recommend including a spindle limiting command in every program for two reasons.

First, the spindle limiting command is *modal.* It will carry over from the last time the command is executed. While this is desirable in a program (from tool to tool), it can cause unexpected problems from job to job. If you machined a workpiece this morning that required spindle limiting, but this afternoon the next program does not, including the spindle limiting command and setting the rpm back to the machine's maximum will ensure that the spindle runs at the appropriate speeds.

Second, you may not expect a problem as the program is written, but as the program is actually run, the setup person may find it necessary to limit spindle rpm. Having the spindle limiting command in the program will make it very easy to modify the program.

Though we don't do so in our tool startup format, there may even be times when you'll need to change the maximum spindle speed from tool to tool. A workpiece in its raw state may be quite unbalanced and unstable prior to the roughing operations. But after roughing, it might be quite balanced and stable. You may be able to speed up the machining operations by allowing the spindle to go faster for finishing operations.

Choosing the appropriate spindle range

Some current model turning centers have but one spindle range. In this case, you will not be selecting spindle ranges. It's likely the machine has a high torque spindle drive motor that can drive the spindle with adequate force.

Most current model turning centers do have two or more spindle ranges. You can think of these ranges as like the transmission of an automobile. You get power (but limited speed) in the low range/s (gears) and you get speed (but limited power) in the high range/s (gears).

One rule of thumb is to perform roughing operations in the low range and finishing operations in the high range. While this is a good rule-of-thumb, there are times when following this rule can be somewhat inefficient. For your first few programs, use this rule-of thumb. In Lesson Sixteen, we do show how to more appropriately determine which spindle range you should select.

Which direction do you run the spindle?

You must, of course, have the spindle running in the appropriate direction in order for machining to occur. While this might sound like a very basic statement, turning center tooling is available in both right-hand and left-hand versions. Spindle direction is directly related to which hand of tooling you're using. In most cases, right hand tools require that the spindle be running in a forward (M03) direction and that left hand tools require the reverse (M04) direction. It is important that you specify the hand of tooling you're using on the setup sheet to eliminate the possibly for confusion in this regard. It's also quite important to confirm that the spindle is running in the right direction for each tool during the program's verification.

How do you check what each tool has done?

The optional stop command, programmed with M01, gives the setup person or operator the ability to stop the machine by turning on an on/off switch (labeled M01 or optional stop) on the operation panel. If the optional stop switch is currently on when the machine executes an M01 in the program, the machine will stop. For most machines, the spindle, coolant, and anything else currently activated will be stopped. If the switch is turned off, the machine will ignore M01 commands in the program (not stop the machine).

Our formats recommend that you include an M01 at the end of every tool. This will allow the setup person or operator to stop the machine after each tool by simply turning on the optional stop switch. This will be necessary during the program's verification and whenever trial machining.

A possible problem with initialized modes

You know that many modal CNC words are *initialized* (automatically instated at power-up). If you work in the inch mode, for example, G20 is initialized. And if you work exclusively in the inch mode, you shouldn't need to include a G20 word in any of your programs. You may have noticed that none of the example programs shown in this text include G20.

We're *assuming* that the inch mode is currently instated when these programs are run. As long as no one has changed the measurement system mode to metric mode (G21), these programs will run properly.

(By the way, what do you think will happen if the machine is somehow in the *metric mode* when a program written for the inch mode is run? The machine will interpret all measurement-system-related words (like coordinates, feedrate, and program zero assignment values) as being in *millimeters*. An X position of X1.0

[which is supposed to be 1.0 inch from program zero in X] will be interpreted as 1.0 millimeter from the program zero position. This has the effect of scaling down the program's motions by a scale factor of 25.4 [the conversion constant from inch to metric]. Obviously, this will not machine the workpiece properly – and will probably be very confusing to the setup person and operator.)

Most programmers don't like to assume that the machine is still in all of its initialized modes when their programs are run. So they include commands in the program (commonly called *safety commands*) to re-instate all of the initialized modes – even though the related words may be redundant. Consider this command:

N005 G99 G20 G23 (Select per-revolution feedrate mode, inch mode, cancel stored stroke limit)

While we have not discussed G23, rest assured that it involves a mode (safety zone canceled state) that must be in effect when your programs run.

How to use our given formats

In Lesson Sixteen, we show four types of program format:

1) Program startup format

2) Tool ending format

3) Tool startup format

4) Program ending format

Using the formats is easy. When you need to begin writing a new program, follow the program-startup format. Once the first tool is ready to machine, you're on your own to program its cutting motions. When you are finished programming the motion commands for the first tool, follow the tool-ending format. Then follow the tool-startup format to begin the second tool. You're on your own again to program the motion commands for the second tool. Continue toggling among tool-ending format, tool-startup format, and cutting motion commands until you are finished programming the motion commands for the last tool in the program. Then follow the program-ending format. Here is a flow chart of this process for a five-tool program.

Program startup
  Machining operations
Tool ending
Tool startup
  Machining operations
Tool ending
Tool startup
  Machining operations
Tool ending
Tool startup
  Machining operations
Tool ending
Tool startup
  Machining operations
Program ending

Again, the formats will work as a crutch until you have them memorized. While you'll have to plug in appropriate speeds, feeds, axis positions, and other pertinent information, the *basic structure* of the program will be easy – and it will remain the same for every program you write.

Key points for Lesson Fifteen:

- Safety and ease-of-use should take priority over efficiency for entry-level programmers.
- You must use a strict structure in your programs to help you become familiar with programming, to ensure consistency among programs, and to gain the ability to re-run tools.
- When you gain proficiency, you can modify the given formats to achieve more efficiency.
- Turning centers vary, meaning you may have to modify the given formats for your particular machine tools.
- You must be careful when determining the safe index position for turret indexing.
- Safety commands in your program can be used to ensure that the machine is still in all of its initialized modes.
- Using the formats shown in Lesson Sixteen will require that you toggle among program-startup format, cutting commands, tool-ending format, and program-ending format.

Talk to experienced people in your company...

... to learn more about the structure used in your CNC programs.

1) Ask to see a few programs actually being used on your company's turning centers. Compare them to the example programs shown in this text. Question an experienced person about any CNC words you don't recognize. Did the programmer use safety commands? Where is the turret index position?

2) Ask to see the procedure used to re-run a tool.

STOP! Do practice exercise number Fifteen in the workbook.

Lesson 16

Four Types Of Program Format

The formats shown in this lesson will keep you from having to memorize most of the words and commands needed in CNC programming. As you'll see, a large percentage of most programs is related to structure.

You know the reasons why programs must be formatted using a strict structure. In Lesson Sixteen, we're going to show the actual formats. We will show two sets of format, one for use when geometry offsets are used to assign program zero – the other for use when program zero is assigned in the program. We will also explain every word in each format in detail. And we'll show example programs that stress the use of our given formats. Finally, we'll show some efficiency-related limitations of our given formats.

There are four types of program format:

- Program startup format
- Tool ending format
- Tool startup format
- Program ending format

Any time you begin writing a new program, follow the program startup format. You can copy this *structure* to begin your program. The actual values of some words will change based on what you wish to do in your own program, but the structure will remain the same every time you begin writing a new program.

After writing the program startup format, you write the motions for the first tool's machining operations. When finished with the first tool motions, you follow the tool ending format. You then follow the tool startup format for the second tool and write the second tool's machining motion commands. From this point, you toggle among tool ending format, tool startup format, and machining motion commands until you are finished machining with the last tool. You then follow the program ending format.

One of the most important benefits of using these formats is that you *will not have to memorize anything*. You simply copy the structure of the format.

Remember that you should use the formats related to assigning program zero with geometry offsets unless your machine does not have geometry offsets.

Format for assigning program zero with geometry offsets

Once again, this should be your format set of choice. The only reason not to use this format is if your turning center does not have geometry offsets (as is the case with older machines).

Program startup format (using geometry offsets)

If you've been doing the programming exercises in this text and/or in the workbook, you may be surprised at how many of these words and commands you are already familiar with. There really won't be all that many new words to learn.

O**0001** (Program number)

N005 G99 G20 G23 (Select feed per revolution feedrate mode, select inch mode, cancel stored stroke limit)

N010 G50 S**4000** (Limit spindle speed to 4,000 rpm)

N015 **T0101 M41** (Index to first tool, instate geometry and wear offset number one, select spindle speed range)

N020 G**96** **S350** M**03** (Select spindle mode, speed, and activate spindle in the forward direction)

N025 G00 X**3.0** Z**0.1** M08 (Move to approach position, and turn on the coolant)

N030 G01 X... Z... F**0.015** (Select feed per revolution mode, first cutting motion must include feedrate)

This is the format you'll use whenever you begin writing a new program. The values shown in bold will vary from program to program. The program number (O0001 in our case) will change from program to program. So will the spindle speed limiting value (S4000 in line N010). In line N015, is possible that your program will start with a tool station other than number one – and you may need a different spindle range. In line N020, you may need a different speed mode, a different speed, and you may sometimes need to start the spindle in the reverse direction. In line N025, you'll likely send the tool to a different approach position. And in line N030, your first machining command, you'll probably need a different feedrate.

While there will surely be variations from program to program, the basic *structure* will remain the same for every program you write. Again, you'll follow this program startup format whenever you begin writing a new program. Use it as a crutch until you have it memorized.

After the program number (specified by the O word), line N005 is the safety command that ensures that important initialized modes are still active.

N010 specifies spindle limiting. Since most workpieces don't require spindle limiting, you normally specify the machine's maximum rpm (4,000 rpm in our case).

N015 specifies the turret index and selects the desired spindle range. The T word selects the appropriate turret station (again, your first tool may not always be station number one) and instates the geometry and wear offsets. We've assumed this tool will be run in the low spindle range, but you may want to begin with the high spindle range (M42 with most turning centers). We're also assuming that your machine has more than one spindle range. If it does not, you must omit the M41 (or M42) from this command.

N020 selects the spindle mode (css or rpm – G96 or G97), the appropriate speed with the S word, and turns the spindle on in the appropriate direction. We're assuming that you're using right hand tools. If you use a left hand tool, the spindle must be started in the reverse direction (M04).

N025 makes the rapid approach movement to within a small distance (usually 0.1 inch) from the surface being machined. We're assuming you want to run the tool with flood coolant. If you do not, leave the M08 word out of this command.

N030 performs the first machining command. The feedrate word (F) is part of the program startup format and must be included in the tool's first machining command.

Tool ending format (using geometry offsets)

N075 G00 **X8.0 Z6.0** (Move to turret index position)

N080 M01 (Optional stop)

These commands will be used to end every tool (but the last one) in your program.

N075 rapids the tool back to a safe turret index position. There is a presentation in Lesson Fifteen that shows how to select a safe turret index position.

N080 is the optional stop (M01) that gives the setup person or operator the ability to stop the machine (with the optional stop on/off switch) to see what this tool has done. This is very helpful during the program's verification – and whenever trial machining must be done.

Tool startup format (using geometry offsets)

N140 T**0202** M**42** (Index to next tool station, instate geometry and wear offset, select spindle speed range)

N145 G**97** S**600** M**03** (Select spindle mode, speed, and turn on spindle in forward direction)

N150 G00 **X0 Z0.1** M08 (Move to first position, turn on the coolant)

N155 G01 Z... **F0.010** (In first cutting movement, include a feedrate word)

This is the format you'll follow whenever beginning a tool in your program. Notice how similar the tool startup format is to program startup format. We've simply eliminated the program number, the safety command, and the spindle limiting command.

Program ending format (using geometry offsets)

N210 G00 **X8.0 Z6.0** (Move to tool change position)

N215 M30 (End of program)

Program ending format is almost identical to tool ending format. The only difference is that you end with M30 (end of program) instead of M01. M30 will turn off anything that's still running (spindle & coolant), rewind the program to the beginning for the next workpiece, and stop the cycle.

Format for assigning program zero in the program with G50

Again, this is the older way of assigning program zero. The only reason for using this format is that your machine does not have geometry offsets.

A reminder about the program's starting point and tool change position

When assigning program zero in the program, *the machine must be properly aligned with the program.* We recommend that novice programmers use the machine's zero return position as the program's starting position and turret index position – and our format reflects this.

While the machine's zero return position may not be the most efficient location at which to change tools (due to the distance the turret may have to travel), it is a very safe place to change tools. Two axis origin lights will come on, indicating that the machine is positioned properly to start the program.

Program startup format (assigning program zero in the program)

O0001 (Program number)

N005 G99 G20 G23 G40 (Select feed per revolution feedrate mode, select inch mode, cancel stored stroke limit)

N010 G28 U0 W0 (Send X and Z axes to zero return position)

N015 G50 **X**______ **Z**______ S**4000** (Assign program zero, limit spindle to 4000 rpm)

N020 T**0100** M**41** (Index to the first tool, select spindle speed range)

N025 G**96** S**350** M**03** (Select spindle mode, speed, and turn spindle on in forward direction)

N030 G00 X**3.0** Z**0.1** T**0101** M08 (Move to approach position, instate wear offset, and turn on the coolant)

N035 G01 X... Z... F**0.015** (First cutting motion must include feedrate)

This is the format you'll use whenever you begin writing a new program. The values shown in bold will vary from program to program. The program number (O0001 in our case) will change from program to program. So will the spindle speed limiting value (S4000 in line N015). In line N020, is possible that your program will start with a tool station other than number one – and you may need a different spindle range. In line N025, you may need rpm mode, a different speed, and you may want to start the spindle in the reverse direction. In line N030, you'll likely send the tool to a different approach position. And in line N035, your first machining command, you'll probably need a different feedrate.

While there will surely be variations from program to program, the basic *structure* will remain the same for every program you write. Again, you'll follow this program startup format whenever you begin writing a new program. Use it as a crutch until you have it memorized.

After the program number (specified by the O word), line N005 is the safety command that ensures that important initialized modes are still active.

N010 commands the machine to go to its zero return position in both axes (this is the program's starting position). This is safety command. We're confirming that the machine is where it should be before the program starts. Note that if the machine is already at the zero return position, this command will be ignored.

N015 does two things. It tells the machine the location of program zero (with X and Z) and it specifies the spindle limiter (with the S word). Notice that we've left the X and Z values blank. You will not know these values until the setup is made and the setup person measures the program zero assignment values for this tool (see Lessons Six and Seven for more information about program zero assignment). Once program zero assignment values have been determined, the X and Z values of this command must be modified (at the machine) to include this tool's program zero assignment values. N015 also specifies spindle limiting. Since most workpieces don't require spindle limiting, you normally specify the machine's maximum rpm in this command (4,000 rpm in our case).

N020 specifies the turret index and selects the desired spindle range. Notice that the T word in line N020 is simply indexing the turret – it is not instating the wear offset. We will wait until the first motion command to instate the wear offset (coming up in line N030). Line N020 also specifies the desired spindle range. We've assumed this tool will be run in the low spindle range, but you may want to begin with the high spindle range (M42 with most turning centers). We're also assuming that your machine has more than one spindle range. If it does not, you must omit the M41 (or M42) from this command.

N025 selects the spindle mode (css or rpm – G96 or G97), the appropriate speed with the S word, and turns the spindle on in the appropriate direction. We're assuming that you're using right hand tools. If you use a left hand tool, the spindle must be started in the reverse direction (M04).

N030 makes the rapid approach movement to within a small distance (usually 0.1 inch) from the surface being machined. We're also instating the wear offset for this tool in line N030. We're assuming you want to run the tool with flood coolant. If you do not, leave the M08 word out of this command.

N035 performs the first machining command. The feedrate word (F) is part of the program startup format and must be included in the tool's first machining command.

Tool ending format (assigning program zero in the program)

N075 G28 U0 W0 T0 (Go back to zero return position and cancel offset)

N080 M01 (Optional stop)

These commands will be used to end every tool (but the last one) in your program.

N075 commands the machine to go to its zero return position (the starting position for the program) – and for each tool. It also cancels the wear offset. The word T0 will cancel any tool's wear offset.

N080 is the optional stop (M01) that gives the setup person or operator the ability to stop the machine (with the optional stop on/off switch) to see what this tool has done. This is very helpful during the program's verification – and whenever trial machining must be done.

Tool startup format (assigning program zero in the program)

N140 G50 **X**______ **Z**______ (Set program zero)

N145 T**0200** **M42** (Index to station number two, select spindle speed range)

N150 G**97** S**600** M**03** (Select spindle mode, speed, and turn on spindle in the forward direction)

N155 G00 **X0** **Z0.1** T**0202** M08 (Move to first position, instate tool offset, and turn coolant on)

N160 G01 Z... F**0.010** (In first cutting movement, include feedrate)

This is the format you'll follow whenever beginning a tool in your program. Notice how similar the tool startup format is to program startup format. We've simply eliminated the program number, the safety command, the zero return command, and spindle limiting.

Program ending format (assigning program zero in the program)

N210 G28 U0 W0 T0 (Rapid back to starting point and cancel offset)

N215 M30 (End of program)

Program ending format is almost identical to tool ending format. The only difference is that you end with M30 (end of program) instead of M01. M30 will turn off anything that's still running (spindle & coolant), rewind the program to the beginning for the next workpiece, and stop the cycle.

Understanding the formats

While we place messages in parentheses for each command and give a short description of each command to help you understand what was going on in each format, you may still have some questions about some of the words used in the formats. Here we list the various words involved in the format and describe them. Note that they are in numerical order for quick reference.

G Words

G00 - This is the rapid positioning word. Whenever you wish motion to occur at the machine's fastest possible rate, the G00 is the command to use. G00 is discussed at length in Lesson Ten.

G01 F___ - Though it may not be readily apparent, the feedrate in the first cutting command is part of the program start-up and tool start-up format. To make each tool independent from the rest of the program, the feedrate must be included for each tool. G01 is discussed in Lesson Ten.

G28 U0 W0 - Used only when program zero is assigned in the program, this command sends the machine to its zero return position in both X and Z. G28 is a two step command. First, it will send the turret to an *intermediate position.* Second, it will send the turret to its zero return position in X and Z (together). With U0 and W0, we're specifying that the intermediate position is incrementally zero in X and zero in Z from the machine's current position (the machine simply stays where it is in the first step of this G28 command). For all intents and purposes, this command as written tells the machine to *go straight to the zero return position in X and Z.*

G50 - This word has two applications. First, it specifies spindle limiting (see description of spindle limiting in Lesson Fifteen). Second, and only when program zero is assigned in the program, G50 is used to specify each tool's program zero assignment values. Program zero assignment is discussed in Lessons Six and Seven.

G96 and G97 - These words specify the spindle mode. G96 selects the constant surface speed mode, and G97 selects the rpm mode. Any S word following a G96 will be used as surface feet per minute (or meters per minute in the metric mode). Any S word following a G97 will be taken as an rpm. (See Lesson Two for more information.)

M Words

M01 - This word is an optional stop. It works in conjunction with an on/off switch on the control panel labeled optional stop (or M01). When the machine executes the M01 word, it looks to the position of the on/off switch. If the switch is on, the machine will stop executing the program. The spindle, coolant, and anything else still running will be turned off. The machine will remain in this state until the cycle is activated again, by pressing the cycle start button. Then the control will continue executing the program from the point of the M01. If the switch is off when the machine executes an M01, the machine will ignore the M01 command entirely. It will be as if the M01 was not in the program at all.

The purpose for including an M01 in the tool ending format is to give the setup person or operator a way to stop the machine at the completion of each tool. Many times, especially during a program's verification and when trial machining, it will be necessary to stop the machine after each tool to confirm that the tool has done its job correctly. In some cases, if the tool has not machined as intended, the subsequent tools in the program may be damaged. For example, say a hole is to be drilled and tapped. Of course, the drilled hole must be deep enough for the tap to reach its desired depth. After the drill had machined the workpiece, if the optional stop switch is on, the operator can rest assured the machine will stop when the drill has finished. They can then check to confirm the drill had machined the hole deep enough for the tap.

Note that there is another program stopping M code called program stop. The program stop command is specified by an M00 word and forces the machine to stop no matter what. There is no on/off switch involved and the operator will have no option; the machine will stop when an M00 is executed.

M03 and M04 - These M codes are used on spindle on in a forward (M03) or reverse (M04) direction. Generally speaking, the M03 is used with right hand tools as long as the machining direction is toward the chuck in the Z axis. If using left hand tools, the spindle must be started in the reverse direction (M04).

You may have noticed that we never commanded the spindle to stop. Though there is an M code available to turn the spindle off (**M05**), it is wise to leave the spindle running, even during turret indexing. There is no reason to turn the spindle off. To do so would be a waste of cycle time and electricity. The **M30** command at the end of the program will turn the spindle off when the program is finished.

M08 - This M code is used to turn on the (flood) coolant. Note that, like the spindle, coolant is never turned off. Coolant will be turned off automatically during a turret index. Also, the **M30** program ending word will turn the coolant off at the completion of the program. If for some reason you must program the coolant to be turned off, an **M09** is used for this purpose.

M30 - This is a program ending word. It tells the control to turn off anything still running (spindle coolant, etc.) and rewind the memory back to the beginning. Then the control stops.

M41 and M42 - Many turning centers have more than one spindle range. This allows power for heavy machining operations in the low range and high speed for finishing operations in the high range. (See Lessons Two and Fifteen for more information.) Two very common M codes to handle the spindle range changes are **M41** and **M42** (**M41** for low range and **M42** for high range). We must warn you, however, that some machines use different M codes to control spindle range changes. You must check in your machine tool builder's programming manual to find the M codes related to spindle ranges for your machine/s.

Other M Words Related to Turning Centers

While not directly related to program formatting, there are other M codes of concern to a turning center programmer. There are several machine functions that can be activated by M codes on turning centers. These things include tailstock (body and quill), chuck jaws (open and close), bar-feeder (if so equipped), and possibly even chip conveyer. Again, you must check the machine tool builder's programming manual to find the list of M codes.

Other Words In The Format

End of Block Word - As discussed in Lesson Eight, every CNC command must end with an *end of block* (EOB) word. This character is entered on a computer when you press the *enter key* and appears on the display screen on most Fanuc controls as a semi-colon.

O Word - Most controls allow more than one program to be stored in the control's memory. The O word assigns the program's number. The operator can easily scan from one program to another by the O word. It will always be the very first word of the program. See Lesson Eight for more information.

S Word - The S word tells the control the desired rpm or sfm (meters per minute in metric mode) for the spindle. The S word, by itself, does not actually turn on the spindle, the **M03** or **M04** does that. See Lesson Two for more information.

T Word - The T word tells the turning center three things, the tool station number, the geometry offset number, and the wear offset number to be used with the tool. The T word is a four digit word. The first two digits tell the control the turret station number and the geometry offset number. The second two digits tell the control the offset number to instate.

F Word - The F word tells the machine the desired feedrate. There are two ways of commanding feedrate, feed per minute and feed per revolution. By far, feed per revolution is the more appropriate way to specify feedrate on turning centers. The only time we recommend using the feed per minute mode is when you intend to make a feedrate movement with the spindle stopped. **G98** specifies the feed per minute mode and **G99** specifies the feed per revolution mode. **G99** is initialized when the power is turned on, so it is not absolutely necessary in the program – but we do include it in the program startup format. See Lesson Two for more information.

Example programs showing format for turning centers

Figure 5.2 is the drawing we use to stress turning center format.

Drawing for example programs that stress the use of program formatting

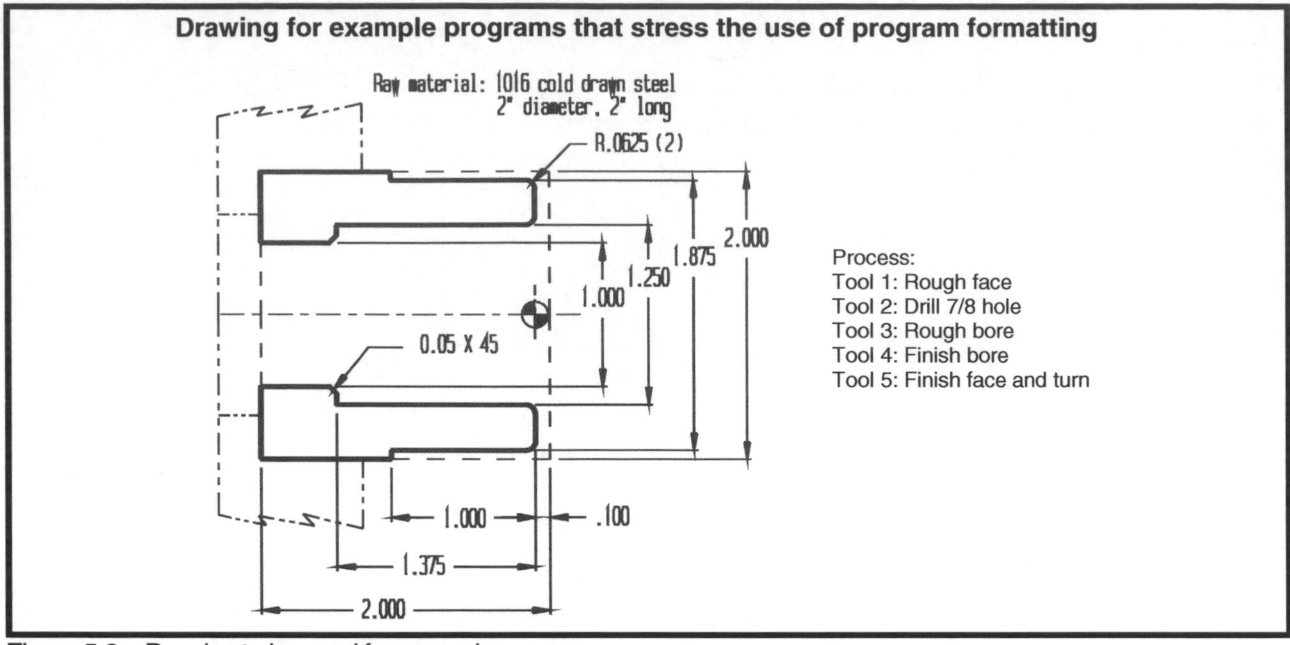

Process:
Tool 1: Rough face
Tool 2: Drill 7/8 hole
Tool 3: Rough bore
Tool 4: Finish bore
Tool 5: Finish face and turn

Figure 5.2 – Drawing to be used for example programs

Example when assigning program zero with geometry offsets

Program:

O0002 (Program number)
(ROUGH FACING TOOL)
N002 G99 G20 G23 (Ensure that initialized modes are still in effect)
N004 G50 S5000 (Limit spindle speed to machine's maximum)
N005 T0101 M41 (Index turret, select spindle range)
N010 G96 S400 M03 (Start spindle in forward direction at 400 sfm)
N015 G00 X2.2 Z0.005 M08 (1) (Rapid to starting position, start coolant)
N020 G01 X-0.062 **F0.012** (2) (Select per revolution feedrate mode, face workpiece at 0.012 ipr)
N025 G00 Z0.1 (3) (Rapid away)
N030 X6.0 Z5.0 (Rapid to tool change position)
N035 M01 (Optional stop)

(7/8 DRILL)
N040 T0202 M41 (Index turret, select spindle range)
N045 G97 S354 M03 (Start spindle forward at 354 rpm)
N050 G00 X0 Z0.1 M08 (4) (Rapid into position, start coolant)
N055 G01 Z-2.2 **F0.008** (5) (Drill hole at 0.008 ipr)
N060 G00 Z0.1 (4) (Rapid out of hole)
N065 X6.0 Z5.0 (Rapid to tool change position)
N070 M01 (Optional stop)

(3/4 ROUGH BORING BAR)
N075 T0303 M42 (Index turret, select spindle range)
N080 G96 S350 M03 (Start spindle forward at 350 sfm)
N085 G00 X1.19 Z0.1 M08 (6) (Rapid into position, start coolant)

N090 G01 Z-1.37 **F0.007** (7) (Begin boring operation at 0.007 ipr)
N095 X0.875 (8)
N100 G00 Z0.1 (9) (Rapid out of hole)
N105 X5.0 Z6.0 (Rapid to tool change position)
N110 M01 (Optional stop)

(3/4 FINISH BORING BAR)
N115 T0404 M42 (Index turret, select spindle range)
N120 G96 S400 M03 (Start spindle forward at 400 sfm)
N125 G00 X1.375 Z0.1 M08 (10) (Rapid into position, start coolant)
N130 G01 Z0 **F0.005** (11) (Begin finish boring at 0.005 ipr)
N135 G02 X1.25 Z-0.0625 R0.0625 (12)
N140 G01 Z-1.375 (13)
N145 X1.1 (14)
N150 X1.0 Z-1.425 (15)
N155 Z-2.0 (16)
N160 G00 X0.8 (17)
N165 Z0.1 (18) (Rapid out of hole)
N170 X6.0 Z5.0 (Rapid to tool change position)
N175 M01 (Optional stop)

(FINISH FACE AND TURN TOOL)
N180 T0505 M42 (Index turret, select spindle range)
N185 G96 S450 M03 (Start spindle forward at 450 sfm)
N190 G00 X2.075 Z0 M08 (19) (Rapid into position, start coolant)
N195 G01 X1.05 **F0.006** (20) (Start finish facing and turning at 0.006 ipr)
N200 G00 Z0.1 (21)
N205 X1.75 (22)
N210 G01 Z0 (23)
N215 G03 X1.875 Z-0.0625 R0.0625 (24)
N220 G01 Z-1.0 (25)
N225 X2.2 (26)
N230 G00 X6.0 Z5.0 (Rapid back to tool change position)
N235 M30 (End of program)

All of the commands and words shown in bold are related to format. The structure for these commands can be copied from our given formats. You'll only be on your own to develop the cutting commands for each tool. Truly, a great percentage of most programs is related to format.

Where are the restart commands?

We've discussed re-running tools several times. And you know that the program must be properly structured in order to re-run tools (which programs written with our given formats are). In order to re-run a cutting tool, you must know the *restart command* for the tool (which is also called the *pickup block*). You will scan to this command before activating the cycle.

To re-run the first tool, of course, you can simply run the program from the beginning. For other tools, finding the restart command with our format is still pretty simple. It is the command that includes the turret index (T word). In the program above, the restart commands are N040 for tool two, N075 for tool three, N115 for tool four, and N180 for tool five.

Example program when assigning program zero in the program

Program:

O0002 (Program number)

(ROUGH FACING TOOL)

N005 G99 G20 G23 (Ensure that initialize modes are still in effect)

N010 G28 U0 W0 (Send machine to zero return position)

N015 G50 X______ Z______ S5000 (Assign program zero, limit spindle speed to machine's maximum)

N020 T0100 M41 (Index turret, select spindle range)

N025 G96 S400 M03 (Start spindle in forward direction at 400 sfm)

N030 G00 X2.2 Z0.005 T0101 M08 (1) (Rapid to position, instate offset, start coolant)

N035 G99 G01 X-0.062 **F0.012** (2) (Select per revolution feedrate mode, face workpiece at 0.012 ipr)

N040 G00 Z0.1 (3) (Rapid away)

N045 G28 U0 W0 T0 (Rapid to zero return position, cancel offset)

N050 M01 (Optional stop)

(7/8 DRILL)

N055 G50 X______ Z______ (Assign program zero)

N060 T0200 M41 (Index turret, select spindle range)

N065 G97 S354 M03 (Start spindle forward at 354 rpm)

N070 G00 X0 Z0.1 T0202 M08 (4) (Rapid into position, instate offset, start coolant)

N075 G01 Z-2.2 **F0.008** (5) (Drill hole at 0.008 ipr)

N080 G00 Z0.1 (4) (Rapid out of hole)

N085 G28 U0 W0 T0 (Rapid to zero return position, cancel offset)

N090 M01 (Optional stop)

(3/4 ROUGH BORING BAR)

N095 G50 X______ Z______ (Assign program zero)

N100 T0300 M42 (Index turret, select spindle range)

N105 G96 S350 M03 (Start spindle forward at 350 sfm)

N110 G00 X1.19 Z0.1 T0303 M08 (6) (Rapid into position, instate offset, start coolant)

N115 G01 Z-1.37 **F0.007** (7) (Begin boring operation at 0.007 ipr)

N120 X0.875 (8)

N125 G00 Z0.1 (9) (Rapid out of hole)

N130 G28 U0 W0 T0 (Rapid to zero return position, cancel offset)

N135 M01 (Optional stop)

(3/4 FINISH BORING BAR)

N140 G50 X______ Z______ (Assign program zero)

N145 T0400 M42 (Index turret, select spindle range)

N150 G96 S400 M03 (Start spindle forward at 400 sfm)

N155 G00 X1.375 Z0.1 T0404 M08 (10) (Rapid into position, instate offset, start coolant)

N160 G01 Z0 **F0.005** (11) (Begin finish boring at 0.005 ipr)

N165 G02 X1.25 Z-0.0625 R0.0625 (12)

N170 G01 Z-1.375 (13)

N175 X1.1 (14)

N180 X1.0 Z-1.425 (15)
N185 Z-2.0 (16)
N190 G00 X0.8 (17)
N195 Z0.1 (18) (Rapid out of hole)
N200 G28 U0 W0 T0 (Rapid to zero return position, cancel offset)
N205 M01 (Optional stop)

(FINISH FACE AND TURN TOOL)
N210 G50 X______ Z______ (Assign program zero)
N215 T0500 M42 (Index turret, select spindle range)
N220 G96 S450 M03 (Start spindle forward at 450 sfm)
N225 G00 X2.075 Z0 T0505 M08 (19) (Rapid into position, instate offset, start coolant)
N230 G01 X1.05 **F0.006** (20) (Start finish facing and turning at 0.006 ipr)
N235 G00 Z0.1 (21)
N240 X1.75 (22)
N245 G01 Z0 (23)
N250 G03 X1.875 Z-0.0625 R0.0625 (24)
N255 G01 Z-1.0 (25)
N260 X2.2 (26)
N265 G28 U0 W0 T0 (Rapid to zero return position, cancel offset)
N270 M30 (End of program)

All of the commands and words shown in bold are related to format. The structure for these commands can be copied from our given formats. You'll only be on your own to develop the cutting commands for each tool. Truly, a great percentage of most programs is related to format.

Where are the restart commands?

We've discussed re-running tools several times. And you know that the program must be properly structured in order to re-run tools (which programs written with our given formats are). In order to re-run a cutting tool, you must know the *restart command* for the tool (which is also called the *pickup block*). You will scan to this command before activating the cycle.

To re-run the first tool, of course, you can simply run the program from the beginning. For other tools, finding the restart command with our format is still pretty simple. It is the command that includes the G50 word. In the program above, the restart commands are N055 for tool two, N095 for tool three, N140 for tool four, and N210 for tool five. The machine must, of course, be at the zero return position when a tool is ru-run.

Suggestions for cycle time improvements.

If you are being exposed to these turning center formats for the very first time, you'll need to gain some experience writing programs before you'll be able to fully master the techniques we show in this presentation. Please skip this presentation for now, and rest assured that what you've learned so far will allow you to program and machine acceptable workpieces.

But as we've pointed out, our recommended formats will not be very efficient. Here we offer some suggestions to speed up the machine's execution of your CNC programs. After you've gained some experience on the machine, come back and read this presentation.

Combine M codes in motion commands

Our recommended formats have you doing one thing at a time per command. First you start the spindle, then you rapid into position. When feasible, combining functions into one command will reduce program execution time. For most turning centers, the command

```
N005 G00 G96 X1.0 Z0.1 S500 M03
```

will execute faster than the commands

```
N005 G96 S500 M03
N010 G00 X1.0 Z0.0
```

With the latter, most turning centers will wait until the spindle has come up to speed in line N005 before they start the motion line N010.

In similar fashion, it's wise to turn off the spindle during the last tool's retract movement to the tool change position (with M05). Why wait for the M30 to begin turning off the spindle?

Minimize spindle dead time

Constant surface speed is a great feature. It makes programming spindle speed easy (in sfm or meters per minute), it improves workpiece finish, and it maximizes tool life. For as good a feature as constant surface speed is, it can be a dreadful cycle time waster if no concern is given to its use and programming. Many programmers are taught in basic texts (including this one) to format their turning center programs something like this.

```
O0001 (Example for constant surface speed)
N002 G99 G23 G40 (Ensure initialized modes are still in effect)
N004 G50 S5000 (Limit spindle speed to machine's maximum)
N005 T0101 (Index turret to rough face and turn tool)
N010 G96 S600 M03 (Start spindle at 600 sfm)
N015 G00 X1.35 Z0.005 M08 (Rapid to approach position, turn on coolant)
N020 G99 G01 X-0.06 F0.012 (Rough face)
N025 G00 Z0.1 (Rapid away)
N030 X1.25 (Rapid to roughing approach position)
N035 G71 P040 Q075 U0.04 W0.005 F0.012 (Rough turn part)
N040 G00 X0.44 (Rapid to finish pass definition first position)
N045G01 Z0 (Come flush with face)
N050 X0.5 Z-0.03 (Chamfer end)
N055 Z-0.5 (Turn first diameter)
N060 X0.69 (Come up second face)
N065 X0.75 Z-0.53 (Form chamfer)
N070 Z-1.0 (Turn second diameter)
N075 X1.25 (Come up to stock diameter)
N080 X8.0 Z5.0 (Rapid to tool change position)
N085 M01 (Optional stop)
N090 T0202 (Index to finish face and turn tool)
N095 G96 S700 M03 (Start spindle at 700 sfm)
N100 G00 X0.6 Z0 M08 (Rapid up to first face)
N105 G01 X-0.06 F0.005 (Finish face)
N110 G00 Z0.1 (Rapid away)
N115 X1.25 (Rapid to beginning of finish turn)
N120 G70 P040 Q075 F0.005 (Finish turn)
N125 G00 X8.0 Z5.0 (Rapid to tool change position)
```


N130 M30 (End of program)

While we have not yet described multiple repetitive cycles G71 and G70 (these cycles are presented in Lesson Eighteen), you should be able to follow the points we make about this program having to do with spindle speed control.

Though this program's format is quite easy for a beginning programmer to understand (one speed specification per tool), it is not very efficient - especially when it comes to spindle activation. No turning center spindle can change speed *instantaneously*. Notice that the programmer has chosen an 8.0 diameter as the tool change position. In line N010, when the spindle is activated, it will start at 287 rpm (3.82 times 600 sfm divided by the 8.0 diameter). Even with today's fastest machines, this will take a second or two.

In line N015 the machine will move into position (a motion of about 5.0 inches). During this motion, the spindle will accelerate to 1,697 rpm (3.82 times 600 sfm divided by the 1.35 diameter). Depending upon your machine's spindle response time, it is quite likely that the motion will occur before the spindle finishes accelerating. Say the machine has a rapid rate of 1,200 ipm. In this case, the approach motion will take less than one second. It is likely that the spindle acceleration time will be well over four seconds (though spindle acceleration/deceleration response times vary dramatically from one turning center to another). The machine will wait while the spindle accelerates up to speed.

During the retract motion in line N080, the spindle will slow to 287 rpm, taking another 3-5 seconds. And the same time consuming techniques are used in lines N095, N100, and N125. With this format the programmer even waits for the M30 (end of program command) to turn the spindle off. Again, while this format is very easy to program and understand, it is somewhat wasteful. Here is another version of this program that is much more efficient when it comes to spindle acceleration and deceleration.

O0001 (Example for constant surface speed)
N002 G20 G23 G40 (Ensure initialized modes are still in effect)
N004 G50 S5000 (Limit spindle to machine's maximum)
N005 T0101 (Index turret to rough face and turn tool)
N010 **G96** G00 X1.35 Z0.005 **S600 M03** (Start spindle at 600 sfm and rapid to approach position)
N015 M08 (Turn on coolant)
N020 G01 X-0.06 F0.012 (Rough face)
N025 G00 Z0.1 (Rapid away)
N030 X1.25 (Rapid to roughing approach position)
N035 G71 P040 Q075 U0.04 W0.005 F0.012 (Rough turn part)
N040 G00 X0.44 (Rapid to finish pass definition first position)
N045G01 Z0 (Come flush with face)
N050 X0.5 Z-0.03 (Chamfer end)
N055 Z-0.5 (Turn first diameter)
N060 X0.69 (Come up second face)
N065 X0.75 Z-0.53 (Form chamfer)
N070 Z-1.0 (Turn second diameter)
N075 X1.25 (Come up to stock diameter)
N080 **G97 S4456** X8.0 Z5.0 (Rapid to tool change position)
N085 M01 (Optional stop)
N090 T0202 (Index to finish face and turn tool)
N095 **G97** G00 X0.6 Z0 **S4456 M03** (Start spindle at 700 sfm and rapid up to position)
N100 **G96 S700 M08** (Turn on coolant)

```
N105 G01 X-0.06 F0.005 (Finish face)
N110 G00 Z0.1 (Rapid away)
N115 X1.25 (Rapid to beginning of finish turn)
N120 G70 P040 Q075 F0.005 (Finish turn)
N125 G00 X8.0 Z5.0 M05 (Rapid to tool change position, turn off spindle)
N130 M30 (End of program)
```

This time in line **N010**, the spindle is started *during* the approach, making the movement internal to the spindle acceleration. In line **N080**, the rpm mode (**G97**) is temporarily selected to keep the spindle from slowing during the machines movement to the tool change position. Note that instead of slowing down, it is actually going to speed up to 4,456 rpm, the speed needed for the *next tool* (3.82 times 700 sfm divided by 0.6 diameter). And in line **N125**, the spindle is being stopped during the retract motion (the motion will be internal to the spindle stopping).

While this program is much more efficient than the first attempt, it is still not as efficient as it could be. Note that there will still be some dead time during the enormous speed changes in lines **N010**, **N080**, and **N125**. Unfortunately, you almost have to be monitoring the cycle at the machine to further improve this cycle. Here is the third and final version of this program after fine tuning was done to virtually eliminate the effects of spindle acceleration/deceleration on cycle time.

```
O0001 (Example for constant surface speed)
N002 G20 G23 G40 (Ensure initialized modes)
N004 G50 S5000 (Limit speed to machine's maximum)
N005 G97 S1400 T0101 M03 (Index turret to rough face and turn tool)
N010 G00 X1.35 Z0.005 S1600 M03 (Start spindle at 600 sfm and rapid to approach position)
N015 G96 S600 M08 (Turn on coolant)
N020 G01 X-0.06 F0.012 (Rough face)
N025 G00 Z0.1 (Rapid away)
N030 X1.25 (Rapid to roughing approach position)
N035 G71 P040 Q075 U0.04 W0.005 F0.012 (Rough turn part)
N040 G00 X0.44 (Rapid to finish pass definition first position)
N045G01 Z0 (Come flush with face)
N050 X0.5 Z-0.03 (Chamfer end)
N055 Z-0.5 (Turn first diameter)
N060 X0.69 (Come up second face)
N065 X0.75 Z-0.53 (Form chamfer)
N070 Z-1.0 (Turn second diameter)
N075 X125 (Come up to stock diameter)
N080 G97 S2400 X8.0 Z5.0 (Rapid to tool change position)
N085 M01 (Optional stop)
N090 G97 T0202 S4100 (Index to finish face and turn tool)
N095 G00 X0.6 Z0 S4456 M03 (Start spindle at 700 sfm and rapid up to position)
N100 G96 S700 M08 (Turn on coolant)
N105 G01 X-0.06 F0.005 (Finish face)
N110 G00 Z0.1 (Rapid away)
N115 X1.25 (Rapid to beginning of finish turn)
```

N120 G70 P040 Q075 F0.005 (Finish turn)

N125 G00 X8.0 Z5.0 **M05** (Rapid to tool change position, turn off spindle)

N130 M30 (End of program)

Notice that in line N005 (the turret index), we begin the spindle acceleration. Depending upon your turret index time, you will be able to change speeds during turret indexes. In our case, the turret indexes just slightly faster than the spindle can accelerate to its required speed. So in line N010, as the machine moves into position, we accelerate the spindle the rest of the way. Notice that even the coolant on function (in line N015) is including some of the spindle acceleration time. The same technique is used during tool changing. During the retract movement in line N080, the turret index in line N090, and the approach in line N095, the costly effect of spindle acceleration on cycle time is eliminated – or at least dramatically reduced.

As stated, you almost have to be monitoring your cycle before you can take full advantage of this technique. You can use lot sizes to help you judge the wisdom of applying this technique. If you have small lots (fewer than 20 parts) with relatively short cycle times, you may want to continue programming as shown in the very first example (it's easiest and can be programmed the quickest). If you have higher quantities, the second technique will shorten cycle time and will not add too much to programming time. If your quantities are very high, you'll want to use the last method and do whatever it takes to get every last second out of your cycle.

Note that there are two more detrimental effects of allowing your turning center's spindle to change speeds so often that have nothing to do with cycle time. First, excessive spindle speed changes can be harmful to your machine's spindle drive system, meaning you can reduce normal wear and tear on the machine by applying these techniques. Second, spindle speed changes require electricity (to decelerate as well as to accelerate). You can save electricity by using this technique.

Efficiently programming spindle range changes

In order to select spindle ranges efficiently, you must understand your machine's spindle power characteristics. In your machine tool builder's programming, operation, or maintenance manual, you will find a power curve chart similar to the one shown in Figure 5.3.

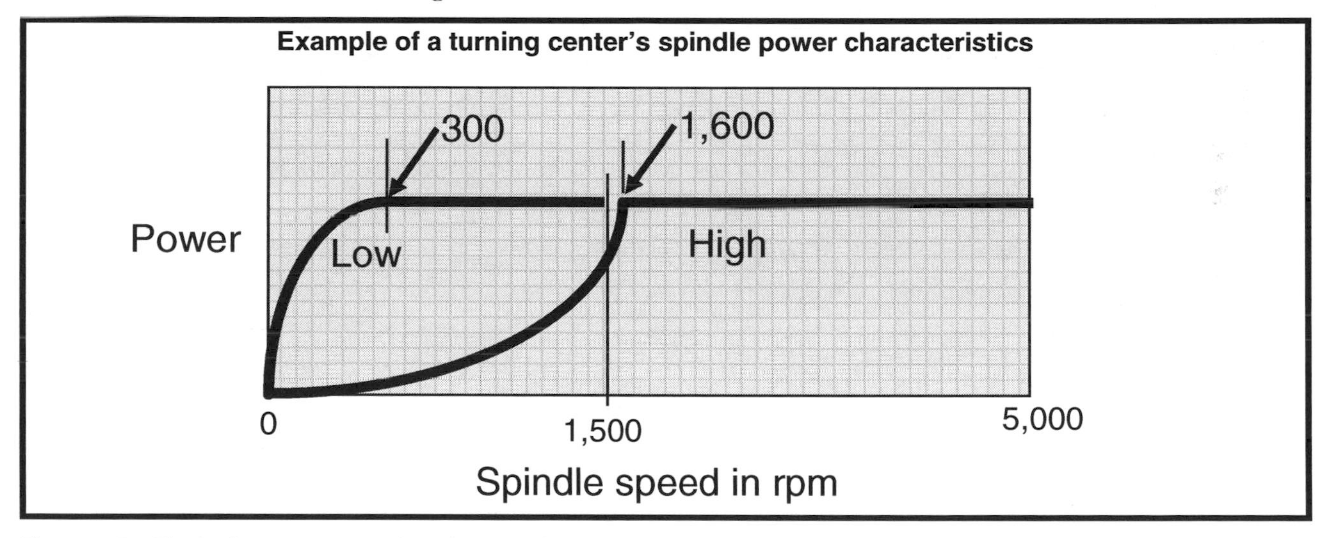

Figure 5.3 – Typical power curve chart for a turning center's main spindle

Notice that this turning center has two spindle ranges (low and high) and a maximum spindle speed of 5,000 rpm. The low range runs from 0-1,500 rpm. The high range runs from 0-5,000 rpm. With most turning centers, the high range completely overlaps the low range since many finishing operations require the full rpm spectrum.

From the chart, notice that this particular machine reaches full power in the low range at 300 rpm. It reaches full power in the high range at 1,600 rpm. When the spindle runs at under 1,600 rpm in the high range, the machine won't have much power. Consider this scenario for the machine depicted in Figure 5.3:

You have a small workpiece being machined from a piece of 1 inch diameter bar stock. All of your tools are made from carbide, and your slowest speed is 600 sfm. When you apply the rpm calculating formula (rpm = 3.82 times sfm divided by diameter), you find that all machining operations require over 2290 rpm. For this workpiece, applying the rule of thumb to rough in the low range will be wasteful. Since the low range peaks out at 1,500 rpm, the appropriate speed for all roughing operations will not be reached. And since time is inversely proportional to spindle speed (because feedrate is specified in per revolution fashion), program execution time will be longer. Since the machine reaches full power in the high range at 1,600 rpm, there will be sufficient power to rough this workpiece in the *high range* and the appropriate spindle speeds will be used.

For some workpieces, just the opposite may be true. Actually switching spindle ranges during a roughing operation may reduce cycle time. Consider, for example, a large 8 inch diameter shaft that must be rough turned down to 1 inch. Say you wish to rough the workpiece at 800 sfm. At the 8 inch diameter, this equates to 382 rpm. Since there is very little power available in the high range at this slow speed, the low range must be selected to get the needed power for the initial roughing passes. But as the workpiece gets smaller in diameter, the required rpm will increase. When the diameter reaches about 2 inches, the spindle will peak out at the top end of the low range (1,500 rpm). Yet there is still more roughing to do. In this case, switching to the high range (where now there is sufficient power), may actually minimize program execution time. Whether this saves any time has a lot to do with how long the turning center takes to change ranges. Time saved with efficient machining could be lost by the act of switching ranges.

As you can see, you really have to know each turning center's specific spindle power characteristics as well as how long it takes to change ranges to make wise decisions when it comes to spindle range selection. This may be difficult for beginners, and you may wish to simply apply the rule-of-thumb (rough in the low range, finish in the high range) until you become more proficient. Just remember that your spindle range selections can have an impact on program execution time.

Minimize spindle reversals

Spindle reversals take time. Yet many programmers and setup people pay little attention to the style of cutting tools they use (right or left hand). As stated earlier, right hand tools commonly require forward direction (M03) while left hand tools commonly require reverse spindle direction (M04). If your tools toggle back and forth between right and left hand, there will be several time-consuming spindle reversals in the program. When feasible, stick to one style of tooling.

Key points for Lesson Sixteen:

- There are four kinds of format to help you write programs: program startup format, tool ending format, tool startup format, and program ending format.
- You are only on your own to come up with the cutting motion commands for each tool – a large percentage of most programs is related to the program's structure.
- The startup formats include all CNC words to make each tool independent of the rest of the program – so tools can be re-run.
- As long as you assign program zero with geometry offsets, the restart command for a tool is the turret index command (with the T word).
- Tool ending format includes an optional stop word (M01) so setup people and operators can easily stop the machine after each tool to see what the tool has done (by turning on the optional stop switch).
- The given formats are not very efficient. Once you've gained some experience you can come back to this lesson to learn how to optimize program execution time.

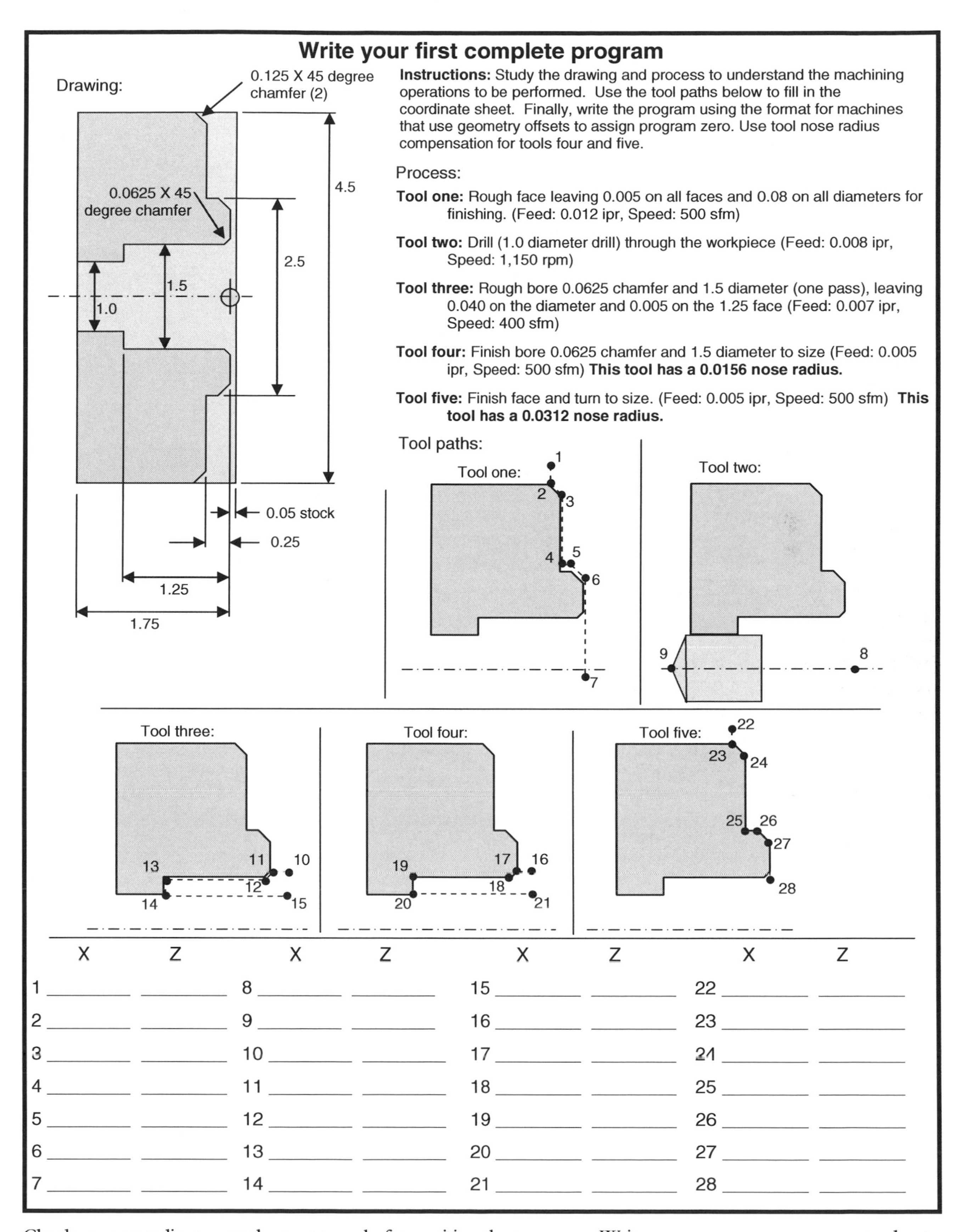

Write your first complete program

Drawing:

Instructions: Study the drawing and process to understand the machining operations to be performed. Use the tool paths below to fill in the coordinate sheet. Finally, write the program using the format for machines that use geometry offsets to assign program zero. Use tool nose radius compensation for tools four and five.

Process:

Tool one: Rough face leaving 0.005 on all faces and 0.08 on all diameters for finishing. (Feed: 0.012 ipr, Speed: 500 sfm)

Tool two: Drill (1.0 diameter drill) through the workpiece (Feed: 0.008 ipr, Speed: 1,150 rpm)

Tool three: Rough bore 0.0625 chamfer and 1.5 diameter (one pass), leaving 0.040 on the diameter and 0.005 on the 1.25 face (Feed: 0.007 ipr, Speed: 400 sfm)

Tool four: Finish bore 0.0625 chamfer and 1.5 diameter to size (Feed: 0.005 ipr, Speed: 500 sfm) **This tool has a 0.0156 nose radius.**

Tool five: Finish face and turn to size. (Feed: 0.005 ipr, Speed: 500 sfm) **This tool has a 0.0312 nose radius.**

Tool paths:

	X	Z		X	Z		X	Z		X	Z
1			8			15			22		
2			9			16			23		
3			10			17			24		
4			11			18			25		
5			12			19			26		
6			13			20			27		
7			14			21			28		

Check your coordinates on the next page before writing the program. Write your program on a separate sheet of paper.

Coordinates for your program

1: X4.7 Z-0.37
2: X4.5 Z-0.37
3: X4.33 Z-0.245
4: X2.58 Z-0.245
5: X2.58 Z-0.12
6: X2.33 Z0.005
7: X-0.062 Z0.005
8: X0 Z0.1
9: X0 Z-2.08
10: X1.585 Z0.1
11: X1.585 Z0.005
12: X1.46 Z-0.0575
13: X1.46 Z-1.245
14: X0.95 Z-1.245
15: X0.95 Z0.1
16: X1.625 Z0.1
17: X1.625 Z0
18: X1.5 Z-0.0625
19: X1.5 Z-1.25
20: X0.95 Z-1.25
21: X0.95 Z0.1
22: X4.7 Z-0.375
23: X4.5 Z-0.375
24: X4.25 Z-0.25
25: X2.5 Z-0.25
26: X2.5 Z-0.125
27: X2.25 Z0
28: X1.5 Z0

Answer program

```
O0010
N002 G99 G20 G23
N004 G50 S5000

(ROUGH FACING TOOL)
N005 T0101 M41
N010 G96 S500 M03
N015 G00 X4.7 Z-0.37 M08 (1)
N020 G01 X4.5 F0.012 (2)
N025 X4.33 Z-0.245 (3)
N030 X2.58 (4)
N035 Z-0.12 (5)
N040 X2.33 Z0.005 (6)
N045 X-0.062 (7)
N050 X6.0 Z5.0
N055 M01

(1.0" DRILL)
N060 T0202 M41
N065 G97 S1150 M03
N070 G00 X0 Z0.1 M08 (8)
N075 G01 Z-2.2 F0.008 (9)
N080 G00 Z0.1 (8)
N085 X6.0 Z5.0
N090 M01

(ROUGH BORING BAR)
N095 T0303 M42
N100 G96 S400 M03
N105 G00 X1.585 Z0.1 M08 (10)
N115 G01 Z0 F0.007 (11)
N120 X1.46 Z-0.0575 (12)
N125 Z-1.245 (13)
N130 X0.95 (14)
N135 G00 Z0.1 (15)
N140 X5.0 Z6.0
N145 M01

(FINISH BORING BAR)
N150 T0404 M42
N155 G96 S500 M03
N160 G41 G00 X1.625 Z0.1 M08 (16)
N165 G01 Z0 F0.005 (17)
N170 X1.5 Z-0.0625 (18)
N175 Z-1.25 (19)
N180 X0.95 (20)
N185 G00 G40 Z0.1 (21)
N190 X6.0 Z5.0
N195 M01

(FINISH FACING TOOL)
N200 T0505 M42
N205 G96 S500 M03
N210 G41 G00 X4.7 Z-0.375 M08 (22)
N215 G01 X4.5 F0.005 (23)
N220 X4.25 Z-0.25 (24)
N225 X2.5 (25)
N230 Z-0.125 (26)
N235 X2.25 Z0 (27)
N240 X1.5 (28)
N245 G00 G40 X6.0 Z5.0
N250 M30
```

STOP! Do practice exercise number sixteen in the workbook.

Special Features That Help With Programming

You now have the basic tools you need to write CNC programs. However, writing programs with only the CNC features we have shown will be very tedious. In Key Concept Number Six, we'll be showing features that make programming easier, shorten the program's length, and in general, facilitate your ability to write programs.

Key Concept Number Six contains seven lessons and an appendix:

17: One-pass canned cycles
18: G71 and G70 – multiple repetitive cycles: rough turning and boring followed by finishing
19: G72-G75 – other multiple repetitive cycles
20: G76 – threading multiple repetitive cycle
21: Working with subprograms
22: Special considerations for Fanuc 0T and 3T controls
23: Other special features of programming
---: Appendix to programming – special machine types and accessories

This key concept should be of great interest to you. While you have learned how to write simple CNC programs, you may be wondering if there are easier ways to get things done. As you'll see in this key concept, there are.

You know, for example, that programming any kind of roughing operation (rough turning, rough facing, or rough boring) can be quite tedious. With only G00, G01, G02, and G03 to work with, every roughing pass will require at least four commands. Consider rough turning a workpiece down from 8.0 inches to 2.0 inches in diameter, taking a depth-of-cut of 0.125 inches. This will require twenty-four passes – at least ninety-six commands. And if a roughing pass intersects a chamfer, taper, or radius, it will be difficult to calculate the end point for the roughing pass. You will learn in Lesson Eighteen that as long as you can program the finish pass, the machine will completely rough turn or bore the workpiece for you – based on one simple command – regardless of how many passes are required. In Lesson Nineteen, you'll see that there is a similar command for rough facing.

In Lesson Twenty, we'll present a simple way to machine threads – one command will machine the entire thread regardless of how many passes it takes. In Lesson Twenty-One, you'll learn how to use a program-shortening feature called *sub-programming*. In Lesson Twenty-Three, we'll present several special features that can help you create programs. While they won't all be of immediate use, you'll surely find at least some of them to be very helpful.

We also include a programming appendix in Key Concept Six. In this appendix, we'll be showing how to handle special machines and machine accessories. For instance, live tooling, bar feeders, and tailstocks will be presented in the appendix.

Almost every CNC function presented thus far is *essential* to your ability to program CNC turning centers. You will be using the features and techniques presented in Key Concepts One through Five in every program you write. Think of what you have learned so far as the rudimentary tools of CNC programming.

You may be wondering why we waited so long to show these features, since they do make programming *so* much easier. Do not underestimate the importance of what you have learned so far. Just as it helps to understand how arithmetic calculations are done manually before you use a calculator, so does it help to understand the long (more difficult) way of programming machining operations before you move on to the more advanced methods. At the very least, the programming activities you have done so far should give you a good appreciation for what is presented in Key Concept Six. More importantly, you should now have a *very* good foundation in CNC that will let you easily grasp the importance of the features we show in this key concept.

Some of the features shown in Key Concept Six are considered by Fanuc to be *options*. But most machine tool builders will include them with the machines they sell. If you come across one of these features that your machine does not have, remember that all of them are *field-installable*, meaning if you have a need for them, they can be added to your control at any time (for an additional price).

As stated, you will not have immediate need for some of the features discussed in this key concept. In fact, some you may never need. But we urge you to study this information if for no other reason than to gain a better understanding of what is possible in CNC programming. You can always come back and review a needed feature when the need arises, *if you know it is available.*

Control series differences

Some of these special programming features (especially multiple repetitive cycles and subprograms) require slightly different programming words within commands based upon the Fanuc control model you are programming. The basic functions of each cycle are exactly the same from one control model to the next. Only the programming format changes slightly from one control to another.

To handle this minor format problem during this text, we will first look at how these functions are programmed for the majority of Fanuc controls. These include the Fanuc 0TC, the 6T, 6TB, 10T, 11T, 15T, 16T, and 20T, and 21T. Most controls that claim to be Fanuc-compatible also use this format. Then we will discuss the minor format differences for Fanuc 0TA, 0TB, 3T, and 18T. These minor program formatting differences exist because the 0TA, some 0TBs, 3T, and 18T controls do not have all of the letter addresses needed to program certain cycles in one command.

If you happen to have a 0TA, 0TB, 3T, or 18T, you must first follow all presentations. You will be exposed to the various functions and features of each cycle and see how they work. When we discuss control model differences (in Lesson Twenty-Two), we'll summarize the command differences.

Lesson 17

One-Pass Canned Cycles

These cycles have been replaced with multiple repetitive cycles. While they are still available – and they can sometimes be helpful – you'll probably find little use for them. But for the sake of completeness, we include them in this text.

There are two types of canned cycles equipped on Fanuc and Fanuc-compatible turning centers. The first, one-pass canned cycles, are quite limited. As the name implies, they will simply *make one pass* (turning, boring, facing, or threading). While there are some good applications for these cycles and you should understand their use, there is a much more helpful series of canned cycles called *multiple repetitive cycles*. These cycles are much better than one-pass canned cycles, especially when it comes to roughing operations (rough turning, rough facing, or rough boring) and threading. Some multiple repetitive cycles actually rival what a good computer aided manufacturing (CAM) system can do when it comes to simplifying a programmer's work.

Cycle consistencies

There are three things that *all* canned cycles have in common (including one-pass canned cycles and multiple repetitive cycles).

1) Every canned cycle requires you to first send the tool to a *convenient starting position*. The convenient starting position will vary from one cycle to the next, but it will always be the tool's current position in X and Z just prior to the canned cycle command. It is the tool's rapid approach position for the canned cycle.

2) At the completion of any canned cycle, *the tool will be left back at the convenient starting position.*

3) All canned cycles can be used for both internal diameter machining (boring) as well as external diameter work (turning). Though the convenient starting position will change from external to internal work, all points about the canned cycle will apply to either.

One-pass canned cycles allow you to program one pass of a machining operation, including turning or boring, facing, and threading. This will minimize the number of commands required in the program since they cause the tool to make four motions from one command. One-pass canned cycles are all *modal*, remaining in effect until another motion type is programmed (like G00 or G01). Additionally, each one-pass canned cycle requires a feedrate, specifying the machining rate to be used for machining during the cycle.

G90 - One pass turning and boring cycle

This cycle is used to make one turning or boring pass. Figure 6.1 shows the motions caused by this command (in a turning mode).

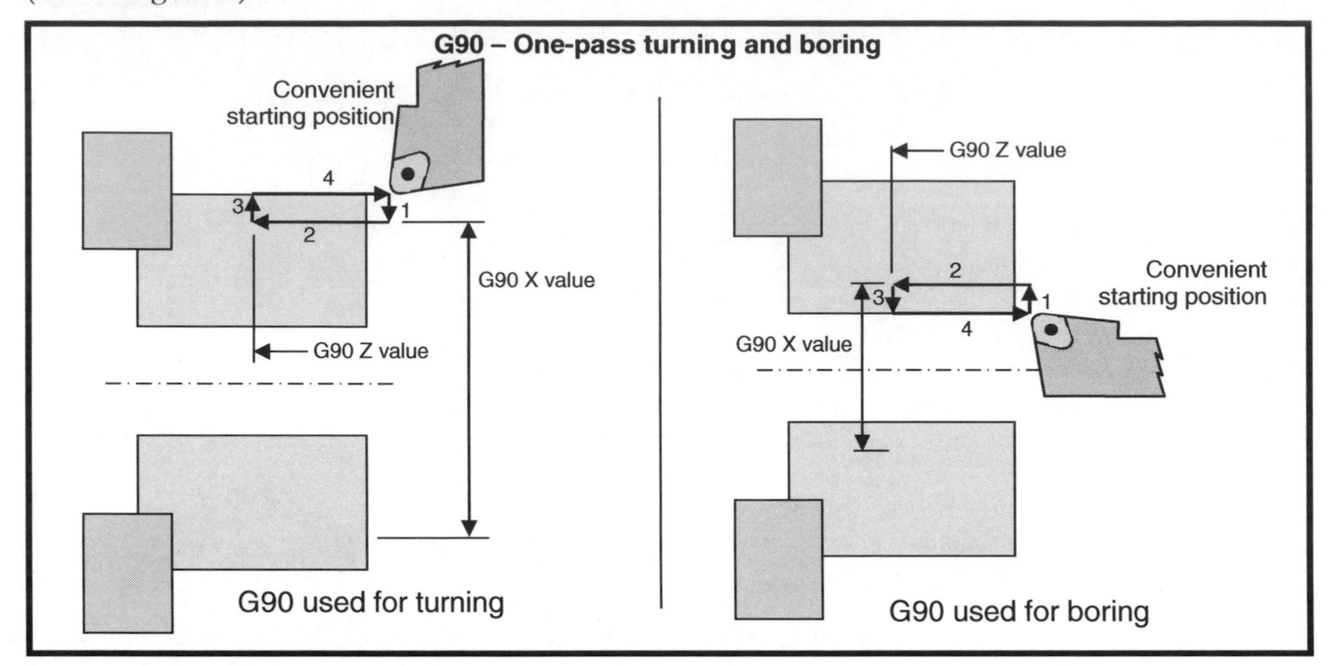

Figure 6.1 – G90 used for one-pass turning and boring

Notice the four motions caused by G90. Prior to the G90 command, you send the cutting tool to its convenient starting position – usually with a rapid motion command. The convenient starting position is flush with the raw material diameter and clear of the workpiece in the Z axis. The G90 command will first cause the tool to rapid to the X diameter specified by the G90 X word. Second, the tool will then feed to the G90 Z position. Third, the tool will feed back to the X position of the convenient starting position. And fourth, the tool will rapid back to the convenient starting location in Z.

For turning (example on the left of Figure 6.1), the convenient starting position in the X axis must be larger than the G90 X value. For boring (example on the right), the convenient starting position in X must be smaller than the G90 X value.

G92 - One pass threading cycle

Like G90, G92 causes the machine to make one pass (four motions). During a threading pass with G92, the machine will synchronize the spindle speed and feedrate in a way that allows thread chasing (several passes over the same thread lead). Spindle speed must be programmed in the rpm mode (G97) when threading. Feedrate for threading is equal to the *pitch* of the thread (for inch threads, thread pitch is equal to one divided by the number of threads per inch). Figure 6.2 shows the motions caused by a G92 command.

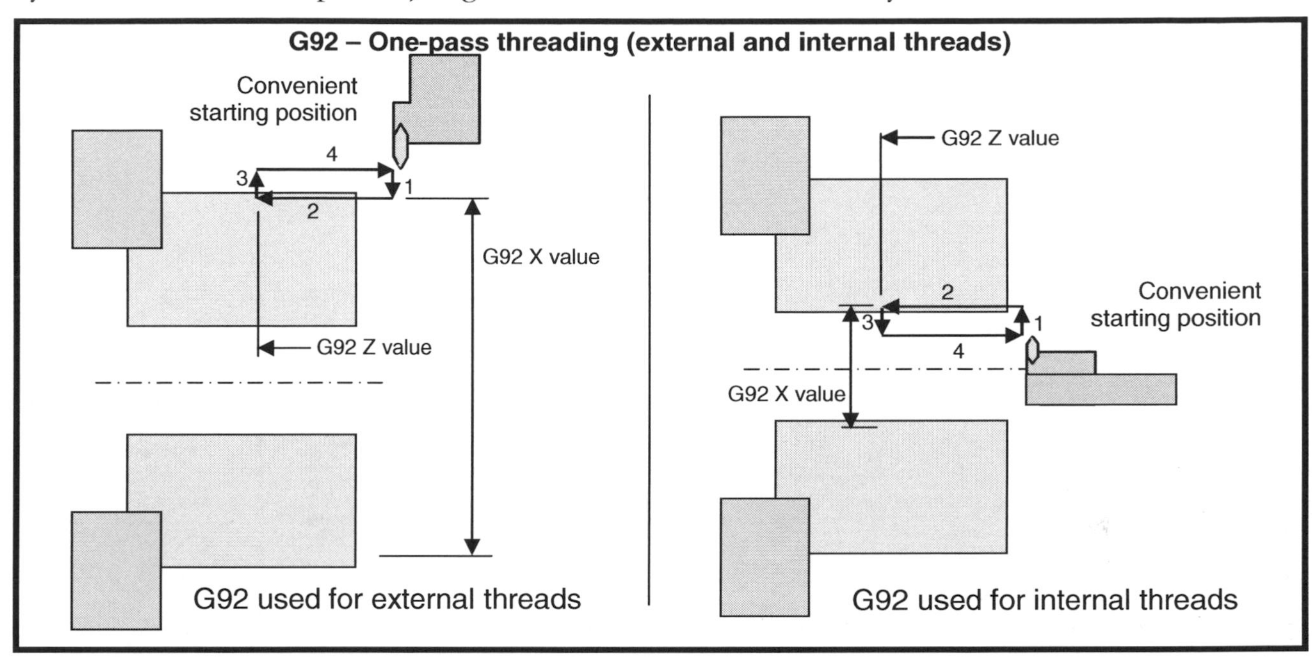

Figure 6.2 – Motions caused by G92

Also like the G90, G92 can be used for internal as well as external threads. For external threads, the convenient starting position in the X axis must be larger than the G92 X value. For internal threads, the convenient starting position in the X axis must be smaller than the G92 X value.

The convenient starting position for G92 must be clear of the diameter to be threaded in X and clear of the workpiece in Z. When G92 is executed, the tool will first rapid to the G92 X position. Then it will feed (with feedrate synchronized with spindle speed) to the G92 Z position. The tool will then rapid to the convenient starting position in X. Finally the tool will rapid back to the convenient starting position in Z.

Threading with G92 can be somewhat tedious and cumbersome (as you'll see in an upcoming example). It requires you to determine the number of passes, the depth of each pass, and the X position for each pass. For this reason, and since the G76 multiple repetitive cycle makes threading *much* easier, G92 is seldom used to generate an entire thread if the program is being prepared manually.

Another limitation of G92 is that it makes it difficult to specify an in-feed angle during threading (allowing the tool to cut only on its leading edge). The multiple repetitive cycle G76 eliminates this problem.

G94 - One pass facing command

G94 is very similar to G90. The only difference is that machining will be done in a facing mode. Figure 6.3 shows the motions caused by G94.

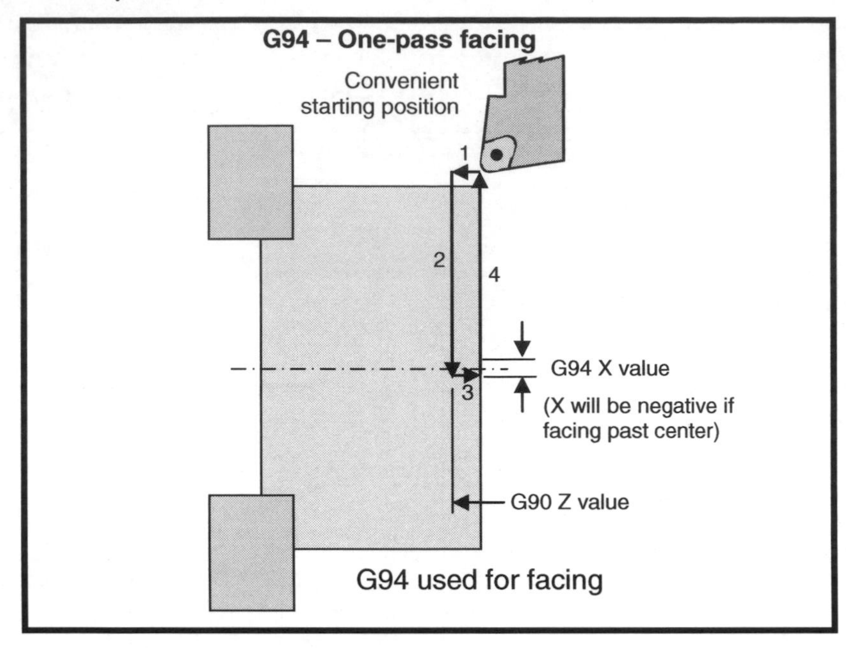

Figure 6.3 – Motions caused by G94

With G94, the convenient starting position must be clear of the workpiece in the X axis and flush with the raw material in the Z axis. G94 will cause the tool to first rapid to the G94 Z position. Second, the tool will feed to the G94 X position. The tool will then feed to the convenient starting position in Z. Finally the tool will rapid back to the convenient starting position in X.

Example of G90 and G94

Figure 6.4 shows the drawing used for our simple example. Only one tool is used. It will first face the workpiece and then turn a diameter.

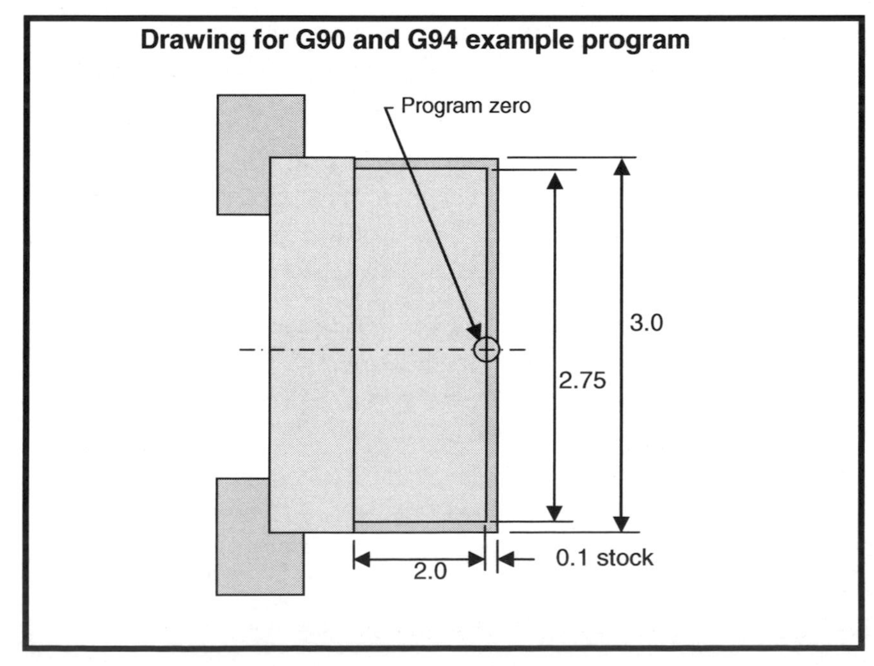

Figure 6.4 – Drawing for G90 and G94 example program

Program:

```
O0008 (Program number)
N003 G99 G20 G23 (Ensure initialized states are still in effect)
N004 G50 S4000 (No need for limiting, limit to machine's maximum speed)
N005 T0101 M42 (Select tool and spindle range)
N010 G96 S400 M03 (Turn spindle on fwd at 400 sfm)
N015 G00 X3.2 Z0.1 M08 (Rapid to convenient starting position for facing, start coolant)
N020 G94 X-0.06 Z0 F0.007 (Face workpiece)
N025 G00 X3.0 (Rapid to convenient starting position for turning)
N030 G90 X2.75 Z-2.0 F0.008 (Turn workpiece)
N035 G00 X6.0 Z6.0 (Rapid to safe index point)
N040 M30 (End of program)
```

Notice that, in line N025, it is necessary to position the tool to the new convenient starting position for the turning operation. While this program will work without this command, it will waste some time as the tool feeds back to the convenient starting position in X during the G90 command.

At first glance, these two cycles may appear to be quite helpful, and they will be if you commonly machine straight faces and turns as our example shows. However, most turned workpieces require the machining of several diameters having tapers, radii, and chamfers. Programming a more complex workpiece with these one-pass canned cycles is almost as difficult as doing so with G00, G01, G02, and G03.

Example of G92 command

Figure 6.5 shows a workpiece to be used for a threading example. As mentioned earlier, G92 forces the programmer to plan each threading pass. In Figure 6.5, we include a chart that shows this prior planning.

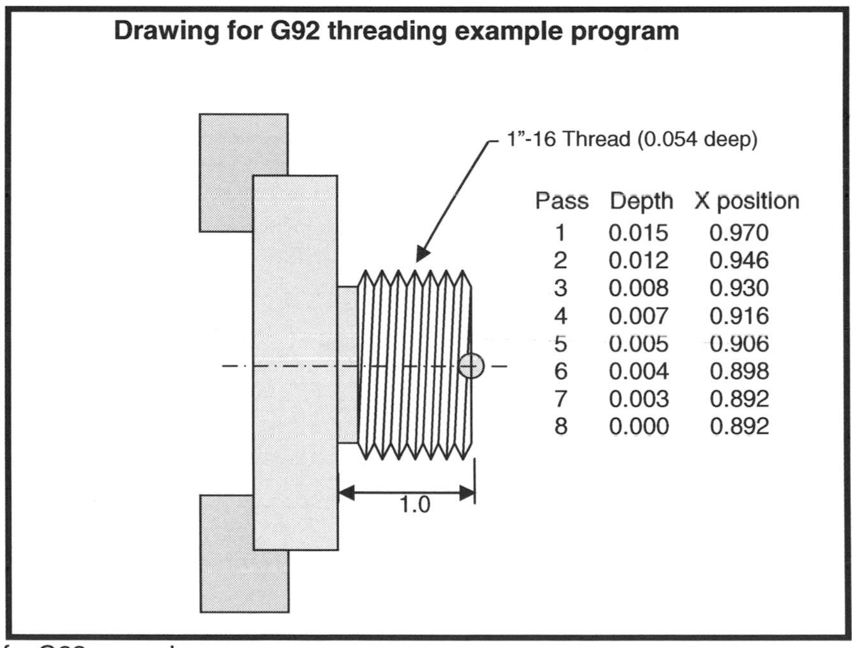

Pass	Depth	X position
1	0.015	0.970
2	0.012	0.946
3	0.008	0.930
4	0.007	0.916
5	0.005	0.906
6	0.004	0.898
7	0.003	0.892
8	0.000	0.892

Figure 6.5 – Drawing for G92 example program

Program:

```
O0009 (Program number)
N003 G99 G20 G23 (Ensure that initialized states are still in effect)
N004 G50 S4000 (No need for limiting, limit to machine's maximum speed)
```

N005 T0101 M41 (Select tool and spindle range)
N010 G97 S1200 M03 (Turn spindle on fwd at 1200 rpm)
N015 G00 X1.2 Z0.2 M08 (Rapid to convenient starting position, start coolant)
N020 G92 **X0.970 Z-0.95 F0.0625** (Make first threading pass)
N025 **X0.946** (Make second pass)
N030 **X0.930** (Make third pass)
N035 **X0.916** (Make fourth pass)
N040 **X0.906** (Make fifth pass)
N045 **X0.898** (Make sixth pass)
N050 **X0.892** (Make seventh pass)
N055 **X0.892** (Make eighth [spring] pass)
N060 **G00** X6.0 Z6.0 (Rapid to safe index point, this cancels G92 mode)
N065 M30 (End of program)

As you can see, G92 is modal. Once instated (in line N020), the machine will continue in the threading mode until it is canceled by another motion type (in line N060 with G00). Notice in line N015 that the convenient starting position in Z is 0.200 in away from the threaded face. This distance allows the machine to accelerate to the proper motion rate for threading before the cutting tool actually contacts the workpiece. Threading should always be done in the lowest spindle range to ensure speed stability during each threading pass.

Again, at first glance G92 may seem like a very helpful canned cycle. But as you can see from Figure 6.5, you must first determine the number of passes, the depth of each pass, and the appropriate X position for each pass. And our example did not incorporate a compound in-feed to keep the threading tool cutting on its front cutting edge as many threading tools require. Doing so will make programming a thread much more difficult. As we discuss the G76 multiple repetitive cycle, you'll see that it's very easy to specify compound in-feed. Also, *the entire thread can be machined by one command.*

Key points for Lesson Seventeen:

- All canned cycles and multiple repetitive cycles share three things in common. Each requires that the tool be first sent to a convenient starting position. Each can be used for external work as well as internal work. And all will leave the cutting tool back at the convenient starting position at the completion of the cycle.
- One-pass canned cycles will cause the machine to make four movements per command.
- One-pass canned cycles can be used for turning and boring (G90), threading (G92) and facing (G94).
- Each one-pass canned cycle is modal – and is canceled with any motion command (like G00).
- While the one-pass canned cycles can sometimes be helpful, multiple repetitive cycles are much more helpful.

STOP! Do practice exercise number seventeen in the workbook.

Lesson 18

G71 And G70 – Rough Turning And Boring Followed By Finishing

These cycles are extremely helpful. They allow you to completely rough turn or rough bore a workpiece based upon a finish pass definition and one special command. In this command you specify important information like depth-of-cut and the amount of stock you want to leave for finishing. Based upon this simple information, the machine will completely rough the workpiece – regardless of the number of passes required.

You know that one-pass canned cycles can be used to minimize the number of commands needed to machine a workpiece. But as you have seen, they are somewhat cumbersome to use. By comparison, the multiple repetitive cycles we discuss in this lesson are *much* more helpful. They make it relatively simple to program complex rough turning and rough boring operations for even the most difficult workpieces. As you'll see, they rival what can be done with a computer aided manufacturing (CAM) system.

As mentioned earlier, first we will show how these multiple repetitive cycles are programmed for the majority of Fanuc turning center controls. Then we will discuss the slight differences required for 0TA, 0TB, 3T, and 18T Fanuc controls.

G71 - Rough turning and boring

If you've been doing the programming exercises in this text and in the workbook, you've seen that rough turning and rough boring operations can be very tedious and error prone to program with only the three basic motion types (rapid, straight line, and circular). Every rough turning or boring pass requires at least four commands. And in our programming exercises so far, we've kept things as simple as possible. In reality, roughing a workpiece is usually much more difficult.

For example, all of the roughing passes you've programmed so far end at a known position (at a face or diameter). If a roughing pass ends at a chamfer, taper, or radius, calculating the end point for the pass will require more difficult math calculations (including trigonometry), even for relatively simple workpieces. Figure 6.6 shows an example workpiece for which programming rough turning will be more difficult.

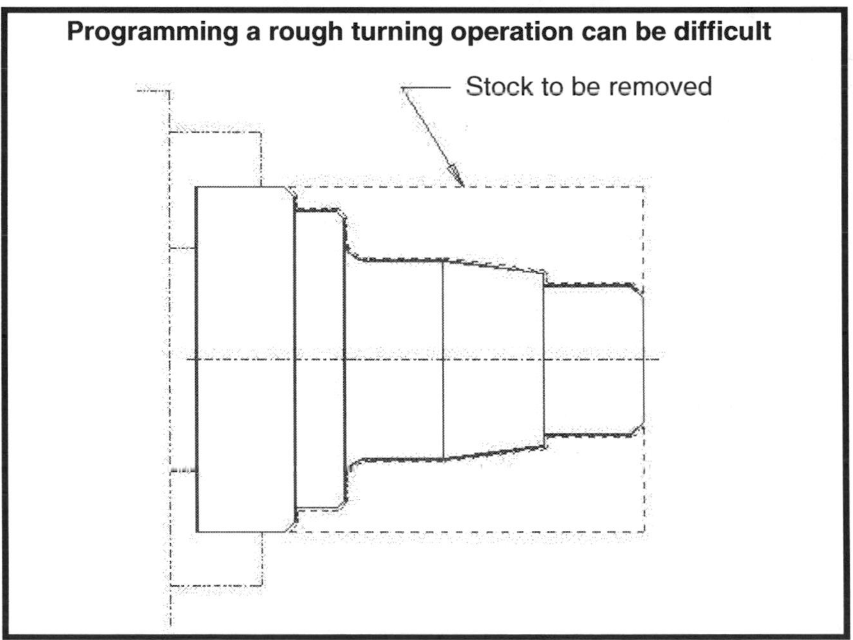

Figure 6.6 – Drawing that stresses the difficulty of programming roughing passes long-hand

Though this is a relatively simple workpiece, notice how much stock must be removed by the rough turning operation. To program this long hand (without any special cycles), a programmer will have to plan and program several roughing passes based on a previously determined *depth-of-cut*. This creates a real problem for manual programmers. And consider the additional difficulty if one or more of the roughing passes ends in the middle of the large taper or fillet radius. Trigonometry will be required in both cases.

Knowing how difficult it can be to program individual turning or boring passes, Fanuc has designed three very helpful multiple repetitive cycles for roughing:

- G71 - Rough turning and rough boring
- G72 - Rough facing
- G73 - Pattern repeating

In this lesson, we limit our discussions to G71, which is used for rough turning and boring. In the Lesson Nineteen, we'll describe G72 and G73. You will see many similarities among these multiple repetitive cycles. All roughing multiple repetitive cycles are followed by the multiple repetitive cycle for finishing. G70 specifies the finishing cycle.

With roughing multiple repetitive cycles, you only need to describe the finish pass of the workpiece surface being roughed. Based on this *finish pass definition* and one very simple command, the control will completely rough the entire workpiece.

The two phases of G71

G71 and G72 commands will be completed in two *phases*. In the phase one, the machine will make a series of roughing passes based on a specified depth-of-cut. But as soon as the tool comes close to an end point for a roughing pass, the machine will immediately retract the tool for another pass. Figure 6.7 shows the motions generated during the first phase of the G71 command.

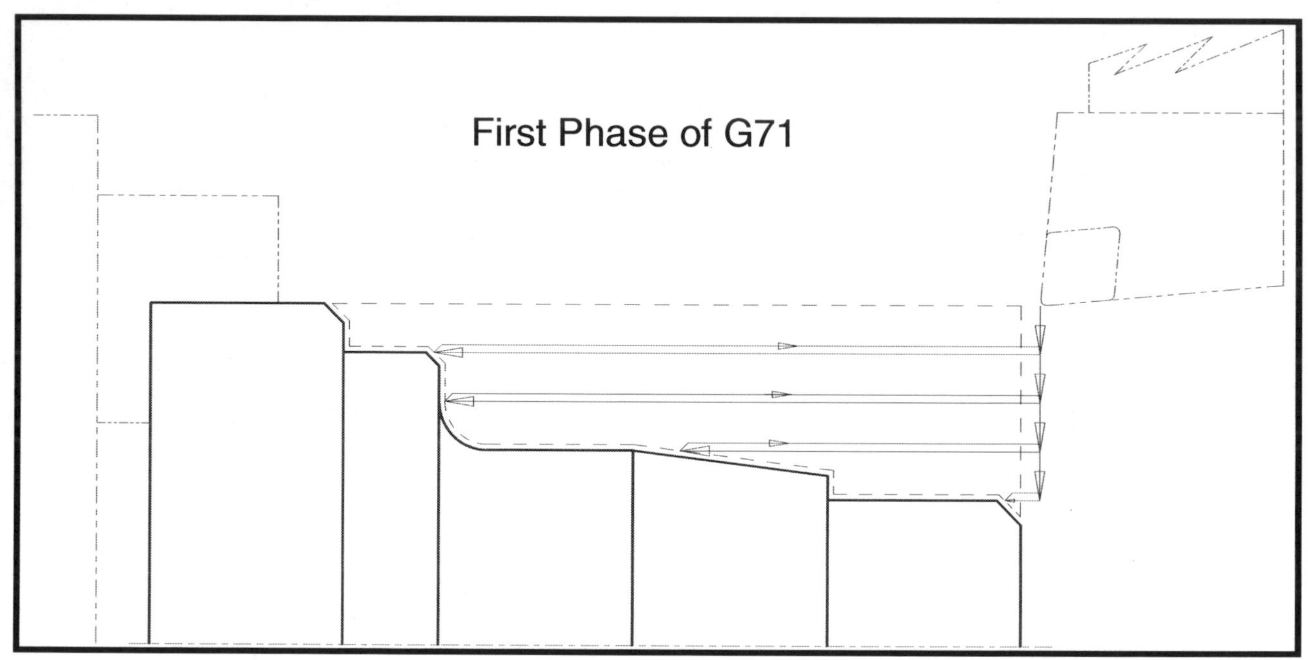

Figure 6.7 – Motions during phase one of a G71 command

After the control finishes the first phase of G71, the workpiece will have a series of steps (like a staircase) and it will not be close to its proper size (it will not have the appropriate amount of finishing stock on all surfaces). None of the steps left by the first phase of G71 will be larger than the depth-of-cut specified in the G71 command. Figure 6.8 shows what the workpiece will look like *after* the first phase of the G71 command.

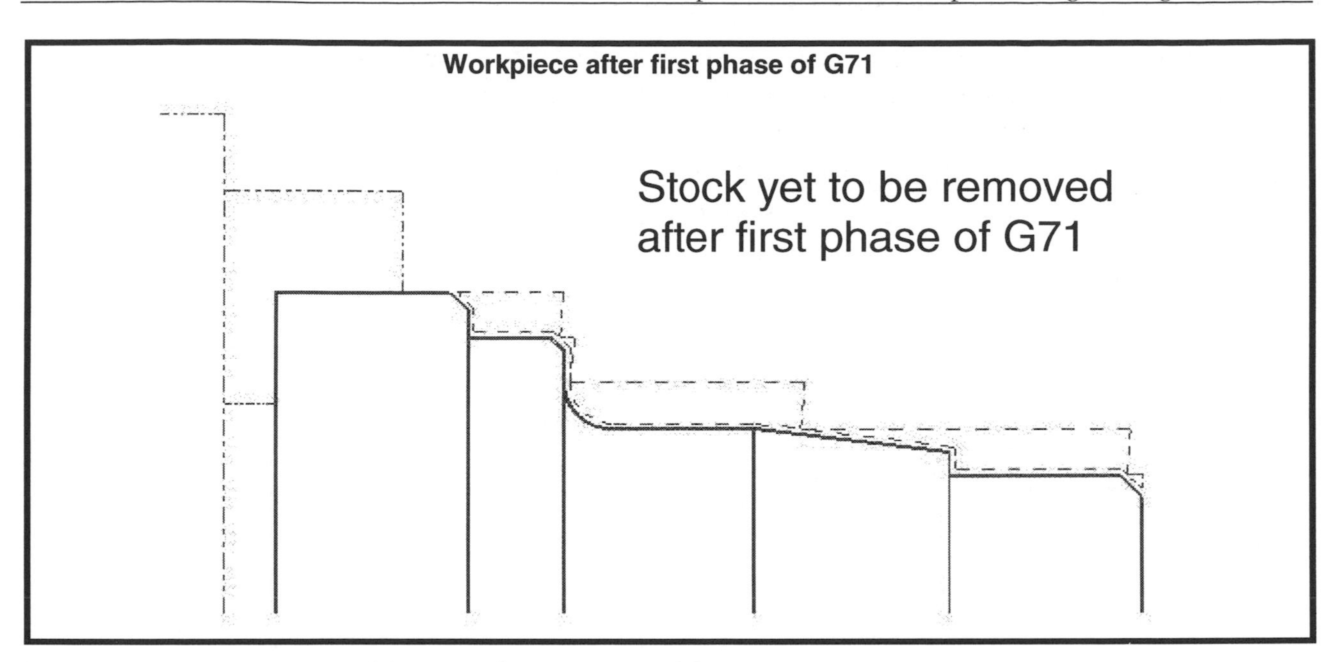

Figure 6.8 – What workpiece will look like after phase one of G71 command

In the second phase of G71, the machine will return to the starting point of the finish pass definition and make one sweeping semi-finish pass over the entire workpiece. It will stay away from the finished surface by the finishing stock values specified in the G71 command. Figure 6.9 shows the motions generated during the second phase of the G71 command.

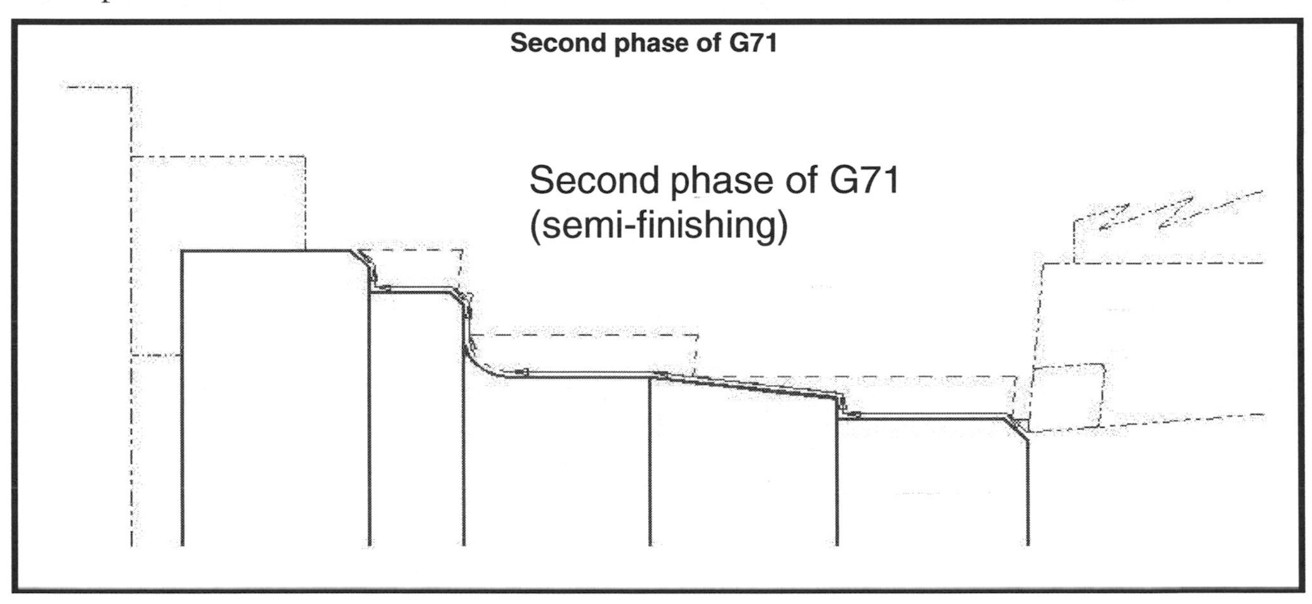

Figure 6.9 – Motions during phase two of G71 command

This is all accomplished by one command (the G71 command) together with the finish pass definition. G71 makes it very easy to program even very lengthy and complex rough turning and boring operations. In essence, the machine will figure out how to make all of the roughing passes for you, in much the same way a computer aided manufacturing (CAM) system does.

There are three steps to programming a roughing operation with G71.

> **First, position the tool to the convenient starting position. For rough turning (external work), this position must be flush with the stock diameter in X and clear of the workpiece in Z. For rough boring (internal work), the convenient starting position must be flush with the current hole size in X and clear of the workpiece in Z.**

Second, specify the G71 command. This command tells the machine *how* to perform the rough turning or boring operation. A series of special words acts as *variables* to specify the required information.

Third, specify the finish pass definition. The finish pass definition begins with a rapid positioning movement in X to the first diameter to be machined. It ends with the tool's movement to the diameter of the convenient starting position. Two of the words in the G71 command (one for diameters and one for faces) specify how much stock is to be left for finishing. All movements during the finish pass definition are specified with finished workpiece dimensions.

At the completion of the G71 command, the workpiece will be entirely rough turned or bored. There will still be stock left on all surfaces for finishing. After a turret index to the finishing tool, another command is used to finish turn or bore after a G71 has been used for roughing (G70).

Understanding G71 command words

Again, a G71 command tells the control how to perform the rough turning or boring operation. Here are the words included within the G71 command.

P word

This word points to the sequence number of the command that starts the finish pass definition. The finish pass definition will always begin with the command immediately following the G71 command. For example, if the G71 command is specified in line N020 (and you skip five numbers for each sequence number), the finish pass definition will begin in line N025. In this case, the word P025 will be included in the G71 command.

Q word

This word points to the sequence number of the command that ends the finish pass definition. The machine will use the commands specified from the P word through the Q word in order to understand what the finished workpiece looks like. As you actually write the G71 command, you won't know how many commands will be in the finish pass definition, meaning you won't know the value of the Q word (yet). So leave the Q word blank, circling it to help you remember to come back and fill it in when the finish pass definition is completed.

Most controls require the P and Q word values to perfectly match the sequence numbers being specified. If N025 is being specified by a P word, for example, the correct word is P025, not P25 or P0025. Also note that the sequence numbers specified by P and Q must only occur once in the program.

U word

The U word tells the control how much stock is to be left for finishing on all diameters. The value U0.040, for example, tells the control to leave 0.040 in on all diameters (0.020 on the side). The U word also points in the direction of stock to be left. For external diameter machining (turning), the U word must be positive. For internal diameter machining (boring), the U word must be negative.

W word

The W word tells the control how much stock is to be left for finishing on all faces. The word W0.005 will leave 0.005 in on all faces. As long as turning or boring is done toward the chuck (in the minus Z direction), as most turning and boring is, the W word value must be positive. If the cutting tool is back turning (machining in the Z plus direction), the W word value must be negative.

D word

The D word tells the control how much stock to remove (on the side) per pass. This is the *depth-of-cut* for the roughing operation. All but the most recent controls do not allow a decimal point to be used with this word. Instead, a four place fixed format must be used in the inch mode (a three place fixed format must be used in the metric mode). For example, the word D2500 specifies a 0.250 inch depth-of-cut. D1250 specifies a 0.125 inch depth-of-cut. D0500 specifies a 0.050 inch depth-of cut. And so on. (In the metric mode, D3000 specifies a three millimeter depth-of-cut.) If you want to know whether a decimal point is allowed with the D word for your particular control, reference the control manufacturer's programming manual in the description of G71.

F word
This is the feedrate for the entire roughing operation. Any feedrate word specified during the finish pass definition will be ignored by during the roughing operation (but will be used during the finishing operation).

What about finishing?
Before we can show a complete example, we need to introduce the **G70** finishing command. As stated, at the completion of the **G71**, the machine will have completely roughed the workpiece based on the **G71** command and what's in the finished pass definition (between the P and Q blocks). But as you know, the workpiece will not yet be finished.

To keep from having to repeat the commands given during the finish pass definition, the machine will allow you to specify the finishing operation with a **G70** command. Like **G71**, the **G70** uses a P word to point to the first command of the finish pass definition and a Q word to point to the last command of the finish pass definition. The P and Q of the **G70** command will always match the P and Q of the related **G71** command. When **G70** is executed, the control will simply go back to the sequence number specified by the P word and do through the sequence number specified by the Q word. This will cause the cutting tool to finish the workpiece. At the completion of the finishing cycle, the tool will be left at the convenient starting position.

Example showing G71 for rough turning and G70 for finish turning
Admittedly, the **G71** command probably sounds a little complicated at this point. When you see an example, things should clear up a bit. And regardless of how difficult you find this to be, it is well worth your time to study until you thoroughly understand the **G71** and **G70** commands. You will save countless programming hours when you master them.

Figure 6.10 shows the workpiece to be used for our first example program. Tool number one is the rough turning tool and tool two is the finish turning tool. The end of this workpiece has been previously faced to size.

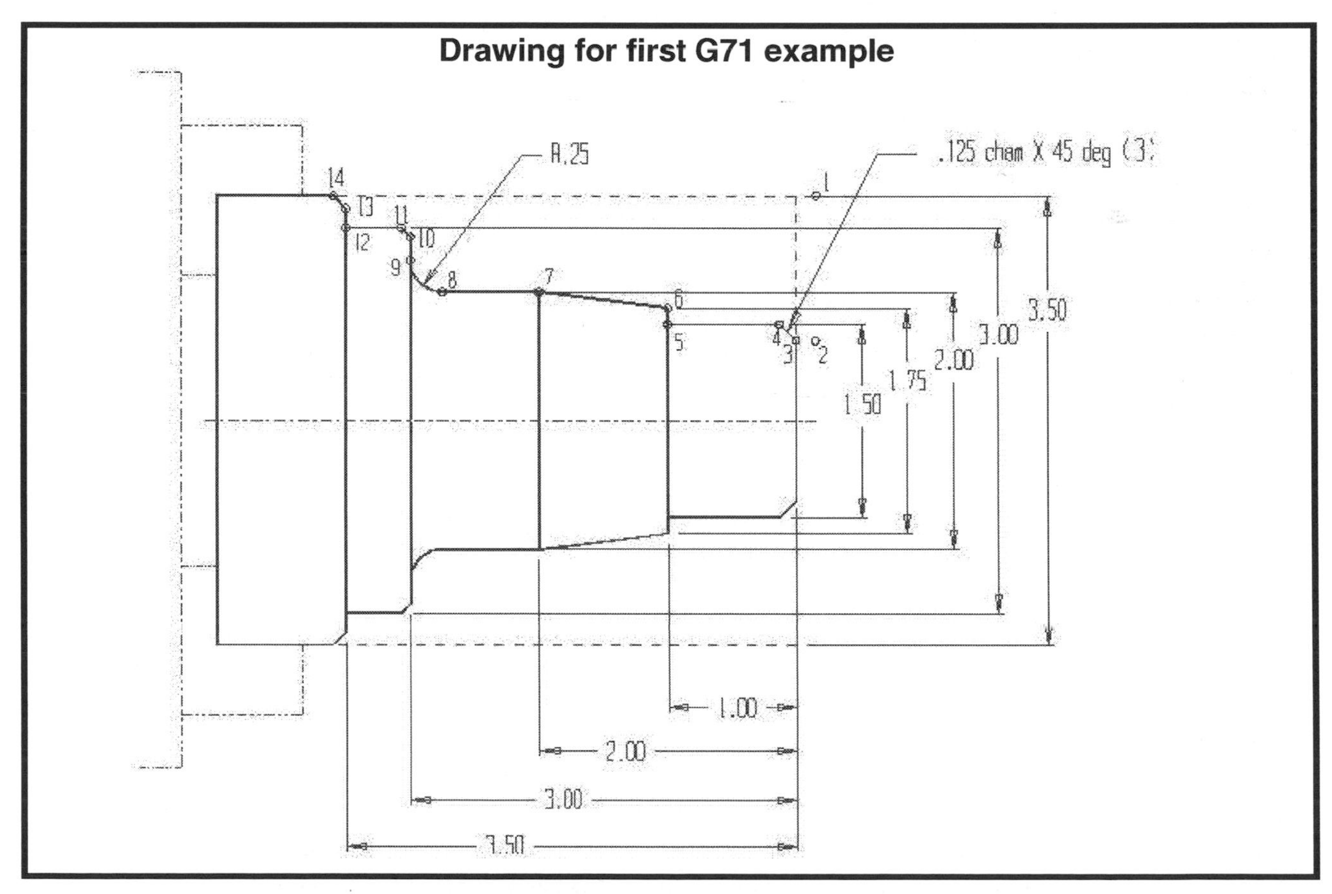

Figure 6.10 – Drawing for first G71 example program

Program:

O0010 (Program number)

(ROUGH TURNING TOOL)

N003 G99 G20 G23 (Ensure that initialized states are still in effect)

N004 G50 S4000 (No need for limiting, limit to machine's maximum speed)

N005 T0101 M41 (Select roughing tool and low spindle range)

N010 G99 G96 S400 M03 (Turn spindle on fwd at 400 sfm)

N015 G00 X3.5 Z0.1 M08 (Rapid to convenient starting position, point 1, start coolant)

N020 **G71 P025 Q085 U0.040 W0.005 D1250 F0.015** (Rough turn entire workpiece based on what is between lines N025 and N085)

N025 G00 X1.25 (First block of finish pass definition, rapid to point 2)

N030 G01 Z0 **F0.008** (Feed to point 3)

N035 X1.5 Z-0.125 (Feed to point 4)

N040 Z-1.0 (Feed to point 5)

N045 X1.75 (Feed to point 6)

N050 X2.0 Z-2.0 **F0.005** (Feed to point 7)

N055 Z-2.75 **F.008** (Feed to point 8)

N060 G02 X2.5 Z-3.0 R.25 (Circular move to point 9)

N065 G01 X2.75 (Feed to point 10)

N070 X3.0 Z-3.125 (Feed to point 11)

N075 Z-3.5 (Feed to point 12)

N080 X3.25 (Feed to point 13)

N085 X3.5 Z-3.625 (Feed to point 14)

N090 G00 X6.0 Z6.0 (Rapid to safe index position)

N095 M01 (Optional stop)

(FINISH TURNING TOOL)

N100 T0202 M42 (Select finish turning tool and high spindle speed range)

N105 G96 S600 M03 (Turn spindle on fwd at 600 sfm)

N110 G00 **G42** X3.5 Z0.1 M08 (Rapid to the same convenient starting position as used for the rough turning operation, start coolant)

N115 **G70 P025 Q085 F0.008** (Go back to line N025 and do through line N085, finish turning workpiece)

N120 G00 **G40** X6.0 Z6.0 (Rapid to safe index point)

N125 M30 (End of program)

Notes about the example program:

1) At the completion of the G71 command (line N020), the workpiece will be entirely rough turned. The control will continue with the program's execution at line N090, after the last command of the finish pass definition. For this reason, it may help to think of all of the commands from N020 through N085 as one command.

2) As you write line N020, you will not know which command ends the finish pass definition. For this reason, the Q word must be left blank until the finish pass definition is completed (in line N085). The finish pass definition will end when the tool reaches the stock diameter (the same X

position as in the convenient starting position). In our example, notice that the tool comes up to a 3.5 position in X in line N085, which is the same X position as is in line N015.

3) Notice that the feedrates specified during the finish pass definition (in lines N030, N050, and N055) will be ignored during the rough turning operation. The entire rough turning operation will be performed at 0.015 ipr (as is commanded in line N020). However, these feedrates within the finish pass definition *will* be used during the finishing operation (commanded by line N115). This allows you to specify different feedrates (meaning different surface finishes) for different surfaces of the workpiece. In our case, the tapered surface will be machined at a slower feedrate than the balance of the workpiece. If no feedrate commands are given during the finish pass definition, the control will use the currently instated feedrate at the time the G70 command is executed. If you wish the entire finishing pass to be made at the same feedrate, you can simply include this feedrate in the G70 command.

4) Notice that the first command of the finish pass definition (line N025) is a motion at rapid to a position flush with the first X surface to be machined (the lower right corner of the chamfer). **There is a special rule about this command.** It goes like this: *The block specified by the P word of a G71 command must include either a G00 or G01 and* ***only*** *an X word to the starting diameter of the finish pass definition.* If you leave out the G00 (or G01) or if you include a Z word in this command, the control will generate a format alarm.

5) Notice (in line N110) that the finishing tool is sent to precisely the same convenient starting location as the rough turning tool. Since all canned cycles leave the tool back at the convenient starting position, this is where the finishing tool will go when the G70 command is finished. If you keep the same convenient starting position for the finishing tool as you used for the roughing tool, you can rest assured that there will be no interference when the finish turning tool returns to its convenient starting position. On the other hand, if you rapid the finish turning tool closer to the starting diameter prior to the G70 command (if the convenient starting position is closer to the first diameter to machine), it is likely that the finish turning tool will crash through the workpiece on its return to the convenient starting position at the completion of the finishing operation.

6) Think of how easy it will be to change the cutting conditions at the machine when G71 is used. If, for example, the depth of cut is too deep and the spindle stalls, only one word (the D word) has to be changed in order to generate a new set of rough turning movements.

7) Notice that it is still possible to use tool nose radius compensation in conjunction with the G70 command. Since tool nose radius compensation is desired for the finishing tool, the proper word (G42 in this case) is included in the tool's first approach to the workpiece (to the convenient starting position) in line N110. It is also being canceled (with G40) on the tool's return to the safe index point in line N120.

Using G71 for rough boring

Only three things change when you use G71 for rough boring. First, the convenient starting position will now be flush with the current hole diameter (not the outside rough stock diameter). Second, the U word of the G71 command will be negative, pointing in the correct direction for stock to be left for finishing. Third, the finish pass definition will be in the form of a finish boring operation instead of a finish turning operation.

Figure 6.11 shows another example workpiece. We will assume that the hole has already been drilled to a 2.0 inch diameter and that the workpiece has been previously faced. Tool number one is the rough boring bar and tool number two is the finish boring bar.

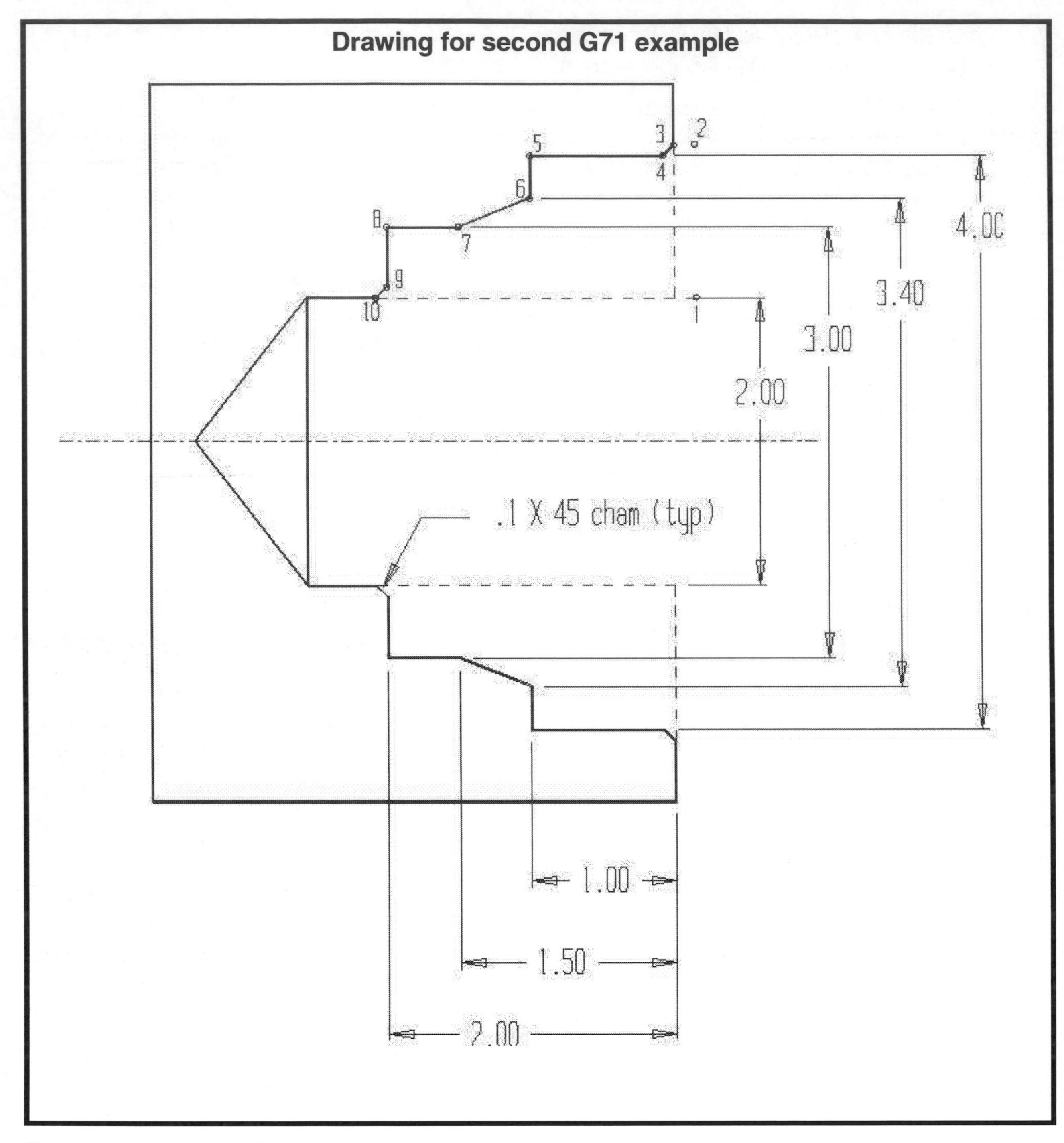

Figure 6.11 – Drawing for G71 rough boring example

Program:

O0012 (Program number)
N003 G20 G23 G99 (Ensure that initialized states are still in effect)
N004 G50 S4000 (No need for limiting, limit to machine's maximum speed)
N005 T0101 M41 (Select rough boring bar and low spindle range)
N010 G99 G96 S300 M03 (Start spindle fwd at 300 sfm)
N015 G00 X2.0 Z0.1 M08 (Rapid to convenient starting position, point number 1, start coolant)
N020 G71 P025 Q065 U-0.040 W0.005 D1250 F0.010 (Rough bore entire workpiece)
N025 G00 X4.2 (Rapid to point number 2)

```
N030 G01 Z0 F0.006 (Feed to point number 3)
N035 X4.0 Z-.01 (Feed to point number 4)
N040 Z-1.0 (Feed to point number 5)
N045 X3.4 (Feed to point number 6)
N050 X3.0 Z-1.5 (Feed to point number 7)
N055 Z-2.0 (Feed to point number 8)
N060 X2.2 (Feed to point number 9)
N065 X2.0 Z-2.1 (Feed to point number 10)
N070 G00 X6.0 Z8.0 (Rapid to safe index position)
N075 M01 (Optional stop)
N080 T0202 M42 (Select finish boring bar and high spindle range)
N085 G96 S400 M03 (Start spindle fwd at 400 sfm)
N090 G00 G41 X2.0 Z0.1 M08 (Rapid to convenient starting position, start coolant)
N095 G70 P025 Q065 F0.006 (Finish bore entire workpiece)
N100 G00 G40 X6.0 Z8.0 (Rapid to safe index point)
N105 M30 (End of program)
```

Notes about the program:

1) Notice that the U word in line N020 is negative. U points in the direction of stock to be left for finishing.

2) You may be questioning how the boring bar gets out of the hole after the roughing and finishing cycles. Remember that all canned cycles and multiple repetitive cycles will leave the tool at the convenient starting position. The convenient starting position for both tools in this program (lines N015 and N090) is at a position of Z0.1.

Limitations of the G71 command

As you can see, the G71 command is extremely helpful. But it does have its limitations. One rather serious limitation, for example, is that unless a special option is purchased (called *type C multiple repetitive cycles*), the G71 command cannot be used to rough machine a recess. After the initial movement to the starting diameter, there can be no reversal in X direction. When it comes to rough turning, this means you will not be able move the tool in a negative X direction during the finish pass definition.

If you must machine a recess with the roughing and/or finishing tools, you must use G00, G01, G02, and G03 commands and program the recess with longhand techniques (after using G71 to rough machine the majority of material).

Another limitation of the G71 command has to do with efficiency. For high production volumes, the method by which the G71 command machines in two phases will waste some time (especially when large faces must be machined). As a cutting tool makes its semi-finish pass (phase two of the G71 command), it must move up each face. In essence, the tool will be air-cutting during these movements.

The advantages of using G71 for most applications dramatically outweigh these limitations (especially for manual programmers). But keep in mind that there are computer aided manufacturing (CAM) systems that will generate roughing motions (with G00, G01, G02, and G03) in a more efficient manner than G71. Of course, CAM system generated CNC programs will be much longer, and it will not be nearly as modify the program at the machine to change cutting conditions (especially depth of cut).

Different format for 0TA, some 0TBs, 3T, and 18T controls

These controls require you to break up each multiple repetitive cycle into two commands because they do not provide the letter address D (for depth of cut).

Here is a G71 rough turning as shown earlier.

```
N020 G71 P025 Q085 U0.040 W0.005 D1250 F0.015
```

This is how the same command must be written for a Fanuc 0TA, some 0TBs, or 3T control:

```
N018 G71 U0.125
N020 G71 P025 Q085 U0.040 W0.005 F0.015
```

As you can see, the only difference is that the depth of cut is be specified in the first G71 command with a U word. Notice that the U word does allow a decimal point. The second G71 specifies the rest of the values, and all are exactly the same as shown earlier.

Key points for Lesson Eighteen:

- With G71, the machine completely roughs the workpiece based upon one command and a finish pass definition.
- G71 can be used for rough turning or rough boring. For rough turning, the convenient starting position is flush with the stock diameter in X and clear of the workpiece in Z. For rough boring, the convenient starting position is flush with the current hole diameter in X and clear of the workpiece in Z.
- The G71 command includes the roughing criteria, including first and last command sequence numbers of the finish pass definition (P and Q), the depth of cut (D), the amount of stock to be left for finishing (U and W) and the roughing feedrate (F).
- For rough turning, the U word (finishing stock left on all diameters) will be positive. For rough boring, the U word will be negative.
- Any feedrate specified during the finish pass definition will be ignored during roughing. It will be used during the finishing operation (specified with G70).
- After roughing and usually with a different cutting tool, a G70 command can be used for finishing – keeping you from having to repeat the commands in the finish pass definition.

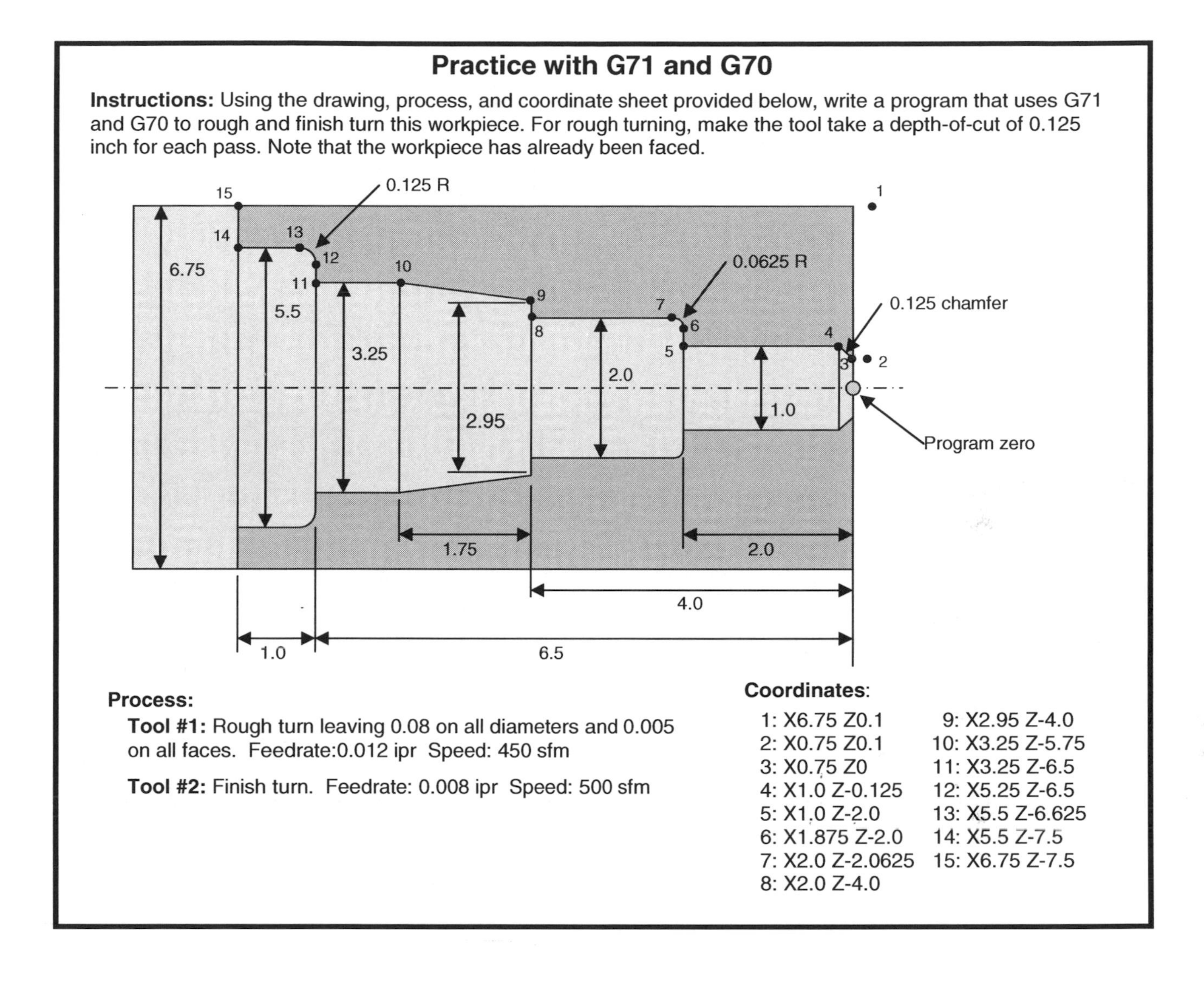

Practice with G71 and G70

Instructions: Using the drawing, process, and coordinate sheet provided below, write a program that uses G71 and G70 to rough and finish turn this workpiece. For rough turning, make the tool take a depth-of-cut of 0.125 inch for each pass. Note that the workpiece has already been faced.

Process:

Tool #1: Rough turn leaving 0.08 on all diameters and 0.005 on all faces. Feedrate:0.012 ipr Speed: 450 sfm

Tool #2: Finish turn. Feedrate: 0.008 ipr Speed: 500 sfm

Coordinates:

1: X6.75 Z0.1
2: X0.75 Z0.1
3: X0.75 Z0
4: X1.0 Z-0.125
5: X1.0 Z-2.0
6: X1.875 Z-2.0
7: X2.0 Z-2.0625
8: X2.0 Z-4.0
9: X2.95 Z-4.0
10: X3.25 Z-5.75
11: X3.25 Z-6.5
12: X5.25 Z-6.5
13: X5.5 Z-6.625
14: X5.5 Z-7.5
15: X6.75 Z-7.5

The answer program is on the next page.

Answer program

```
O0011
N002 G99 G20 G23
N004 G50 S5000

(ROUGH TURNING TOOL)
N005 T0101 M41
N010 G96 S450 M03
N015 G00 X6.75 Z0.1 M08 (1)
N020 G71 P025 Q090 U0.08 W0.005 D1250
  F0.012
N025 G00 X0.75 (2)
N030 G01 Z0 (3)
N035 X1.0 Z-0.125 (4)
N040 Z-2.0 (5)
N045 X1.875 (6)
N050 G03 X2.0 Z-2.0625 R0.0625 (7)
N055 G01 Z-4.0 (8)
N060 X2.95 (9)
N065 X3.25 Z-5.75 (10)
N070 Z-6.5 (11)
N075 X5.25 (12)
N080 G03 X5.5 Z-6.625 R0.125 (13)
N085 G01 Z-7.5 (14)
N090 X6.75 (15)
N095 X6.0 Z5.0
N100 M01

(FINISH TURNING TOOL)
N105 T0202 M42
N110 G96 S500 M03
N110 G00 G42 X6.75 Z0.1 M08 (1)
N115 G70 P025 Q090 F0.008
N120 G00 X5.0 Z6.0
N125 M30
```

STOP! Do practice exercise number eighteen in the workbook.

Lesson 19

G72: Rough Facing, G73: Pattern Repeating, G74: Peck Drilling, And G75: Grooving

The rough facing cycle (G72) works very much like G71 – but roughing is done by facing instead of turning. If your workpiece is large in diameter and short in length, it makes more sense to rough face than to rough turn. The pattern repeating cycle (G73), the peck drilling cycle (G74) and the grooving cycle (G75) are not so helpful. We'll explain why in this lesson.

In this lesson, we cover four more multiple repetitive cycles. But frankly speaking, you'll probably find that – of these four – only the G72 rough facing multiple repetitive cycle is very helpful.

G72 - Rough facing

There are times when it is not efficient to rough machine in a turning manner. When a workpiece is large in diameter and relatively short in length, it is usually better to rough machine in a facing manner since fewer passes will be required. G72 can be used for this purpose.

The G72 command is very similar to the G71 command. Here we will discuss *only the differences*. See Lesson Eighteen for more about how G71 works.

First, the convenient starting position is in a different location with G72. When using G72, the convenient starting position must be a position in X that clears the outside diameter of the raw material. In Z, it will be flush with the face of the raw material. (Point one in the figure 6.12 shows the convenient starting position for G72.) Second, the command specified by the P word must now include only a Z departure (no X word) along with a G00 or G01 (normally G00). Third, the finish pass definition must be done in a facing manner.

That's it. All words in the G72 mean exactly the same thing as they do in G71. And G70 can still be used for finishing. Knowing these differences, you should be able to understand this example program. Figure 6.12 shows the workpiece drawing along with the points needed for programming.

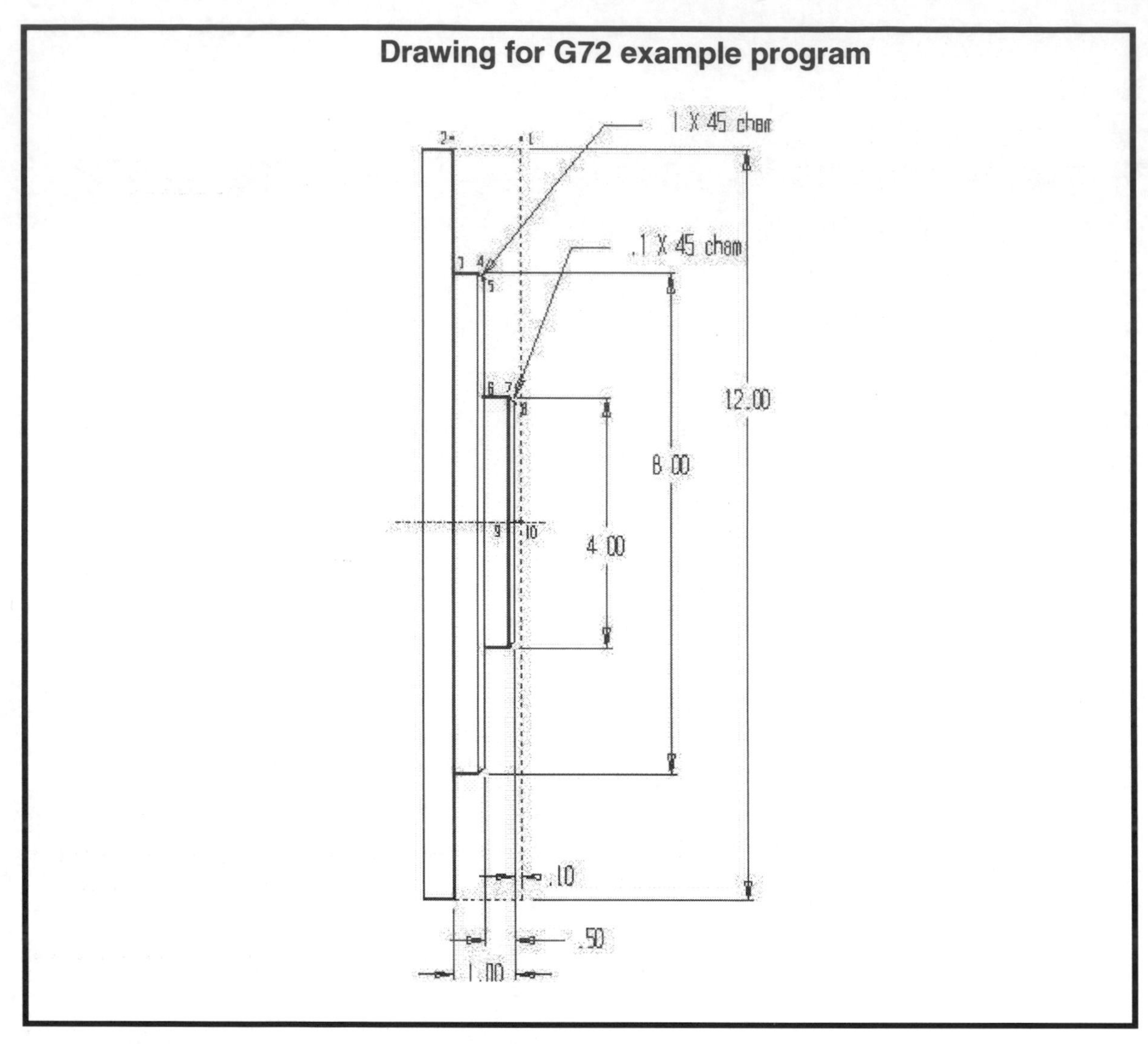

Figure 6.12 – Drawing for a G72 example program

Program:

O0013 (Program number)

N003 G99 G20 G23 (Ensure that all initialized states are still in effect)

N004 G50 S4000 (No need for limiting, limit to machine's maximum speed)

N005 T0101 M41 (Select rough facing tool and low spindle range)

N010 G99 G96 S300 M03 (Start spindle fwd at 300 sfm)

N015 G00 X12.4 Z0.1 M08 (Rapid to convenient starting position, point number 1, start coolant)

N020 **G72 P025 Q075 U0.04 W0.005 D1250 F0.015** (Rough face entire workpiece)

N025 G00 Z-1.1 (Rapid to point 2)

N040 G01 X8.0 F0.006 (Feed to point 3)

N045 Z-0.6 (Feed to point 4)

N050 X7.8 Z-0.5 (Feed to point 5)

N055 X4.0 (Feed to point 6)

N060 Z-0.1 (Feed to point 7)

N065 X3.8 Z0 (Feed to point 8)

N070 X-0.06 (Feed to point 9)

N075 Z0.1 (Feed to point 10)

N080 G00 X14.0 Z5.0 (Rapid to safe index point)

N085 M01 (Optional stop)

N090 T0202 M42 (Select finish facing tool and high spindle speed range)

N095 G96 S400 M03 (Start spindle fwd at 400 sfm)

N100 G00 **G41** X12.4 Z0.1 (Rapid to convenient starting position)

N105 **G70 P025 Q075 F0.006** (Finish face entire workpiece)

N110 G00 **G40** X14.0 Z5.0 (Rapid to safe index point)

N115 M30 (End of program)

Different format for 0TA, some 0TBs, 3T, and 18T controls

Again, these Fanuc controls require you to break up each cycle into two separate commands. As you know, this is because these controls do not have the letter address D to specify the depth-of-cut.

Here is the one-command G72 as executed on most controls:

N020 G72 P025 Q075 U0.04 W0.005 D1250 F0.015

Here is the format for 0TA, some 0TBs, and 3T controls.

N018 G72 U0.1250

N020 G72 P025 Q075 U0.04 W0.005 F.015

Again, only the depth-of-cut word (now a U word in the first G72) is affected.

G73 - pattern repeating

Though not as often used as G71 and G72, there may be times when a programmer wishes to rough machine by repeating a series of motions several times, each time letting the cutting tool get closer to the finished path. Figure 6.13 shows one possible example.

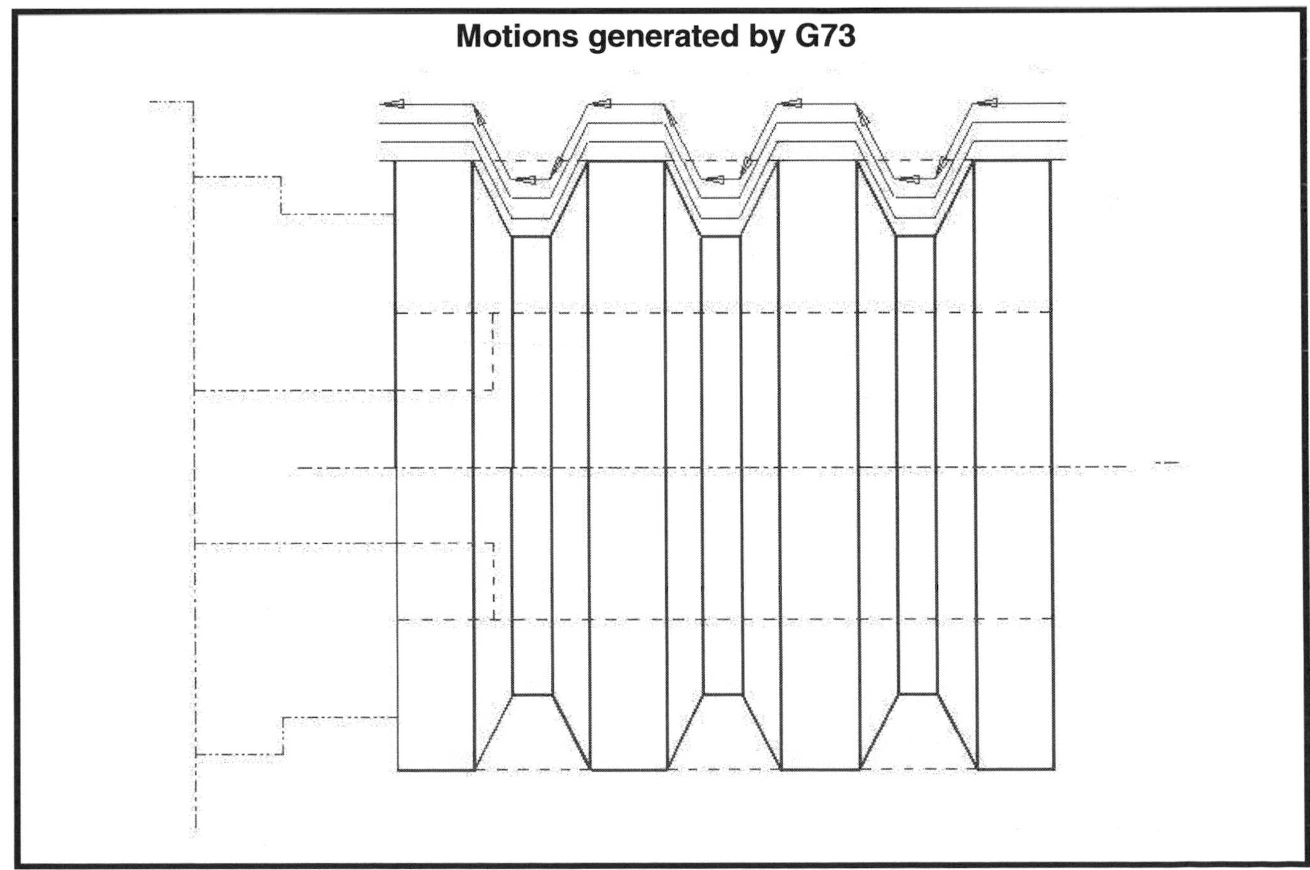

Figure 6.13 – Example of when G73 pattern repeating can be used

Notice that for the first few passes, the G73 command will waste a great deal of time cutting air. The first pass, for example, will not machine anything except during the very bottom of its movements in X. This is a common problem when using G73. For low production quantities (one or two workpieces), this may be acceptable, but for higher production quantities, it will be necessary to program more efficiently (but with much more difficulty) using G00, G01, G02 and G03 commands for roughing.

Many of the words used with G73 mean exactly the same thing as they do in G71 and G72. Only the new or changing word meanings will be presented.

D word

The D word no longer specifies a depth-of-cut in a G73 command. Instead, the D word now specifies the *total number of roughing passes.* D3, for example, specifies three passes. In essence, the control will divide the stock to be removed (specified by I and K in the G73 command) by the value of the D word to determine the depth of each pass in X and Z.

I word

The I word in a G73 command specifies how much stock is to be removed from each diameter (on the side). If left out of the G73 command, the control will assume a value of zero for the I word.

K word

The K word in a G73 command specifies how much stock is to be removed from each face. If left out of the G73 command, the control will assume the K value is zero.

Example of G73 pattern repeating

Figure 6.14 shows the drawing for our example. Tool number one is the rough turning tool and tool number two is the finish turning tool.

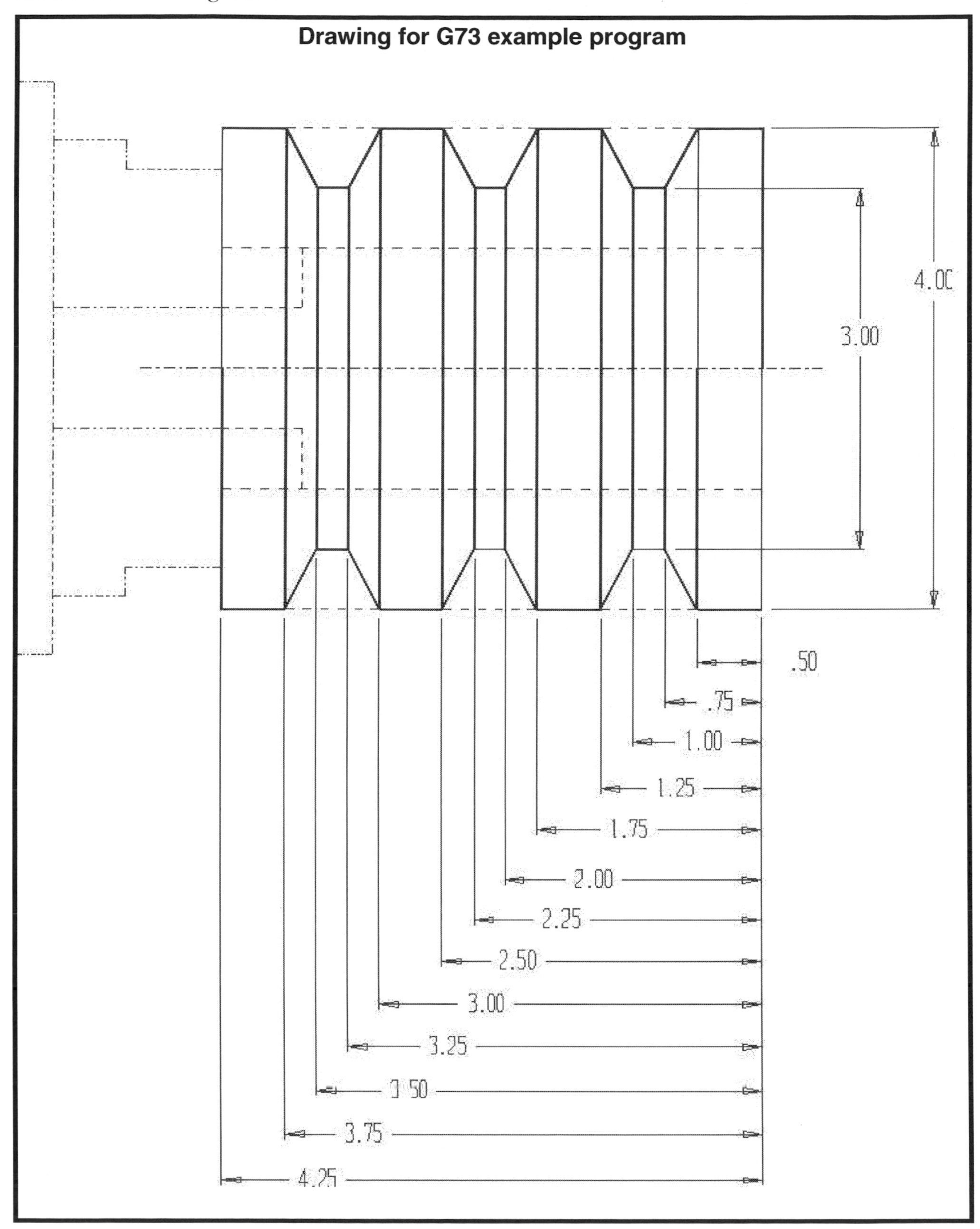

Figure 6.14 – Drawing for G73 pattern repeating example

Program:

```
O0014 (Program number)
N003 G99 G20 G23 (Ensure that initialized states are still in effect)
N004 G50 S4000 (No need for limiting, limit to machine's maximum speed)
N005 T0101 M41 (Select rough profiling tool and low spindle speed range)
N010 G99 G96 S300 M03 (Start spindle fwd at 300 sfm)
N015 G00 X5.2 Z0.1 M08 (Rapid to convenient starting position, start coolant)
N020 G73 P025 Q090 U0.040 W0 I0.5 D4 F0.012 (Rough entire workpiece)
N025 G00 X4.0 (Rapid flush with diameter to machine)
N030 G01 Z-.5 F0.006
N035 X3.0 Z-0.75
N040 Z-1.0
N045 X4.0 Z-1.25
N050 Z-1.75
N055 X3.0 Z-2.0
N060 Z-2.25
N065 X4.0 Z-2.5
N070 Z-3.0
N075 X3.0 Z-3.25
N080 Z-3.5
N085 X4.0 Z-3.75
N090 Z-4.35
N095 G00 X6.0 Z5.0 (Rapid to safe index point)
N100 M01 (Optional stop)
N105 T0202 M42 (Select finish profiling tool and high spindle range)
N110 G96 S400 M03 (Start spindle fwd at 400 sfm)
N115 G00 G42 X5.2 Z0.1 (Rapid to convenient starting position)
N120 G70 P025 Q090 F0.006 (Finish profile outside diameters)
N125 G00 G40 X6.0 Z5.0 (Rapid to safe index point)
N130 M30 (End of program)
```

Notes about the program:

1) A special tool which is capable of reaching into these grooves must be used.
2) In line N015, notice that the convenient starting position must be at least the stock to be removed amount (I) away from the surface to be machined.
3) Four passes will be made with a depth of cut of about 0.125 inch (I0.5 divided by 4)

Can you use G73 for castings and forgings?

Some programmers try to program the roughing operations for castings and forgings with the G73 command. For workpieces that have a balanced amount of stock on all surfaces, and for workpieces that do not vary much from one diameter to the next, the G73 will work just fine.

Unfortunately, there will be few times when the amount of roughing stock on a casting or forging will be balanced. Most design engineers design the casting or forging to be as simple as possible. In many cases, no two finished surfaces will have the same amount of roughing stock.

Also, when it comes to large faces, if the programmer attempts to program the true amount of roughing stock (with the K word), when the control divides the K word by the D word (number of passes), there will usually be a substantial amount of stock to be removed on each face. The turning tool will attempt to machine this amount of stock for the entire length of the face. This will bury the tool (breaking it), and will makes it infeasible to use G73 alone for roughing castings and forgings.

Some programmers get around this problem by roughing all faces (using G00 and G01 – or G94 motions) prior to using the G73 command. Unfortunately, from an efficiency standpoint, nothing beats laying out the entire surface to be machined (rough and finish) on a piece of graph paper and making the movements longhand with G00 and G01. Though this method takes much longer to program, if lot sizes are large, the program will be much efficient than one using G73.

Another alternative when programming castings and forging is to use a computer aided manufacturing (CAM) system. Many CAM systems designed for turning centers have elaborate casting and forging capabilities. They generate efficient roughing commands using G00 and G01.

Different format for 0TA, some 0TBs, 3T, and 18T controls

Again, these controls require you to break up each cycle into two independent commands. This is because these controls do not have the letter address D to specify the number of passes, nor do they use I and K in the command to specify machining stock.

Here is the G73 command as shown earlier:

N020 G73 P025 Q090 U.040 W0 I.5 D4 F0.012

The new format for 0TA, some 0TBs, 3T, and 18T is as follows:

N018 G73 U0.5 R4.0

N020 G73 P025 Q090 U.040 W0 F.012

Notice that the format for G73 is a little different. The number of passes is now specified by the R word in the first G73 command. Also, a U word in the first G73 command specifies the amount of stock to be removed on the side, and a W word is used to specify of stock to be removed in Z (not used in this example).

G74 - peck drilling

Though not an extremely helpful cycle, the G74 can be used to break chips as a hole is being drilled. It causes the tool to peck into the hole a specified distance, back out about 0.005 inch (which breaks the chip), and peck again. This process is repeated until the hole bottom position is reached.

We say it is not very helpful because most programmers would rather have a peck drilling cycle that will *clear chips* after each peck rather than just break them. To clear chips, the drill must be retracted all the way out of the hole between pecks. G74 will not do this.

Figure 6.15 shows the word meanings of the G74 command.

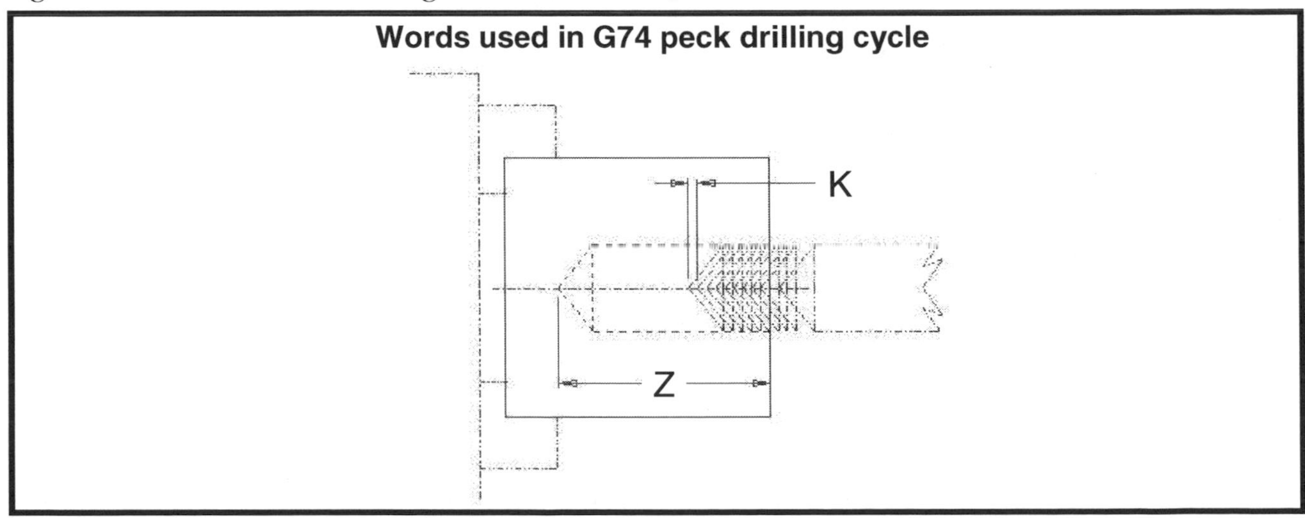

Figure 6.15 – G74 word meanings

Here is an example program that will drill a hole two inches deep with 0.100 inch pecks. As you know, drilling must be done in the rpm mode (G97).

Program:

```
O0015 (Program number)
N003 G99 G20 G23 (Ensure that initialized states are still in effect)
N004 G50 S4000 (No need for limiting, limit to machine's maximum speed)
N005 T0101 M41 (Select drill station and low spindle range)
N010 G97 S300 M03 (Start spindle fwd at 300 rpm)
N015 G00 X0 Z0.1 M08 (Rapid up to workpiece, start coolant)
N020 G74 Z-2.0 K0.1 F0.010 (Peck drill hole)
N025 G00 X4.5 Z6.0 (Rapid to safe index point)
N030 M30 (End of program)
```

What if I must clear chips between pecks?

Any hole that has a depth greater than about four times the drill diameter requires deep hole peck drilling – the drill must retract from the hole to chips between pecks. If you have a two-axis turning center (without live tooling), there will only be one hole to drill – right at the center of the workpiece. If you have a three axis turning center equipped with live tooling, there could be more than one hole to drill – and most machine tool builders will equip their live tooling turning centers with a series of hole machining canned cycles to facilitate the programming of hole machining operations (see the appendix for more on turning centers that have live tooling).

But again, with two axis turning centers there can only be one hole to drill. And it is not at all difficult to program a series of chip-clearing pecks using G00 and G01 motions. Say for example, you have a 0.5 inch diameter hole to drill that is 4.0 inches deep. This hole has a depth of eight times the drill diameter (4.0 divided by 0.5) – and most machinists will agree that this hole requires a chip clearing peck drilling operation. Here is the program segment for the drill. The workpiece has already been faced and the Z axis program zero point is at the right end of the finished workpiece. Each peck is about three times the drill diameter (we're using a balanced peck depth of 1.33 inches – 4.0 inch depth divided by three passes).

```
.
.
N075 T0404 M41 (Index turret to 0.5" drill)
N080 G97 S600 M03 (Start spindle fwd at 600 rpm)
N085 G00 X0 Z0.1 M08 (Rapid to approach position, start coolant)
N090 G01 Z-1.33 F0.01 (Plunge first peck)
N095 G00 Z0.1 (Retract to clear chips)
N100 Z-1.23 (Rapid back into the hole to within 0.1 of where the drill left off)
N105 G01 Z-2.66 (Plunge second peck)
N110 G00 Z0.1 (Retract to clear chips)
N115 Z-2.56 (Rapid back into the hole to within 0.1 of where the drill left off)
N120 G01 Z-4.0 (Plunge third and final peck)
N125 G00 Z0.1 (Retract from hole)
N130 X5.0 Z6.0 (Rapid to turret index position)
N135 M01 (Optional stop)
.
.
```

Different format for 0TA, some 0TBs, 3T, and 18T controls

Here is the format for the G74 peck drilling command given earlier.

```
N020 G74 Z-2.0 K.1 F.010
```

The format for the 0TA, some 0TBs, 3T, and 18T is as follows.

```
N020 G74 Z-2.0 Q0.1 F0.010
```

The only difference is that the Q word is used to specify the depth of each peck instead of a K word.

G75 - grooving cycle

Like G74, the G75 grooving cycle is not very helpful. It simply causes the grooving tool to plunge one time and rapid out of the groove. When grooving is required, the grooving tool must additionally machine chamfers or radii on the outside corners of the groove. Also, the grooving tool may be narrower than the groove itself, meaning multiple passes must be made. Some programmers even like to have the grooving tool sweep the bottom of the groove to machine a nice finish. None of this is done by G75.

For this reason, most programmers rely on longhand programming techniques (G00 and G01) to machine grooves. During our discussion of sub-programming techniques (Lesson Twenty-One), we will show a way to simplify the machining of multiple identical grooves. You'll also see our recommended method to machine grooves.

Figure 6.16 shows the words related to the G75 cycle.

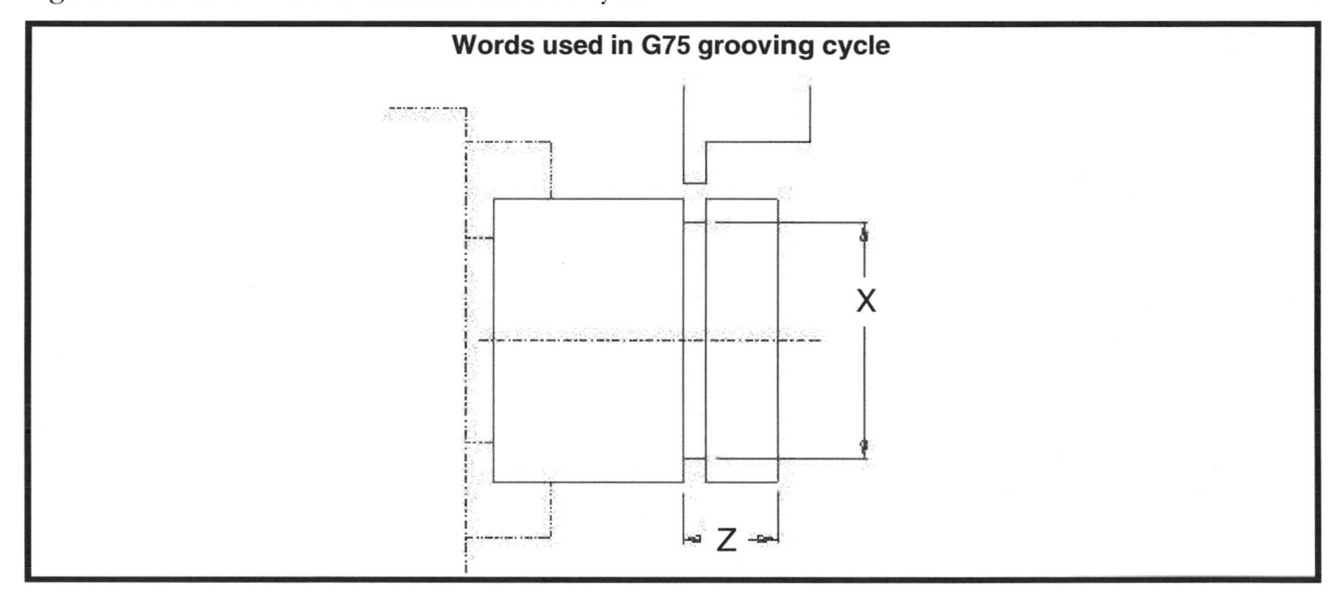

Figure 6.16 – Words related to G75 grooving cycle

Here is a simple program that machines a groove (without corner breaks) into a 3.0 inch diameter to a bottom position of 2.75 in the X axis at a -1.0 inch Z position.

Program:

```
O0016 (Program number)
N003 G99 G20 G23 (Ensure that initialized states are still in effect)
N004 G50 S4000 (No need for limiting, limit to machine's maximum speed)
N005 T0101 M41 (Select grooving tool station and low spindle range)
N010 G97 S300 M03 (Start spindle fwd at 300 rpm)
N015 G00 X3.2 Z-1.0 M08 (Rapid up to workpiece, start coolant)
N020 G75 X2.75 F0.005 (Plunge groove and retract)
```

N025 G00 X4.5 Z6.0 (Rapid to safe index point)

N030 M30 (End of program)

As you can see, the G75 command doesn't do much. It simply causes the tool to feed to the bottom of the groove and rapid out.

Key points for Lesson Nineteen:

- Of the four multiple repetitive cycles shown in this lesson, only the G72 rough facing command is very helpful.
- Like G71, G72 will completely rough (face) the workpiece based upon one command and a finish pass definition.
- There are three differences between G71 and G72. First, the convenient starting position for G72 will be clear of the workpiece in the X axis and flush with the face in the Z axis. Second, the command specified by the P word in the G72 command must contain a G00 and only a Z word. Third, the finish pass definition for G72 must be done in a facing mode.
- G73 is the pattern repeating cycle. It can sometimes be helpful – as long as cycle time is not critical.
- G74 is the peck drilling cycle – but it only breaks chips as the hole is drilled. Most programmers prefer a deep hole peck drilling cycle – one that clears chips between pecks.
- G75 is the grooving cycle – but it does not machine chamfers or radii on the corners of the groove. Nor does it machine a groove that is wider than the grooving tool.

STOP! Do practice exercise number nineteen in the workbook.

Lesson 20

G76 – Threading Command

This command is extremely helpful for machining threads. It machines external threads and internal threads with equal ease. Only one command is needed to machine the thread – regardless of how many passes are required.

Threading involves *chasing* a thread on the workpiece with a single point threading tool. By chasing, we mean the threading tool will make a series of synchronized passes to machine the thread to its required depth. Thread depth will increase with each pass. The threading tool has the form of the thread ground into its insert.

G76 is an extremely helpful multiple repetitive cycle that will machine an entire thread in one command. Like the other multiple repetitive cycles, the G76 command allows you to specify the criteria for the thread to be machined with a series of *variables* that are included right in the G76 command.

Like the G71 command, G76 can be used for outside diameters as well as inside diameters. Also like the G71, the programmer will rapid the threading tool to a *convenient starting position.* This position for threading will be clear of the surface to be threaded in the X axis (by at least 0.03 inch or so), and well off the face of the diameter to be threaded in the Z axis (we recommend a distance of four times the pitch or 0.2 inch, *whichever is smaller*). This large approach distance in Z is required to allow the threading tool to accelerate to the required motion rate (feedrate) before contacting the workpiece. If it does not, the thread's pitch will not be correct at the beginning of the thread.

Figure 6.17 shows the words related to the G76 threading cycle. Note that an external thread is being shown.

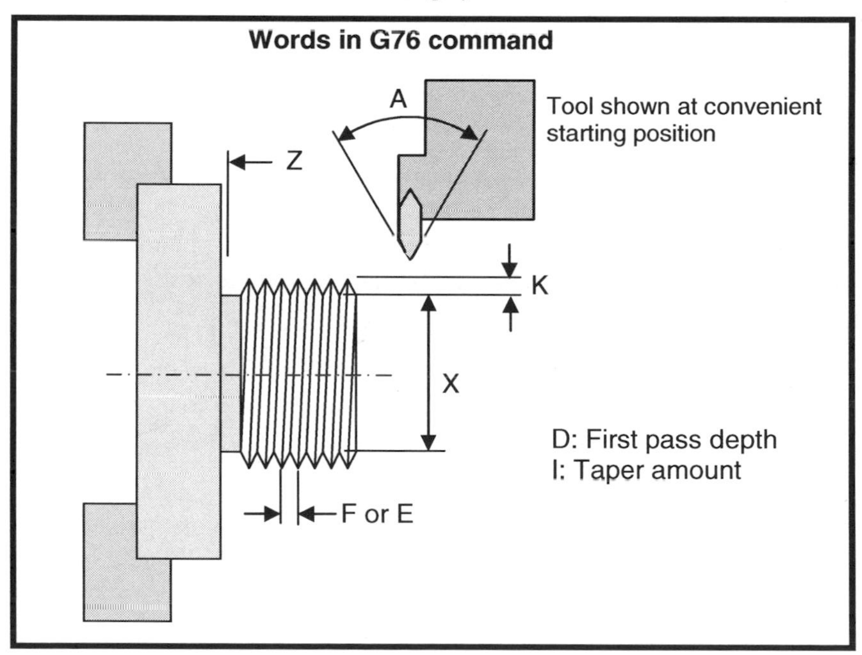

Figure 6.17 – Words related to G76 command

X word

The X word in a G76 command specifies the *final threading diameter.* For external threads, the X word is the minor diameter of the thread. For internal threads, the X word is the major diameter of the thread.

Z word
The Z word in a G76 specifies the end point for every threading pass in the Z axis. When threading into a recess (as Figure 6.17 shows), the Z word value will be in the middle of the recess the recess. When threading a solid surface, the Z word value will be the ending point of the thread. As discussed in Lesson Ten, we recommend programming the leading edge of the threading insert in the Z axis (as opposed to the point of the insert). This makes it easy for the setup person to determine the program zero assignment values for the tool – and when threading up to a shoulder (as is done in Figure 6.17), it allows you to specify a Z word that is very close to the shoulder without fear of crashing the threading tool.

What is thread chamfering?
There is a feature called *thread chamfering* that affects how close to the Z word value the threading tool will actually come before chamfering begins. If thread chamfering is turned on, the tool will come to within about one pitch (1 divided by the number of threads per inch) of the Z value. At this point, the threading tool will begin feeding away in the X axis while it continues moving in the Z axis (forming a forty-five degree angle of pull-out). This causes the threading tool to gradually retract from the thread.

Thread chamfering is most helpful when there is no thread relief groove at the end of the thread. With this kind of thread – and especially with older machines - the thread will *look better* if threading is done with thread chamfering turned on.

If thread chamfering is off (which is the initialized state for most turning centers), the tool will feed precisely to the Z word value of the G76 command and pull directly away from the thread. This is required when threading into a thread relief.

How thread chamfering is turned on and off is determined by the machine tool builder. Some builders provide two M codes, one to turn thread chamfering on and another to turn it off. Others require that a control parameter be changed in order to activate thread chamfering. Again, the initialized state for thread chamfering is usually off.

Since newer machines provide very accurate threading motions, the need for thread chamfering has been reduced over the years. Most current machines can provide accurate and nice looking threads with thread chamfering turned off – even when there is no relief groove at the end of the thread.

K word
The K word specifies the thread depth (on the side). Unfortunately, most design engineers do not provide the thread depth as part of the thread specification. To determine this value, you will need to reference a machining handbook (like Machinery's Handbook). Look for the major and minor diameters of the thread. The thread depth (K word value) can be calculated by subtracting the minor diameter from the major diameter – and then dividing the result by two.

Some programmers approximate thread depth by using 75% of the thread's pitch. This works quite well for National Standard (60 degree included angle) threads. If the thread depth is approximated, the threading tool's wear offset can be used to fine-tune the depth of the thread when the first workpiece is run.

D word
The D word specifies the *depth of the first pass* (on the side). The control will automatically reduce each subsequent threading pass depth from this value – so the depth of successive passes will get progressively shallower.

The D word value controls how many passes will be made. Generally speaking, the larger the D word, the fewer the number of threading passes. And actually determining the D word value will take some practice. If you're in doubt, we recommend starting with a small value (forcing more passes than necessary) and working your way up. Generally speaking, coarse threads will allow a larger D word than fine threads. As a starting point, we recommend a D word of about 15% of the pitch for external threads and about 10% of the pitch for internal threads. For example, if machining a 1"-8 external thread (0.125 pitch), start with a D word of about 0.019 (which is fifteen percent of 0.125).

As with G71 and G72, the D word in G76 can help with optimizing. If the threading cycle is making too many passes, increase the D word. If too few passes are being made, reduce the D word.

Also like G71 and G72, most controls do not allow a decimal point to be used with the D word used with G76. In the inch mode, a four place fixed format must be used. D0150 specifies a 0.015 inch depth of cut. D0100 is 0.010 inch, and so on. In the metric mode, a three place format must be used. D500 specifies a 0.5 mm depth-of-cut.

A word

The A word specifies the *tool angle*. The A word does not allow a decimal point. Many threading tools are designed to machine with only the leading edge of the tool. If A is specified in the G76 command (and not set to 0), the threading tool will machine with only its leading edge.

For National Standard threads, the included angle for the thread is sixty degrees. With an A word specified as A60 in the G76 command, the threading tool will in-feed with each pass at an angle of just under thirty degrees.

Older Fanuc controls require that you specify the A word from these choices: A80, A60, A55, A30, A29, and A0. Newer controls allow any tool angle to be specified. An A word of A29, for example, will be used to machine an ACME thread.

Some threading tools are designed to machine on both sides of the threading insert. These threading tool inserts are commonly referred to as *cresting inserts*. If you use a cresting insert, simply leave the A word out of the G76 command – or set it to a value of zero (A0). This will cause the threading tool to approach in only the X axis for each threading pass. That is, three will be no angular in-feed.

F word and E word

As you know, the F word specifies feedrate in feed per revolution (always thread in feed per revolution mode – G99). And of course, the pitch of any thread is the amount of Z axis motion needed during each spindle revolution – meaning the F word will be equal to the pitch of the thread.

With threads specified in the inch mode, pitch is equal to one divided by the number of threads per inch. With metric threads, the pitch is provided right in the threads specification.

For example, an external metric thread specified as 50-1.5 has a fifty millimeter major diameter and a 1.5 millimeter pitch. An external inch thread specified as 2.0-12 has a two inch major diameter and twelve threads per inch. The pitch for this thread is 0.08333333333333... (1/12).

In the inch mode, you know the F word has a four-place decimal format. This means you can only program the F word to four places. A sixteen-threads-per-inch thread has a pitch of precisely 0.0625, meaning you can perfectly specify this pitch with the F word (with F0.0625). But with a twelve threads per inch thread, the pitch is 0.083333333333... The best you can do with an F word is F0.0833.

This is usually acceptable (as long as mating parts are programmed with the same pitch value). However, you can gain two more digits of pitch accuracy by using the E word instead of the F word in the G76 command. That is, the E word allows a six-place decimal format. With the E word, you can specify the pitch of a twelve threads per inch thread as E0.083333. Again, in most applications, you won't notice the difference. But consider workpieces that require a perfect pitch value for threading. If you are machining a lead screw, for example, it is quite important to make the pitch of the lead screw as perfect as possible.

I word

The I word is only used for *taper threading* to specify the amount of taper in the X axis. We'll show how taper threading is programmed a little later in this lesson.

Q word

Some newer Fanuc controls use the Q word to specify the angular position of the thread start when machining multiple start threads. We'll show how multiple start threads are programmed a little later in this lesson.

Example program for threading

Figure 6.18 shows the workpiece to be threaded. All machining (other than threading) has been previously done. Only the threading tools are shown (one external threading tool and one internal threading tool).

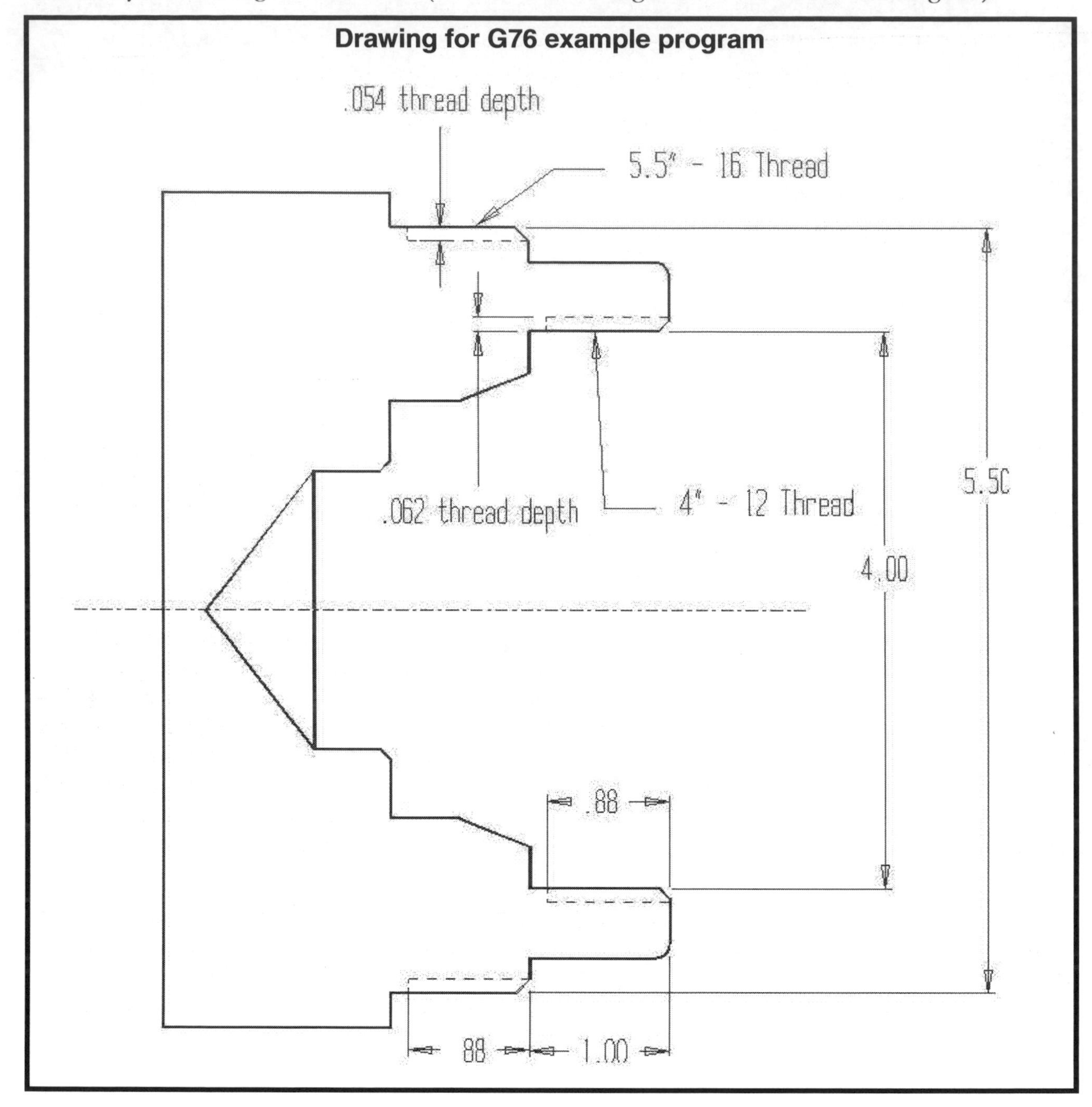

Figure 6.18 – Drawing for threading example program

Program:

O0016 (Program number)
N003 G99 G20 G23 (Ensure that initialized states are still in effect)
N004 G50 S4000 (No need for limiting, limit to machine's maximum speed)
N005 T0505 M41 (Select external threading tool and low spindle range)
N010 G97 S500 M03 (Start spindle fwd at 500 rpm)
N015 G00 X5.7 Z-0.8 M08 (Rapid to convenient starting position, start coolant)
N020 **G76 X5.392 Z-1.88 K0.054 D0100 A60 F0.0625** (Chase 5.5-16 thread)

```
N025 G00 X8.0 Z6.0 (Rapid to safe index position)
N030 M01 (Optional stop)
N035 T0707 M41 (Select internal threading tool and low spindle range)
N040 G97 S700 M03 (Start spindle fwd at 700 rpm)
N045 G00 X3.8 Z0.2 M08 (Rapid to convenient starting position, start coolant)
N050 G76 X4.124 Z-0.88 K0.062 D0080 A60 F0.0833 (Chase 4-12 thread)
N055 G00 X8.0 Z6.0 (Rapid to safe index point)
N060 M30 (End of program)
```

Other tips on threading

When possible, thread in the lowest spindle range

This will ensure that the spindle has the inertia to maintain a constant speed in rpm while each threading pass is done. If the spindle speed varies, of course, the pitch of the thread will not be maintained.

Thread in the rpm mode (G97)

Most machines require that you thread in the rpm mode – specified by G97. Remember that in constant surface speed mode, the machine will change rpm to match the diameter of the cutting tool. When threading, this means the spindle speed will change in rpm with each threading pass if the machine is in the constant surface speed mode. Again, most machines do not allow this because they cannot perfectly synchronize the feedrate with the (changing) spindle speed as each pass is made.

Watch out for maximum allowable feedrate

Remember that all turning centers have a maximum feedrate. With most machines, the maximum feedrate is about half the rapid rate. For a machine having a rapid rate of 400 inches per minute (rather slow by today's standards), the maximum feedrate will be about 200 inches per minute. While 200 ipm may sound pretty fast, you may exceed this value when machining coarse threads (or multiple start threads). Consider machining 1.0-8 thread (one inch diameter having eight threads per inch). The pitch, of course, is 0.125 inches (1/8). Say you run the spindle at a speed of 2300 rpm (about 600 sfm). The required feedrate to machine this thread is 286 ipm (2300 times 0.125). The machine just mentioned cannot feed this fast. Unfortunately, most machines will not generate an alarm in this case. Instead, they'll simply do their best to machine the thread – and the pitch of the thread will be incorrect. This can be a difficult problem to diagnose at the machine!

Thread with thread chamfering turned off

In almost all cases, you will have no problems threading with thread chamfering off (even into a solid diameter with no thread recess). On the other hand, some odd things can happen when thread chamfering is turned on (especially if you don't know it). As an example, say you are machining a very short coarse thread – maybe a thread that is only 0.5 inches long. The threading tool may begin retracting in the X axis to start chamfering long before it should – possibly even before the threading tool comes into contact with the workpiece.

If your machine has M codes to specify thread chamfering on and off, be sure to command the M code that turns thread chamfering off prior to every threading command. If you don't have M codes for this purpose (many machines do not), we recommend setting the thread chamfering parameter so that the amount of thread chamfering is zero. Unfortunately, parameter numbers vary from one Fanuc model to another, so you'll have to look it up. You'll find it documented in the G76 discussion in the Fanuc programming manual.

Finish the thread before removing workpiece

If you remove the workpiece prior to finishing your thread, there will be no (feasible) way to replace it in the machine and continue threading. Additionally, we recommend *not turning the machine's power off* until a thread is completely finished for the same reason.

Right hand threads versus left hand threads

If you are threading toward the chuck with a right hand threading tool (spindle running forward – M03), you will machine a right hand thread. If you need to make left hand threads, you will either have to thread from the chuck side toward the tailstock side (spindle still forward – M03 with right hand tool) or use a left hand threading tool and thread toward the chuck (spindle running reverse – M04).

Offsetting for threading tools

You can use wear offsets to sneak up on the final threading diameter (in X) just as you would for any cutting tool. You can even use trial machining techniques when you're in doubt about how the threading tool will perform (first workpiece). You can offset the X axis and re-machine with the tool as many times as necessary. However, *you cannot offset the Z axis for a threading tool unless you can do so in increments of the thread's pitch.* Offsetting in any other increment will result in *cross threading* (the threading tool will not follow the previously machined lead). This will cause a scrap workpiece.

Start the tool far enough away from the thread being machined

In order to allow the threading tool to accelerate to a constant and stable feedrate, we recommend a larger approach distance for threading tools than for other types of cutting tools. Though turning centers have different axis acceleration characteristics and you may be able to use a smaller approach distance, we recommend an approach distance that is four times the thread pitch or 0.2 inch, whichever is smaller.

Minimum depth-of-cut, final pass depth, and number of spring passes

There are three more important threading-related *parameters* that control the *minimum depth-of-cut*, the *final pass depth*, and how many *spring passes* will be made. The minimum depth of cut is the smallest depth-of-cut during threading the tool will take (other than the final pass depth). For most threads, the minimum depth-of-cut should be set to about 0.005 inch. The final pass depth is the depth-of-cut the threading tool will use for the last pass (other than spring passes) and should be set to about 0.0002 inch. The number of spring passes controls how many passes of zero depth will be made after the threading tool has reached its final depth (spring passes are used to relieve tool pressure). Normally one or two spring passes will suffice.

With most controls, these three important values are set within the control's parameters. That is, you cannot control them from within a program (again with most controls). To check or change these values, you must find the appropriate parameters. But again, parameter numbers vary from one control model to another, meaning you must reference your Fanuc programming manual to find the related parameter numbers. You will find them documented in the G76 description of the Fanuc programming manual.

If these parameters are improperly set, threading operations will behave poorly. We've actually seen programmers who have given up on using G76 because these parameters were improperly set – and they didn't know it.

If working with a 0TA, some 0TBs, 3T, or 18T Fanuc control, you do have control of these three functions right from within your G76 commands. We'll show how a little later in this lesson.

Disabled or modified control functions during threading

There are three control panel functions that will behave differently during a threading operation. *Feedrate override* – which is normally used to modify the programmed feedrate during a machining operation – is disabled during G76 command execution. The operator cannot be allowed to manipulate the programmed feedrate, or course, or else the thread pitch will not be machined correctly.

Feed hold is also disabled during the threading motion. This button normally allows the operator to temporarily stop the machine's axis motions at any time. But if this button is enabled during a threading pass, of course, threads will be damaged when it is pressed. If the operator presses feed hold during a G76 treading cycle, the machine will complete the current pass before it will stop the motion.

In similar fashion, *single block* will cause the machine to make one entire pass when a G76 command is being activated. It will not stop the machine at the end of the current motion as it normally does.

Tapered threads

Taper threading requires that the workpiece be *taper turned* prior to threading. The taper angle for taper threading is usually 3.718 degrees (on the side), though you should confirm this angle with your thread's specification on the workpiece drawing on in a machinist's handbook. Figure 6.19 shows a diagram giving the meaning of each word when the G76 command is used for taper threading.

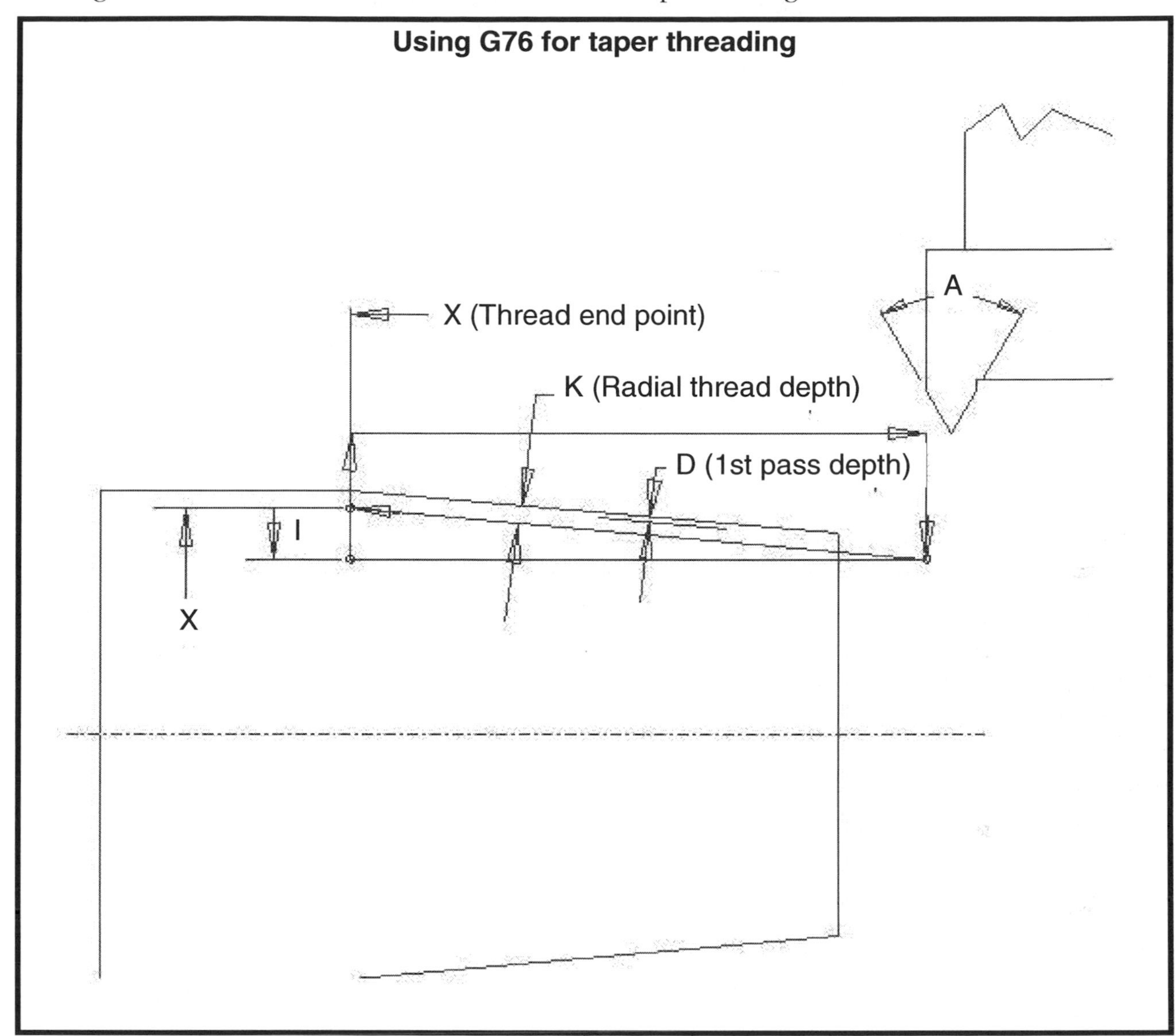

Figure 6.19 – Words involve with G76 when taper threading

Notice the addition of an *I word*. The I word specifies the distance and direction from the end point of the thread to the starting point of the thread along the X axis. For external taper threads, I will be negative. For internal taper threads, I will be positive.

The actual value of I requires trigonometry to calculate. The value of I is equal to the tangent of the taper angle (again, usually 3.718 degrees) times the total length of the threading pass along the Z axis. *The thread pass length must include the approach distance* (usually 0.200 inch). The tangent of 3.718 degrees is 0.06498.

If, for example, your external thread is 1.0 inch long and you have a 0.2 inch approach, the value of the I word will be I-0.0779 (tangent of 3.718 [0.06498] times 1.2). But again, you must confirm the taper angle for the threads you machine.

Multiple start threads

Multiple start threads are required when mating workpieces require the smooth fit of a fine thread with the fast axial motion of a coarse thread. Multiple start threads require more than one lead to be machined. Figure 6.20 shows an example of a *four-start* thread.

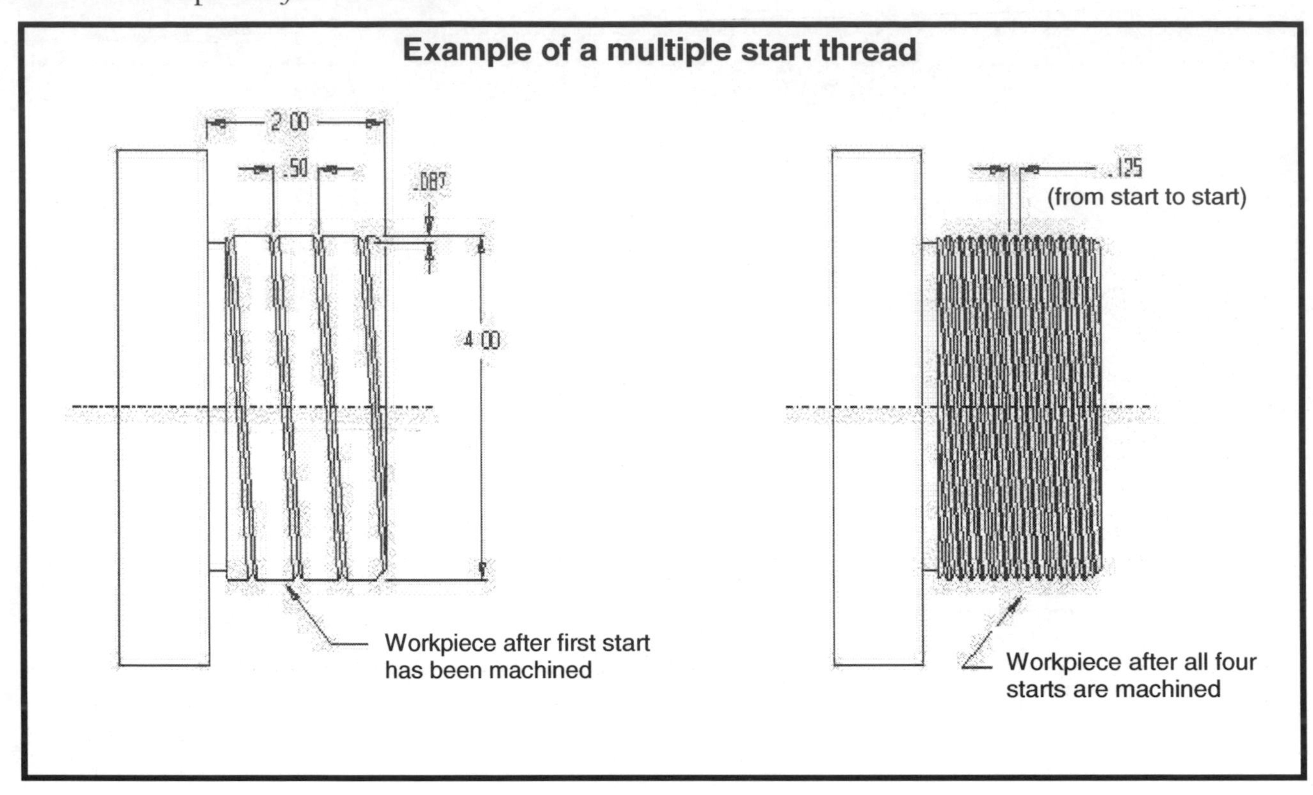

Figure 6.20 – Drawing shows a four-start multiple start thread

The drawing on the left shows what the thread will look like after the first start is machined. The *lead* of the thread (0.5 inch in this case) is determined over one start. This will be the feedrate in the G76 commands. The drawing on the right shows what the thread will look like after all four starts have been machined.

Most machines require that multiple start threads be machined by *cross threading in a controlled manner*. In essence, you will be machining four independent thread starts – meaning four separate G76 commands for a four-start thread. The feedrate used for each thread start must be the overall lead of the thread (0.5 inch in our example).

One way to machine this thread (that will work on all controls) is to physically move the tool in the Z axis after each G76 command by a value equal to the overall lead (0.5 in our case) divided by the number of thread starts (4 in our case). The move-over amount after each start is machined in our example is 0.125 inch (again, 0.5 divided by 4). This will force the tool to cross thread in a controlled manner when subsequent G76 commands are given.

Notice the very large feedrate word. In our case, the threading tool will be moving at a rate of 0.5 inches per revolution. You must be careful not to exceed the machine's maximum feedrate as discussed earlier. You may have to slow the spindle speed to ensure that maximum feedrate is not exceeded. Multiply feedrate (0.5 in our case) times your intended rpm to determine the threading tool's feedrate in inches per minute. This ipm feedrate cannot be more than about half the machine's rapid rate. In the example program that follows, we're using a speed of 500 rpm. The machine must have a maximum feedrate of at least 250 ipm in order to machine our example workpiece (0.5 times 500).

Here is a program that that machines the four start thread shown in Figure 6.20.

Program:

O0018 (Program number)
N003 G99 G20 G23 (Ensure that initialized states are still in effect)
N004 G50 S4000 (No need for limiting, limit to machine's maximum speed)
N005 T0101 M41 (Select threading tool and low spindle range)
N010 G97 S500 M03 (Turn spindle on fwd at 500 rpm)
N015 G00 X4.2 Z0.2 M08 (Rapid to convenient starting position to machine first start, start coolant)
N020 **G76 X3.826 Z-1.9 K0.087 D0160 A60 F0.500** (Machine first start)
N025 **G00 Z0.325** (Move over to starting position for second start)
N030 **G76 X3.826 Z-1.9 K0.087 D0160 A60 F0.500** (Machine second start)
N035 **G00 Z0.450** (Move over to starting position for third start)
N040 **G76 X3.826 Z-1.9 K0.087 D0160 A60 F0.500** (Machine third start)
N045 **G00 Z0.575** (Move over to starting position for fourth start)
N050 **G76 X3.826 Z-1.9 K0.087 D0160 A60 F0.500** (Machine fourth start)
N055 G00 X8.0 Z6.0 (Rapid to safe index point)
N060 M30 (End of program)

In blocks N025, N035, and N045, the tool is being moved more and more plus in the Z axis in 0.125 increments. While this technique works, it has two limitations. First, each subsequent thread start will take longer than the last because the threading tool must move further during each pass. This results in a longer cycle time. Second, and more important, there may be times when it is impossible to move plus in the Z axis without driving the threading tool into an obstruction (like the tailstock). For these reasons, newer controls make it possible to machine multiple start threads without moving in the Z axis between starts.

For these controls, a Q word is used within the G76 command to tell the control the rotational angular position of each thread start. Here is another program that uses the Q word and machines the workpiece shown in Figure 6.20.

Program:

O0019 (Program number)
N003 G99 G20 G23 (Ensure that initialized states are still in effect)
N004 G50 S4000 (No need for limiting, limit to machine's maximum speed)
N005 T0101 M41 (Select threading tool and low spindle range)
N010 G97 S500 M03 (Turn spindle on fwd at 500 rpm)
N015 G00 X4.2 Z0.2 M08 (Rapid to convenient starting position to machine first start, start coolant)
N020 G76 X3.826 Z-1.9 K0.087 D0160 A60 F0.500 **Q0** (Machine first start)
N025 G76 X3.826 Z-1.9 K0.087 D0160 A60 F0.500 **Q90.0** (Machine second start)
N030 G76 X3.826 Z-1.9 K0.087 D0160 A60 F0.500 **Q180.0** (Machine third start)
N035 G76 X3.826 Z-1.9 K0.087 D0160 A60 F0.500 **Q270.0** (Machine fourth start)
N055 G00 X8.0 Z6.0 (Rapid to safe index point)
N060 M30 (End of program)

Notice the Q word in each G76 command. It simply tells the control where (angularly) to begin the thread start. For a four start thread, each thread start is simply 90 degrees apart (360 divided by four).

Different format for 0TA, some 0TBs, and 3T controls

The multiple repetitive cycle for threading G76 has several differences among the Fanuc control models. Actually, the G76 format for 0TA, some 0TBs, 3T, and 18T allows more flexibility than the method shown so far.

Here is an example threading command as given in program number O0016 (drawing in Figure 6.18).

```
N020 G76 X5.392 Z-1.88 K.054 D0100 A60 F0.0625
```

Now here is the same command given in 0TA, some 0TBs, 3T, and 18T format:

```
N018 G76 P000060 Q0.005 R0.0002
N020 G76 X5.392 Z-1.88 P0540 Q0.0100 F0.0625
```

The P word in the first G76 command does three things. The first two digits represent the number of spring passes (currently set to none). The second two digits set the chamfer amount in tenths of a pitch (currently set to nothing). If you want a thread chamfer amount of one pitch, this P word will be specified as P001060. The third two digits of the P word represent the tool angle (60 degrees in our example).

The Q word in the first G76 command specifies the minimum depth-of-cut. The R word in the first G76 command specifies the final pass depth. The values set by P, Q, and R in the first G76 can also be set by parameters as they are on other controls, meaning if you leave these words out of the first G76 command, the machine will use the values that are currently in the related parameters.

Again, Fanuc 0TA, some 0TBs, 3T, and 18T controls give you control of functions during threading that other Fanuc controls do not (other than through parameter settings). If machining a wide variety of threads on the same workpiece, it is often helpful to be able to change these important threading functions from within the program.

The P word in the second G76 command is the overall thread depth on the side (no longer specified with a K word). The P word does not allow a decimal point – a four place fixed format must be used. The Q word in the second G76 command is the depth of first pass (instead of D). The Q word does allow a decimal point.

When taper threading with G76, an R word is used in the second G76 command to specify the taper amount (instead of an I word).

What about tapping?

Though there are exceptions, *most Fanuc controls do not have a tapping cycle.* While this may be considered rather severe limitation, it is relatively easy to command a tapping operation on two axis turning centers (there will only be one hole to tap, and it will always be right in the center of the workpiece). Note that machines equipped with live tooling can tap any number of holes, and most live tooling machines do have a tapping cycle (we show how to program machines with live tooling in the Appendix that follows Lesson Twenty-Three). For now we'll concentrate on tapping with longhand commands.

Tap in the rpm mode

As with any center-cutting tool, you must calculate the speed for tapping in rpm (G97).

Tap in the low range

Especially for larger taps, tapping requires a great amount of torque. Your machine will have more torque in the low range than the high range, especially at slower speeds.

Use G32 as the motion command for tapping

If you tap with G01, the operator must have the feedrate override switch positioned at 100% when the tapping command is executed. They must also *not* press the feed hold button during the tapping motion. In either case, the tap will break if these rules are broken. But G32 will overcome both problems. Like G01, G32 will cause straight line motion at the programmed feedrate (the feedrate will be the pitch of the thread – one divided by the number of threads per inch for inch threads). But unlike G01, G32 will disable feed hold and feedrate override, making it a more fail-safe command for tapping. Note that the *single block switch must be turned off* during tapping in order to avoid breaking the tap at the bottom of the hole. G32 does not disable single block.

Use a tension/compression tap holder

Most machine tool builders recommend using a tap holder that allows the tap to float along the Z axis. This kind of holder is spring-loaded. You can pull the tap to extend it – and when released, it will spring back to its neutral position. You can push the tap to compress it – and when released, it will spring back to its neutral position. This kind of tap holder is required because most turning centers cannot perfectly synchronize the pull-out motion of the tap with the spindle reversal at the bottom of the hole. You must contact your machine tool builder to find out if they recommend tapping with a (less expensive) solid tap holder.

The program below shows the commands you can use for tapping. However, since machine tool builders vary when it comes to how M codes behave, you must cautiously verify these commands. Again, contact your machine tool builder or reference their programming manual to find the commands they recommend for tapping.

Keep the approach position 0.2 inch from the Z surface to tap

This is important when using a tension compression holder. As the spindle reverses at the bottom of the hole, it's possible that the tap will extend (pull out) slightly in its holder. If it does, the tap may not truly be out of the hole (if you use a 0.1 inch approach distance) at the completion of the retract movement.

```
O0001 (Program number)
(Commands for drilling hole prior to tapping)
.
.
.
(½ - 13 tap)
N075 T0202 M41 (Index turret, select low range)
N080 G97 S229 M03 (Turn spindle on fwd at 229 rpm)
N085 G00 X0 Z0.2 M08 (Rapid into position, start coolant)
N090 G32 Z-0.75 F0.0769 (Feed tap into hole, feedrate equals thread pitch)
N095 M04 (Reverse spindle)
N100 G32 Z0.2 (Feed tap out of hole)
N105 G00 X6.0 Z5.0 M03 (Rapid back to tool change position, reverse spindle)
N110 M01 (Optional stop)
.
.
.
```

Key points for Lesson Twenty:

- G76 is an extremely helpful multiple repetitive cycle that will completely machine a thread based upon one command in the program. It can be used for external and internal threads.
- In most threading applications, thread chamfering should be turned off.
- The E word allows two more digits of pitch accuracy than the F word.
- If the A word is specified (and not set to zero), the cutting tool will make each successive pass using an in-feed angle.
- Threading must be done in the rpm mode (G97). You must be sure not to exceed the machine's maximum feedrate when threading. Once you begin threading, you must not remove the workpiece

from the machine before the thread is completed. While you can offset in any way for the threads diameter (in X), you cannot offset in the Z axis unless you do so in increments of the thread's pitch.

- Certain control panel functions are disabled during a threading operation.

STOP! Do practice exercise number twenty in the workbook.

Lesson 21

Working With Subprograms

There are times when a series of CNC commands must be repeated – within one program – and sometimes among several programs. Whenever you find yourself writing a series of commands a second time – you should consider using a subprogram. The longer the series of commands and the more often they must be repeated, the more a subprogram can help.

You know that a CNC turning center will execute a program in sequential order. It will start with the first command in the program: read it – interpret it – and execute it. Then it will move on to the next command – read, interpret, execute. It will continue this process for the entire program.

You also know that there are times when a series of commands in your program must be repeated. We've introduced one time so far – when multiple identical grooves must be machined.

In this lesson, you will learn about a way to change the order of program execution to some extent – and this will be especially helpful when commands must be repeated.

The difference between main- and sub- programs

A *main program* is the program that a setup person or operator will execute when they activate a cycle. *Every program shown to this point in the text has been a main program.* Main programs almost always end with M30 (or M02 with some machines).

A *subprogram* is a program that is invoked by a main program or by another subprogram. A subprogram ends with M99.

In a main program, when you reach a series of commands that you know will be repeated, you can invoke a subprogram that contains the repeated commands (you place the repeated commands in the subprogram instead of in the main program). Each time the commands must be executed, you will invoke the subprogram.

M98 is the word used to invoke a subprogram. A *P word* in the M98 command specifies the program number of the subprogram to be executed. When the machine is finished executing the subprogram, it will come back to the main program to the command after the calling M98.

Subprograms provide two important benefits – one rather obvious one and the other not so obvious. The obvious benefit is that subprograms will minimize the number of commands that must be specified in order to machine a workpiece. So programming becomes simpler and programs become shorter.

A second, less obvious, benefit of using a subprogram has to do with program verification. If the subprogram is correct the first time it is executed, it will be correct *every time* it is executed. Consider machining multiple identical grooves. When you independently program each groove (without using a subprogram), you can make a mistake at any time – when either writing or typing the program. When verifying the program, you must be careful with every command in every groove – again, a mistake can be made at any time.

But when you use a subprogram, the commands to machine one of the grooves will be included in the subprogram. If the first groove comes out correctly, so will all the others. *The same commands are being executed for every groove.*

Loading multiple programs

If you're not careful, using subprograms will require more time and effort on the setup person's part. Any programming time gained by using subprograms may be lost when setups are made. And of course, setup time (machine time) is much more valuable than programming time – at least it is if the machine is running production while programs are written.

When you use subprograms, more than one program will be involved with a given job. If you use them for the kinds of applications we show in this lesson, it is not uncommon for a single job to require *several* programs. And of course, *all* programs related to the job must be loaded into the machine's memory before the job can be run. Your setup documentation must specify all program numbers related to the job – as well as where they can be found – so the setup person can quickly and easily load programs.

While it may require a control parameter setting, there is a way to load all programs that are related to a job from a single computer file. For Fanuc controls, you must specify a percent sign (%) before the first program and after the last program to be loaded.

This will shorten the time it takes to load programs, but it won't help when *saving* programs. With most Fanuc controls, each program must be saved separately – into an individual file.

Words used with subprograms

Once again, here are the words used with subprograms.

> M98 – This word is used to call a subprogram from a main program – or another subprogram
>
> P word – This word specifies the subprogram to be called
>
> M99 – This word is used to end a subprogram – machine will return to the calling program to the command after the calling M98
>
> L word (part of the P word with Fanuc 0M and 3M controls) – if included in the M98 command, this word specifies the number of times the subprogram will be executed.

We've presented all but the *L word*. This word specifies how many times the subprogram will be executed. If left out of the M98 command, the subprogram will be executed once. With Fanuc 0M and 3M controls, the number of executions is part of the P word. Both of these commands will execute subprogram O1000 five times:

```
N050 M98 P1000 L5 (for almost all Fanuc controls)
N050 M98 P0051000 (for 0M and 3M Fanuc controls)
```

For Fanuc 0M and 3M controls, the first three digits of the P word specify the number of executions and the last four digits specify the program number. Note that with *all* Fanuc controls, the command

```
N050 M98 P1000
```

will call subprogram O1000 and execute it once.

A quick example

This example shows the structure of subprogram usage – though we're not going to explain the application yet. Here is the main program. Note that it ends with M30.

```
O0019 (Program number)
N003 G99 G20 G23 (Ensure that initialized states are still in effect)
N004 G50 S4000 (No need for limiting, limit to machine's maximum speed)
N005 T0101 M42 (Select grooving tool and high spindle range)
N010 G96 S500 M03 (Start spindle fwd at 500 sfm)
N015 G99 G00 X3.2 Z-0.5 M08 (Rapid to convenient starting position for grooving, start coolant)
N020 M98 P1000 (Jump to subprogram 1000 and machine 1st groove)
N025 G00 Z-1.0 (Move over to second groove)
N030 M98 P1000 (Jump to subprogram 1000 and machine 2nd groove)
```


N035 G00 Z-1.5 (Move over to third groove)
N040 **M98 P1000** (Jump to subprogram 1000 and machine 3rd groove)
N045 G00 Z-2.0 (Move over to fourth groove)
N050 **M98 P1000** (Jump to subprogram 1000 and machine 4th groove)
N055 G00 Z-2.5 (Move over to fifth groove)
N060 **M98 P1000** (Jump to subprogram 1000 and machine 5th groove)
N065 G00 X6.0 Z5.0 (Rapid to safe index point)
N070 **M30** (End of main program)

Here is the subprogram being called by the main program. Note that it ends with M99.

O1000 (Subprogram number)
N1 G00 W0.0625 (Move over to groove center)
N2 G01 U-.7 F0.005 (Feed to groove bottom)
N3 G04 P500 (Pause for .5 second)
N4 G00 U0.7 (Rapid out of groove)
N5 W-0.0925 (Move to left side of left chamfer)
N6 G01 U-0.2 (Feed flush to diameter)
N7 U-0.06 W0.03 (Form chamfer on left side of groove)
N8 U-0.440 (Feed to groove bottom)
N9 G04 P500 (Pause for .5 second)
N10 G00 U0.7 (Rapid out of groove)
N11 W0.155 (Rapid to right side of right chamfer)
N12 G01 U-0.2 (Feed flush with diameter)
N13 U-0.06 W-0.03 (Form chamfer on right side of groove)
N14 U-0.440 (Feed to groove bottom)
N15 G04 P500 (Pause for .5 second)
N16 G00 U0.7 (Rapid out of groove)
N17 W-0.125 (Rapid back to convenient starting position)
N18 **M99** (End of subprogram)

Again, we'll explain the application in a bit. For now, let's concentrate on the words related to subprograms. When the main program is activated (operator presses the cycle start button), it will be program O0019 that they run. Like any other program, this program will be executed command-by-command – in sequential order.

But when line N020 is executed, the machine will temporarily exit from program O0019 to execute program O1000. And of course, program O1000 will be executed in sequential order. When completed with the subprogram and line N18 of program O1000 is executed, the machine will return to the calling program (the main program in our case) to the command *after* the calling M98 command. At the present time, the machine will return to line N025 of the main program and continue executing this program.

When line N030 of the main program is executed the process is repeated. The only difference is that when the M99 of the subprogram is executed this time, the machine will return to line N035 of the main program.

In our example main program this process is repeated three more times (lines N040, N050, and N060).

Nesting subprograms

Again, you can call a subprogram from a main program – *or from another subprogram.* When you call a subprogram from another subprogram, it is called *nesting subprograms* – and there is a limitation to how *deep* you can nest. With most controls, you can only nest four deep, as the Figure 6.21 shows.

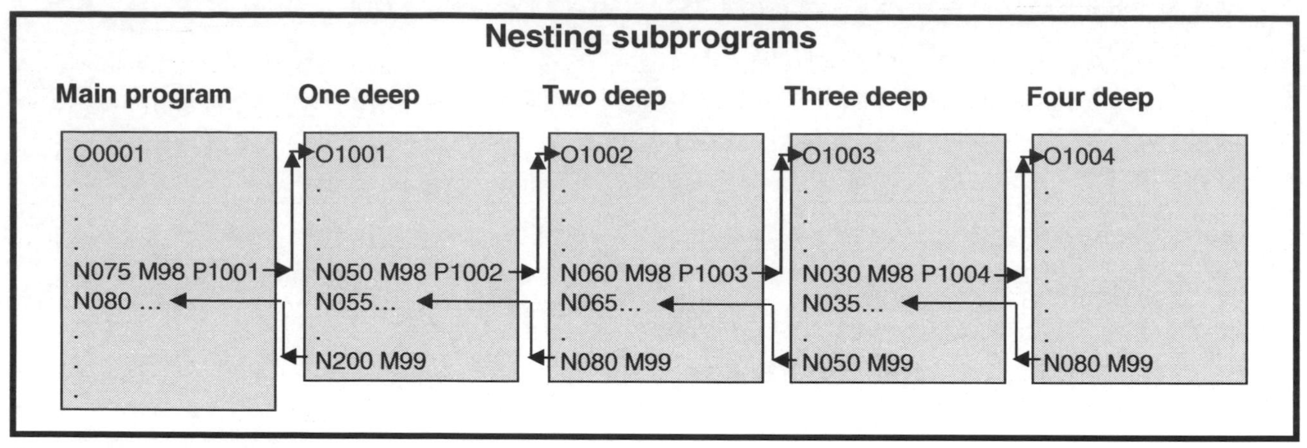

Figure 6.21 – Nesting subprograms four deep

Applications for subprograms

As with all CNC features discussed to this point, we wish to stress the reasons why subprograms are used as importantly as how they are used. Generally speaking, subprogram applications fall into three basic categories. At this point, we only introduce the applications. After introducing the application categories, we'll show specific examples for each.

Repeated machining operations

This is the most common application category for manual programmers. You will often find it necessary to repeat commands for machining operations. For example, you may have a 0.25 inch wide groove with 0.031 inch chamfers on each outside corner that must be machined in several places on a workpiece (or on several different workpieces). A subprogram will let you program the groove once. Whenever the groove must be machined, you simply call the subprogram.

Unfortunately, most applications in this category require that you write the sub-program in the *incremental* positioning mode. Otherwise, the operation will be repeated in the same location every time the subprogram is executed. In our grooving example, the same groove will be machined if the sub-program is written in the absolute mode.

Frankly speaking, this eliminates (or nearly eliminates) the feasibility of using subprograms for many repeated machining operations. In many cases, it just as easy (and often less time consuming) to repeat the commands in the main program using the absolute positioning mode than it is to write the subprogram in the incremental positioning mode.

Control programs

Subprograms are often helpful when it is necessary to control the program's general flow. Consider, for example, a *flip job*. With a flip job, the operator will load the workpiece and activate the cycle. Once the first operation has been completed, the machine will stop. At this point the operator will turn the workpiece around in the chuck. Then they will reactivate the cycle to machine the second operation. This provides a completed workpiece in each cycle.

While a lengthy program containing all operations can be created, it may be better to break up the program into two parts (subprograms). One program (again, a subprogram) will be used for the first operation and another will be used for the second operation. A third program (the *control program*) will cause the activation of the two subprograms. This technique is especially helpful when a computer aided manufacturing (CAM) system is used to prepare the related programs. Most CAM systems will not automatically combine the two programs needed for a flip job into one program. Instead, they create two separate programs.

Utility applications

We call any method or technique that facilitates usage of your CNC machines a utility. Given the broad scope of programming features, and the almost unlimited ideas that stem from human ingenuity, there is very little that is impossible with CNC machine tools. Subprograms give you another helpful tool to solve problems. An example of a utility application for subprograms is related to bar feeding turning centers.

When a bar feeder is being used, the CNC commands needed for advancing the bar are quite redundant. They will be needed in *every* program using the bar feeder. Instead of placing these redundant commands in the main program, they can be placed in a subprogram. Whenever bar feeding is required, the bar-feed subprogram can be called. Though the setup person may still have to modify a value or two in the subprogram during setup (based upon workpiece length changes), sub-programming can minimize the number of commands in the machining program. And if for any reason the bar feeding commands must be changed, *only the subprogram must be changed.* The programming of bar feeders is discussed in the Appendix that follows Lesson Twenty-Three.

Example for repeating machining operations - multiple identical grooves

Figure 6.22 shows five identical grooves that must be machined on a workpiece. Note that without a subprogram, you will have to program all five grooves individually. With a subprogram, you only need to program the groove once. Notice that the grooving tool in Figure 6.22 is narrower than the groove. Multiple passes must be made.

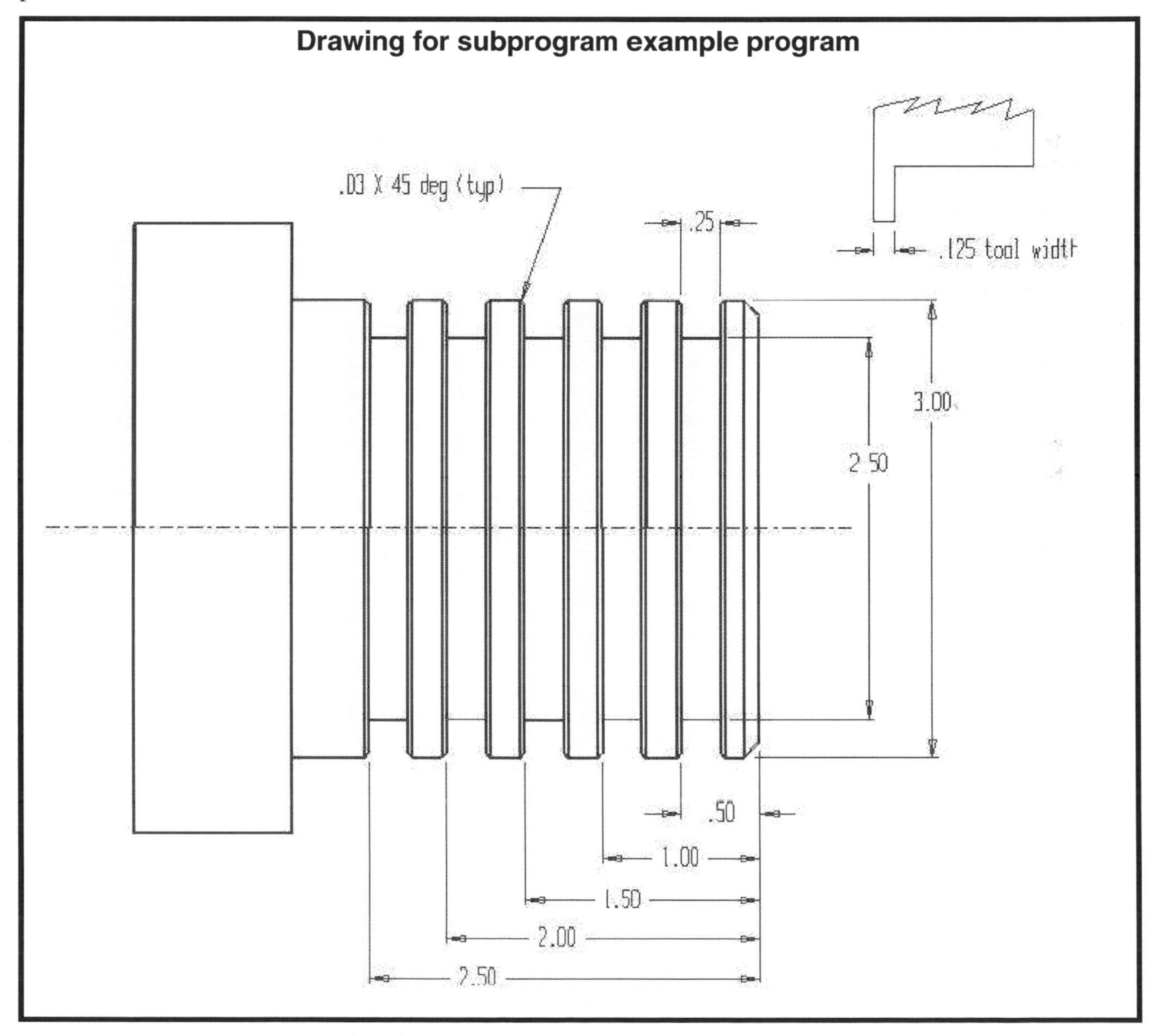

Figure 6.22 – Drawing for multiple grooving subprogram example

This also happens to be one good application for programming in the incremental positioning mode. If the groove is programmed in the absolute positioning mode, the location of the groove cannot be changed (all grooves will be machined in the same location). Frankly speaking, programming a turning center in the incremental mode can be a little confusing. As you know, X axis departures must be specified with U words for most controls. And the U word specifies a *diameter change*. Z axis departures must be specified with W words.

Main program:

O0019 (Program number)
N003 G99 G20 G23 (Ensure that initialized states are still in effect)
N004 G50 S4000 (No need for limiting, limit to machine's maximum speed)
N005 T0101 M42 (Select grooving tool and high spindle range)
N010 G96 S500 M03 (Start spindle fwd at 500 sfm)
N015 G99 G00 X3.2 **Z-0.5** M08 (Rapid to convenient starting position for grooving, start coolant)
N020 **M98 P1000** (Jump to subprogram 1000 and machine 1st groove)
N025 G00 **Z-1.0** (Move over to second groove)
N030 **M98 P1000** (Jump to subprogram 1000 and machine 2nd groove)
N035 G00 **Z-1.5** (Move over to third groove)
N040 **M98 P1000** (Jump to subprogram 1000 and machine 3rd groove)
N045 G00 **Z-2.0** (Move over to fourth groove)
N050 **M98 P1000** (Jump to subprogram 1000 and machine 4th groove)
N055 G00 **Z-2.5** (Move over to fifth groove)
N060 **M98 P1000** (Jump to subprogram 1000 and machine 5th groove)
N065 G00 X6.0 Z5.0 (Rapid to safe index point)
N070 M30 (End of main program)

Subprogram:

O1000 (Subprogram number)
N1 G00 W0.0625 (Move over to groove center)
N2 G01 U-.7 F0.005 (Feed to groove bottom)
N3 G04 P500 (Pause for .5 second)
N4 G00 U0.7 (Rapid out of groove)
N5 W-0.0925 (Move to left side of left chamfer)
N6 G01 U-0.2 (Feed flush to diameter)
N7 U-0.06 W0.03 (Form chamfer on left side of groove)
N8 U-0.440 (Feed to groove bottom)
N9 G04 P500 (Pause for .5 second)
N10 G00 U0.7 (Rapid out of groove)
N11 W0.155 (Rapid to right side of right chamfer)
N12 G01 U-0.2 (Feed flush with diameter)
N13 U-0.06 W-0.03 (Form chamfer on right side of groove)
N14 U-0.440 (Feed to groove bottom)
N15 G04 P500 (Pause for .5 second)
N16 G00 U0.7 (Rapid out of groove)
N17 W-0.125 (Rapid back to convenient starting position)
N18 **M99** (End of subprogram)

In line N020 of the main program, the control execute program O1000 and machine the first groove. In line N18 of the subprogram, the control returns to line N025 of the main program. Up to this point, the programmer has not saved any programming commands (or effort).

But in line N030 of the main program, the control is told to execute the subprogram a second time. At this point, the programmer has kept from having to write about 20 commands. This is repeated in lines N040, N050, and N060. In all, the programmer has saved 80 commands.

While programming incrementally is cumbersome, remember the second benefit of using subprograms for multiple machining operations. During program verification, if the first groove is machined correctly, so will all of the other grooves (again, the control is using the same set of commands to machine each groove).

There is yet another benefit for using subprograms for multiple identical machining operations. If this groove is required on *several workpieces*, the subprogram (O1000 in our case) can be left in the machine for use any time it is needed. Whenever you must machine a 0.25 wide groove that has 0.03 chamfers with a 0.125 grooving tool, you simply call program O1000. Again, this not only minimizes programming effort, it also minimizes program verification time since a proven program is being used to machine the grooves.

Example for control program applications - flip jobs

There are times when the operator will run the first operation of a workpiece – turn the workpiece around in the chuck at a program stop commanded by M00 – and then run the second operation of the workpiece. This usually requires double-bored jaws that can hold on two different diameters. One workpiece will be completed in each cycle. It can be helpful to separate the programs needed to machine the workpiece into subprograms, especially when a computer aided manufacturing (CAM) system is used to create the programs. Figure 6.23 shows an example of a flip job.

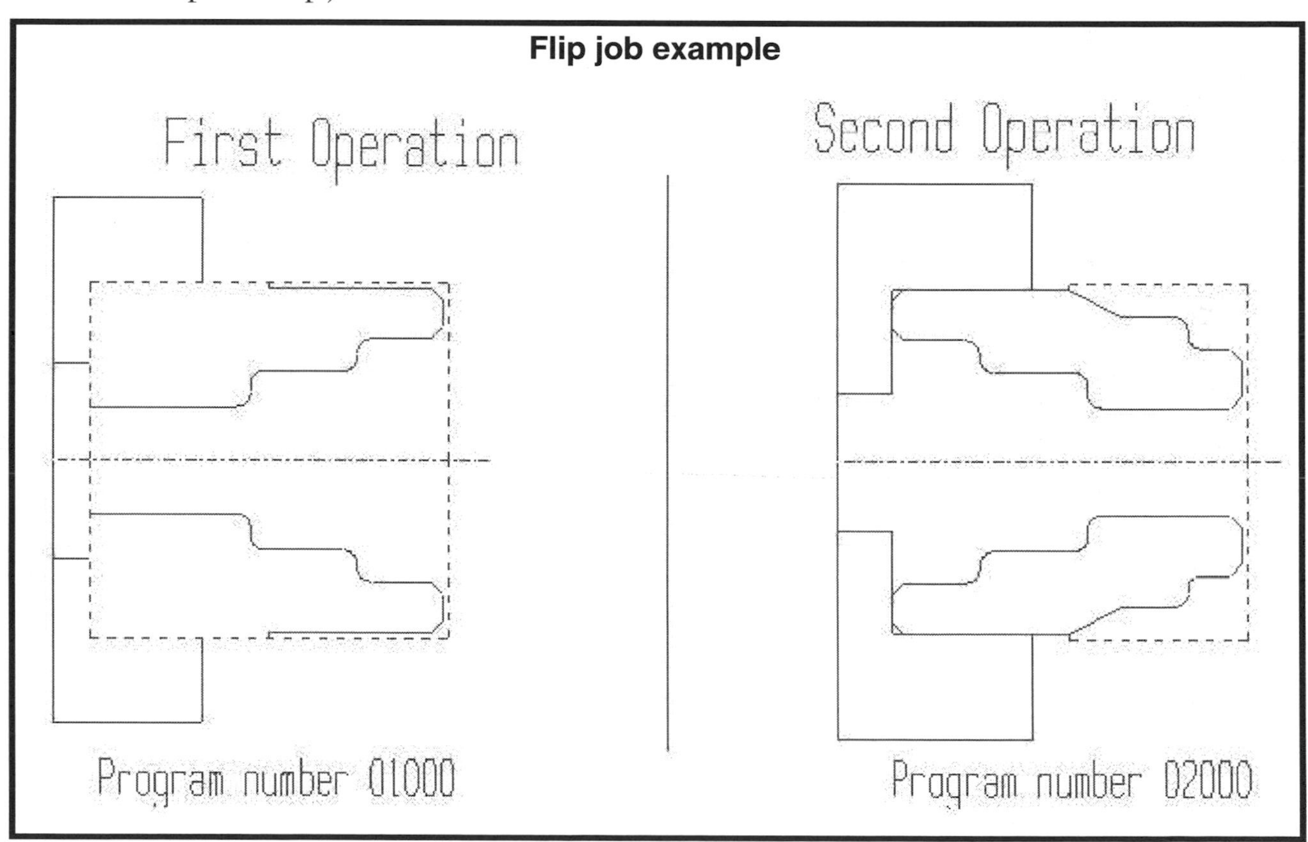

Figure 6.23 – Drawing shows an example of a flip job

Notice that program O1000 is used to machine the first operation and program O2000 is used to machine the second operation. Here is the main program needed to machine both ends followed by a portion of each subprogram needed to do the actual machining.

Main program:

O0020 (Program number)
N003 G99 G20 G23 (Ensure that initialized states are still in effect)
N004 G50 S4000 (No need for limiting, limit to machine's maximum speed)
N005 **M98 P1000** (Machine first side)
N010 **M00** (Program stop to turn workpiece around in jaws)
N015 **M98 P2000** (Machine second side)
N020 M30 (End of main program)

Subprograms:

O1000 (Subprogram to machine first side)
N005... (Machine entire first side)
.
.
M99 (End of subprogram)

O2000 (Subprogram to machine second side)
N005 ... (Machine entire second side)
.
.
M99 (End of subprogram)

When the operator activates program number O0020, the machine is told to execute program O1000 (in line N005). This machines the first operation of the workpiece. When finished with program O1000, the control will return to line N010 of the main program. The M00 in this command will cause the machine to stop. The operator can safely turn the workpiece around in the chuck and then reactivate the cycle. Line N015 tells the control to execute program O2000, which machines the second operation of the workpiece. When finished, the control will return to the main program to line number N020, which rewinds the program and stops the machine. The completed workpiece is removed at this point and another piece of raw material is loaded.

As stated in Lesson Thirteen, when you program flip jobs, it's wise to use two wear offsets for each tool used in both programs (O1000 and O2000). Use an offset number corresponding to the tool station number for tools used in program O1000 (for example, tool five, wear offset five, just as you normally do). For tools used in the second operation (program O2000), add twenty to the tool station number to come up with the wear offset number to be used with the tool (tool five, wear offset twenty-five).

Example for utility applications - bar feeder activation

If you'll be working with a machine that has a bar feeder, you'll find that the commands necessary to advance the bar are very redundant from job to job. This makes a good application for sub-programming (note that bar feeder programming is discussed in greater detail later in the Appendix that follows Lesson Twenty-Three).

Here is the main program that machines a workpiece being run on a bar feeding turning center. Notice that only one command, line N005 is related to the bar feed activation:

O0004 (Program number)
N002 G99 G20 G23 (Ensure that initialized states are still effective)


```
(BAR STOP)
N005 M98 P4000 (Activate the bar feeding subprogram)
N010 G50 S5000 (No spindle limitation necessary, limit to machine maximum)
(Finish face and turn tool)
N015 T0101 M42
N020 G96 S550 M03
N025 G00 X1.45 Z0 M08 (1)
N030 G01 X-0.062 F0.008 (2)
N035 G00 Z0.1 (3)
N040 G42 X1.025 (4)
N045 G01 Z0 (5)
N050 X1.125 Z-0.05 (6)
N055 Z-1.05 (7)
N060 X1.45 (8)
N065 G40 G00 X6.0 Z5.0
N070 M01
(1/12 DRILL)
N075 T0202 M42
N080 G97 S740 M03
N085 G00 X0 Z0.1 M08 (9)
N090 G01 Z-1.21 F0.007 (10)
N095 G00 Z0.1 (9)
N100 X6.0 Z5.0
N105 M01
(FINISH BORING BAR)
N110 T0303 M42
N115 G96 S300 M03
N120 G41 G00 X0.6625 Z0.1 M08 (11)
N125 G01 Z0 F0.004 (12)
N130 X0.5625 Z-0.05 (13)
N135 Z-1.03 (14)
N140 X0.4 (15)
N145 G00 Z0.1 (16)
N150 G40 X6.0 Z5.0
N155 M01
(1/8 WIDE CUT-OFF TOOL)
N160 T0404 M42
N165 G96 S450 M03
N170 G00 X1.45 Z-1.125 M08 (17)
N180 G01 X1.025 F0.005 (18)
N185 G00 X1.45 (17)
N190 Z-1.075 (19)
N195 G01 X1.125 (20)
```

```
N200 X1.025 Z-1.125 (18)
N205 M18 (Activate part catcher)
N210 X0.44 (21)
N215 M19 (Retract part catcher)
N220 G00 X1.45 (17)
N225 X6.0 Z5.0
N230 M99 (Return to start and continue)
```

Bar feeding subprogram:

```
O4000 (Subprogram for bar feeding)
N1 T1212 M05 (Index, stop spindle)
N2 G00 X0 Z-1.025 (This Z position MUST change for each setup)
N3 M15 (Open collet chuck)
N4 G98 G01 Z0.05 F30.0 (Bar advance)
N5 M14 (Close collet chuck)
N6 G99 G00 X6.0 Z5.0
N7 M01
N8 M99
```

Special notes about M99

As stated, M99 is used to end a subprogram. This command sends program execution back to the main program. However, there are two other usages for M99.

Ending a main program with M99

First, as you know, M30 is normally used to end a main program. This command turns off anything still running (spindle, coolant, etc.) and returns the program back to its beginning. Then the machine stops for the operator to load another workpiece.

But look at line N230 of the bar feeding main program (program O0004). We're ending a *main program* with M99. If M99 is used to end a main program, the machine will return to the beginning of the main program and continue without stopping. While some bar feeding turning centers are equipped with special functions that allow continuous machining (and main program still end with M30), many do require the use of M99 as is shown in program O0004 above. Again, we provide more information about bar feeder programming in the Appendix that follows Lesson Twenty-Three.

Changing the order of program execution with M99

Second, M99 can be used in a main program (not a subprogram) to cause an *unconditional branch*. The control can be told to jump from one point in the main program to another with M99. A P word within the M99 specifies the sequence number to which you wish the control to jump. For example, if the command

```
N010 M99 P050
```

is included in a *main program*, it will cause to machine to jump to line N050 and continue executing the program from there.

There may be times during a program's verification that severe mistakes in process are discovered. For example, say a program is prepared for a turning center that follows this process.

1) Rough turn
2) Finish turn
3) Drill 2" hole
4) Rough bore
5) Finish bore

Experienced machinists will agree that you should *rough everything before you finish anything.* In this process, the finish turning is being prior to the drilling and rough boring. It is quite likely that during these operations, the workpiece will be deformed slightly – or it may shift in the chuck – resulting in the turned diameter not being concentric with the bored diameter.

While this may be a relatively obvious processing mistake, there may be times when your CNC machine is be down waiting for you to change a program to reflect a correction in machining order. In these cases, the programmer must usually to go back to the office and call up the program within a CNC text editor. Using cut and paste techniques, they will modify the original program to reflect the process change and download the corrected program to the CNC machine tool. Depending on the availability of the programmer, the programmer's aptitude, whether the computer is available, and how quickly programs can be transferred, this task may take quite a bit of time.

Newer CNC controls do have text editors which allow cut and paste techniques to be done right on the CNC control. When available, these editors should be used to minimize the CNC machine's downtime. Most are simple enough to use that the setup person can make the required program changes without needing the programmer's help.

For those machines that do not have cut and paste features, the M99 command can be used to change machining order. The setup person can quickly and easily modify the program to match the desired machining order. Here is an example to stress how it's done. First we show the *incorrect* program based on the previous process.

```
O0008 (Program with incorrect process)
N003 G20 G23 G40 (Ensure that initialized states are still in effect)
N004 G50 S4000 (No limiting necessary, limit to machine's maximum)

N005 T0101 M41 (Rough turning tool)
N010 G96 S400 M03 (Start spindle CW at 400 SFM)
N015 G00 X3.040 Z0.1 (Rapid to rough turn diameter)
N020 G99 G01 Z-1.995 F0.017 (Rough turn)
N025 X3.25 (Feed up face)
N030 G00 X6.0 Z5.0 (Rapid to tool change position)
N035 M01 (Optional stop)

N040 T0202 M42 (Finish turning tool)
N045 G96 S600 M03 (Start spindle CW at 600 SFM)
N050 G00 X3.0 Z0.1 (Rapid to diameter to be turned)
N055 G01 Z-2.0 F0.006 (Finish turn)
N060 X3.25 (Feed up face)
N065 G00 X6.0 Z5.0 (Rapid to tool change position)
N070 M01 (Optional stop)

N075 T0303 M41 (2" drill)
N080 G97 S300 M03 (Start spindle CW at 300 RPM)
N085 G00 X0 Z0.1 (Rapid to position)
N090 G01 Z-2.6 F.009 (Feed to depth)
N095 G00 Z0.1 (Rapid out of hole)
N100 G00 X6.0 Z5.0 (Rapid to tool change position)
N105 M01 (Optional stop)
```

```
N110 T0404 M41 (1.5" rough boring bar)
N115 G96 S400 M03 (Start spindle CW at 400 SFM)
N120 G00 X2.085 Z0.1 (Rapid to rough bore diameter)
N125 G01 Z-1.995 F0.010 (Rough bore)
N130 X2.0 (Feed down face)
N135 G00 Z0.1 (Rapid out of hole)
N140 X6.0 Z5.0 (Rapid to tool change position)
N145 M01 (Optional stop)
N150 T0505 M42 (1.5" finish boring bar)
N155 G96 S600 M03 (Start spindle CW at 600 SFM)
N160 G00 X2.125 Z0.1 (Rapid to position)
N165 G01 Z-2.0 F0.006 (Finish bore)
N170 X2.0 (Feed down face)
N175 G00 Z0.1 (Rapid out of hole)
N180 G00 X6.0 Z5.0 (Rapid to tool change position)
N185 M30 (End of program)
```

Here is the corrected version of the program that uses the desired process.

```
O0008 (Program with correct process)
N003 G99 G20 G23 (Ensure that initialized states are still in effect)
N004 G50 S4000 (No limiting necessary, limit to machine's maximum)
N005 T0101 M41 (Rough turning tool)
N010 G96 S400 M03 (Start spindle fwd at 400 sfm)
N015 G00 X3.040 Z0.1 (Rapid to rough turn diameter)
N020 G99 G01 Z-1.995 F0.017 (Rough turn)
N025 X3.25 (Feed up face)
N030 G00 X6.0 Z5.0 (Rapid to tool change position)
N035 M01 (Optional stop)
N038 M99 P075 (Jump to line N075)
N040 T0202 M42 (Finish turning tool)
N045 G96 S600 M03 (Start spindle fwd at 600 sfm)
N050 G00 X3.0 Z0.1 (Rapid to diameter to be turned)
N055 G01 Z-2.0 F0.006 (Finish turn)
N060 X3.25 (Feed up face)
N065 G00 X6.0 Z5.0 (Rapid to tool change position)
N070 M01 (Optional stop)
N073 M99 P185 (Jump to end of program)
N075 T0303 M41 (2" drill)
N080 G97 S300 M03 (Start spindle fwd at 300 sfm)
N085 G00 X0 Z0.1 (Rapid to position)
N090 G01 Z-2.6 F.009 (Feed to depth)
N095 G00 Z0.1 (Rapid out of hole)
```



```
N100 G00 X6.0 Z5.0 (Rapid to tool change position)
N105 M01 (Optional stop)
N110 T0404 M41 (1.5" rough boring bar)
N115 G96 S400 M03 (Start spindle fwd at 400 sfm)
N120 G00 X2.085 Z0.1 (Rapid to rough bore diameter)
N125 G01 Z-1.995 F0.010 (Rough bore)
N130 X2.0 (Feed down face)
N135 G00 Z0.1 (Rapid out of hole)
N140 X6.0 Z5.0 (Rapid to tool change position)
N145 M01 (Optional stop)
N150 T0505 M42 (1.5" finish boring bar)
N155 G96 S600 M03 (Start spindle fwd at 600 sfm)
N160 G00 X2.125 Z0.1 (Rapid to position)
N165 G01 Z-2.0 F0.006 (Finish bore)
N170 X2.0 (Feed down face)
N175 G00 Z0.1 (Rapid out of hole)
N180 G00 X6.0 Z5.0 (Rapid to tool change position)
N183 M99 P040 (Go back to line N040 to finish turn)
N185 M30 (End of program)
```

With three commands (lines N038, N073, and N183) we have changed the machining order to match our desired process. Admittedly, this program is rather difficult to follow since it is not executed in sequential order. For this reason, we still recommend that the programmer eventually changes it in the conventional manner. But since the machine is running production, this can be done at the programmer's leisure.

What is parametric programming (custom macro B)?

As you have seen, subprograms dramatically expand what can be done with manual programming functions. But as you know, they have one major limitation. Subprograms must be *totally redundant.* If *anything* changes from one execution to the next, subprograms cannot be used.

Parametric programming gives you the ability to overcome this limitation. (Fanuc's version of parametric programming is called *custom macro B.*) In essence, parametric programming gives you the ability to write general-purpose subprograms.

While parametric programming is beyond the scope of this text (we have other materials available to help you learn parametric programming), all CNC people should at least be able to recognize applications for this very powerful programming tool. If your company has applications that fall in to one of these five categories, you'll want to learn more about parametric programming.

Part families

Many companies machine workpieces that fit into a close family. While the workpieces are different, they have very similar attributes and are machined with the same process (sequence of machining operations) and tooling. For example, one company machines *air cylinders* in a variety of sizes. Each component making up the air cylinder (end caps, piston, cylinder, etc.) falls into a close part family. A program used to machine one of the workpieces in the family will be very similar to programs for other workpieces in the family.

With conventional programming methods, the programmer will commonly modify one program to create another program for a different workpiece in the family. When finished, they will have one hard-and-fixed CNC program for each workpiece in the family. If a process or engineering change is made that affects all workpieces in the family, *all* of the CNC programs must be modified. Think of a part family that contains over one hundred workpieces. This could involve a great deal of work.

With parametric programming, one program can be developed to machine *all* workpieces in the family. In essence, a *general purpose program* is created that uses *variables* to specify the changing attributes for each workpiece in the family. The machine will behave differently based upon the current settings of these variables.

User defined canned cycles

As you know from Lesson Nineteen, most Fanuc controls do not have a deep hole peck drilling cycle. However, with parametric programming, you can create your own deep hole peck drilling cycle. Or maybe you want to create a grooving cycle that allows you to easily specify the groove location in X and Z, the tool width, the groove width, and the size of the chamfers on the groove. Based upon these *variables* in your main program, the machine will machine your intended groove. The potential for creating your own canned cycles is unlimited.

Utilities

Parametric programming opens the door to countless utility applications that can reduce setup time, cycle time, and in general, make your machines easier to work with. Examples of utilities include part counters, tool life managers, and cycle time meters.

We could, for example, easily improve upon the utility subprogram shown earlier for bar feeding. As you know, if this application is handled with a simple subprogram, the setup person must manually modify the subprogram with each new setup. They must do so due to the changing lengths of workpieces being machined. If they forget to modify the subprogram, the results could be disastrous. With parametric programming, you can actually specify the workpiece length, facing stock, and cutoff tool width as part of the *call statement* to the (general purpose) subprogram. Here is an example that shows how.

Main program:

```
O0004 (Program number)
N002 G99 G20 G23 (Ensure that initialized states are still effective)
(BAR STOP)
N005 G65 P4000 C0.125 W1.0 S0.05 (Execute bar feeding parametric program)
N010 G50 S5000 (No spindle limitation necessary, limit to machine maximum)
(FINISH FACE AND TURN TOOL)
N015 T0101 M42
N020 G96 S550 M03
N025 G00 X1.45 Z0 M08 (1)
N030 G01 X-0.062 F0.008 (2)
N035 G00 Z0.1 (3)
.
.
.
```

In line N005, M98 has been replaced with G65. G65 is used to call the bar feeding custom macro (custom macro B is Fanuc's version of parametric programming). Letter address P is used to specify the custom macro's program number. The C word is being used to specify the cutoff tool width. W is being used to specify the workpiece length. And S is being used to specify the stock on the face to be machined in the operation. These letter addresses are chosen by the CNC programmer to represent what ever needs representing. But in the body of the custom macro, you cannot reference their values with the letter addresses. While this may seem a little complicated, you must reference them with a series of *local variables*. Local variable #3 is used to reference the value of C. #23 is used to reference the value of W. And #19 is used to reference the value of S. Knowing this, consider the bar feeding parametric program:


```
O4000 (Custom macro for bar feeding)
N1 T1212 M05 (Index, stop spindle)
N2 G00 X0 Z-[#3 + #23 – 0.1] (#3 contains the value of C, #23 contains the value of W)
N3 M15 (Open collet chuck)
N4 G98 G01 Z#19 F30.0 (Bar advance, #19 contains the value of S)
N5 M14 (Close collet chuck)
N6 G99 G00 X6.0 Z5.0
N7 M01
N8 M99
```

In line N2, the machine will rapid to a position in Z based upon the current values of C and W coming from the call statement in the main program. So will the bar advance movement (line N4) be appropriate based upon the current value of S. Since every main program that uses this custom macro will have these values plugged in, the setup person no longer needs to modify the subprogram during setup. While this is a good application for custom macro, you may find it a little complicated. Our main point is to stress what's possible with parametric programming, not to teach you how to do it. Again, we have other training materials available to help you learn parametric programming. With parametric programming, *you can create general purpose subprograms.*

Complex motions and shapes

Since arithmetic calculations can be done in parametric programs, any shape that can be calculated with an arithmetic formula can be machined by a parametric program. Consider, for example, the kind of motion necessary for machining an ellipse. The XZ motion will no longer be circular; it will be in the form of an elliptical motion. Believe it or not, parametric programming actually allows you to create your own motion types.

> **Talk to experienced people in your company...**
>
> ... to learn more about how your company uses subprograms.
>
> 1) Ask an experienced person if your company uses subprograms. If the answer is yes, ask to see the applications they're used for.
>
> 2) If subprograms are used, ask how program numbers are organized. How quickly can the setup person load them (and remove them when the job is finished)?
>
> 3) Ask if your company uses bar feeders. If so, is a subprogram used to advance the bar?

Key points for Lesson Seventeen:

- Subprograms are used to repeat commands. Just about any time you find yourself repeating commands in a program, you should consider using a subprogram. The more commands that must be repeated, the fewer commands you'll need to for the job.
- One benefit of subprograms is obvious – shortened program length. But the other, easier program verification is not so obvious. When verifying your program, if subprogram runs properly the first time it is executed, it will run properly every time it is executed.

- M98 is the word used to call a subprogram. A P word in the M98 command specifies the subprogram number. M99 is the word used to end the subprogram.
- Applications for subprograms fall into three basic categories – multiple identical machining operations, control programs, and utilities.
- Parametric programming, which is an option on most controls, gives you the ability to write multi-purpose subprograms.

STOP! Do practice exercise number twenty-one in the workbook.

Lesson 22

Differences To 0TA, 0TB, 3T, And 18T Controls

Though we've been showing the differences related to multiple repetitive cycles and subprograms for these controls throughout the lessons in Key Concept Six, this lesson quickly summarizes them. If you do not have one of these control models, you can skip this lesson.

Remember that the only reason for these minor differences is that the Fanuc 0TA, 0TB, 3T, and 18T controls do not have the letter addresses D and L. While these letter addresses may happen to be on the keyboard, they have no programming function with the Fanuc 0TA, some 0TBs, 3T, and 18T controls. For multiple repetitive cycles, these controls force the programmer to break up each cycle into two separate commands.

G71

Here is a G71 rough turning command for most Fanuc controls (0TC, 6T, 6TB, 10T, 11T, 15T, 16T, 20T, and 21T):

```
N020 G71 P025 Q085 U0.040 W0.005 D1250 F0.015
```

Here is how the same command must be specified for a Fanuc 0TA, some 0TBs, 3T, or 18T control:

```
N018 G71 U0.125
N020 G71 P025 Q085 U0.040 W0.005 F0.015
```

The only difference is that the depth-of-cut must be specified in the first G71 command with a U word. Notice that the U word does allow a decimal point. You may wish to go back through this key concept and modify all examples given for G71 to reflect this minor difference (if you have a 0TA, some 0TBs, 3T, or 18T control).

G72

G72 uses the same technique. Here is the format for most Fanuc controls.

```
N020 G72 P025 Q075 U0.04 W0.005 D1250 F0.015
```

Here is the format for 0TA, 0TB, 3T, and 18T.

```
N018 G72 U0.1250
N020 G72 P025 Q075 U0.04 W0.005 F0.015
```

Again, only the depth of cut word (now a U word in the first G72) is affected.

G73

Here is how the G73 command is given for most Fanuc controls.

```
N020 G73 P025 Q090 U.040 W0 I.5 D4 F.012
```

The format for 0TA, some 0TBs, 3T, and 18T is as follows:

```
N018 G73 U0.5 R4.0
N020 G73 P025 Q090 U0.040 W0 F.012
```

The format for G73 is a little different. The number of passes is now specified by the R word in the first G73. Also, U in the first G73 represents the amount of stock to be removed on the side, and W is the amount of stock to be removed in Z (not used in this example).

G74

Here is the format for the G74 peck drilling (chip breaking only) for most Fanuc controls.

N020 G74 Z-2.0 K0.1 F0.010

The format for the 0TA, some 0TBs, 3T, and 18T is as follows.

N020 G74 Z-2.0 Q0.1 F0.010

The only difference for 0TA, some 0TBs, 3T, and 18T is that the Q word is used to specify the depth of each peck instead of a K word.

G75

For the G75 grooving cycle, the format for the 0TA, 0TB, 3T, and 18T is the same as for most Fanuc controls.

G76

The multiple repetitive cycle for threading G76 has yet a few more differences. In fact, the G76 format for 0TA, 0TB, 3T, and 18T actually allows more programming flexibility than it does for most Fanuc controls.

Here is an example threading command for most Fanuc controls.

N020 G76 X5.392 Z-1.88 K0.054 D0100 A60 F0.0625

Now here is the same command given in 0TA, 0TB, 3T, and 18T format:

N018 G76 P000060 Q0.005 R0.0002
N020 G76 X5.392 Z-1.88 P0540 Q0.0100 F0.0625

The P word in the first G76 command actually does three things. The first two digits represent the number of spring passes (currently set to none). The second two digits set the chamfer amount (currently set to nothing). The third two digits represent the tool angle (60 degrees).

The Q word in the first G76 command allows you to specify the minimum depth-of-cut. The R word in the first G76 allows you to specify the final thread pass depth. The P and Q in the first G76 are also set by parameters, meaning if you leave them out of the first G76 command, the control will assume the values set by parameters.

The P word in the second G76 is the overall thread depth on the side. Notice that no decimal point is allowed and a four place fixed format must be used. The Q word in the second G76 is the depth of first pass.

If taper threading must be done, an R word is placed in the second G76 to specify the taper amount. This R word replaces the I word used with most Fanuc controls.

Subprograms

The only difference in programming format will present itself only if more than one execution of the subprogram is required (seldom needed for turning center applications). The command:

N050 M98 P1000

will execute program O1000 one time on all controls.

However, the command:

N050 M98 P1000 L5

will execute program O1000 five times for most Fanuc controls.

This command will do so for 0TA, some 0TBs, 3T, and 18T controls.

N050 M98 P0051000

The first three digits is the number of times the subprogram will be executed. The final four digits is the program number to be executed.

Again, if you have only 0TAs, 0TBs, 3ts, and/or 18Ts, it might be wise to go back and mark up all example programs and practice exercises to match your control's required format for multiple repetitive cycles and subprograms.

What about controls that are not made by Fanuc?

As you know, we have been using a Key Concepts approach to help you learn how to program CNC turning centers. We've been stressing the reasons why things are done as importantly as how they are done. And we've been applying the Key Concepts to one specific type of control. All specific examples have been shown for Fanuc and Fanuc-compatible controls.

If one or more of your turning centers does not have a Fanuc or Fanuc-compatible control, you must of course, learn how to program it. The Key Concepts you have learned will still apply. But some of the specific commands we have shown will not work perfectly for other controls (if they work at all). We want to prepare you for some of the most common differences between Fanuc controls and controls that are not compatible with Fanuc. While this short discussion will not present every programming difference for every control, it will show you the kinds of things you must watch for.

G code differences

As you begin working with a control that is not Fanuc-compatible, begin by referencing the control manufacturer's list of G codes to find differences. With basic G codes, like those used for motion (G00 through G03), it is likely that you won't find many differences. But with less common G codes, you'll likely find some differences. Figure 6.24 shows a list of G codes for an Okuma OSP control.

Partial list of G codes used on an Okuma OSP control (7000L)

G00 Rapid motion	G42 Tool nose radius compensation right	G77 Tapping cycle (right hand)
G01 Straight line cutting motion	G50 Spindle limiter	G78 Tapping cycle (left hand)
G02 Circular motion (CW)	G64 Droop control off	G80 through G88 Lap cycles
G03 Circular motion (CCW)	G65 Droop control on	G90 Absolute positioning
G04 Dwell command	G71 X axis threading cycle	G91 Incremental positioning
G31 Fixed threading canned cycle	G72 Z axis threading cycle	G94 Feed per minute mode
G34 Variable lead thread cutting cycle	G73 X axis grooving cycle	G95 Feed per revolution mode
G40 Tool nose radius compensation	G74 Z axis grooving cycle	G96 Constant surface speed mode
G41 Tool nose radius compensation left	G75 Automatic chamfering	G97 RPM mode
	G76 Automatic rounding	

Figure 6.24 – List of G codes for an Okuma OSP control

As you can see, many of the G codes remain the same as those used with Fanuc controls. The basic motion type G codes (G00 through G03) remain the same. G04 is still a dwell command. Tool nose radius compensation is still programmed with G40 through G42. The spindle limiter is still G50. G96 and G97 still specify the spindle modes. Even for those G codes you do recognize, you'll need to confirm how they work. With some Okuma OSP controls, for example, an L word is used in circular motion commands (instead of R) to specify the radius of the arc being machined.

Some G codes are simply number conversions. For example, with Okuma OSP, G94 is used to specify per minute feed mode and G95 specifies feed per revolution feedrate mode (instead of G98 and G99). G90 and G91 are used to specify absolute and incremental positioning (instead of X and Z and U and W words). A series *Lap cycles* (G80 through G88) is used instead of multiple repetitive cycles (G70 through G76) – and most programmers that have experience with both types of controls prefer the Lap cycles from Okuma.

There are several G codes that you don't recognize. For example, you'll have to reference Okuma's programming manual to learn the function of *droop control* (G64 and G65).

Admittedly, we're just scratching the surface. Learning another control's programming methods can be challenging – especially if you're on your own. Most machine tool builders provide applications people that are available for phone assistance when you have questions. And most also provide training classes. If you already have experience programming one control type (like Fanuc), learning how to program another control type is relatively easy (compared to learning how to program your first control type).

STOP! Do practice exercise number twenty-two in the workbook.

Lesson 23

Other Special Programming Features

Current model CNC turning centers come with many features to help with special applications. While some of these features will be of little need to you in the immediate future, it is important to know they exist. You cannot begin to apply any feature that you don't know about.

CNC control manufacturers strive to equip their controls with as many helpful programming features as possible. Those mentioned so far (multiple repetitive cycles and subprograms) are used on a very regular basis — and you should strive to master them. However, there are some special programming features that are not used nearly as regularly. Indeed, there are some features that are extremely important to one company but of no value to another.

As you study this lesson, you need to consider your own company's CNC applications. If you are in doubt about the value of a given feature, ask an experienced person in your company about its value to your company. You can minimize your studies about those features your company does not currently need. You can always come back and study this lesson in greater detail should the need arise.

As you study this lesson, remember that *your ingenuity is based predominantly upon your knowledge of what is possible.* You cannot apply a feature of which you are unaware. At the very least, this lesson will acquaint you with what is possible with special CNC programming features.

The organization of this lesson is not as tutorial as previous lessons. While we will explain each feature in detail and in tutorial format, we don't present them in a special order. Here are the topics contained in this lesson:

- Block delete (also called optional block skip)
- Special techniques with sequence numbers
- G codes that have not yet been introduced (in numerical order)

Block delete (also called optional block skip)

The block delete function is used to give the CNC operator a choice between one of two possibilities. An on/off switch on the control panel (commonly labeled block delete or optional block skip) is used to actually make the choice. Since applications for block delete vary, the programmer must make each use of block delete *very clear* to the operator. This should be done in the setup- and production-run-documentation.

A slash code (/) in the program tells the control to look to the position of the block delete switch. If the switch is on, the control will skip any words to the right of the slash code. If the switch is off, the control will execute these words.

Here is a simple example. Say your turning center does not have an adequate coolant switch. Your setup person has no way to turn off the coolant when programs are being verified (whenever an M08 is executed, coolant will come on). To solve this problem, you can place a slash code at the beginning of every coolant command. Here is one way to do so.

/ N015 M08 (If the block delete switch is on, the M08 will be skipped and coolant will stay off)

During program verification, the setup person can turn *on* the block delete switch. This will force the control to *skip* commands that turn coolant on (leaving the coolant off).

The slash code does not have to be placed at the very beginning of a command. If, for example, you have a totally enclosed work area (as most turning centers do), you may want the coolant to come on during each cutting tool's approach to the workpiece. Consider this command.

N015 G00 X3.2 Z0.005 **/ M08** (Only M08 is affected by slash code)

In this case, only the M08 is influenced by the slash code. The rest of the command will be executed regardless of the position of the block delete switch.

Applications for block delete

As stated, block delete can help whenever you wish to give the operator a choice between one of two possibilities. In the coolant example, either the setup person wants coolant or they do not. There are *many* times when a programmer wants to give the operator a choice between one of two possibilities.

Another optional stop

As you know, the optional stop function (M01) lets the operator stop the machine at key times in the program. In the format shown in Lesson Sixteen, we have you place an M01 at the end of every tool (except the last one) to give the setup person and operator the ability to check and see what each tool has done prior to going on to the next tool. This is very important during the program's verification – and is especially important when trial machining. But if you use our recommended programming format, you cannot (feasibly) use the optional stop function for any *other* purpose that might come up.

Say you want the CNC operator to perform sampling inspections on the workpiece at a critical point during the machining cycle. It might be a critical finish boring operation. After this operation in *every tenth workpiece*, you want the operator to take the measurement. This is an excellent application for optional stop - but if you have already placed an M01 at the end of every tool, you cannot use M01 to additionally make the machine stop only after the critical milling operation. If the operator turns on the optional stop switch, the machine will stop *at the end of every tool* as well as at the sampling measurement. This will be distracting and will not provide the desired result.

Consider this program stop command.

N060 M00

M00 is a *program stop*. It causes the machine to stop until the operator reactivates the cycle. There is no option to it – *the machine will stop*. This command can be used if you want your operator to measure *every* workpiece. Still, this is not the desired result. But if used in conjunction with the slash code, you can use the *block delete switch* as a second optional stop function. Consider this command.

/ N060 M00

If the block delete switch is turned on, the control will *skip* the M00 command and the machine will not stop. For this application, *on* will be the normal setting for the block delete switch. When the operator turns off the block delete switch (at every tenth workpiece), the machine will stop after the finish boring operation to allow the measurement.

A warning about block delete: In this application, the block delete switch will function in just the opposite fashion as *optional stop switch* (when the optional stop switch is turned on, the machine will stop). This can cause some confusion. Remember that when the block delete switch is turned *on*, words to the right of the slash code will be skipped. We're accustomed to something happening when a switch is turned on. With block delete, something *doesn't* happen when the switch is turned on.

Trial machining

Throughout this text we have been stressing the need for trial machining. Remember the five steps required to trial machine:

1: Recognition of a tight tolerance that worries you

2: Make an initial adjustment that causes additional stock to be left on the workpiece

3: Let the tool machine under the influence of the trial machining adjustment and stop the cycle after the tool is finished

4: Measure the machined dimension and adjust accordingly

5: Re-run the tool under the influence of the new adjustment

As shown so far, trial machining requires a great deal of manual intervention. After the initial adjustment and trial machining, for example, the setup person or operator must manually restart the cycle.

Think about step one. If a setup person or operator can recognize a tight tolerance that requires trial machining, so can a programmer. There are things a programmer can do in a program to simplify the trial machining operation – and the block delete function can help.

Whenever using block delete for trial machining, the block delete switch will be on during normal production (skipping the trial machining commands). When the setup person or operator wants to trial machine, they'll turn *off* the block delete switch.

Our example shows the use of block delete to help with trial machining for lengthy rough turning operations. From Lesson Thirteen, you know that the amount of finishing stock a rough turning tool leaves is often critical. The setup person can adjust the rough turning tool's wear offset in such a way that the tool will leave excess stock (even more than is left for finishing). They then run the tool and stop the machine when the tool is finished. They measure the workpiece and determine just how much the wear offset must be adjusted. After adjusting the offset, they must rerun the tool. For short roughing operations (under two minutes), this may be just fine. But consider a lengthy rough turning operation for a very large shaft. Say the roughing tool takes over twenty minutes to rough turn the workpiece. In this case, forty minutes (twenty minutes for the original roughing and twenty minutes for the second try) will be required to get the rough turning tool machining properly.

By combining block delete with trial machining, this time can be dramatically reduced. Knowing that the setup person commonly requires trial machining techniques for rough turning, the programmer can include one trial machining pass under the influence of slash codes. If the setup person or operator wants to trial machine for the tool, they turn off the block delete switch. If they do not wish to trial machine (normal operation), they turn on the block delete switch. Here's an example program that shows how this can be done.

O0003 (Program number)

N003 G99 G20 G23 (Ensure that initialized modes are still effective)

N004 G50 S4000 (No spindle limitation necessary)

N005 T0101 M41 (Select rough turning tool, offset, and spindle range)

N010 G99 G96 S400 M03 (Start spindle fwd at 400 sfm)

N015 G00 X6.0 Z0.1 M08 (Rapid up to the workpiece)

/N020 X5.5 (Begin trial machining operation)

/N025 G01 Z-0.3 F0.020 (Trial machine)

/N030 X6.0 (Feed up face)

/N035 G00 X8.0 Z3.0 (Rapid to convenient measuring position)

/N040 M00 (Stop for measurement, DIAMETER SHOULD BE 5.50 IN)

/N045 T0101 M03 (Reinstate offset, restart spindle)

/N050 G00 X6.0 Z.1 (Rapid back to starting point)

N055 G71 P060 Q160 D2500 U0.040 W0.005 F0.020 (Rough turning cycle)

N060 . . .

.

.

In line N020, we begin the trial turning operation. Notice the slash codes. In lines N025, N030, and N035, the roughing tool makes the trial turning pass and rapids to a convenient measuring position. At this location the setup person can easily measure the workpiece. In line N040, the machine stops due to the M00. We strongly recommend that you include a message in the program at this point telling the setup person what the diameter (and if necessary, the Z face position) the workpiece should currently be. The setup person measures and adjusts the offset accordingly. Line N045 reinstates the offset, based on the setup person's recent offset change. In line N050, the tool rapids back to its starting point. From line N055, the program continues in its normal manner. After setting the offset, the operator turns on the block delete switch until they need to trial machine again.

Again, the block delete switch can be turned off whenever the setup person wishes to use the trial machining sequence, meaning the CNC operator will also be able to trial machine when changing (or indexing) the rough turning tool's insert during the production run.

Warning about block delete applications

You must always consider what will happen if the operator has the block delete switch in the wrong position when they run the program. In the case of the coolant example, the operator might get wet. In the case of the trial machining operation, trial machining may be ignored when it is required. While these are not pleasant situations, at least the operator is not placed in extreme danger. Consider an application when the operator (and/or machine) may be in danger if the block delete switch is not correctly positioned.

Say you have some castings that are varying substantially from one workpiece to the next. A rough face and turn tool is being used to machine a large portion of the workpiece, and the raw material is varying as much as 0.5 inch from one casting to another (0.5 in excess stock on some workpieces). In this case, you might decide to make a series of roughing passes just to remove the excess stock under the influence of the slash code. Then you program the last roughing pass in the normal manner without the slash codes.

If a workpiece does not have excess stock, the operator will turn on the block delete switch to ignore the extra passes (saving time). But if the workpiece has excess stock, the operator will turn the block delete switch off. If they do not do so, the rough face and turn tool will machine much more stock than intended. This will cause damage to the tool, workpiece, and possibly the machine – and if the workpiece is thrown from the work area, possibly injury to the operator.

When faced with this problem, most programmers will do one of two things. Either the workpieces will be separated (those that have excess stock from those that do not) or the program will be written to make the extra passes, whether they are needed or not. While cycle time may be compromised, safety is enhanced.

Sequence number (N word) techniques)

As you know, sequence numbers provide a way to number each line in the program in an organized manner. Most programs in this text, for example, skip five numbers between each sequence number (N005, N010, N015, etc.). This allows everyone to find important commands, and allows room between commands should you need to add more.

Eliminating sequence numbers

While sequence numbers are very helpful, especially to beginning programmers, they do require space in the machine's memory. Fanuc controls have the reputation of having rather small memory capacities. The time may come when you want to load a lengthy program into the machine, but there is insufficient storage capacity. Either some (possibly important) programs must be deleted from memory to make room, or your programs must be made shorter. The first technique most programmers apply to reduce program length is to eliminate sequence numbers. While this makes it more difficult to reference important commands in each program, more and longer programs can be loaded into the machine.

Using special sequence numbers in program restart commands

Though you may consider it rather easy to find the beginning of each tool by scanning T words (with our format used with geometry offsets, the restart command for each tool is the command including the T word), you can help your operators more easily find the restart command for each tool if you use a special sequence number to begin each tool. Place this special sequence number in the T word command. Consider this program.

```
O0025 (Program number)
G99 G20 G23
G50 S5000
(ROUGH FACE AND TURN TOOL)
N1 T0101 M41
G96 S450 M03
G00 X2.2 Z0.005 M08 (1)
G01 X-0.062 F0.012 (2)
G00 Z0.1 (3) X1.810 (4)
G01 Z0.005 (5)
X1.935 Z-0.245 (6)
Z-0.495 (7)
X2.06 Z-0.745 (8)
G00 X6.0 Z5.0
M01
(1" DRILL)
N2 T0202 M41
G97 S380 M03
G00 X0 Z0.1 M08 (9)
G01 Z-2.8175 F0.009 (10)
G00 Z0.1 (11) X6.0 Z5.0
M01
(ROUGH BORING BAR)
N3 T0303
G96 S400 M03
G00 X1.19 Z0.1 M08 (11)
G01 Z-0.4325 F0.008 (12)
X0.94 Z-1.1825(13)
G00 Z0.1 (14)
X6.0 Z5.0
M01
(FINISH BORING BAR)
N4 T0404 M41
G96 S450 M03
G41 G00 X1.5 Z0.1 (15)
G01 Z0 F0.005 M08 (16)
G02 X1.25 Z-0.125 R0.125 (17)
G01 Z-0.4375 (18)
X1.0 Z-1.1875 (19)
```

```
X0.8 (20)
G00 Z0.1 (21)
G40 X6.0 Z5.0 (Cancel tnr comp)
M01
(FINISH TURN)
N5 T0505 M42
G96 S500 M03
G00 X1.95 Z0 M08 (22)
G01 X1.15 F0.007 (23)
G00 Z0.1 (24)
G42 X1.75 (25, instate comp)
G01 Z0 (26)
X1.875 Z-0.25 (27)
Z-0.5 (28)
X2.0 Z-0.75 (29)
X2.2 (30)
G40 G00 X6.0Z5.0 (cancel comp)
M30
```

If this technique is used consistently, everyone will know how to easily find the beginning of each tool. Even if you eliminate sequence numbers for the purpose of conserving memory space, we recommend retaining these special tool-restarting sequence numbers.

Documenting your programs with messages in parentheses

Almost all current model Fanuc controls allow you to include messages within the CNC program with parentheses (exceptions to this statement are the 3T and some 0Ts). Any information within parentheses will be shown on the display screen of the control but ignored during the execution of the CNC program. This gives the CNC programmer an excellent way to provide documentation for the CNC setup person and operator.

While we've been using messages excessively (documenting every line), you'll often need to clarify things for setup people and operators.

General information about the job

A well documented CNC program begins with a series of messages that remove any doubt about the program's use. Here is a sample program beginning that includes sufficient documentation for this purpose. (Note that many controls require *upper case characters* for messages.)

```
O0001
(          MACHINE: MORI SEIKE SL4)
(  PART NUMBER: A-2355-2C)
(      PART NAME: BEARING FLANGE)
(         REVISION: F)
(       CUSTOMER: ABC COMPANY)
(       OPERATION: 20, MACHINE BORED END OF PART)
(   PROGRAMMER: MLL)
( DATE FIRST RUN: 4/11/04)
(  LAST REVISION: 6/30/04 BY CRD)
```

```
(           RUN TIME: 00:05:25)
N005 G99 G20 G23
N010 G50 S5000
(ROUGH FACE AND TURN TOOL)
N015 T0101 M41
.
.
.
```

Notice that anyone viewing this header information can easily tell which CNC machine the program is for, what workpiece and operation the program is machining, who wrote the program, who changed it last, and some important dates. Though this kind of information may seem quite basic, remember that many companies eventually accumulate thousands of CNC programs. Without this basic documentation in each program, it can be very difficult to keep track of which programs are used for a given job.

Of special importance in our example header is the *current revision* for the workpiece. Remember that the designs for production workpieces are commonly changed, meaning changes to CNC programs. These changes can wreak havoc with the organization and maintenance of CNC programs. In the event that a revision may not be permanent (the design engineer may delete the revision), many companies maintain a copy of each CNC program for every revision, meaning a given operation for one workpiece may eventually have many CNC programs. The setup person and/or CNC operator must be very cautious when starting each new job, confirming that the CNC program they are about to run will machine the workpiece to its most recent revision. Documenting and maintaining the revision information at the beginning of every CNC program makes this checking easy.

Also notice the specification of run time. Once the program has run once, it can be helpful to document its run time right in the CNC program. Anyone looking at the program in the future (while the program is not currently running) can easily determine how long the program takes to run.

Tool information

We've been using this technique in every example program. Place a descriptive message in a line by itself at the beginning of each tool (skipping a line before this message makes it even easier to isolate a tool's set of commands). In this message, you can name the tool, the operation being performed, tool station, etc.

At every program stop

There will be times when the CNC operator must perform a special (manual) task during the CNC program's execution at a program stop, commanded by an M00. In a flip-job, for example, the turning center will stop after the first operation is completed for the operator to turn the workpiece around in the chuck. Whenever you program an M00, *always* place a message in the program telling the operator what is expected. Here are some examples:

```
N205 M00 (TURN WORKPIECE AROUND)
N315 M00 (BLOW CHIPS OUT OF HOLE)
N545 M00 (REDUCE CLAMPING PRESSURE FOR FINISHING OPERATIONS)
```

To document anything out of the ordinary

It only makes sense to strive to do things in a consistent and logical manner. CNC programmers should keep their programming techniques as consistent as possible to ensure that everyone gets comfortable with their method of programming. However, there may be special considerations that force a CNC programmer to do something unusual. Whenever this happens, a message in the program can clarify the reason why common practices are not being followed and give instructions about how to proceed.

For example, most CNC turning center programs require only one wear offset per tool. For this reason, most CNC programmers make the wear offset number for each tool the same value as the tool station number.

This way the CNC operator can easily get comfortable with how to determine which wear offset is related to each tool. Tool number one will use wear offset number one. Tool number two uses wear offset number two, and so on.

In Lesson Thirteen, we introduce times when more than one wear offset is required for a given tool (or set of tools) within a CNC program. For example, say a grooving tool is necking two identical grooves. One groove is being machined in an area of the workpiece that has good support. The other groove is being machined in a different area of the part that has poor support. This difference in support and tool pressure is causing one groove (the one in the supported area of the workpiece) to come out deeper than the other. So the programmer uses two wear offsets for the grooving tool. One wear offset determines how grooving will be machined in the weak area of the workpiece. The other determines how the groove will be machined in the well supported area. Since this technique is rarely used, the programmer should include messages in the program to alert the setup person and operator about what is being done.

```
O0010
(SPECIAL NOTE! TWO WEAR OFFSETS ARE USED FOR GROOVING TOOL!)
(USE WEAR OFFSET 5 FOR RIGHT-MOST GROOVE AND WEAR OFFSET 25 FOR)
(LEFT-MOST GROOVE.)
N003 G99 G20 G23
N004 G50 S5000
(ROUGH TURNING TOOL)
N005 T0101 M41
.
.
.
```

For changes made after a dispute

There are times when a CNC programmer is asked to make changes to a CNC program with which they do not agree. A change that improves machining time may (in the programmer's opinion) open the door to safety related problems. Whenever a CNC program change is made after any dispute, it is a good idea to document the circumstances that lead to the change. Within the CNC program makes the most obvious place to document such a change.

Automatic corner rounding and chamfering

This *optional* feature can make programming corner radii and chamfers easier. This feature will only work when square shoulders are involved. It allows you to program chamfers and radii as part of the previous facing or turning command. This can keep the programmer from having to specify chamfers and radii in separate commands.

The three words involved are I, K, and R. I and K are used for commanding 45 degree chamfers. R is used for commanding radii. One of these words will be included in G01 commands with the X or Z axis departure. They must point in the direction of the upcoming chamfer or radius (plus or minus) as well as specify the size of the chamfer or radius.

Figure 6.25 shows a drawing to be used for an example. Only the finish turning operation is being shown.

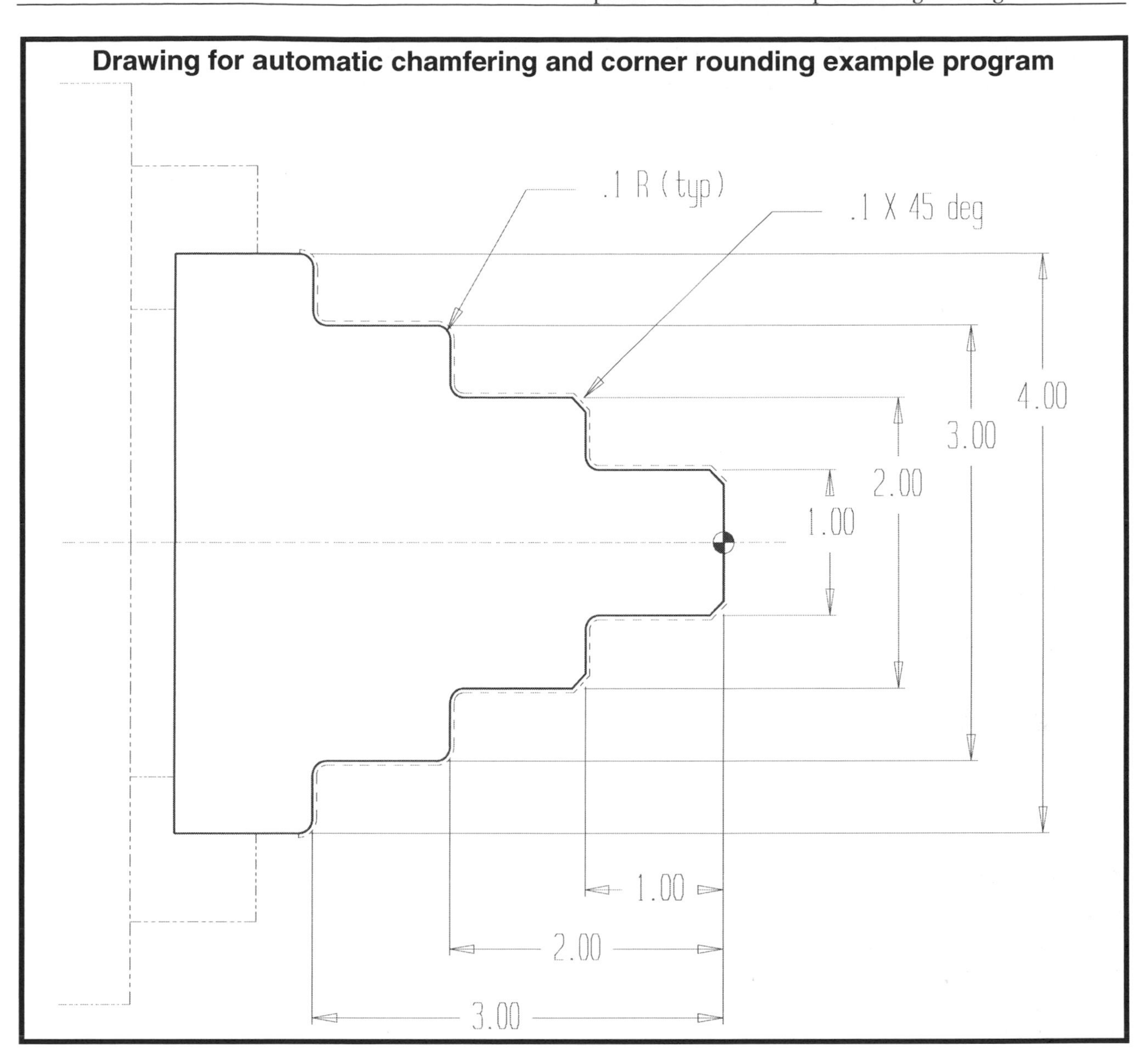

Figure 6.25 – Drawing for automatic corner rounding and chamfering example

Program:

O0023 (Program number)

N003 G99 G20 G23 (Ensure that initialized states are still in effect)

N004 G50 S5000 (No limiting necessary, limit spindle speed to machine's maximum)

N005 T0101 M42 (Select finish turning tool and high spindle range)

N010 G96 S500 M03 (Start spindle fwd at 500 sfm)

N015 G00 X0.8 Z0.1 M08 (Rapid up to workpiece, start coolant)

N020 G01 Z0 F0.007 (Feed flush with face)

N025 X1.0 Z-0.1 (Form first chamfer)

N030 Z-1.0 **R0.1** (Turn diameter and form fillet)

N035 X2.0 **K-0.1** (Come up face and form chamfer)

N040 Z-2.0 **R0.1** (Turn diameter and form fillet)

N045 X3.0 **R-0.1** (Come up face and form radius)

N050 Z-3.0 **R0.1** (Turn diameter and form fillet)

N055 X4.0 **R-0.1** (Come up face and form radius)

N060 G00 X6.0 Z5.0 (Rapid to safe index position)

N065 M30 (End of program)

In line **N025**, the first chamfer is being programmed in the normal (longhand) manner. Since the approach movement is not at ninety degrees to the direction of the next motion, automatic corner chamfering cannot be used in this command.

In line **N030**, notice the program is commanding a motion in Z to a minus one inch end point. But due to the R word in this command, the machine will begin forming a 0.100 radius as soon as it reaches within 0.100 of the programmed end point. Since the R word is plus, the direction of the radius is formed in the X+ direction.

In line **N035**, the next face is being machined. Notice the **K-0.1** in this command. This tells the machine that a chamfer is to be formed in the Z minus direction as soon as the tool comes within 0.100 of the X end point. And so on. In each straight line cutting command, since I, K, or R is included, chamfers and radii are automatically formed.

Remember, this is an optional feature that may not be equipped with your control. However, since it makes programming much simpler, it will be worth the time you spend testing whether it is available (one way to do so is to load and run the previous example program).

Other G codes of interest

If we have not yet discussed a given G code, it probably has a very special application. And frankly speaking, it is not be required by all CNC users. While some of these G codes will be very important to you and your company's needs, others will not. Talk with experienced people in your company to see which of these G codes your company uses.

G04 - Dwell command

A dwell command will cause all axes (X and Z) to pause for a specified period. With Fanuc and Fanuc-compatible controls, the dwell period is specified in seconds (some controls allow the period of dwell to be specified in number of spindle revolutions).

With Fanuc controlled turning centers, you can actually use three different letter addresses to specify the dwell time in seconds (X, U, and P). X and U allow a decimal point, while P does not and requires a three-place fixed format. Here are three examples. All specify a one second dwell.

N045 G04 X1.0 (Pause for one second)

N045 G04 U1.0 (Pause for one second)

N045 G04 P1000 (Pause for one second)

The most basic application for the dwell command is to allow time for *tool pressure* to be relieved. This is commonly required when necking a groove. When the grooving tool reaches the bottom of the groove, if it is immediately retracted from the groove, the bottom of the groove will not be perfectly round (especially at larger diameters). Pausing at the bottom of the groove will allow the spindle to rotate far enough for all material at the groove bottom to be removed. It will also allow tool pressure to be relieved. Here's an example:

.

.

.

(1/8 wide grooving tool)

N250 T0505 M42 (Index turret, start spindle)


```
N255 G96 S500 M03 (Start spindle fwd at 500 sfm)
N260 G00 X4.2 Z01.5 M08 (Rapid to approach position, start coolant)
N265 G01 X3.7 F0.005 (Plunge groove)
N270 G04 X0.5 (Pause for one-half second)
N075 G00 X4.2 (Rapid out of groove)
.
.
.
```

Though we've used an X word to specify the dwell time, remember that *all axes* will pause whenever a **G04** command is used (not just X).

Though we don't agree, we've seen some machine tool builders actually recommend that you program a dwell command to bypass some machine problems related to M code activation. Fully interfaced M codes should have a confirmation signal to inform the control that the M code function is completed and that it's all right to continue (the machine should *automatically* pause during the M code's activation).

Consider, for example, the jaw-open and jaw-close function for bar feed turning centers. An **M14** may be used to close the jaws and **M15** may be used to open them. After the bar has advanced, and the programmer specifies an **M14** to close the chuck jaws, *the machine should automatically wait* until the jaws are completely closed before it allows the program to continue. The next command is a motion back to the tool change position. If the machine does not wait, the bar will continue feeding out during the retract motion. This, of course, will cause severe problems when machining the next workpiece.

We've seen some turning center manufacturers that recommend programming a dwell command after closing the chuck jaws to allow time for the jaws to close prior to the rapid movement. While this usually works, if for any reason, the chuck jaws have not closed completely during the dwell (a variation causing jaw closing time to vary), you may still have the problem. For this reason, you need to specify an excessive dwell time (wasting cycle time) to ensure that the jaws will always be closed before the motion occurs.

Another machine problem that can be handled with a dwell command is related to your coolant system. Many turning center cutting tools require that the coolant be flowing full force in order for the operation to be successful. A through-the-tool, coolant-fed drill is one such tool. When the drill enters the hole, the coolant must be flowing at full force. If it is not, the drill will be damaged due to excessive heat buildup. If the coolant system is poor (maybe it has a failing check-valve to hold the coolant in the coolant lines), it may take a while for the coolant to get up to full pressure. Many programmers simply program a dwell command after the drill approaches to allow time for the coolant to get up to full force.

Frankly speaking, *we do not recommend programming around machine problems.* The best way to handle these problems is to *fix the machine.* In the case of the chuck jaw activation, ask the machine tool builder to fully interface the M code that closes the chuck jaws. In the case of the poor coolant system, fix the check-valve. While the **G04** may be a reasonably good *temporary* solution, you can minimize the potential for long-term problems by eliminating the source of the problem.

Other G codes

We'll discuss a few more G codes in the Appendix as we present the programming of certain accessory devices. If your turning centers have live tooling capabilities, for example, you likely have a series of hole machining canned cycles (**G80-G89**) to help you machine holes. We'll also be discussing *polar coordinate interpolation* commanded by **G112** and **G113** during this discussion. As you look in the G codes list of your Fanuc programming manual, you'll see a few more G codes that we have not introduced, but for the most part, their application is so limited that the typical CNC user will never have a need for them. If you feel you have a need for one of these commands, you can contact us or your machine tool builder to learn more about its application.

Key points for Lesson Twenty-Three:

- Control manufacturers provide a number of special features to help with programming. While some won't be of immediate need, it is important to know what is available.
- Block delete can be used to give the operator a choice between one of two possibilities.
- Block delete can help you program trial machining operations.
- Use a special sequence number to specify the restart command for each tool.
- Messages in parentheses can minimize confusion.
- Automatic corner rounding and chamfering can help when machining chamfers and radii.

STOP! Do practice exercise number twenty-three in the workbook.

Appendix

Special Machine Types And Accessories

To this point, we've limited our presentations to programming methods for basic two axis turning centers. In this Appendix, we'll be showing how to handle the programming of special machines as well as special accessory devices.

As you know, turning centers come in many configurations – and they can be equipped with many special accessories that change the way programs must be written. If you'll be working with a basic two axis turning center, you my may be able to skip much of this appendix. Just skim this material to see what's here. You can always come back and read more thoroughly should the need arise. Be sure to check with an experience person in your company to find out which accessories and machine types you must study.

All accessories and machine types discussed in this appendix require that you have a firm understanding of what has been presented thus far. In all cases, we'll be building upon what you already know.

Unfortunately, the actual programming and usage of certain accessories is highly machine-builder-dependent. While we can easily explain the application for a given device, it will be impossible to show specific usage techniques for all turning centers. In these cases, you'll at least be exposed to one builder's way of handling the accessory's usage.

Work holding and work support devices

How a workpiece is held and supported during its turning center machining operations is one of the most critical factors contributing to the success of the turning center application. If a workpiece is held and supported properly, all other facets of the machining operation will be easy to control. But an improperly held and/or supported workpiece will, at best, make the machining operation difficult to perform. In worse cases, an inadequate work holding and/or support device will make it impossible to hold size on the workpiece. That is, you won't be able to make good workpieces. And at worst, a poor selection of workholding and support devices can make for a very dangerous working environment.

In this section we will take a close look at the most common work holding and support devices available for use with CNC turning centers. While we cannot possibly discuss every detail and implication of each device, this presentation should give you a reasonably well rounded-view of the devices available.

Work holding devices

During any turning operation, the workpiece will be under two kinds of stress. One cause of workpiece stress is the centrifugal force generated with workpiece rotation. At the high spindle speeds available from today's turning centers, an out-of-round workpiece will have the tendency to shake itself apart. The second cause of workpiece stress is caused by the machining operation itself. As any cutting tool comes into contact with the workpiece, it can have the tendency to push the workpiece out of the workholding device.

During the typical purchase of a turning center, often too little consideration is given to possible work holding and work support alternatives. And, as stated, a poorly chosen work holding and/or work support device will make for a poor machining environment. In this section, you will be exposed to the most common work holding alternatives, how they are used, and for what kinds of work they are best.

Try not to view a turning center's work holding device as an integral part of the machine tool itself. Rather, try to view it as only part of the machine's setup, just as a common table vise is only a part of a vertical machining center's setup. Though a table vise is a very common work holding tool, it will not properly hold all workpieces to be machined by a vertical machining center. Just as a table vise will not suffice for all machining center setups, no single turning center work holding device will suffice for all kinds of workpieces that can be machined on a turning center.

Three jaw chucks

The three jaw chuck can be thought of as the table vise of turning center work holding devices. This form of work holding device is, by far, the most popular device used to hold workpieces during machining on turning centers. Almost every company that utilizes CNC turning centers owns at least one form of three jaw chuck. Here we will present an overview of the types available as well as their basic applications.

Figure A.1 shows a common three jaw chuck used with CNC turning centers. The chuck in this drawing has no top tooling attached. We are currently stressing the basic components of the chuck body itself.

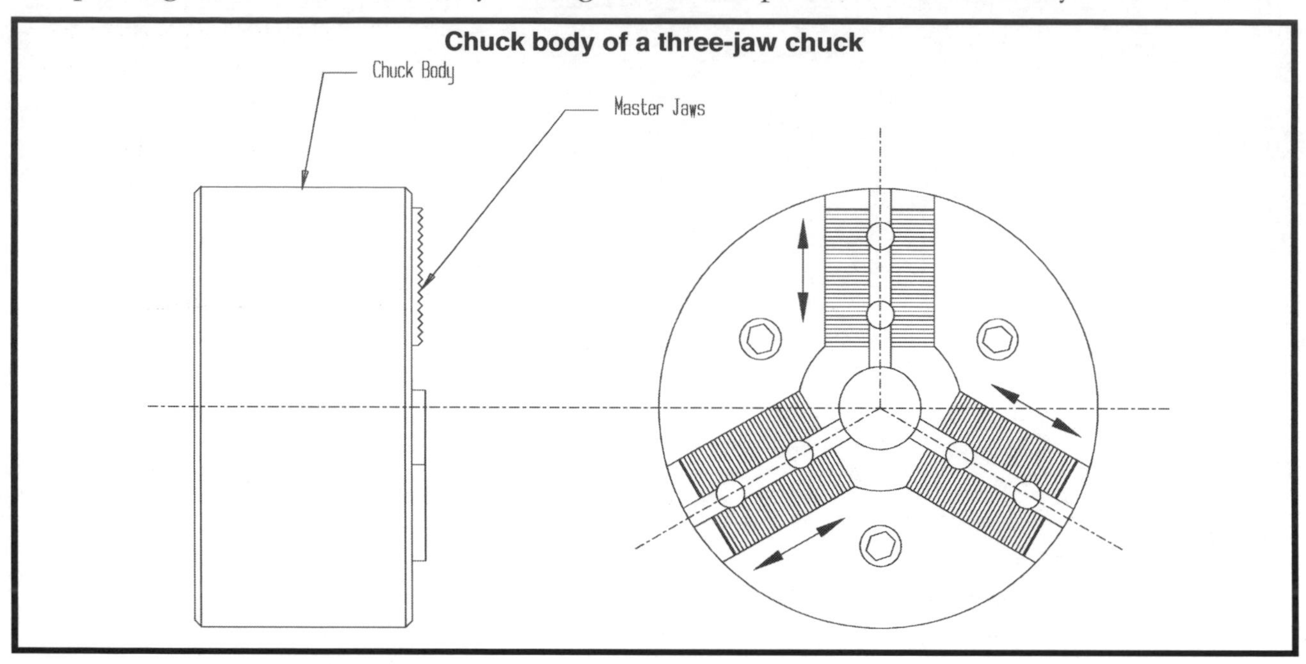

Figure A.1 – Three jaw chuck without top tooling

A three jaw chuck has three moving master jaws. These jaws are serrated to accommodate the top tooling used to clamp onto the workpiece. The master jaws are mounted 120 degrees apart, allowing symmetrical clamping around the workpiece. As the chuck is activated, the three jaws move into the clamped position. Note that the three master jaw chucks are designed to clamp in both directions, giving it the ability to hold a workpiece on outside diameters as well as on inside diameters.

Most three jaw chucks used with CNC turning centers are activated by a foot switch. By pressing the foot switch once, the jaws close. Pressing it again makes the jaws open. The two most popular styles of three jaw chuck are *hydraulic* and *pneumatic*.

Hydraulic chucks require an activator to be mounted at the rear of the spindle. A *draw tube* (or draw bar) runs through the spindle and moves fore and aft as the chuck is activated to provide the driving force to open and close the chuck. Note that the draw tube reduces the maximum size of bar stock that can be fed through the spindle. In fact, for very small chucks, a *draw bar* is used for activation, eliminating the possibility to feed bar through the spindle.

The major advantage of hydraulic chucks is that they provide a tremendous amount of chucking pressure at the jaw. This gives the hydraulic style chuck the ability to securely hold very heavy workpieces.

The major drawback of hydraulic chucks is that they have a very limited range of pressure settings. In fact, it can be very difficult to truly tell exactly how much chucking pressure is being applied to each jaw. Since chuck pressure is changed by changing the flow of oil into the chuck's activator, precise pressure changes are next to impossible. Also, the amount of pressure supplied by a hydraulic chuck may vary throughout the day, as the temperature of the oil within the activator device changes. For this reason, hydraulic style chucks are best suited for applications that require a great deal of chucking force, but no chucking finesse.

Pneumatic chucks are activated by air. Most are mounted to the front of the spindle, with no draw tube or draw bar required. Pneumatic chucks cannot provide the clamping pressure available from hydraulic chucks. However, their range of chucking pressure adjustment is much better. When clamping on somewhat flimsy workpieces (like tubing), it is necessary to more precisely regulate the amount of pressure at the jaw. If too much pressure is applied, the workpiece will be deformed during chucking. The fine chuck pressure adjustment allowed by pneumatic chucks make holding flimsy workpieces much easier.

Hard jaws

Hard jaws are so named because they are hardened and ground. This hardening process makes the hard jaw harder than the workpiece it is designed to hold. For this reason, hard jaws bite into the workpiece being held.

Figure A.2 shows one type of hard jaw. Hard jaws are used when you must get the best possible grip on the workpiece. The sacrifice for this powerful grip is that hard jaws may leave severe witness marks on the workpiece. As long as the witness marks will cause no damage to the workpiece, or as long as the surface gripped by the hard jaws will be machined in a later operation, hard jaws make the best possible choice if gripping potential is of primary concern, as would be the case with powerful roughing operations.

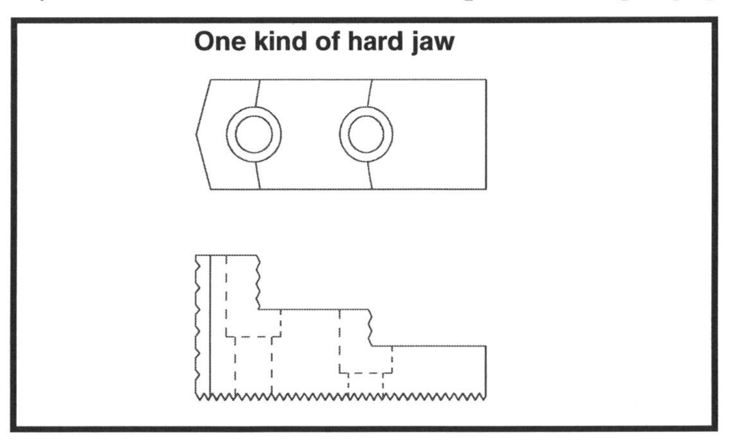

A.2 – One type of hard jaw

Most hard jaws are designed to grip on more than on surface of the jaw. As the drawing shows, most hard jaws can be placed in the chuck one way to allow relatively small diameters to be gripped, or turned around in the chuck to grip on larger diameters. Figure A.2 shows one style of hard jaw. Hard jaws can be designed to suit a wide variety of workpiece configurations.

Hard jaws do not make a very good choice for clamping on finished surfaces for two reasons. First, and as mentioned, they tend to leave nasty witness marks on the workpiece. These witness marks are usually unacceptable if left on finished surfaces. Second, due the inconsistency of biting depth when hard jaws are used, it is not possible to hold concentricity from the gripped end of the workpiece to the end of the workpiece being machined. If concentricity is required from one end of the workpiece to the other, another form of top jaw must be used. In many applications, hard jaws are used for the first operation (when raw material is being held) and *soft jaws* are used for the second operation.

Soft jaws

Soft jaws are best applied when leaving a witness mark on the workpiece is unacceptable, and/or if concentricity is critical from one end of the workpiece to the other. Unlike hard jaws, soft jaws are not hardened and ground. They are made from steel. If concentricity is important, soft jaws must be machined in the chuck for every setup in which they are used. The diameter two which jaws are machined is very close to the diameter of the workpiece that is to be held.

Figure A.3 shows a typical soft jaw.

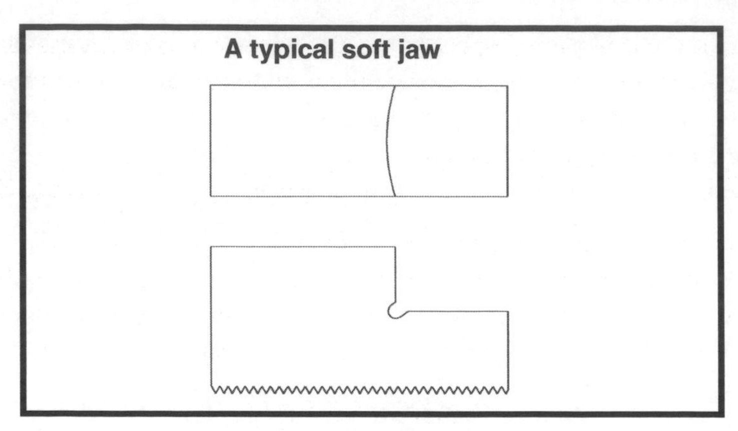

Figure A.3 – A typical soft jaw

Say for example, soft jaws must clamp on a three inch external workpiece diameter. This diameter has been machined in a prior operation, so it is perfectly round and quite smooth. The setup person will mount the soft jaws in the three jaw chuck and bore the jaws to a diameter that is very close to a three inch diameter. Ideally, the soft jaws will be bored just slightly smaller than three inches (2.995 inch or so). Once bored, the jaws will have the tendency to expand a small amount in diameter. When bored slightly smaller than the chucking diameter, the jaws will nicely expand to the workpiece diameter to be held. If bored too large, the jaws will not properly hold the workpiece since only the very center of the jaw will contact the workpiece. Jaw boring is discussed in greater detail during Lesson Twenty-Four.

Programmable features of three jaw chucks

Almost all three jaw chucks designed for use with CNC turning centers are automatic, activated by a foot pedal or some form of switch. Additionally, there are other features of three jaw chucks that are also automatic. Some of these automatic features are programmable.

Chuck jaws open and close

The operator must have the ability to manually open and close the chuck jaws in one manner or another. Usually this is accomplished by a foot switch, making it easy to hold the workpiece with both hands during the loading of a workpiece.

Most CNC turning centers also allow the chuck jaws to be opened and closed by program commands. Usually two M codes are used – one for open and another for close. The actual M code numbers used for this purpose vary dramatically from one turning center to the next and can be found in the turning center's programming manual.

For safety reasons, most chucks require the spindle to be stopped prior to opening or closing the chuck. This is true of both manual activation as well as programmed activation.

The most common application for programmable chuck open and close is bar feeding. Though collet chucks (discussed later) make a better choice for bar feeding applications, a programmable three jaw chuck will also work. During a bar feed operation, it is necessary for the bar stock to feed through the center of the spindle. Before this is possible, the chuck jaws must be opened. After the bar is fed, the chuck jaws must be closed before the next workpiece can be machined. Bar feeders will be discussed in detail a little later in this Appendix.

Chucking pressure

In most turning center applications, the workpiece is loaded into the chuck and clamped once. The chucking pressure (manually selected during setup by a valve or rheostat) will be used to hold the workpiece during the entire machining operation. For a one-time chucking operation, the chuck pressure must be adequate to hold the workpiece rigidly during machining operations, yet not be so great so as to deform the workpiece. If the workpiece is deformed, it will spring once the chuck is opened.

With structurally strong workpieces, the operator or setup person will be able to easily make this chucking pressure compromise. However, weaker workpieces cannot be completely machined without sufficient clamping force to deform the workpiece, especially if powerful roughing operations must be performed. If only one clamping pressure is used for the entire machining cycle, it may be impossible to successfully machine the workpiece without deformation.

Some companies, when faced with this problem, have the operator manually change chucking pressure during the machining cycle. With the chucking pressure set quite high, the workpiece is loaded and the rough machining operations are performed. After roughing, the cycle is stopped (with a program stop **M00**). The operator then lowers the chucking pressure and reactivates the cycle. The finish machining operations are then performed at the lower chucking pressure. This ensures that the workpiece will not be deformed during the finishing operations. When the cycle is completed, the operator *must* remember to increase the chucking pressure before the next workpiece is loaded (if they do not, the results could be disastrous).

Keep in mind that if a hydraulic chuck is being used to hold the workpiece, it can be very difficult for the operator to precisely set the chucking pressure consistently from time to time. In most cases, a separate *chuck pressure gauge* must be used to confirm the chuck pressure. This gauge is actually clamped in the chuck to provide the chuck pressure reading.

Using this method to confirm chuck pressure is tedious and time consuming. The operator may have to adjust the chuck pressure valve two or three times in order to get the right pressure. And this must be done twice during each cycle. If the operator forgets to change the chucking pressure, at best the workpiece will be scrapped. At worst, the workpiece could be thrown from the chuck.

A programmable high/low pressure three jaw chuck improves this situation. Two M codes, one for high pressure chucking and another for low pressure chucking are used to change the chucking pressure. The operator sets two rheostats (one for high clamping pressure and another for low) during setup. If a chuck pressure gage is used to confirm chucking pressure, the operator need only use it one time (not twice for each workpiece). In the program, two M codes tell the control which chucking pressure to use. Operation becomes much more automatic.

Almost all three jaw chucks require that the jaws be re-clamped during the changing of the high/low valve. This is even true of most programmable high/low pressure chucks. During the re-clamping process, the workpiece must be held securely in order to maintain concentricity and to keep it from falling out of the chuck. To allow totally automatic operation, most companies incorporate some form of turret-mounted workpiece pusher (possibly spring-loaded) that holds the workpiece firmly in place during the re-clamping required for changes in chuck pressure.

Chucking direction

Three jaw chucks can clamp in both the opened and closed position, allowing the chuck to hold on inside as well as outside diameters. For external diameter clamping, of course, the clamping direction is inward (toward the center of the chuck). For internal diameter clamping, the clamping direction is outward. Almost all CNC turning centers allow an easy way of changing the clamping direction. Usually a toggle switch or manually activated valve controls which way the chuck clamps. Some even allow the chuck direction to be changed by programmed command (usually two more M codes).

Note that for safety reasons, most CNC turning centers will not allow the spindle be activated if the chuck has not been brought to the clamped position, so the chucking direction must be correctly selected before a program can be activated. Also for safety reasons, most turning centers do not allow the chucking direction to be changed while the spindle is running.

Collet chucks

Figure A.4 shows one kind of collet chuck. Instead of incorporating top jaws with which to grip the workpiece, a collet chuck utilizes an internal hardened collet. The external configuration of collets will remain identical from one collet to the next. Only the inside diameter of the collet (the diameter that holds the workpiece) will change. This diameter is machined just slightly larger than the workpiece diameter to be held. Collets can be purchased as standard items from most tooling supply companies and come in standard sizes ranging from 0.0312 in to well over 2.0 inches in diameter.

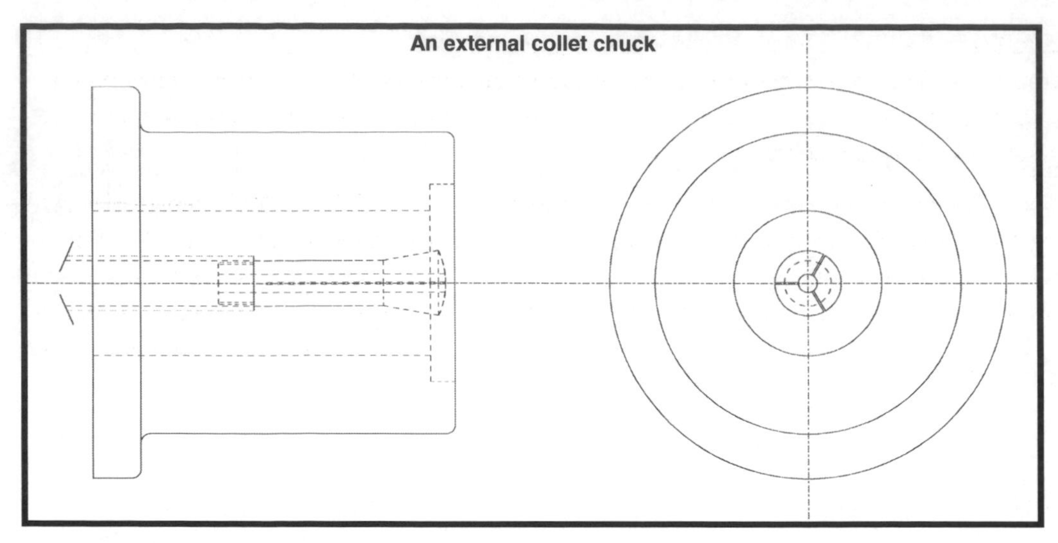

Figure A.4 – A typical collet chuck

The first benefit of a collet chuck is collet-change time. Off-the-shelf collets can be used. It is very easy to exchange collets so setup time is dramatically reduced compared to three jaw chucks. Generally, a collet can be changed in less than two minutes.

The second benefit is related to concentricity. A collet holds the workpiece nearly all the way around the periphery of the workpiece. This give a collet chuck excellent concentricity qualities.

A third benefit is reduced activation time. Since collets are made to spring to slightly larger than the diameter they are designed to hold, they must collapse only about 0.020 in to 0.030 in when clamping. Clamping time is very short.

Yet another benefit is reduced wind resistance. Most collet chucks are smooth around their outside diameter. Since they need not hold top jaws that protrude from the face of the chuck, there is no wind generated during rotation. A three jaw chuck can sometimes resemble a fan in this regard if long top jaws are used. For this reason, collet chucks take much less energy to rotate.

While the drawing in Figure A.4 shows an external collet chuck, many CNC turning centers that are designed exclusively for bar feed applications incorporate the collet chuck as an integral part of the machine's spindle. In essence, these machines appear to have no chuck.

A collet chuck makes an excellent choice with bar feeding applications. Bar feeders will be discussed later.

Figure A.5 shows some of the collet styles available.

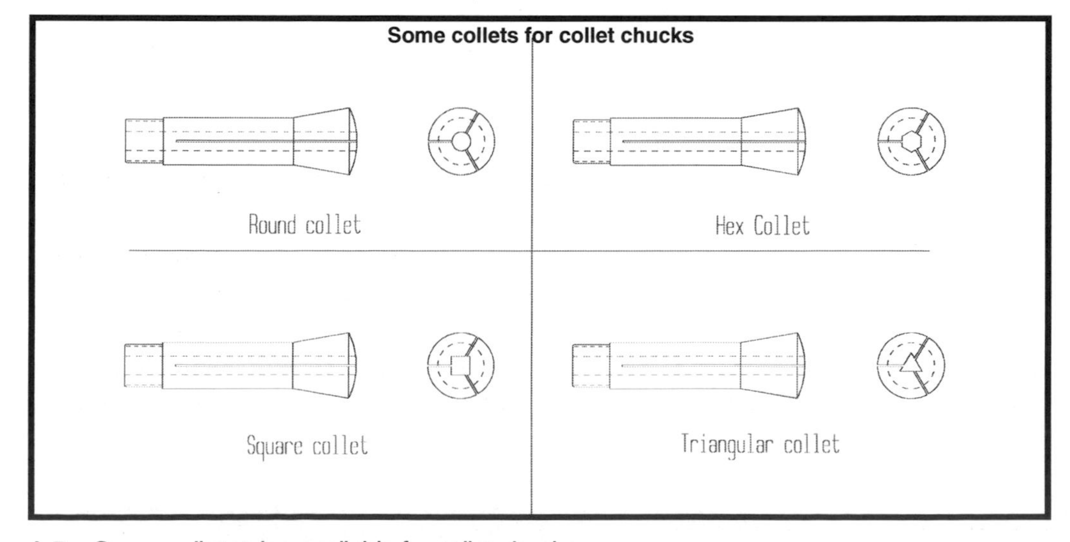

A.5 – Some collet styles available for collet chucks

Work support devices

Properly holding the workpiece is often only part of a good turning center setup. While there are times when the workholding device by itself is sufficient to hold and support the workpiece, there are also times when additional support is required for stable machining.

Generally speaking, the longer the workpiece, the more probable it will be that the workholding device by itself will not be sufficient to properly hold and support the workpiece. To help determine whether some kind of work support device is required, we offer a simple rule-of-thumb: *the length to diameter ratio.* If the workpiece extends from the workholding device by a distance over about three times its diameter, and if optimum cutting conditions must be achieved, some form of work support device will be necessary. For example, if a one inch diameter workpiece must be machined, it can extend about three inches from the workholding device before some form of work support device should be used.

Tailstocks

The most common form of work support device is a tailstock. A tailstock is used to support the end of the workpiece opposite the work holding device. In essence, when a tailstock is used, both ends of the workpiece will be supported. Figure A.6 shows an example setup incorporating a tailstock.

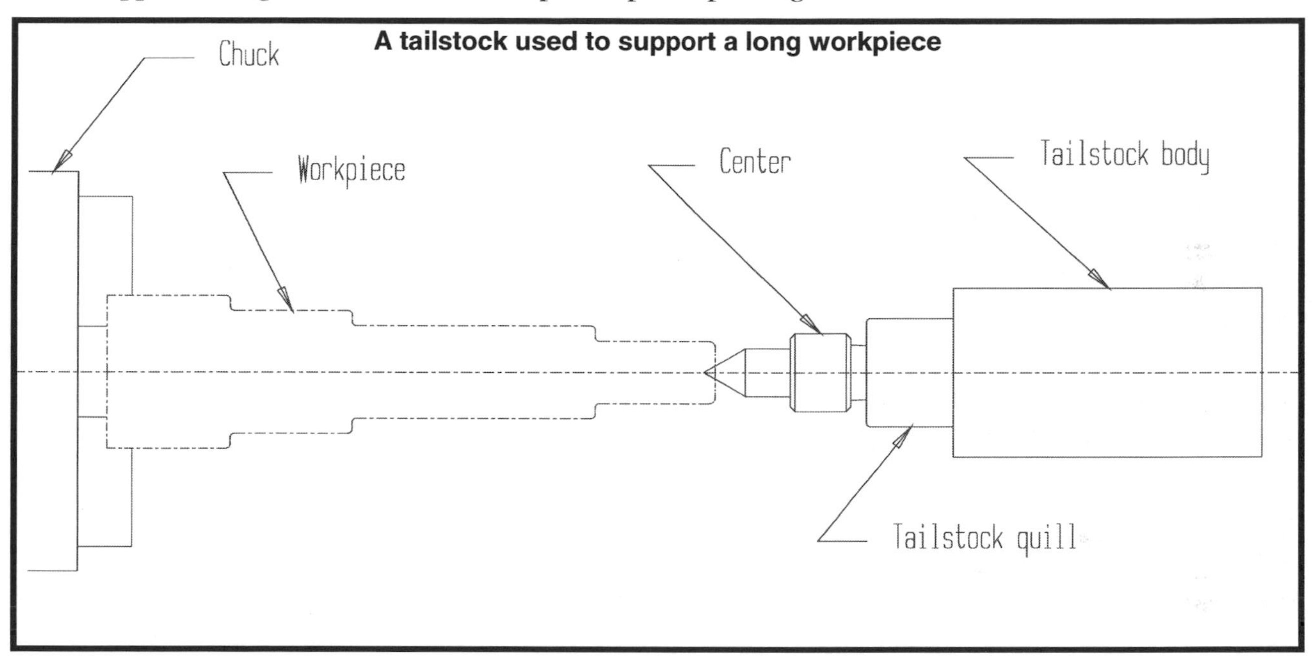

Figure A.6 – A tailstock used to support a long workpiece

The tailstock body

The body of a tailstock provides the tailstock with its rigidity. All CNC turning centers that incorporate a tailstock allow the tailstock's body to move. This allows for workpieces of different lengths. With most machines, the tailstock can be moved manually as well as by programmed commands.

The setup person can activate a push button, foot switch, or lever to manually move the tailstock body. They will move the tailstock body into position as part of the workpiece setup.

Machine tool builders vary when it comes to how the tailstock body is commanded to move within a program. Most utilize two moveable end stops in conjunction with limit switches mounted on the tailstock body. As part of the setup, the operator moves the end stops to the desired positions. Two M codes are used to command tailstock body movement. One M code tells the machine to move the tailstock body forward until the front end stop is contacted. Another tells the machine to move the tailstock body back until the rear end stop is contacted.

Tailstock quill

The tailstock quill is the device that actually applies the pressure to the workpiece. Like the tailstock body, it can move forward to contact the workpiece and back to retract from the workpiece. The quill pressure is usually adjustable (with a valve or rheostat) to allow for variances in workpiece rigidity. All CNC turning center manufacturers allow the tailstock quill motion to be activated manually (by a foot switch or push button), and most allow automatic quill motion (by programmed commands). When activating the tailstock quill motion manually, the operator must be very careful during the tailstock quill forward motion. *An operator should always grip the workpiece from below and never hold the workpiece in a way that allows a hand to be in the way of the tailstock quill.*

The programmed commands required for tailstock quill activation vary from one turning center manufacturer to the next. Most use two M codes. One M code brings the quill forward until it contacts the workpiece. The other retracts the quill.

Tailstock center

The center actually comes into contact with the workpiece. Centers come in a variety of styles, allowing for a wide variety of work support configurations. For small workpieces, the center comes in the form of a small pointed cone. The angle of this cone is sixty degrees to match a center drilled hole in the workpiece.

Keep in mind that there must be some machining operation on the workpiece prior to engaging the center. Most commonly, a center drilled hole is used for this purpose. However, larger holes in the center of the workpiece can also be used.

The portion of the center that actually contacts the workpiece is usually allowed to rotate. This form of center is called a *live center*. Another form that does not rotate (usually more common on manual engine lathes) is called a *dead center*. This rotation capability is sometimes designed into the machine's quill, but is more commonly designed into the center itself.

Tailstock alignment problems

During machining, the center must be perfectly aligned with the spindle. If it is not, the tailstock will have the tendency to actually bend the workpiece and will cause a taper to be machined on all diameters.

During setup, the setup person must check the alignment of the tailstock for taper. After the workholding setup is made, and after the tailstock has been properly positioned, a test cut is made (manually) on the workpiece. This test cut is made with the tailstock center engaged with the workpiece. The operator will skim cut a relatively long length of the workpiece and measure the diameters at both ends. If the two diameters are precisely the same, the tailstock is properly aligned. If they are not, the tailstock is out of alignment and must be adjusted.

How tailstock alignment is adjusted varies from one manufacturer to another, as does the difficulty involved. Most incorporate a series of locking bolts which must be released. At this point, a set screw is turned to physically move the tailstock center position. If a dial indicator is attached to the turret and made to contact the tailstock center, the setup person can monitor the amount of motion at the tailstock center when the adjusting set screw is turned. Unfortunately, this tends to be an inexact procedure. In most cases, the setup person must adjust several times, and test cut after each adjustment.

Programming considerations

Though the tailstock body and quill movements are programmable on most CNC turning centers, there are relatively few applications when programmability is required. In most applications, the operator will simply activate the tailstock manually as part of the workpiece loading process. For example, if loading a long shaft (with a previously machined center drilled hole), the operator will first place the workpiece in the jaws of the chuck and clamp the chuck. Next they will activate the tailstock quill (usually by a foot switch) to engage the quill into the workpiece. Finally, the chuck jaws will be released and re-clamped to ensure a proper seating of the tailstock center.

One application for a programmable tailstock is when bar feeding. Right after the cut-off operation, the bar can be center drilled. After the bar feeding operation, the tailstock will be engaged. Since this must occur during the CNC cycle, the tailstock motions must be automatically engaged (programmed).

Steady rests

As you know, when a tailstock is used, the workpiece will be supported at both ends, and machining at either end will be quite stable. However, if the workpiece is quite long, and if machining is to occur throughout the entire length of the workpiece, machining may not be so stable in the *middle* of the workpiece. The workpiece may have the tendency to push away from the cutting tool. If this occurs, diameter dimensions in the middle of the workpiece will be difficult to hold. A steady rest can be used to provide additional support. Figure A-6 shows a steady-rest being used to support the middle of a workpiece.

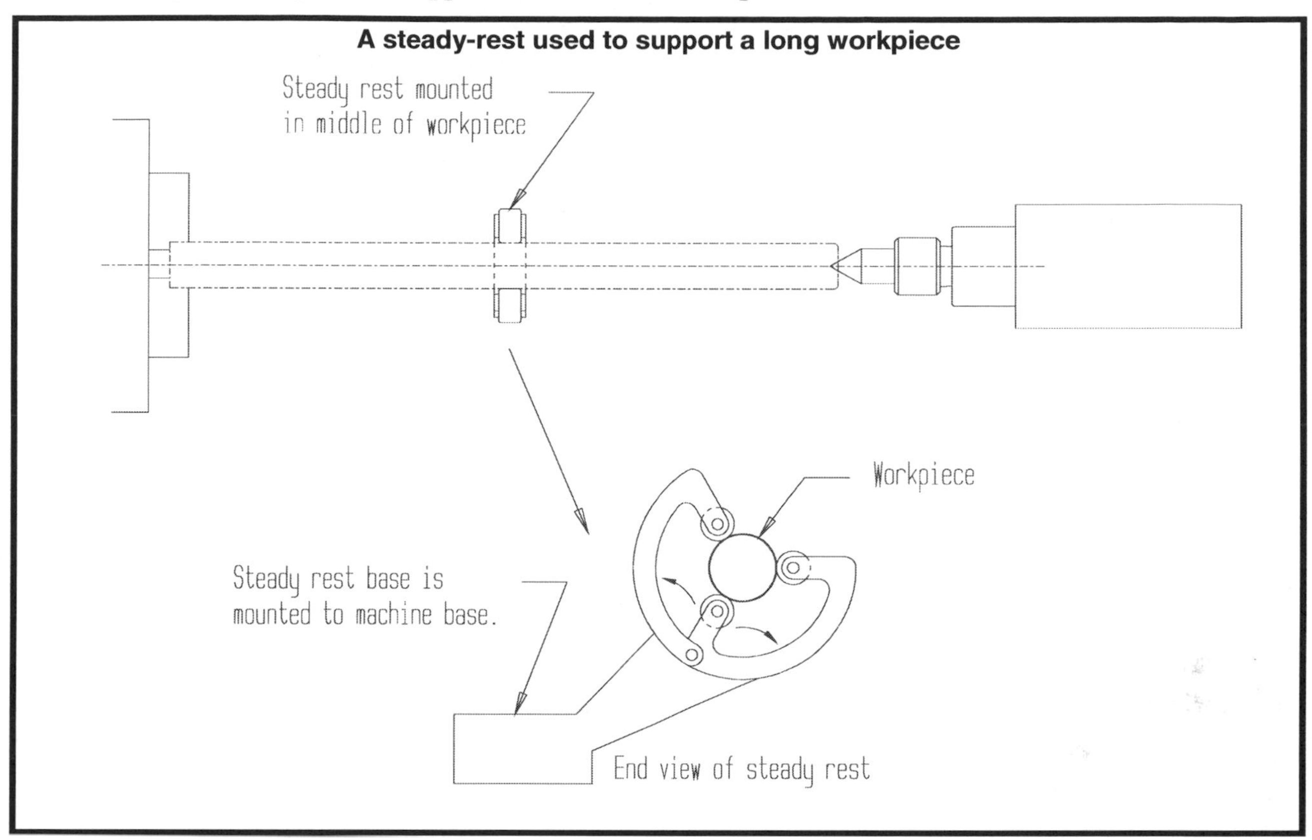

Figure A.6 – A steady rest

As you can see, this form of steady rest incorporates three rollers which come into contact the workpiece. This action provides the needed support for machining close to the center of the workpiece. Note that two or more steady rests can be placed in tandem if a cutting tool must turn from one end of the workpiece to the other. For this application, one steady rest will remain open while the tool passes through. Once the tool has passed the steady rest, it will be closed and the one in close proximity will be opened.

Turning centers that incorporate this form of steady rest allow the opening and closing of the rollers to be done by programmed commands. Usually two M codes are used for this purpose, one to open and another to close. Some turning centers even allow the positioning of the steady rest along the Z axis to be activated by programmed command.

Another application for a steady rest is related to machining the very end of a long workpiece. If only external machining (turning) is to be done, a tailstock can be easily used to accomplish the necessary support. However, if internal machining (boring) is to be done, a tailstock cannot be used. In this case, a steady rest can be used to support the workpiece in a way that allows machining inside the end of the workpiece. It will locate from a previously machined outside diameter and provide the needed support while machining the inside diameters.

Bar feeders

Bar feeders provide the simplest form of automation for turning centers. They feed a long bar through the spindle and provide the rough stock for workpieces to be machined. For each cycle, a workpiece is machined and cut off. The bar is fed again and another workpiece is machined. This process is repeated for the entire length of the bar. Some bar feeders even have automatic bar loading systems that keep the operator from having to load bars into the bar feeder itself. Instead, they simply place the bars in the rack of the bar feeder.

The advantage of a bar feeder, of course, is that a long period of unattended operation is possible while the machine runs all the workpieces from the bar/s.

How a bar feeder works

Generally speaking, bar feeders can accommodate bars from under 0.25 inches in diameter to over 2.0 inches in diameter. Depending on the style of bar feeder, there may be bushings mounted within the bar feeder that guide the bar through the center of the spindle. The hole in each bushing is just larger than the outside diameter of the bar.

Most bar feeders require the bar to be quite straight, free from bends or bows. You can imagine trying to rotate a crooked 2.0 inch diameter that is twelve feet long bar at high spindle speeds. The resulting vibration will make machining impossible. Many companies that use bar feeding turning centers have bar straightening devices capable of removing bends from long bars.

Some bar feeders require that the end of the workpiece being fed into the machine be chamfered. This allows the bar to easily find its way through the spindle. Those bar feeders that require chamfering usually come with a chamfering device that makes it easy and convenient for the CNC operator to machine the chamfer right at the machine. Unfortunately, once the bar is fed through the spindle, the chamfer on its end must be removed before machining can begin (one or more facing passes will be required).

Most bar feeders apply a constant pushing force to the bar. As soon as the bar feeder is activated, it will try to push the bar in the direction of the spindle. As the collet (or chuck) is opened, the bar will immediately feed through the spindle. For this reason, a bar stop mounted in the turret will be programmed to provide the proper stopping position for the bar (more on how a little later). Also, the operator must be careful not to open the chuck manually once the bar feeder is turned on.

The bar feeder must be interfaced with the machine tool for the purpose of providing an *end-of-bar signal* to the CNC control. As stated, the machine will continue to machine workpieces for the entire length of the bar. If left unattended, the bar will eventually be run out. At some point, there will be insufficient material left on the bar to provide adequate gripping support in the collet or chuck jaws. At (or before) this point, the bar feeder must tell the control to stop the cycle.

Workholding considerations

While a three jaw chuck will work and is sometimes used, most companies that are serious about bar feeding will equip the turning center with a collet chuck. The collet chuck makes the best choice for bar feeding for the reasons mentioned earlier – during our discussion of collet chucks.

Styles of bar feeders

There are two basic types of bar feeders. The difference between these two types has to do with how the bar is supported within the bar feeder itself. One form of bar feeder supports the bar within a series of bushings. The bushings are placed at even intervals for the length of the bar feeder. As the bar rotates, the bushings provide a bearing surface for the bar and keep the bar running true.

This relatively inexpensive form of bar feeder works nicely for short bars. For short bars, the bushings within the bar feeder can easily provide enough support. However, as bar length grows, this form of bar feeder is very prone to vibration. When these bar feeders vibrate, a great deal of noise is generated, and the quality of machining will be affected.

For long bars, a better (and more expensive) form of bar feeder is the hydraulic-style bar feeder. With this style of bar feeder, the bar is supported within the bar feeder in a cushion of oil, instead of by solid bushings. The cushion of oil provides a nice dampening effect, minimizing vibration and noise, even for large diameter and somewhat crooked bars.

A hydraulic style bar feeder can rotate a long bar at much higher speeds without vibration than a bushing style bar feeder. Though this speed improvement varies based on the quality of the bar feeder, most hydraulic style bar feeders can out-perform bushing style bar feeders by a margin of at least three to one when it comes to maximum spindle speeds.

Several bar feeder manufacturers offer a compromise. They provide automatic loading bar feeders for shorter bars (up to about four feet in length). They incorporate bushings to support the workpiece, which keeps the cost of the bar feeder down. And because several bars can be loaded, they provide a long period of unattended operation. The only disadvantage is that long bars must be cut prior to use in the CNC turning center (most bars come in lengths of twelve or fifteen feet).

How to program for bar feeders

Again, most bar feeders will apply constant pressure to the bar in the direction of the spindle. For this type of bar feeder, there is no actual command needed to activate the bar feeder itself. The entire bar feeding operation is programmed based solely synchronizing the opening and closing the chuck with the positioning of a *stock stop* that is in close proximity to the bar end. As the chuck opens, the bar feeds out a small amount and contacts the stock stop. The stock stop advances, causing the bar to feed out of the spindle. When it is in position, the chuck is closed, clamping on the bar. The stock stop is then retracted. We'll show the programming of these of movements a little later.

Some bar feeders do allow the programmer to turn on and off the pushing pressure of the bar feeder by programmed commands. This minimizes the amount of pressure the bar is influenced by during machining. For bar feeders that allow this, usually two M codes are used to control the function. One M code (commonly an M13) activates the pushing pressure and another stops it (possibly M14).

Note that some bar feeders even work with a time delay. For these bar feeders, only one M code is required (to activate the pushing). After a certain amount of time (commonly ten to twenty seconds), the pushing pressure will cease. The amount of time delay is usually an adjustable feature of the bar feeder.

Determining how much to feed the bar

Before you can write the bar feeding portion of your program, you must determine how far the bar must advance. Figure A.7 shows how to calculate the amount of bar advance needed. As you can see, you simply add the workpiece length plus the cutoff tool width plus any facing stock to be machined from the workpiece in order to come up with the bar advance amount (in the example, 0.125 + 0.75 + 0.03 = 0.905 advance amount).

Some collet chucks tend to suck the bar back into the collet as the collet is closed. If this happens, the length of bar feed pull out must be adjusted accordingly. Fortunately, the pull-back amount will remain fairly consistent from one setup to the next.

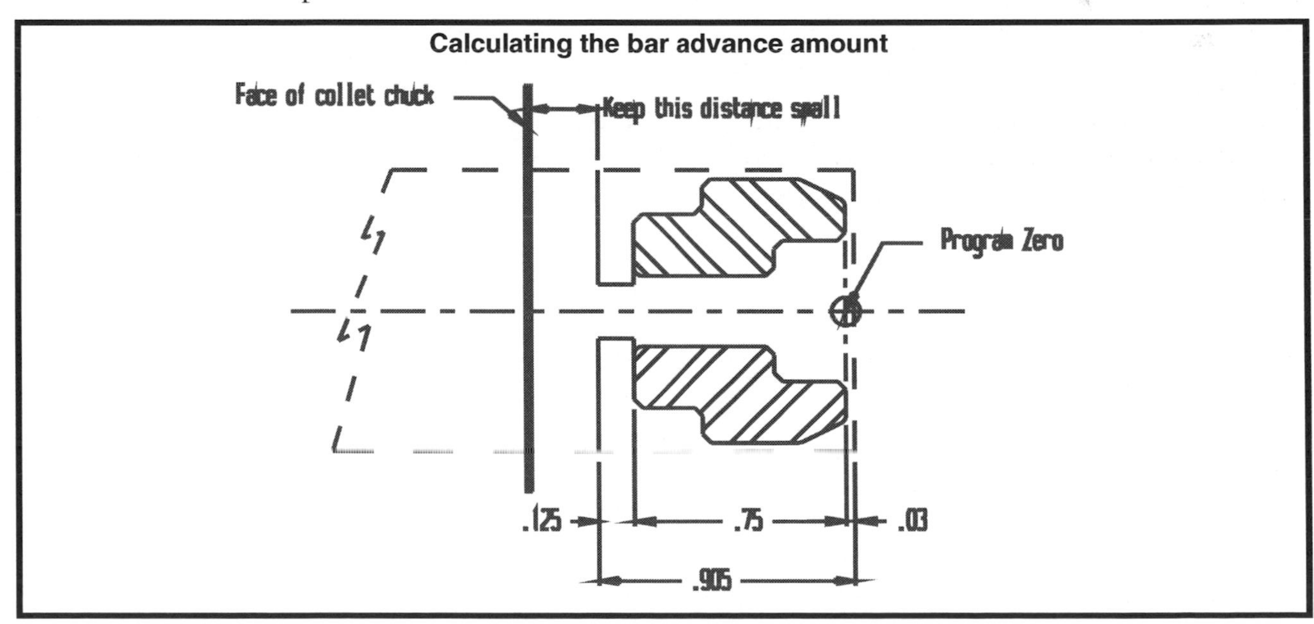

A.7 – How to calculate advance amount

The workpiece should be kept very close to the face of the collet chuck. This minimizes the amount of overhang required for machining, which in turn minimizes the potential for vibration and maximizes the potential for machining with optimum cutting conditions. Most companies will try to maintain a constant and safe distance here. About 0.25 inch will usually do nicely.

The steps to bar feeding

Again, the bar feeder will apply a constant pressure to the bar, pushing it toward the spindle. But it does not control how far the bar is fed during each cycle. As soon as the chuck opens, the bar will continue to advance – until it hits something – or until the chuck is closed.

When bar feeding, some form of *stock stop* that is mounted in a turret station must be used to provide a stopping surface for the bar end. During the bar feed, this stock stop will advance a distance equal to the amount of the bar feed. This, in effect, draws the bar out from the spindle.

Some companies use a special stock stop mounted in its own turret station. However, most programmers consider this to be a waste of a turret station. The shank end of the cut off tool makes an excellent stock stop and eliminates the need for a special turret station. Since the cut off tool will, of course, be very close to the workpiece after cutting it off, it is in an excellent position to provide the stock stop function. So this also minimizes cycle time. Figure A.8 shows the steps to bar feeding.

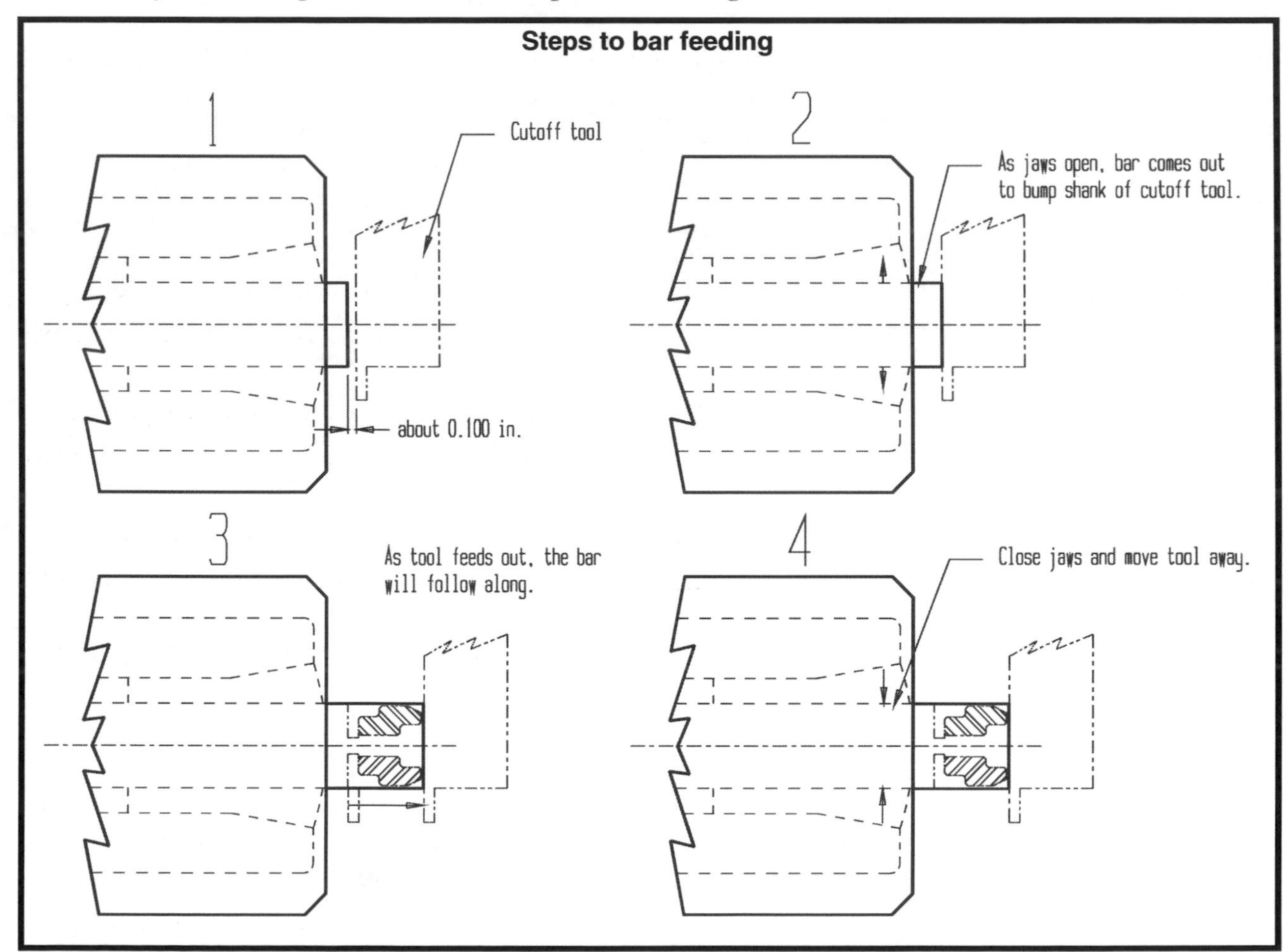

A.8 – The steps needed for bar feeding

1: Position the stock stop close to the bar end

As the upper left drawing shows, the stock stop is first brought to a position that is very close to the end of the bar. This position must be close enough to the bar end so as not to allow the bar to slam into the stock stop when the chuck is opened. For larger bar diameters, there will be a great deal of inertia built up as the bar moves forward. Most programmers keep the stock stop about 0.100 in away from the current bar. If the cut off tool is used as the stock stop (as we're showing), it must be brought down in X to a position that allows the shank of the tool to be the stock stop (a large negative X position).

If the bar feeder is one that allows pushing pressure to be turned off (or turns off automatically after a time delay), the proper M code must be commanded to activate the bar feeder at this time

2: Open the chuck to allow the bar to come out

Second, the chuck is opened (by an M code) to release the bar. Since the bar feeder constantly applies pushing pressure, the bar will immediately advance and contact the stock stop. At the end of this step, the bar will be resting flush with the stock stop.

3: Draw the bar out

Third, the stock stop is advanced (Z plus movement) the required amount. The rate of motion for this movement should be slightly slower than the bar feeder's advance rate to ensure that the stock stop will remain in constant contact with the bar. If the stock stop gets ahead of the bar end, the bar will slam into the stock stop at the end of its motion. Most programmers use a feedrate of about 30.0 to 40.0 inches per minute for this motion. Find out how fast your bar feeder pushes and use a slightly slower feedrate.

With most machines, the spindle is stopped at this point, meaning the feedrate of the pull-out motion cannot be specified in per revolution mode fashion. This motion must be specified in per minute fashion (G98). After the bar feed operation, the programmer must remember to re-select the per revolution mode (G99) if machining feedrates are to be given in per revolution fashion.

4: Close the chuck and move the stock stop away

The chuck is closed to stop the bar from further motion. The stock stop is then moved away from the workpiece (first in the plus Z direction – then to the turret index position).

The redundancy of bar feed programming

As you can imagine, the series of commands required for bar feeding will be very similar from one job to the next. As long as the cut off point remains the same from workpiece to workpiece (usually about 0.250 in from the collet chuck face), *only one command in the entire bar feed routine will change.*

Since bar feeding commands are so similar from one job to the next, most companies that bar feed regularly use a subprogram in which to store the bar feeding routine. This subprogram remains in memory from one job to the next, and the setup person changes it as part of each new job's setup.

If the CNC control allows *parametric programming* techniques, programming the bar feed operation becomes even simpler. Once the parametric bar feeding program is written and verified, the programmer can simply call the parametric program from the machining program, passing a variable equal to the bar pull out amount for use during the execution of the parametric program. (An example bar-feeding parametric program is given at the end of Lesson Twenty-One.)

When to program the bar feed

Some companies like to program the bar feed cycle at the very beginning of the program, before any machining takes place. Others like to program it at the end of the program, after the cut off operation. By and large, this is a matter of personal preference. However, the method chosen has a lot to do with how the operator will position the bar when it is loaded into the machine at the very beginning of the bar.

If the programmer elects to bar feed at the very beginning of the program, the operator will load the bar into the spindle with the bar end the same distance from the collet chuck face as it will be just after cut off (or close to it). When the cycle is activated, the bar feed will occur in the normal manner. With this method, the operator need not position the bar perfectly. If anything, they will want to position the bar slightly on the short side to ensure that the stock stop will have sufficient clearance when it approaches.

If the programmer elects to bar feed at the end of the program, the operator must load the bar into the spindle with the bar end extending to the same position as after the bar feed (ready for the first workpiece to be machined). In this case, it is a much more critical that the operator position the bar properly. If extending out of the chuck too far, the first tool will have more stock to machine from the face of the workpiece. If not extending from the chuck far enough, the face of the workpiece may not clean up.

Since bar positioning ins not so critical, we recommend programming the bar feed operation at the beginning of the program, prior to machining.

Ending a bar feed program

A bar feeding program must be totally automatic. Therefore, at the end of the machining program the control must not stop. It must return to the program's beginning and execute the program again for the next workpiece. And, of course, this process must be repeated for the entire length of the bar. The CNC program must be executed as many times as there are workpieces in the bar. It will be the end-of-bar signal of the bar feeder that actually stops the cycle.

CNC control manufacturers vary with regard to how the program is repeated. All Fanuc and Fanuc-compatible controls can use an **M99** with which to end the bar feeding program. When used to end a main program, the **M99** word tells the machine to return to the beginning of the program and continue without stopping. In essence, the machine is put into an endless loop. It is not until the control receives the end-of-bar signal from the bar feeder that the cycle is halted.

Some machine tool builders use a switch mounted on the control panel to choose between automatic or single cycle operation. How this switch is labeled varies. Some manufactures call it *continuous operation.* Others call it *automatic operation.* If this switch is set to continuous operation, the program can still end with **M30** or **M02**.

An example bar feeding program

Figure A.9 shows the workpiece to be used for our example. For the purpose of this example, we'll say that **M14** closes the chuck and **M15** opens the chuck. We'll also say that the bar feeder for this example applies constant pushing pressure so no special command is required to turn on the bar feeder's pushing pressure. Also, this bar feeder utilizes an end-of-bar signal with which to stop the cycle at the end of the bar. Note that our example shows bar feeding being done at the beginning of the program, meaning the operator will load the bar with the bar end close to the collet chuck, and ready for the bar advance. Program zero is at the very front of the finished workpiece.

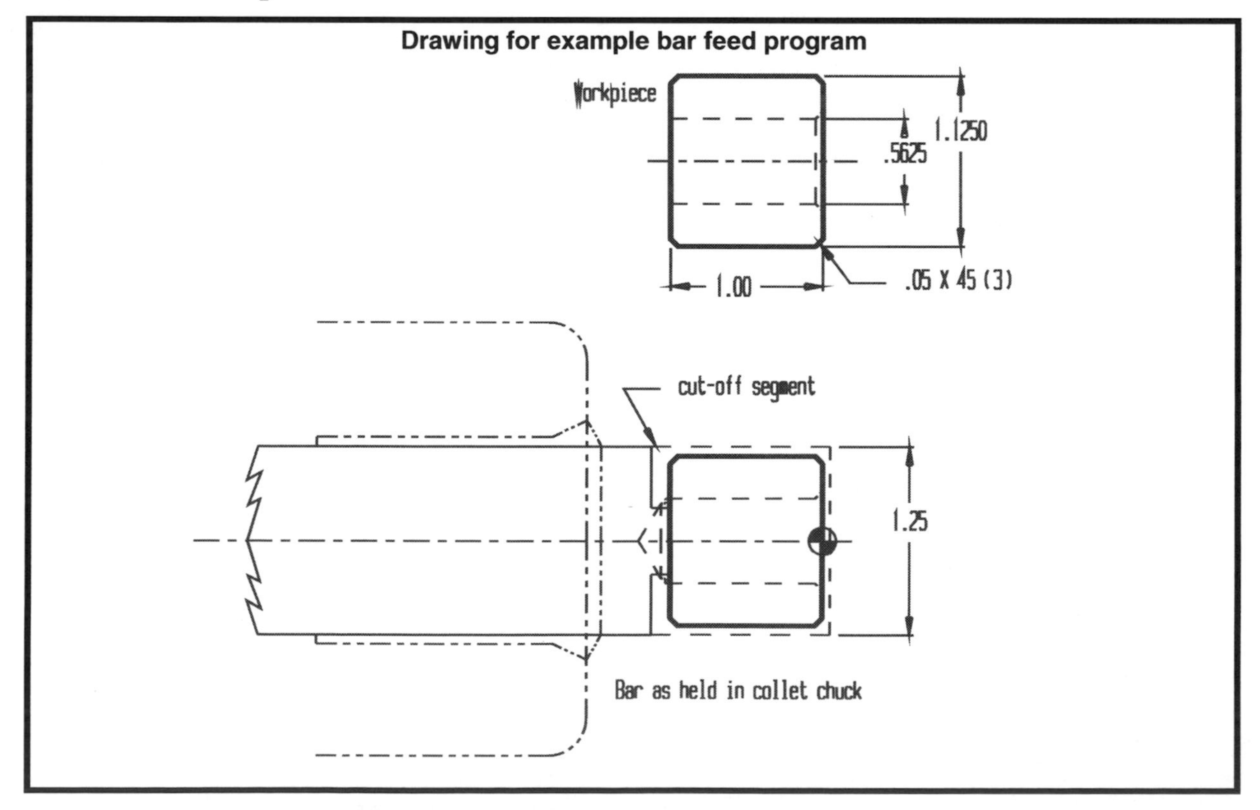

Figure A.9 – Drawing for bar feed example program

Program:

```
O0006 (Program number)
N005 G99 G20 G23
N010 G50 S5000
(BAR STOP)
N015 T1212 M05 (Index, stop spindle)
N020 G00 X0 Z-1.025 (This is the only command that changes from one job to the next)
N025 M15 (Open collet chuck)
N030 G98 G01 Z0.05 F30.0 (Bar advance)
N035 M14 (Close collet chuck)
N040 G99 G00 X6.0 Z5.0
N045 M01

(FINISH FACE AND TURN TOOL)
N050 T0101 M42
N055 G96 S550 M03
N060 G00 X1.45 Z0 M08 (1)
N065 G01 X-0.062 F0.008 (2)
N070 G00 Z0.1 (3)
N075 G42 X1.025 (4)
N080 G01 Z0 (5)
N085 X1.125 Z-0.05 (6)
N090 Z-1.05 (7)
N095 X1.45 (8)
N100 G40 G00 X6.0 Z5.0
N105 M01
(1/2 DRILL)
N110 T0202 M42
N115 G97 S740 M03
N120 G00 X0 Z0.1 M08 (9)
N125 G01 Z-1.21 F0.007 (10)
N130 G00 Z0.1 (9)
N135 X6.0 Z5.0
N140 M01
(FINISH BORING BAR)
N145 T0303 M42
N150 G96 S300 M03
N155 G41 G00 X0.6625 Z0.1 M08 (11)
N160 G01 Z0 F0.004 (12)
N165 X0.5625 Z-0.05 (13)
N168 Z-1.03 (14)
N170 X0.4 (15)
N175 G00 Z0.1 (16)
```

```
N180 G40 X6.0 Z5.0
N185 M01
(1/8 WIDE CUT OFF TOOL)
N190 T0404 M42
N195 G96 S450 M03
N200 G00 X1.45 Z-1.125 M08 (17)
N205 G01 X1.025 F0.005 (18)
N210 G00 X1.45 (17)
N215 Z-1.075 (19)
N220 G01 X1.125 (20)
N225 X1.025 Z-1.125 (18)
N230 M18 (Advance part catcher)
N235 X0.44 (21)
N240 M19 (Retract part catcher)
N245 G00 X1.45 (17)
N250 X6.0 Z5.0
N255 M99 (Return to beginning of program and continue)
```

Part catchers

When a workpiece is parted (cut off) during a bar feed operation, it will fall away from the bar. If no special provisions are made, the workpiece (still rotating at a very fast rate) will fall to the bottom of the machine's bed, and possibly into the chip conveyor. During the fall the workpiece will bounce around quite a bit due to the inertia of its rotation and it will likely be damaged before it finally comes to a stop.

Some companies attempt to solve this problem by padding the lower internal surfaces of the machine. Though this may help keep the workpiece from becoming damaged, the operator will have to dig around among the chips in order to find the workpieces.

A part catcher eliminates this problem. Part catchers swing into position to catch the workpiece just before the parting operation. After it is cut off, the workpiece falls into the part catcher. After the part catcher retracts from the work area, it drops the workpiece into a collecting bin.

The activation of a part catcher is programmable by M codes. One M code commands the part catcher to advance into position and another commands it to retract. The previous bar feeding example program activates the part catcher just before cutoff.

Part catchers vary dramatically in design. Their effectiveness also varies. Some will effectively catch over 99% of workpieces cut off, while others catch a much smaller percentage.

Live tooling

The primary application for any turning center is to machine workpieces in a *turning* fashion. A stationary cutting tool is brought into contact with a rotating workpiece. Operations such as rough and finish turning, rough and finish boring, grooving, knurling, and threading are performed in this fashion. All presentations made to this point have assumed your turning center is limited to machining in this manner.

However, many workpieces to be machined on turning centers require *secondary operations* after the turning operation. A flange, for example, may require that a series of holes be machined around the outside of the flange after the turning is done. A shaft may require a keyway to be milled or a cross hole to be drilled. In some cases, a flat or contour must be milled on the periphery of a diameter.

Traditionally, secondary operations have been performed on separate machine tools – like manual milling machines, drill presses, and even CNC machining centers *after the turning center operation.* Each workpiece requiring secondary operations is brought to the other machine for additional machining. As you can imagine, secondary operations mean more setups and more workpiece handling. This equates to higher production costs. For this reason, more and more companies are minimizing (or eliminating) secondary operations by performing secondary operations *right on the CNC turning center*. When this is possible, not only are production costs reduced, but workpiece quality also improves because fewer setups are required.

Features of live tooling turning centers (also called mill/turn machines)

In order for the turning center to perform secondary operations, the turning center must have the ability to perform milling, drilling, tapping, and other operations not commonly considered to be turning center operations. Secondary operations usually require relatively light-duty machining.

Rotating tools

The turning center must be able to rotate a tool held in the turret (hence the name *live tooling*). A special spindle drive motor and transmission mounted within the turret provides this rotation. The speed and activation of the rotating tool are programmable functions.

Special tool holders

The tool holders that contain rotating tools must have the ability to point the tool along the X axis (as would be required for drilling cross holes) as well as along the Z axis (as would be needed for drilling flange holes). Figures A.10 shows examples of how rotating tools are held in the turret.

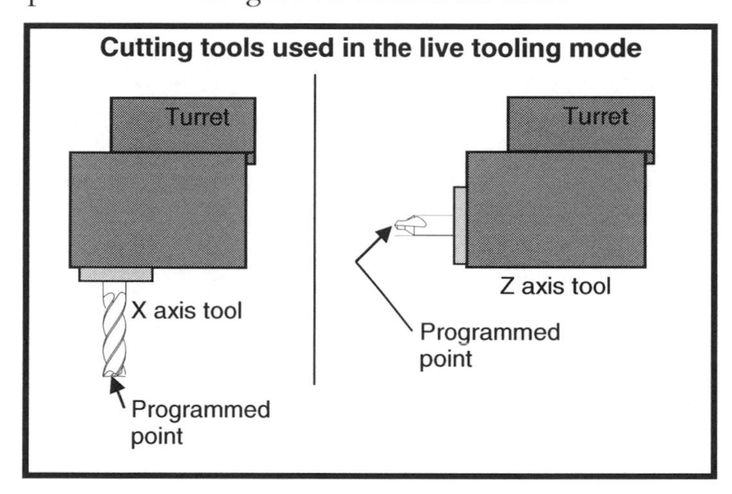

Figure A.10 – Cutting tools used in the live tooling mode

Precise control of main spindle rotation

You have precise control of the main spindle's rotation. In the normal turning mode, the machine's spindle is rotated at a (rather fast) specified speed in either revolutions or surface feet per minute. When the main spindle is stopped, it is in neutral, and still free to rotate. When in the live tooling mode, the main spindle must be rotated in a more controlled manner. You must, for example, be able to rotate to workpiece to a precise angular position at which a hole must be drilled.

There are two common ways turning center manufacturers handle this kind of main spindle rotation. One way is to utilize an *indexer* within the headstock. The other way is to incorporate a true *rotary axis* (called the C axis) within the headstock. Either way, the machine tool builder must provide a way (usually by M codes) to specify which spindle mode is desired, the normal turning mode for turning operations or the live tooling mode for live tooling operations. The most desirable (and most common) form or rotary device is a rotary axis (C axis).

Only one way to specify speed and feedrate

Whenever live tools are used, spindle speed (for the live tool) *must* be specified in rpm. Since a rotating cutting tool's diameter does not change during the machining operation, constant surface speed is not applicable. Along the same lines, feedrate cannot be specified in per revolution fashion with most machines. Since the main spindle is not in the turning mode, the machine will consider the main spindle to be off. If a motion is commanded with a per-revolution feedrate, the axes will not move.

How feedrate is programmed depends upon what kind of rotary device is being used within the spindle. If an indexer is used, all feedrates will be specified in per minute fashion. If a C axis is used, and if only X and/or Z motion is required, feedrate must still be specified in per minute fashion. However, *any motion that includes a rotary axis departure must be programmed in degrees per minute (dpm)* – unless the feature polar coordinate interpolation is being used.

Hole machining canned cycles

Most turning centers with live tools are equipped with a full set of hole machining canned cycles. These canned cycles help you perform center drilling, drilling, tapping, reaming, and counter boring operations.

Polar coordinate interpolation

If the turning center will be performing milling operations around the periphery of the workpiece with a milling cutter pointing along the Z axis (machining flats or special contours), it will require a motion that combines the X and C axes. This kind of motion is programmed with a special motion type called *polar coordinate interpolation.* This feature allows you to treat the C axis as a linear axis

Selecting the main spindle mode

All turning centers equipped with live tooling have a way to specify *which* spindle mode is desired, the normal turning mode or the live tooling mode. Two M codes are usually used for this purpose. One M code selects the normal turning mode. This mode is usually initialized, meaning at power up the normal turning mode is automatically selected. Again, everything shown in this text so far has used the normal turning mode.

Another M code selects the live tooling mode. When this M code is executed, a mechanical device within the headstock will engage the rotary device. Note that some machines with C axis actually use the same drive motor for both modes, but M codes are still used to select which main spindle mode is desired. The M code selecting the live tooling mode will perform some other functions as well. When the live tooling mode is selected, for example, all spindle related commands (speed, activation, and direction) will be transferred to the live tooling spindle motor in the turret.

When the live tooling mode is selected, any spindle speed (specified by an S word) will be taken as the speed in rpm to be used for the live tooling spindle drive motor. M03 will turn the live tooling spindle drive motor on in the forward direction. M04 turns it on in the reverse direction, and M05 stops the live tooling spindle motor. Again, only rpm mode is allowed with live tooling.

We'll be using M81 to select the normal turning mode and M82 to select the live tooling mode, but you must determine the M code numbers for your machine/s by referencing your machine tool builder's programming manual.

Programming an indexer

We include presentations for both indexers and rotary axes. If your turning center has live tooling, it will have only one of these devices. By far, a rotary axis is the more popular rotary device.

An **indexer** is used to quickly rotate the workpiece to an attitude that allows a machining operation to be performed. No machining can occur during the index. Figure A.11 shows a good application for an indexer.

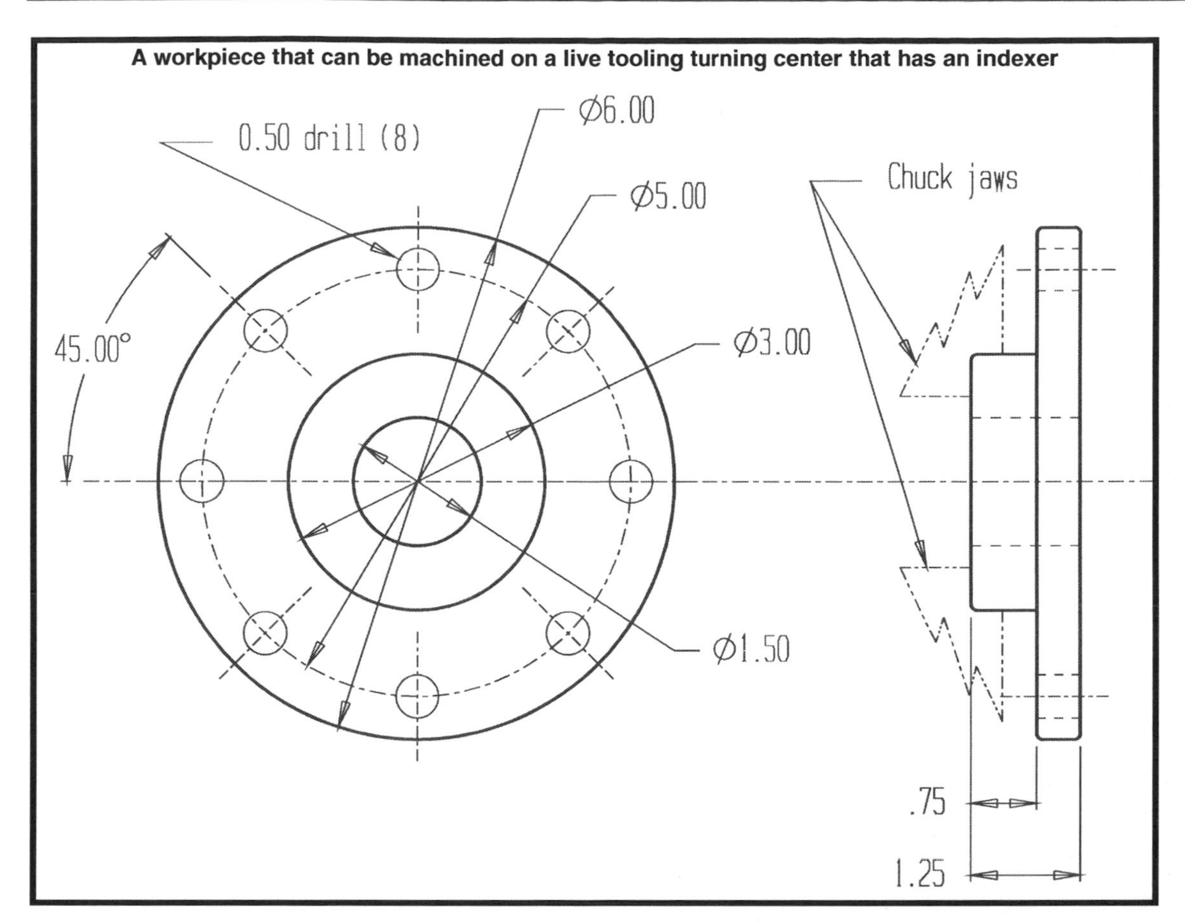

Figure A.11 – An application requiring an indexer (remember, a C axis can also be used for indexer applications)

Notice the series of holes to be machined in the face of this workpiece. The indexer within the headstock will be commanded to rotate the workpiece to the first angular hole location. A hole will then be machined. Then the indexer is commanded to rotate 45.0 degrees to the next hole location. Another hole is machined. This process is repeated for all eight holes.

Keep in mind that with an indexer, machining can only occur *after* the rotation is completed (the workpiece is stopped).

The smallest amount of index for most indexers is one degree. One degree indexers use a special programming word to command the angular departure distance. The letter address C is often used for this purpose. (Do not confuse this with a true C axis.) With this kind of indexer, to index forty five degrees, the command

```
N040 C45
```

is given. Notice that for use with one degree indexers, the C word does not allow a decimal point. You simply specify the incremental amount of angular departure from the indexer's current position.

Most turning centers with one-degree indexers allow the programmer to specify the direction of rotation within the indexing command. Most use two M for this purpose. One M code specifies clockwise rotation while another specifies counter clockwise rotation.

Example program for an indexer

This example program will perform the drilling operation for the workpiece shown in Figure A.11. For the purpose of this example, we'll say that M81 is the M code that selects the normal spindle mode and M82 is the M code that selects the live tooling mode. We'll also say the machine has a one degree indexer and that it is commanded by a C word. In this application, the direction of rotation is not important (the indexer can rotate in either direction), so we're not including any commands to specify rotation direction.

Program:

```
O0001 (Program number)
N002 G99 G20 G23
N003 G50 S5000
N005 M81 (Select turning mode)
.
.
.
(Completely turn workpiece)
.
.
.
N205 T0101 (Rotate turret to 0.500 in drill)
N210 M82 (Select live tooling mode)
N215 G97 S650 M03 (Turn the live tool spindle motor on at 650 rpm in the forward direction)
N220 G00 X5.0 Z0.1 (Rapid to first hole location)
N225 G98 G01 Z-0.65 F5.0 (select feed per minute mode, drill first hole)
N230 G00 Z0.1 (Rapid out of hole)
N235 C45 (Index 45 degrees to machine another hole)
N240 G01 Z-0.65 (Drill second hole)
N245 G00 Z0.1 (Rapid out of hole)
N250 C45 (Index another 45 degrees to machine another hole)
N255 G01 Z-0.65
N260 G00 Z0.1
N265 C45
N270 G01 Z-0.65
N275 G00 Z0.1
N280 C45
N285 G01 Z-0.65
N290 G00 Z0.1
N295 C45
N300 G01 Z-0.65
N305 G00 Z0.1
N310 C45
N315 G01 Z-0.65
N320 G00 Z0.1
N325 C45 (Last hole)
N330 G01 Z-0.65
N335 G00 Z0.1
N340 G00 X10.0 Z5.0 (Return to a safe index position)
N345 G99 M81 (Re-select per revolution mode and normal turning mode)
N350 M30 (End of program)
```

Notice that this program uses G01 to machine each hole and G00 to rapid out of the hole. For drilling, this requires three commands for each hole (index, machine the hole, rapid out). For other hole machining operations (like tapping and peck drilling) there could be even more commands per hole. For this reason, most turning centers that have live tooling are equipped with a series of *canned cycles* to help with hole machining operations. These cycles make it possible to machine one hole per command, dramatically shortening the program's length and making programming easier. We'll present more about canned cycles a little later.

The holes are being machined in the feed per minute mode (notice the G98 in line N225). Since the main spindle is in the live tooling mode, the control considers it to be turned off. The axes will not move if the feed per revolution mode is selected in the live tooling mode.

Programming a rotary axis (C axis)

For turning center applications, the rotary axis within the headstock is called the C axis. Though the C axis can be still used as an indexer (rotate, then machine), you will also have the ability to rotate the workpiece *while machining.*

When used to specify rotary axis motions, the C word allows a decimal point and has a three-place format. The smallest increment is 0.001 degrees. You can think of a rotary axis as being a 360,000 position indexer! Again, the rotary axis has the additional benefit of allowing rotation at a controlled feedrate.

Angular values

The C axis requires angular position to be programmed in degrees. And angular positions less than one degree must be specified in decimal portion of a degree. If you happen across a print that dimensions an angular position in minutes and seconds, you must convert to decimal portions of a degree.

Decimal potion of a degree = minutes/60 + seconds/3600

For the angular position 3 degrees, 40 minutes, 32 seconds, first divide forty by sixty (0.666) and thirty-two by thirty-six-hundred (0.009). Add the two together (0.675) and the result is the decimal portion of a degree. This position or departure will be programmed with the C word:

C3.675

Zero return position

Like X and Z, the C axis has a zero return position. And for many machines, the power-up procedure will require that you zero return the C axis, just like you do X and Z.

Rapid versus straight line motion

Like X and Z, you can specify motion types with the C axis. G00 (rapid motion) will cause the rotary axis to move as fast as it can. Generally speaking, you use the G00 mode whenever you're using the C axis as an indexer (no machining during rotation).

As with X and Z, G01 can be used to rotate the rotary axis at a controlled feedrate. With polar coordinate interpolation (discussed later), you can even command circular (G02 and G03) motions with the C axis.

Unless you are using polar coordinate interpolation, feedrate can be a little complicated when commanding the C axis during in a cutting motion. Any feedrate motion involving the C axis must be programmed in *degrees per minute.* Calculating feedrate in degrees per minute can be little difficult. Here are the formulae.

Time (in minutes) = length of cut divided by desired inches per minute feedrate

Degrees per minute = incremental rotation amount divided by time in minutes

To calculate degrees per minute feedrate, we recommend first calculating how much time it will take to make the motion (each motion may have a different feedrate in degrees per minute if the angular rotation amount is different). Then you can calculate the degrees-per-minute feedrate based upon how much rotation is required. Note that if you're using polar coordinate interpolation (as is commonly the case when machining during rotation) most turning centers do allow you to specify feedrate in inches or millimeters per minute.

Program zero assignment

Like X and Z, you can assign a program zero point for the C axis. Though it's not always required (when indexing, it's easier to program the C axis incrementally), once program zero is assigned, you can specify angular positions from the program zero program zero point. For most machines, the assignment of program zero is rather *transparent.* The program zero point for the C axis will be automatically assigned when the rotary axis is sent to its zero return position. Any time you perform a zero return for the C axis, you're also assigning program zero (the program zero point being the C axis zero return position). While turning center manufacturers vary when it comes to the polarity of the C axis, clockwise (as viewed from in front of the chuck) is usually the plus direction and counter clockwise is usually the minus direction.

Absolute versus incremental

Like X and Z, you are can program the C axis with either positioning method. When using absolute positioning, you can specify angular positions from program zero (normally the C axis position at zero return). Just as letter addresses X and Z specify absolute positions for the X and Z axes for most machines, so does letter address C specify absolute angular positioning in the C axis. Most machines use the letter address H to specify incremental departures in the C axis.

When used as a simple indexer, it's usually easier to program the C axis in the incremental mode. Here is the same operation shown earlier (flange workpiece) programmed for a C axis instead of a one degree indexer.

Program:

```
O0001 (Program number)
N002 G99 G20 G23
N003 G50 S5000
N005 M81 (Select turning mode)
.
.
.
(Completely turn workpiece)
.
.
N205 T0101 (Rotate turret to 0.500 in drill)
N210 M82 (Select live tooling mode)
N215 G97 S650 M03 (Turn the live tool spindle motor on at 650 rpm in the forward direction)
N220 G00 X5.0 Z0.1 (Rapid to first hole location)
N225 G98 G01 Z-0.65 F5.0 (Select per minute feedrate mode and drill first hole)
N230 G00 Z0.1 (Rapid out of hole)
N235 H45.0 (Incrementally rotate C axis 45degrees, note rapid mode is instated)
N240 G01 Z-0.65 (Drill second hole)
N245 G00 Z0.1 (Rapid out of hole)
N250 H45.0 (Incrementally rotate C axis 45degrees, note rapid mode is instated)
N255 G01 Z-0.65
N260 G00 Z0.1
N265 H45.0 (Incrementally rotate C axis 45degrees, note rapid mode is instated)
N270 G01 Z-0.65
N275 G00 Z0.1
N280 H45.0 (Incrementally rotate C axis 45degrees, note rapid mode is instated)
N285 G01 Z-0.65
N290 G00 Z0.1
N295 H45.0 (Incrementally rotate C axis 45degrees, note rapid mode is instated)
N300 G01 Z-0.65
```



```
N305 G00 Z0.1
N310 H45.0 (Incrementally rotate C axis 45degrees, note rapid mode is instated)
N315 G01 Z-0.65
N320 G00 Z0.1
N325 H45.0 (Incrementally rotate C axis 45degrees, note rapid mode is instated)
N330 G01 Z-0.65
N335 G00 Z0.1
N340 G00 X10.0 Z5.0 (Return to tool change position)
N345 G99 M81 (Re-select feed per revolution feedrate mode and normal turning mode)
N350 M30 (End of program)
```

Canned cycles for hole machining

Most turning centers with live tooling are equipped with a series of hole machining canned cycles to simplify the programming of hole machining operations. They allow each hole to be machined with one command, regardless of the type of operation (drill, peck drill, tap, etc.). They're also modal, meaning once a cycle is instated, you simply list the hole locations for holes to be machined.

All canned cycles cause these basic movements:

1) Rapid to the hole location. Depending upon the plane in which you are machining (along X or along Z), you specify the hole's coordinates in either X/C or Z/C. The machine will first rapid the tool to this location.

2) Rapid to the R plane. This is also dependent upon the plane in which you are machining. If machining along the Z axis (end holes), the R plane will be a location along the Z axis that clears the work surface. If you will be machining along the X axis (cross holes) the R plane will be a location along the X axis that clears the work surface.

3) Machine the hole. Based upon the cycle type selected (drill, tap, etc.), the hole will be machined.

4) Come out of the hole. When finished the tool will be retracted from the hole and left at the R plane.

How do you specify the machining direction?

In order for the machine to correctly interpret the canned cycle command (knowing which way to machine), you must specify the *plane* in which you are machining. If machining in the X/C plane (machining holes along the Z axis), most machines require that you specify a G18 (X/C plane selection) before you machine holes. If machining in the Z/C plane (machining holes along the X axis), most controls require that you specify G19 (Z/C plane selection).

Canned cycle types

Here is a list of the most common canned cycles used on turning centers and their function.

G80: cancel cycle - This word must be used when you're finished machining holes, canceling the cycle.

G81: standard drilling cycle - This cycle feeds the drill to the bottom of the hole and rapids out.

G82: counter boring cycle - This cycle is used for flat bottom holes. It feeds the counter bore to the hole bottom, pauses for a specified length of time, and then rapids the tool out of the hole.

G83: deep hole peck drilling cycle - When holes are too deep to be machined in one pass, this cycle can be used to clear chips from the hole during machining. It will cause the drill to peck a specified amount, and rapid out of the hole to clear chips. It will then cause the tool to rapid back into the hole to within a small distance from where it left off. It will then peck again and rapid out. This is repeated until the drill reaches its final depth. When finished, the drill will rapid out of the hole.

G84: tapping cycle - This cycle machines right hand threads. With the spindle running in the forward direction (M03), the tap will be fed into the hole. At the hole bottom, the spindle will reverse and the tap will feed back out of the hole. To come up with the proper feedrate (in per minute fashion), multiply the pitch of the thread being machined times the rpm being used to tap the hole.

G85: reaming cycle - Most programmers use G81 to ream holes. G85 will cause the tool to feed to the hole bottom and *feed* out of the hole. Since the reamer will not be machining on its way out of the hole, most programmers feel it is waste of cycle time to do so.

Words used in canned cycles

The meanings of X, Z, and R will vary based upon which plane is selected.

With X/C plane selection (G18 on most controls), **X and C** will specify the hole center position (its coordinates). The **R word** specifies a position along the Z axis that clears the work surface (the rapid plane). Z will be the hole bottom position. Note that X, C, R, and Z are specified from the program zero point. With this scenario, you are machining holes along the Z axis (end holes) with a tool that is pointing along the Z axis.

With Z/C plane selection (G19 on most controls), **Z and C** will specify the hole center position (its coordinates). The **R word** specifies a position along the X axis that clears the work surface (the rapid plane). X will be the hole bottom position. Note that X, C, R, and Z are specified from the program zero point. With this scenario, you are machining holes along the X axis (cross holes) with a tool that is pointing along the X axis.

The **F word** specifies the feedrate for machining. Again, if you are in the live tooling mode, the feedrate must be specified in per minute fashion.

A **P word** is used in the counter boring cycle (G82) to specify the pause time at the hole bottom. For most controls, P does not allow a decimal point. You must specify P with a three place fixed format. P500, for example, specifies a one-half second pause.

A **Q word** is used in the deep hole peck drilling cycle (G83) to specify the peck depth.

While it's a little off the subject of live tooling, note that you can use these canned cycles when in the normal turning mode. This can be very helpful, especially for tapping and peck drilling to minimize the number of commands in your program.

An example program

Here is the same program shown earlier (for the flange workpiece), but this time the holes are being machined with the G81 drilling cycle. Though we're showing this example for a rotary axis (C axis machine), the same principles apply if your machine has a one degree indexer.

Program:

```
O0001 (Program number)
N002 G99 G20 G23
N003 G50 S5000
N005 M81 (Select turning mode)
.
.
.
(Completely turn workpiece)
.
.
.
N205 T0101 (Rotate turret to 0.500 in drill)
```

```
N210 M82 (Select live tooling mode)
N215 G97 S650 M03 (Turn the live tool spindle motor on at 650 rpm in the forward direction)
N220 G00 X5.0 Z0.1 (Rapid to first hole location)
N225 G98 G18 G81 Z-0.65 R0.1 F5.0 (Select per minute feedrate mode, drill first hole)
N230 H45.0 (Incrementally rotate C axis 45degrees, machine hole)
N235 H45.0 (Incrementally rotate C axis 45degrees, machine hole)
N240 H45.0 (Incrementally rotate C axis 45degrees, machine hole)
N245 H45.0 (Incrementally rotate C axis 45degrees, machine hole)
N250 H45.0 (Incrementally rotate C axis 45degrees, machine hole)
N255 H45.0 (Incrementally rotate C axis 45degrees, machine hole)
N260 H45.0 (Incrementally rotate C axis 45degrees, machine hole)
N265 G80 (Cancel cycle)
N270 G00 X10.0 Z5.0 (Return to tool change position)
N275 G99 M81 (Re-select per revolution feedrate mode and normal turning mode)
N280 M30 (End of program)
```

In line N225, the canned cycle is instated and the first hole is machined (wherever the C axis happens to be). From this point on, we simply list hole positions. Since the holes are angularly 45 degrees apart, the H45.0 is used to incrementally rotate the C axis. After each rotation, another hole is drilled. After the last hole (in line N260), the cycle is canceled in line N265 with G80.

Understanding polar coordinate interpolation

Most machining operations with live tools are quite simple to program. Canned cycles, for example, make it very easy to machine holes. If you must machine slots (face slots or diameter slots) on the center of the workpiece, it will be a simple matter of positioning the end mill close to the surface to be milled (with G00), plunging into the slot (with G01) and machining along an axis (X or Z, whichever is appropriate).

There is one kind of machining operation that at first glance, may appear easy. But in reality, is quite difficult. Consider the two flats on the workpiece shown in Figure A.12.

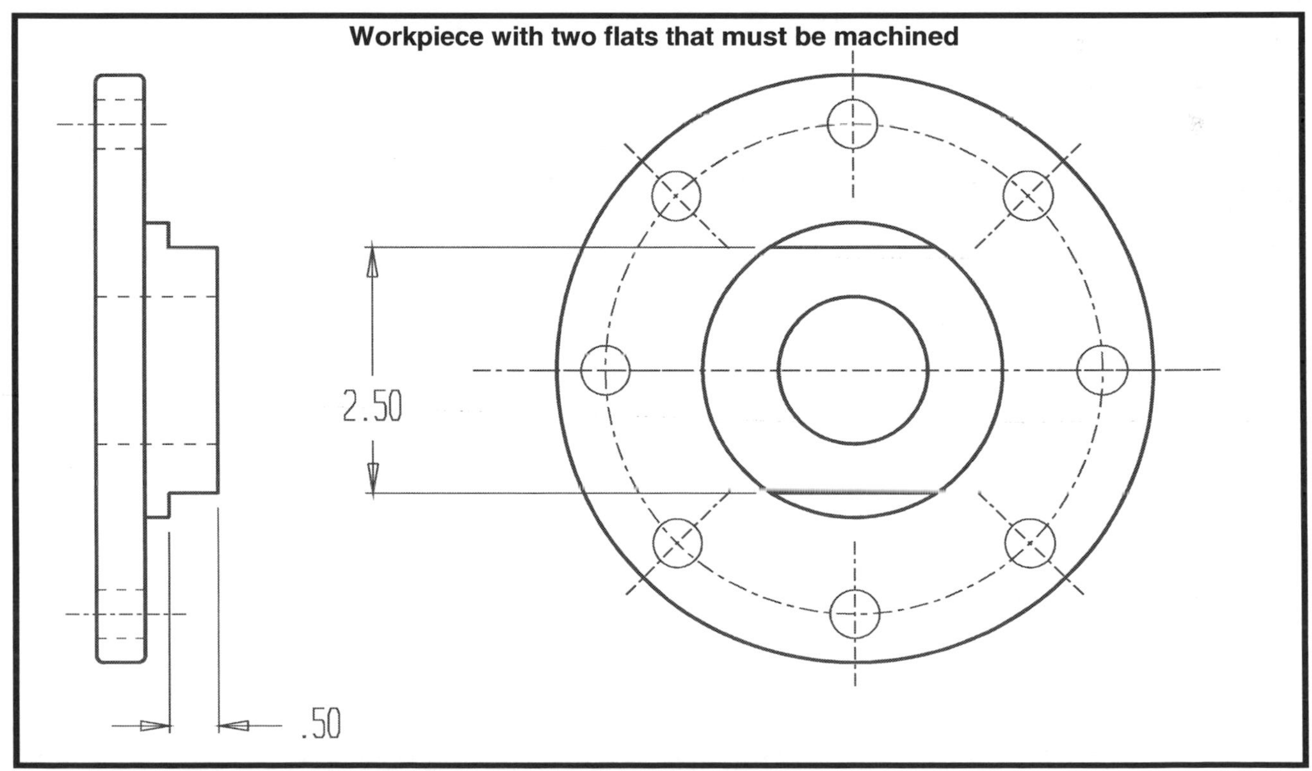

Figure A.12 – Machining the flats requires a combination of X and C motion

These flats must be machined with an end mill pointed along the Z axis. However, unless the turning center has a *Y axis* (perpendicular to the XZ plane), machining the flats must be done by combining X axis motion with C axis motion. This will be a more difficult than it looks. As the milling cutter begins machining, the C axis will be rotating and the X axis will be getting smaller in diameter. As the tool reaches the center of the flat, the X axis will reverse (getting bigger in diameter). At first glance, you may think that two simple **G01** commands (one to center, one the rest of the way) will suffice. But two **G01** motions will not render a flat surface. Instead the surface will be *heart-shaped.*

To get the desired flat shape will require motion in a much more *articulated* manner. Prior to polar coordinate interpolation, this kind of motion required the use of a computer aided manufacturing (CAM) system to generate. And the result was a very long series of very tiny movements along X and C. And this is a simple flat. Consider what it will take to machine a more elaborate shape on the outside diameter of the workpiece, like the hex shape in Figure A.13.

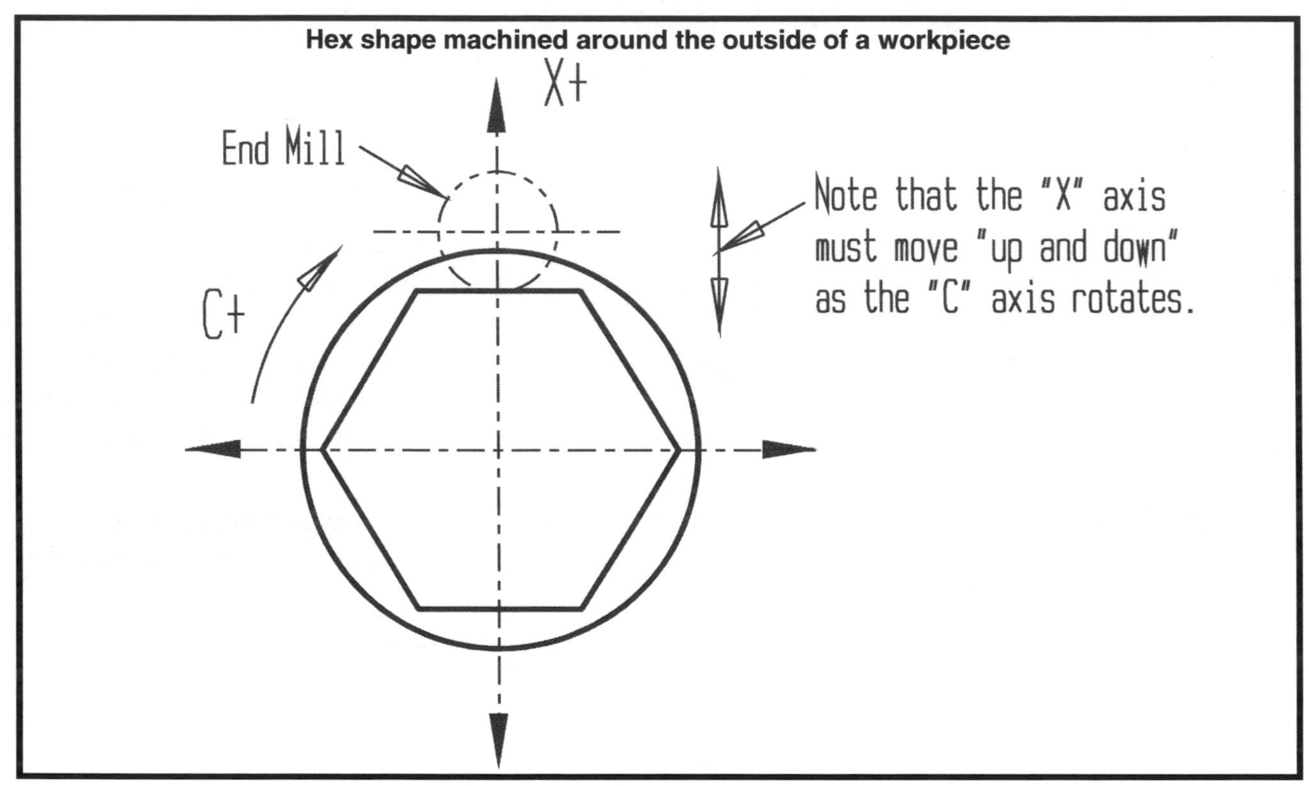

Figure A.13 – Hex shape is difficult to program without polar coordinate interpolation

With *polar coordinate interpolation*, this hex shape can be machined with six simple commands.

In essence, polar coordinate interpolation allows you to treat the C axis it like a linear axis. This dramatically simplifies the commands needed to machine contours around the outside of the workpiece. You'll even be allowed to machine circular workpiece attributes with **G02** and **G03**. And you can specify feedrate in per minute fashion instead of degrees per minute (inches or millimeters per minute).

Figure A.14 shows is the coordinate system for polar coordinate interpolation.

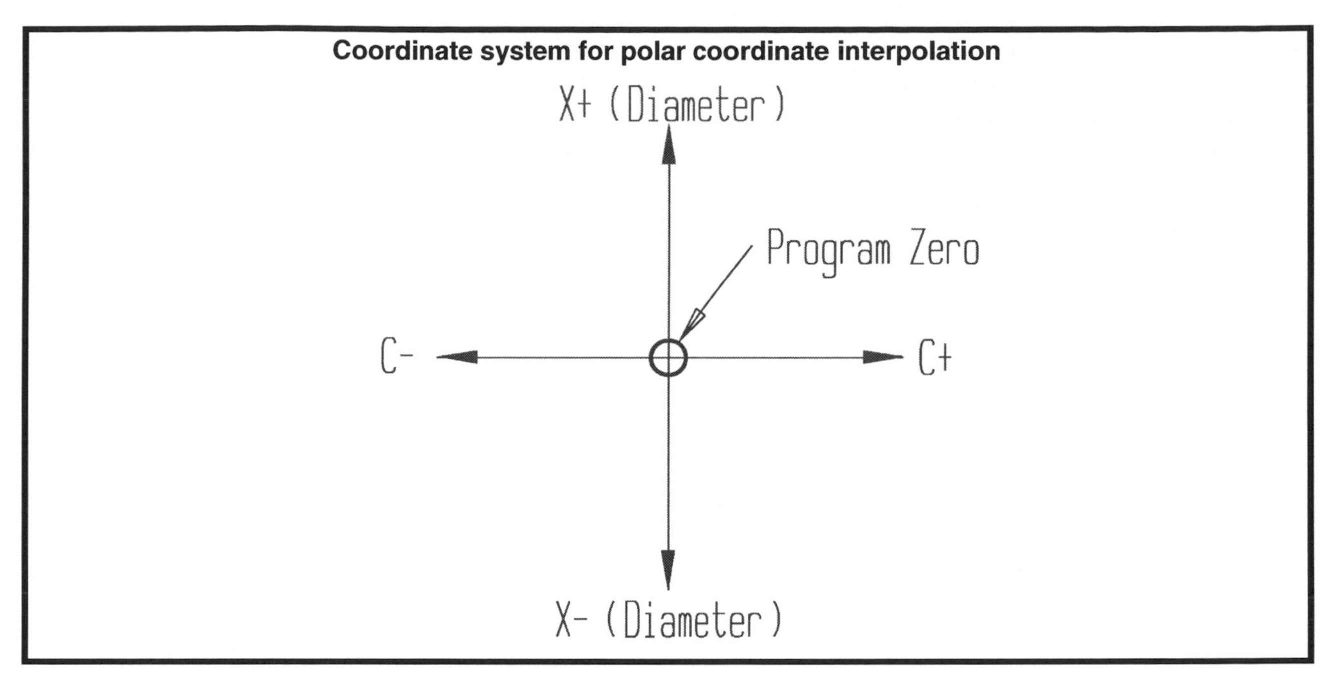

Figure A.14 – Coordinate system for polar coordinate interpolation

In the X axis (vertical), all coordinates are specified in *diameter* (just as they always are). In the C axis (horizontal), coordinates are specified in radius. The program zero point is the center of the workpiece for both axes. Also note the polarity of each axis. Anything above program zero in X is plus, anything below program zero is minus. (In the normal turning mode, you seldom see negative coordinates for X. With polar coordinate interpolation, it is quite common.) Anything to the right of the origin in C is plus, anything to the left is minus.

Unless your control is equipped with cutter radius compensation, all coordinates must be specified to the center of the milling cutter. Since few turning centers have cutter radius compensation, our examples show the cutter's centerline coordinates.

Figure A.15 shows the example drawing and coordinate sheet for our example program.

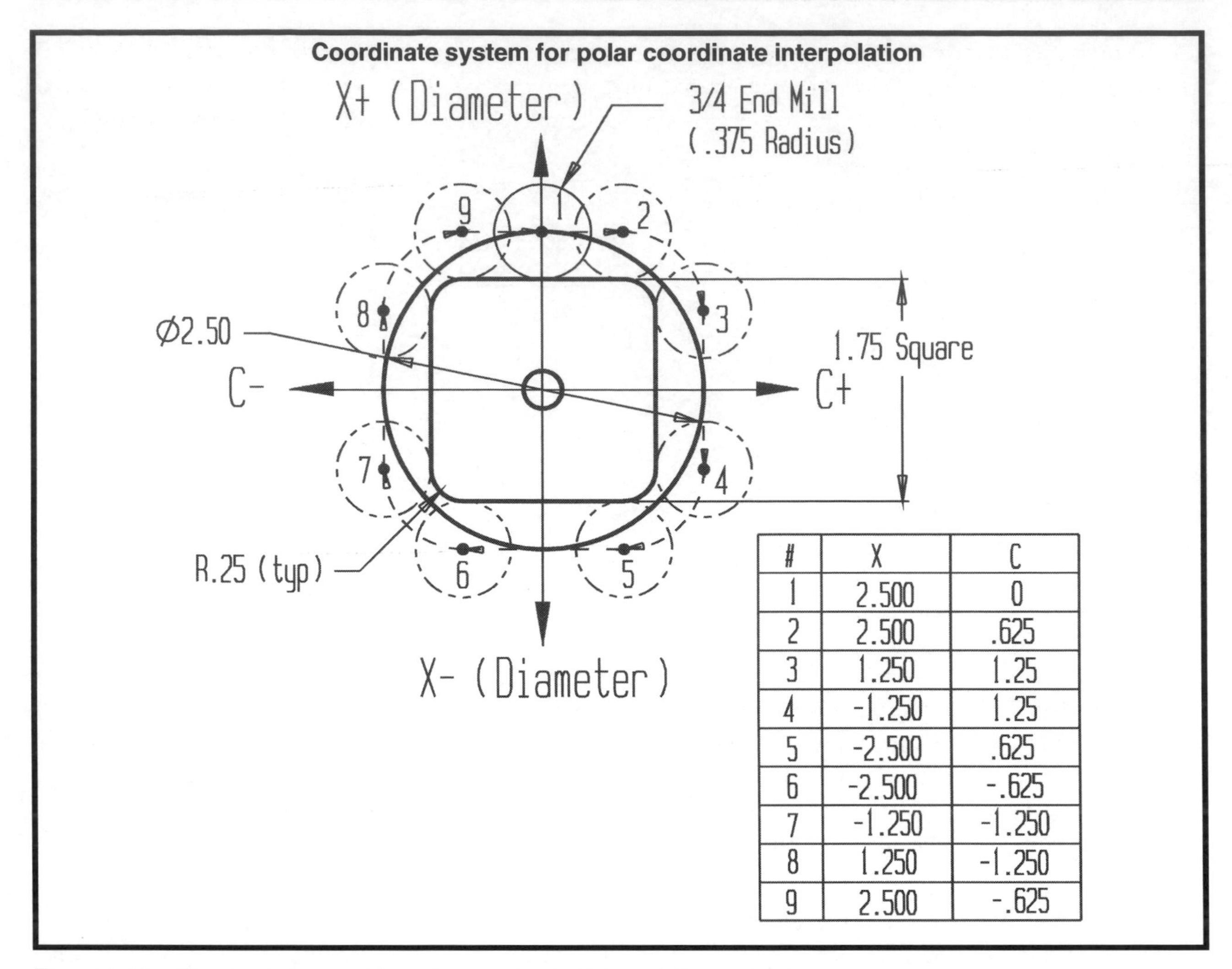

#	X	C
1	2.500	0
2	2.500	.625
3	1.250	1.25
4	-1.250	1.25
5	-2.500	.625
6	-2.500	-.625
7	-1.250	-1.250
8	1.250	-1.250
9	2.500	-.625

Figure A.15 – Drawing to be used for polar coordinated interpolation example program

As you can see, a 1.75 inch square shape with a 0.25 inch radius in each corner is to be machined by a 0.75 inch diameter cutter. The tool path (after approaching) runs from point one through point nine – then back to point one. If you have CNC machining center experience, this resembles how you treat the X and Y axes when contour milling is required.

The X value of each coordinate is specified in *diameter* to the center of the milling cutter. For point number one, for example, the distance from the center to the surface being machined is 0.875 inch (half of the 1.75 square). Add this value to the cutter radius (0.375 inch). When you double the result (1.25), you come up with the needed X axis position of 2.5. The center of the cutter is right on the program zero point in the C axis at point one, so the C axis coordinate of point number one is zero.

At point number six, the X position is the same value, but on the negative side of program zero (notice the negative X value -2.5 on the coordinate sheet)

For the X value of point number three, the distance from the program zero point to the center of the cutter is 0.625 (1.875 minus 0.25). Double this value, and you come up with the needed X axis coordinate for this point, X1.25. In the C axis for point three, the distance from the program zero point to the center of the milling cutter is 1.25 (half of 1.75 plus the cutter radius 0.375).

While calculating coordinates is a little difficult, once you determine them, commanding motions is very easy. Simply make your motion commands using **G01**, **G02**, and **G03** in the normal fashion. For circular commands, the radius you program (the R word) must reflect the cutter's centerline path. The radius for each circular motion in Figure A.15, for example, will be 0.625 (the 0.25 workpiece radius plus the 0.375 cutter radius).

The G codes used to instate and cancel polar coordinate interpolation vary, even among the various Fanuc control models. For most, it is invoked with G112 and canceled with G113. But you'll have to check in your Fanuc programming manual to determine which G codes are used for your machine/s (some controls use G12.1 and G13.1).

Here is the program that machines the square shape in Figure A.15

```
O0001 (Program number)
N003 G20 G23 G40 (Ensure that initialized modes are still effective)
N004 G50 S4000 (No spindle limitation necessary)
N005 T0101 (Index the turret to the 3/4 in end mill, a live tool pointing along the Z axis)
N010 M82 (Instate the live tooling mode)
N013 G97 S500 M03 (Start live tooling spindle fwd at 500 rpm)
N015 G00 Z-0.5 (Rapid to work surface in Z axis)
N020 G18 G112 (Instate polar coordinate interpolation mode)
N025 G00 X3.65 C0 (Rapid to clearance position)
N030 G98 G01 X2.5 F10.0 (Select per minute feedrate mode, feed to point 1)
N035 G01 C.625 (Feed to point 2)
N040 G02 X1.25 C1.25 R.625 (Form radius to point 3)
N045 G01 X-1.25 (Feed to point 4)
N050 G02 X-2.5 C.625 R.625 (Form radius to point 5)
N055 G01 C-.625 (Feed to point 6)
N060 G02 X-1.25 C-1.25 R.625 (Form radius to point 7)
N065 G01 X1.25 (Feed to point 8)
N070 G02 X2.5 C-.625 R.625 (Form radius to point 9)
N075 G00 C0 (Feed back to point 1)
N080 G00 X3.65 (Rapid away in X)
N085 G99 G113 M81 (Cancel polar coordinate interpolation mode, reinstate normal turning mode)
N090 G00 X10.0 Z5.0 (Return to tool change position)
N095 M30 (End of program)
```

Other machine types

While we will not describe them in great detail, there are other turning center configurations you should be aware of. Most are off-shoots of two axis turning centers and should be relatively easy to adapt to if you find yourself having to work with them.

Twin spindle turning centers

As the name implies, a twin spindle turning center is like having two machines in one. Just about everything on the machine (spindle, turret, axes, etc.) is duplicated. Some even have two CNC controls. These machines are, in fact, handled (programmed, setup, and operated) as two separate machines. The two spindles are totally independent of one another, meaning two different workpieces (or first and second setting of the same job) can be running at the same time. The biggest advantage of these turning centers (over having two separate machines) is the close proximity of the spindles. It is very easy for one operator to run both spindles. Another advantage is their small foot print. They take up less floor space than two separate machines.

Sub-spindle turning centers

Like twin spindle turning centers, sub-spindle machines have two spindles. One, the main spindle, is usually on the left (as viewed from the front of the machine). The sub-spindle, which is in-line with the main spindle, is on the right (in the same position a tailstock would be). These machines will have at least four axes (two turrets with X and Z). But it is not uncommon for at least one of the spindles to have live tooling capabilities and a C axis for spindle rotation (a five axis machine). If both spindles have live tooling and C axis, the machine will have six axes.

While all of this may sound a little complicated (having six axes to deal with may seem a little intimidating), you can treat this machine like having two separate machines. As with twin spindle machines, sub-spindle machines require two separate programs, one for the main spindle and one for the sub-spindle.

Sub-spindle turning centers are often equipped with bar feeders, making them quite automated. The main advantage is that they can complete a workpiece (including secondary operations if live tooling is used) from raw material state (a long bar) to finished workpiece.

The bar is fed into the main spindle (as it is for any bar feed turning center). The first operation is performed on one end of the workpiece. The sub spindle is then advanced to grab the workpiece, taking it from the main spindle (many sub-spindle turning centers do this during cut-off, while the spindle is rotating). While the main spindle goes to work on the first operation for the next workpiece, the sub-spindle machines the second operation for the first workpiece. When dropped into the part catcher from the sub-spindle, the workpiece will be completely finished.

About the only new programming consideration for sub-spindle turning centers has to do with the transfer of the workpiece from the main spindle to the sub-spindle. Since it is unlikely that the two operations (main and sub spindle) will take exactly the same amount of time (the main spindle usually takes longer), there must be some way to command the spindle that finishes first (commonly the sub-spindle) wait until the other spindle has finished. Most Fanuc controlled turning centers use a series of M codes (sometimes three digit M codes) for synchronizing purposes. When the control executes an **M110** in one of the spindle's programs, for example, it will wait until the same M code (**M110** in our case) is executed in the other spindle's program. When it is, the programs for both spindles will continue. This allows the programmer to make one spindle wait until the other is finished.

The actual transfer mechanism varies from one machine tool builder to another. Some builders actually make the sub-spindle's extension and contraction an axis of motion (sub-spindle turning centers having live tooling on both spindles with this feature have seven axes of motion). Others have simpler mechanical devices activated by M codes for workpiece transfer. We must bow to your machine tool builder to explain how the workpiece transferring system is programmed.

Swiss-type turning centers (also called sliding headstock turning centers)

The design of this machine is radically different from what we have been showing throughout this text. Indeed, it will require more instruction than we provide. We provide only a simple introduction.

For every machine type addressed by this text, the cutting tool moves in the X and Z axes. This is acceptable and desirable for most turning applications. However, with small diameter workpieces that are very long, the tool pressure during machining will cause too much deflection – excessive enough to render machining impossible.

Consider, for example, a needle valve. A pin less than 0.05 inches in diameter must be turned down to a point over a length of one or two inches. This cannot be (feasibly) done with a conventional turning center. But it's a piece of cake for a Swiss-type turning center.

With a Swiss-type turning center, the tool remains stationary in the Z axis (it still moves in the X axis to control diameter). While the tools are still commonly held in a turret, the turret cannot move along the Z axis. The turret is mounted very close to the spindle nose, so as machining takes place, the workpiece will be well supported by the work holding device (normally a collet or spacer).

The Z axis is now *workpiece motion.* The headstock and spindle actually move with the workpiece to drive it into the cutting tool. When you watch this kind of machine, it will appear that the workpiece is being *extruded* from the spindle. This rigidity of setup allows tiny diameters to be machined with ease.

These machines have two major limitations. First, the amount the headstock can move will determine the maximum workpiece length. While sometimes it's possible to *double grab* the stock to allow longer workpieces, it's not always feasible. Second, it is impossible to make a series of roughing passes (as is commonly done with more conventional turning centers). Once the stock is machined once, there will be no support at the spindle nose for subsequent passes.

This second limitation has made it necessary for tooling manufacturers to develop some rather unique cutting tools. These tools are applied only to Swiss-type turning centers. Grooving tools, for example, that traditionally cannot machine in a turning mode (they can only plunge) are available for both necking and turning. This makes it possible to easily machine recesses of any length.

Know Your Machine From An Operator's Viewpoint

Key Concept Number Seven begins the setup and operation portion of this text. It parallels Key Concept Number One - but in this Key Concept, we'll look at a CNC turning center from the perspective of a setup person or operator.

Key Concept Number Seven contains two lessons:

- 24: Tasks related to setup and running production
- 25: Buttons and switches on the operation panels

We now begin discussions related to setup and operation. Throughout these discussions, we will be assuming that you have read and understand certain presentations from the programming-related lessons (in Lessons One through Twenty-Three). There are *many* setup- and operation-points made during the programming-related lessons. Without an understanding of these important points, much of what is presented from now on will make little sense.

Are you only interested in setup and operation?

Our goal with this text is to provide you with the knowledge you need to master *all three tasks* related to CNC turning center usage – programming, setup, and operation. And if you are to become proficient with all three tasks, you must read this text from beginning to end. But as we point out in Lesson Three, companies vary with regard to what they expect of their CNC people.

Some companies expect one person to do everything related to turning center usage, including program, setup, and run production. This is common in workpiece producing companies (job shops) and tooling producing companies. Other companies divide the related tasks and assign them to several people (one person programs, another makes setups, and yet another runs production). This is common in product producing companies.

If you work for a product producing company, you may not be expected to write programs. Your responsibilities may be limited to making setups and/or running out the job once a setup has been made. While we feel you can still benefit from reading the *all* of the programming related presentations, your immediate interests may be solely related to setup and/or operation.

If you are interested only in setup and/or operation, you should, *at the very least*, read the programming-related presentations listed below. This will allow you to understand material presented from this point on. Even if you *have read* all of the programming-related lessons, this list will help you quickly review some important material before continuing. *We will not be repeating these important setup- and operation-related discussions* during Key Concepts Seven through Ten:

From Lesson One: Machine configurations

- Read from the beginning to the heading "Programmable functions of turning centers"

From Lesson Two: Understanding turning center speeds and feeds

- Read the entire lesson

From Lesson Three: General flow of the programming process
- Read the entire lesson

From Lesson Five: Program zero and the rectangular coordinate system
- Read from the beginning to the heading "Wisely choosing the program zero point location"

From Lesson Six: Determining program zero point assignment values
- Read the entire lesson

From Lesson Seven: Four ways to assign program zero
- Read the entire lesson

From Lesson Eleven: Introduction to compensation
- Read the entire lesson (also read the introduction to Key Concept Four that precedes this lesson)

From Lesson Thirteen: Wear offsets
- Read the entire lesson

From Lesson Fourteen: Tool nose radius compensation
- Read from the beginning to the heading "Steps to programming tool nose radius compensation"
- Read from the heading "Tool nose radius compensation from the setup person's point of view" to the heading "Programming tool nose radius compensation values"

From Lesson Fifteen: Introduction to program structure
- Read from the beginning to the topic "Machine variations that affect program structure"

You may be wondering why so many setup- and operation-related topics are presented during *programming-related* presentations. Again, many companies expect one person to program, setup, and run production – these people, must master all three tasks related to turning center use.

Even in companies that do separate these tasks and have different people perform them, a programmer must know enough about setup and running production to direct setup people and operators. Indeed, it is the programmer in these companies that creates the setup and production run documentation. *A programmer must know at least as much about setup and operation as setup people and operators.* It is a natural progression to discuss certain setup- and operation-related topics *during programming* when they apply to programming-related features and functions.

If you have read all material to this point, you may be surprised at how much you already know about setting up and operating a CNC turning center.

The need for hands-on experience

Hands-on experience is extremely important to fully mastering CNC turning center setup and operation. But unfortunately, no text can provide hands-on experience. If you are using this text as part of a course you are taking at a technical school, hopefully there are machines available for practicing what you learn in class and from this text. *We cannot overstate the importance of hands-on experience* – experience we cannot provide in this text.

That stated, we can provide *the principles* needed to set up and run CNC turning centers. As a comparison, think of what it takes to learn to fly an airplane. A future pilot must, of course, spend a great deal of time in the cockpit at the flight controls in order to master the skills needed to become a pilot. But before a person begins any serious hands-on flight training, they *must* understand the basics of aerodynamics and flight. Without this understanding, controls in the cockpit will have no meaning. At some point, the student must attend *ground school.* In ground school they will learn concepts that help them understand the controls in the cockpit.

In similar fashion, a CNC setup person and/or operator *must* spend time at the CNC machine to fully master tasks related to setting and running production on a CNC turning center. But before a person can spend any meaningful time at the machine, they *must* understand concepts related to setting up and running the CNC machine. Without an understanding of the material presented from this point in the text, machine functions will have no meaning.

Think of Key Concepts Seven through Ten in this text as a kind of *ground school for learning to setup and run CNC turning centers.* While you will not gain the important hands-on experience you need to fully master running a CNC turning center from this text alone, you will learn the *concepts* required to begin practicing (with the supervision of your instructor or an experienced person in your company). An understanding of these concepts is mandatory *before* you begin working at the machine.

About Key Concept Number Seven

Again, this Key Concept parallels Key Concept Number One. But now we'll now be looking at a CNC turning center from a setup person's or operator's viewpoint. Much of what is presented in Key Concept Number One is important to setup people and operators. Again, be sure to read from the beginning of Lesson One to the heading *Programmable functions of turning centers* before continuing.

Procedures you must know about

Setting up and operating a CNC turning center requires that you master some basic *procedures.* While we show many *step-by-step* procedures in Lesson Twenty-Seven, we want to acquaint you with some of the most rudimentary machine-usage procedures and describe when they are required. We'll be referring to these procedures throughout Key Concepts Seven through Ten.

Machine power-up and power-down

You must, of course, be able to turn on your turning center/s. This procedure usually involves turning on a main power breaker (a switch that is usually located on the back of the machine), then pressing the control *power on* button, and finally, pressing a button to turn on the machine's hydraulic system (if it has one). To turn most machines off, you simply reverse the order of these steps.

Sending the machine to its zero return position

Most turning centers require that you perform this procedure when the machine is first powered up.

The zero return procedure varies from one machine tool builder to another. With some machines, it is as simple as selecting the zero return mode and pressing one button which causes each axis (first X, then Z) to move to its zero return position. With other (indeed most) machines, you must send each axis to it zero return position independently. After selecting the zero return mode, you select the axis you want to zero return first (usually X). Next you press the plus button and hold it (though many turning centers have a *joystick* for this purpose). The selected axis will rapid until it comes to within an inch or so of its zero return position. Then it will slow down, eventually reaching the zero return position. You can tell when this occurs, since an *axis origin light* will come for the axis as it reaches the zero return position. This must be repeated for the other axes.

Manually moving each axis

Manually moving an axis is often required. Some examples of when you must manually move an axis include measuring program zero assignment values, manually machining soft jaws, and manually positioning a cutting tool into a position so that it can be inspected or replaced. Most machine tool builders provide you with two ways to manually move an axis (most machines allow you to manually move only one axis at a time).

One way to move an axis is to use the *jog* function of the machine. This allows rather crude control of motion, and will be used when there is no danger of an axis component colliding into an obstruction. To jog an axis, first you select the jog mode (sometimes called the manual mode). Next you select the axis you want to move (X or Z). You then select the motion rate with a multi-position switch (commonly called *jog feedrate*). And finally, you activate the axis motion in the desired direction (plus or minus). With most machines there are two buttons used for this purpose, one labeled plus – the other labeled minus (though again, some machines have a joystick for this purpose). As long as you hold the button or joystick, the axis will move. Also, you can manipulate the jog feedrate switch during motion to speed up or slow down the motion.

Another way to move an axis is to use the *handwheel.* This provides a much more precise way to move an axis. It is commonly required when there is danger of an axis component (like a cutting tool) colliding with an obstruction. With the handwheel, you can control how quickly the machine will move when you turn the handwheel. In the times-one (X1) position, each increment of the handwheel will be 0.0001 inch (or 0.001

mm). In the times-ten (X10) position, each increment will be 0.001 inch (or 0.010 mm). In the times one-hundred position (X100), each increment will be 0.010 inch (or 0.1 mm). To use the handwheel, first you select the handwheel mode, then you select the axis you want to move (X or Z). Finally, you select the rate of motion (X1, X10, or X100). When you turn the handwheel in the desired direction (plus or minus), the axis will move.

Manually starting the spindle

This procedure is required when measuring program zero assignment values if your machine does not have a tool touch off probe. As you know, measuring program zero assignment values often requires that you actually machine a small amount of material from the workpiece.

Machine tool builders vary when it comes to how much manual control they provide for the spindle. With some machines, you can simply select the manual mode, set a multi-position switch to select the desired spindle speed, select the desired direction (forward or reverse), and press a button to activate the spindle. Another button will be pressed to turn the spindle off.

If your machine does not provide manual control for the spindle, you must use a function called *manual data input* (MDI) and enter and execute a program-like command to start the spindle (like G97 S500 M03). Another command must be entered and executed to stop the spindle (M05). We'll discuss the use of manual data input in Lesson Twenty-Six.

Manually indexing the turret

Machine tool builders vary when it comes to how much manual control they provide for turret indexing. Your machine may have a twelve-position switch that allows you to select the turret station to which you want the turret to index (assuming the machine has a twelve position turret). Once you select the desired turret station, you press a button to activate the turret index.

Unfortunately, many turning centers do not have this multi-position switch. Most will require that you use manual data input (MDI) mode to enter and execute a program-like command to index the turret (like T0100). The turret must, of course, be clear of obstructions before a turret index can be commanded. Again, we'll discuss the use of manual data input in Lesson Twenty-Six.

Manipulating the display screen

The display screen will show you a great deal of important information. For setup people and operators, the four most important display screen modes are *position*, *program*, *offset*, and *program check*.

The position display screen

As you would expect, in position mode, the display screen mode shows information about machine position. Four position pages are involved. The *absolute* position display page shows the machine's current position relative to program zero. The *relative* position display page is used to take measurements and can be set or reset to specify a point of origin for the measurement. The *machine* position display page shows the machine's current position relative to the zero return position. And a fourth display screen page shows a combination of the previous three – along with a special *distance-to-go* position display that lets you see how much further the machine will move in the current command.

More about the relative position display page

Again, *the relative position display page is the page you will use to take measurements.* With this display screen page you can set a point of reference for the measurement. The relative position display is the *only* position display screen page that can be set or reset in this manner.

More about distance-to-go

This is a very important function – one that is also displayed on the *program check* display screen page. When you are verifying a program, it is often helpful to know just how much further the tool will move in the current command. Say a tool is making its first approach movement. You know the tool is *supposed* to stop 0.1 inch away from the work surface. But you're worried. With *dry run* and *single block* turned on (two other program verification functions), you will have total control of the machine's motion rates (including rapid) with a multi-position switch (commonly the *feedrate override* switch). As the tool approaches you slow motion with the feedrate override switch. When the tool is about 0.5 inch away from the work surface, you get worried, so you press the *feed hold* button to stop axis motion (feed hold is yet another program verification function). When you look at the *distance-to-go* display, you know the value shown in the Z axis should be under 0.5 inch. If it is not, the current command will cause the tool to crash into the work surface (and, of course, you've found a problem with the program or setup).

The program display screen

In edit or memory mode, this display screen shows you the active program. If you are running the program, this will allow you to see up-coming commands. If you need to modify the program, this display provides you with a kind of text editor. In manual data input (MDI) mode, this display screen shows the page used to enter MDI commands.

The offset display screen

This display screen allows you to view, enter, and modify offsets. Any kind of offset is accessible, including work shift, wear, and geometry offsets.

The program check display screen

Used for program verification, this display screen shows important information to help you find mistakes in a program. It shows a few upcoming commands of the active program, currently active CNC words (like G and M codes, as well as S word and F word), and the distance-to-go display.

Loading and saving CNC programs

Before programs can be run, of course, they must reside in the machine. And once a job is finished, it may be necessary to save the program for future use, especially if changes have been made to the program during the program's verification. These functions require the use of a *distributive numerical control* (DNC) system. When loading a program, most DNC systems require that the machine be prepared to receive the program and then the DNC system is commanded to send the program. The specific procedures to load and save programs are shown in Lesson Twenty-Seven.

Lesson 24

Tasks Related To Setup And Running Production

We define setup time as the total time a machine is down between production runs. We define cycle time as the total time it takes to complete a production run divided by the number of good workpieces produced. When you think about it, there are only two general tasks that occur on CNC turning centers – machines are either in setup or they are running production.

It is important to understand the distinction between making a setup and running production. The tasks you perform during setup are *getting the machine ready* to run production. Only when the setup is completed and a workpiece has passed inspection is it possible to run production. The person making the setup is called the *setup person* – the person completing the production run is called the *CNC operator* (though in many companies, one person makes the setup *and* runs production – this kind of person is commonly called a *CNC technician*).

Setup-related tasks

- Tear down previous setup an put everything away
- Gather the components needed to make the setup
- Make the workholding setup
- Assemble the cutting tools
- Load cutting tools into the machine's turret
- Assign program zero for each new tool (tool's not in the turret from the previous job)
- Enter tool nose radius compensation values
- Load the CNC program/s
- Verify the correctness of a new or modified program
- Verify the correctness of the setup
- Cautiously run the first workpiece – ensure that it passes inspection
- If necessary, optimize the program for better efficiency (new programs only)
- If changes to the program have been made, save a corrected version of the program

Running production-related tasks

Done during every cycle:

- Load a workpiece
- Activate the cycle
- Monitor the cycle to ensure that cutting tools are machining properly (first few workpieces only)
- Remove the workpiece
- Clean/de-burr the workpiece
- Perform specified measurements (if required)
- Report measurement results to statistical process control (SPC) system

Not required in every cycle:

- Make offset adjustments to maintain size for critical dimensions
- Replace dull tools
- Remove chips from the machine (if required)

In this lesson, we'll be going through this list of tasks and describing each one in detail. Notice how many of these tasks have been described in detail during the programming-related presentations. We will not be repeating these presentations.

There are certain setup-related tasks that must sometimes be repeated during a production run. If for example, a cutting tool gets dull and must be replaced, the same tasks required to initially assemble it, determine its program zero assignment values, and enter its offsets must be repeated. And if trial machining is required during setup for a given tool, it will probably be required when the tool is replaced during the production run.

Many of these tasks, of course, draw upon your basic machining practice skills. Tasks related to making workholding setups and assembling cutting tools, for example, require that you understand workholding- and cutting-tool-components. And these tasks are identical to those that must be performed on conventional (not CNC) machine tools. So, as with programming, setting up and operating a CNC turning center require a firm understanding of basic machining practices. And also as with programming, if you have experience working with conventional machine tools, you have a head start for mastering CNC turning center setup and operation.

As you look through the list of tasks just shown, notice how many have been introduced during our discussions of programming. While we're going to expand our discussions from this point on, you have already been exposed to many setup- and operation-related concepts.

This lesson is truly at the heart of the setup and operation presentations in this text. It presents the most important tasks you must understand in order to be a successful setup person or operator.

A CNC job from start to finish

Figure 7.1 shows the job we're going to use to discuss every task related to setup and maintaining production. We're going to be describing this job as if we're actually running it. While this isn't as good as actually demonstrating a job being run, it's as close as we can come. We'll use a universal style slant bed turning center.

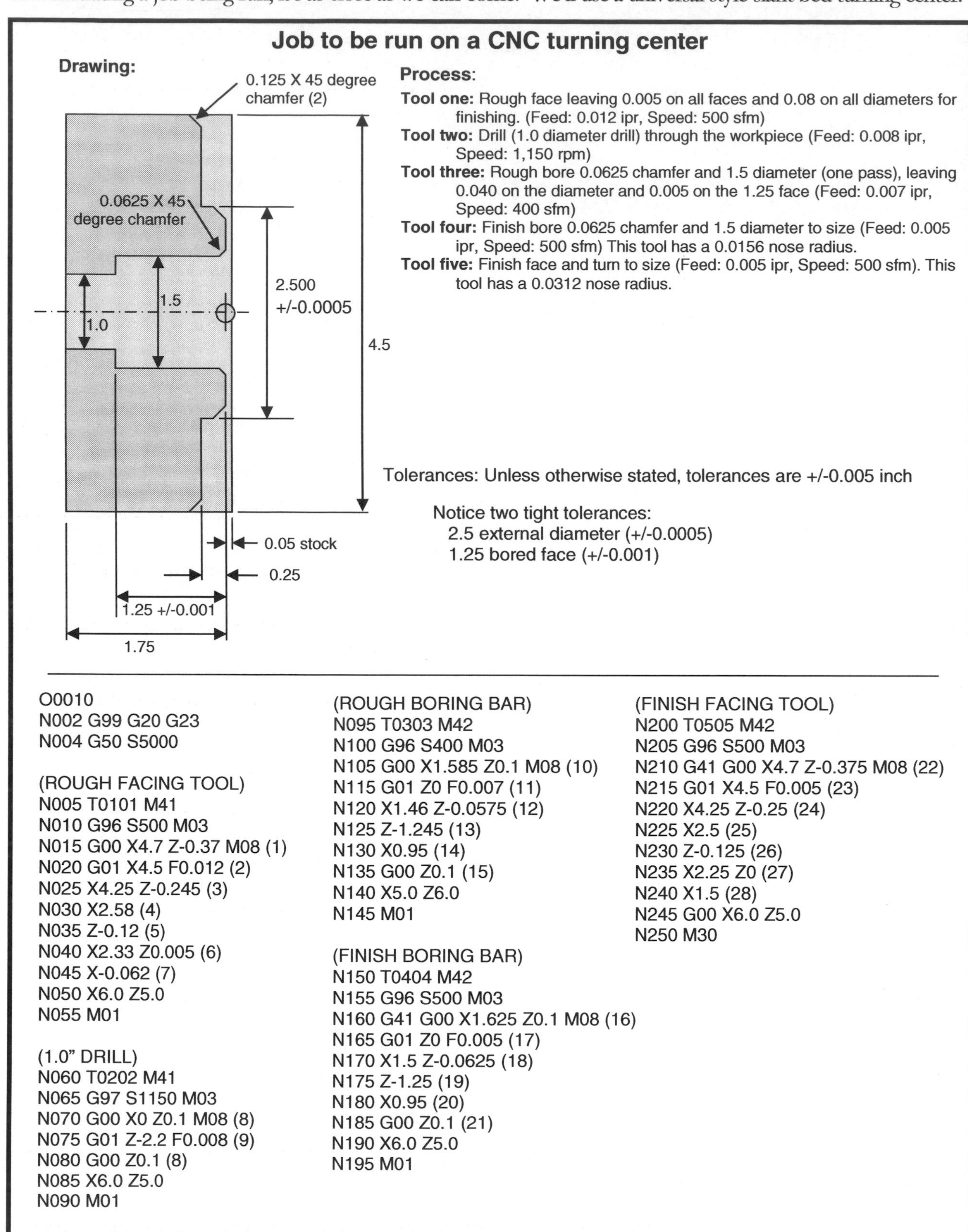

Figure 7.1 – A job that must be run on a CNC turning center

This workpiece is the practice exercise from Lesson Sixteen. Notice that there are now two tight tolerances specified on the print. The process is shown, providing you with the machining order used in the program. All coordinates in the program have been specified as mean values for each tolerance.

Notice that the program is not documented nearly as well as most we have shown in this text. Most programmers don't provide much documentation in their programs (remember – documentation takes up memory space in the machine) – so this is typical of what you'll see with the jobs you'll be running. Tool naming messages are included in the program at the beginning of each tool. The program uses the format shown in Lesson Sixteen (using geometry offsets to assign program zero), which allows the re-running of cutting tools.

Again, we'll use this job to describe the various tasks that must be completed in order to make a setup and complete a production run. Our descriptions will assume, however, that you have some basic machining practice experience. Be sure to question an experienced person in your company if any of these presentations require more basic machining practice skills than you currently possess.

Setup documentation

As described in Lesson Nine, a programmer must provide setup documentation. Most programmers will use a one-page setup sheet to describe everything that must be done to make the setup. Figure 7.2 shows the setup sheet for our example job. Most companies also supply the setup person with a copy of the print and a print-out of the CNC program.

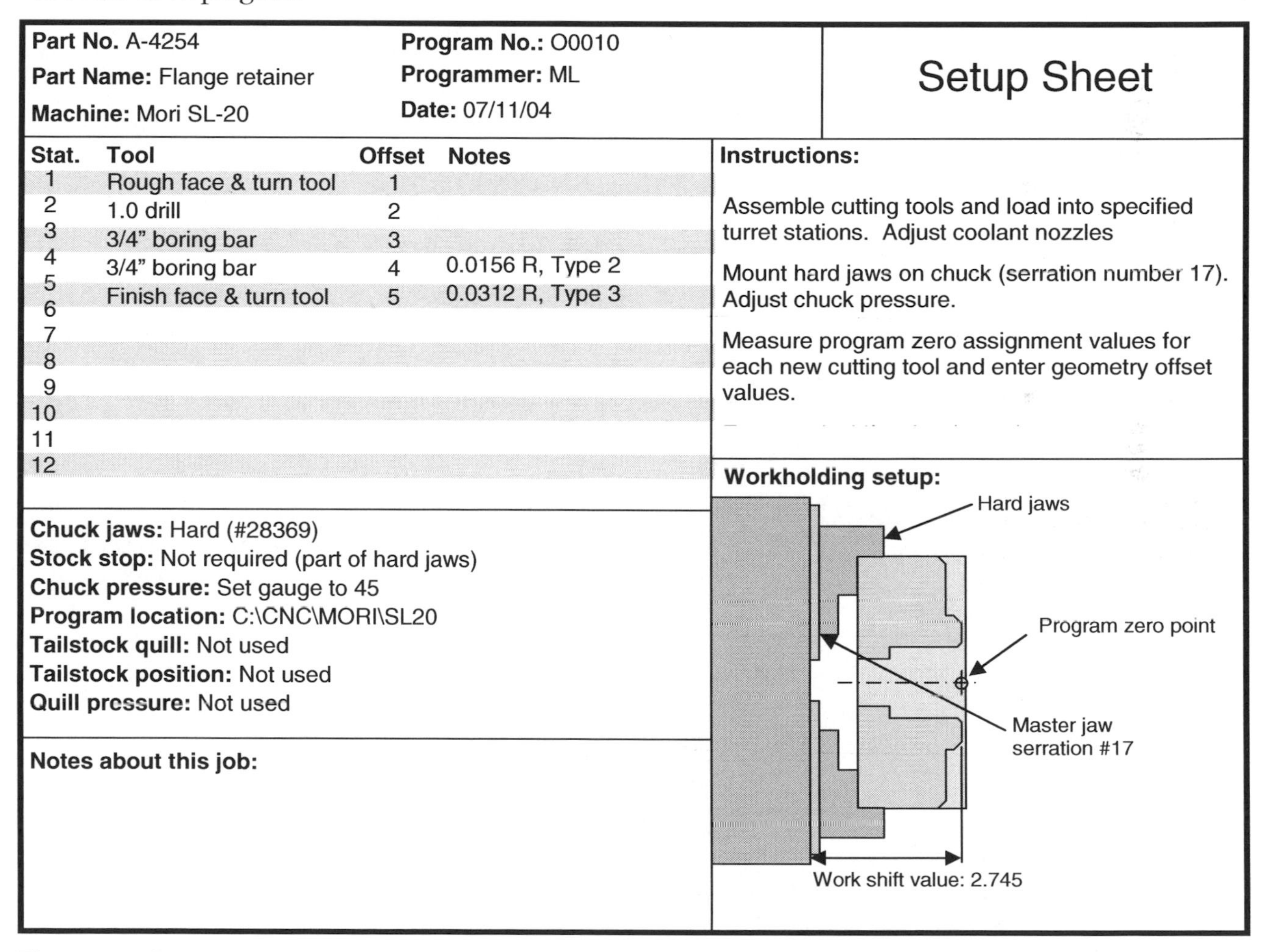

Part No. A-4254
Part Name: Flange retainer
Machine: Mori SL-20

Program No.: O0010
Programmer: ML
Date: 07/11/04

Setup Sheet

Stat.	Tool	Offset	Notes
1	Rough face & turn tool	1	
2	1.0 drill	2	
3	3/4" boring bar	3	
4	3/4" boring bar	4	0.0156 R, Type 2
5	Finish face & turn tool	5	0.0312 R, Type 3
6			
7			
8			
9			
10			
11			
12			

Instructions:

Assemble cutting tools and load into specified turret stations. Adjust coolant nozzles

Mount hard jaws on chuck (serration number 17). Adjust chuck pressure.

Measure program zero assignment values for each new cutting tool and enter geometry offset values.

Chuck jaws: Hard (#28369)
Stock stop: Not required (part of hard jaws)
Chuck pressure: Set gauge to 45
Program location: C:\CNC\MORI\SL20
Tailstock quill: Not used
Tailstock position: Not used
Quill pressure: Not used

Notes about this job:

Workholding setup:

Figure 7.2 – Setup sheet for our example job

As is typical of setup documentation provided by most programmers, our setup sheet does not explain every detail of what must be done to get the job up and running. It assumes quite a bit of the setup person who will be making the setup. Many setup-related tasks are quite redundant from one setup to the next – as are the cutting tools used from job to job – and most programmers won't be very specific about the most redundant tasks. We'll point out common shortcomings of setup documentation as we continue with this discussion.

We explain the tasks related to making setups in the approximate order that setups are made.

Tear down the previous setup and put everything away

Before you can begin making a new setup, you must remove certain components from the previous setup. Tasks include removing the top tooling (jaws) from the work holding device – assuming they not required in the next job, and cleaning chips and debris from the work area within the machine. The old CNC program should also be removed from the machine's memory (after saving a corrected version – if required).

When it comes to cutting tools that will not be required in the next job, setup people vary. As long as their turret stations are not required for cutting tools in the up-coming job, we recommend leaving most cutting tools in the turret, especially if they are common tools – tools that will be needed in upcoming jobs. Remember that as long as a cutting tool remains in the turret, its offset settings (wear, geometry, and tool nose radius compensation) will remain correct. Removing tools (and later replacing them) can result in duplicated effort.

Once removed from the machine, tooling components should be put back where they belong – so you can easily find them in the future.

Gather the components needed to make the setup

The more organized you can be when making a setup, the easier the setup will be to make. Part of being organized is gathering *all* of the components you'll need to make the setup. Gathering everything at one time will keep you from having to walk to and from the tool crib every time you need something.

Companies vary with regard to how well they list the components needed for a given setup. Our setup sheet is quite poor in this regard. For example, we're assuming that the setup person can figure out which components are needed to assemble the cutting tools for the job. Indeed – we're assuming they can determine which cutting tools will be used. This kind of setup sheet is very common in workpiece- and tooling-producing companies – when the setup person is expected possess a high degree of skill.

Some companies do much better, providing a complete list of every needed component, including workholding components, cutting tool components, hardware (fasteners like screws and bolts), gauging tools, and even hand tools. This makes the gathering of needed components much easier. With this kind of list, someone other than the (highly skilled) setup person will be able to gather components. This will free up the setup person to be doing other things while components are being gathered. This kind of component list is common in product-producing companies.

Make the workholding setup

You can learn more about work holding and support devices (three jaw chucks, collet chucks, tailstocks, steady rests, and bar feeders) in the Appendix that follows Lesson Twenty-Three.

Setups will vary dramatically from machine to machine and from job to job. Using the setup instructions prepared by the CNC programmer – and based upon your machining practice skills – you will make the holding setup. If using a collet chuck, it's usually a relatively quick and easy task to mount the new collet. If using a three jaw chuck – as is the case with our example job – top tooling must be placed on the chuck. In our case, hard jaws must be mounted to the chuck.

Depending upon the chuck style, mounting jaws can be a tedious procedure. The serrations in the master jaws of most three jaw chucks are quite small and close together, meaning you must be careful and ensure that all three jaws are properly placed. Jaws will be numbered (one, two, and three), as will be the master jaws on the chuck, so be sure to place each jaw on the correct master jaw. If using hard jaws (for repeated jobs), the setup sheet should specify the precise location for the jaws (as our setup sheet does – serration number 17). If using soft jaws, it's usually left to the setup person to determine in which serrations to place the jaws. This takes practice, and can result in some trial-and-error.

How To Mount Jaws In The Correct Serrations

Turning center setup people eventually master the placement of jaws on three-jaw chucks. But novices tend to spend a great deal of time using trial and error methods. The master jaws of most three-jaw chucks have tiny serrations into which the jaws (hard or soft) must be mounted. Two socket head cap screws are commonly used to mount each jaw (six screws total). And again, the same master jaw serration must be used for all three of the jaws.

The setup person won't know if one of the jaws is misplaced until *after all three jaws* are mounted. If a jaw is misplaced, all of the jaws must be removed and the entire process of mounting jaws must be repeated.

Many setup people count serrations to determine which serration each jaw must be mounted in. But the setup person must first determine the diameter of the serration into which jaws must be placed. This is yet another source of mistakes that leads to repeating the entire jaw-mounting process.

The diameter at which to mount jaws

Since most hydraulic three-jaw chucks have a stroke of 0.25 or more, and since serrations are commonly 0.062 inch (1/16) or less apart, three is a little flexibility when it comes to which master jaw serration is used (as long as all three jaws are mounted in the ***same*** serration). Most chuck manufacturers recommend having the chuck clamp on the workpiece close to the middle of the master jaw stroke. For this reason, knowing the jaw stroke is important for determining the diameter at which you want to mount jaws.

Say, for example, you have a chuck with a 0.25 inch stroke (0.5 diameter increase/decrease). And you're going to mount jaws with the chuck in its closed position (master jaws toward chuck center). Each jaw must be placed in the master jaw in such a way that its workpiece-contacting surface is 0.25 smaller (in diameter) than the diameter the jaws will be clamping on. This allows for half the jaw stroke. When the jaws are actually clamped on a workpiece, they will contact the workpiece in the middle of the chuck's stroke.

If you need to clamp on a 4.5 inch diameter (as is required in our example job) with a chuck having a 0.25 inch jaw stroke, first close the jaws (with no workpiece, of course). The jaw contact surface must be at a 4.25 inch diameter (again 0.25 smaller than the workpiece clamping diameter). With the workpiece-contacting surface of each jaw at this diameter, when you open the chuck, the jaw contact surfaces will be at a 4.75 inch diameter, which will allow the workpiece to be loaded. When you close the chuck on a workpiece, the jaws will contact the workpiece at the 4.5 inch diameter, and half way through their stroke.

Again, trying to figure out which serration to use that allows this can be very difficult, especially when you consider that there is no relationship between the contact surface of the jaw and the end of the jaw. But there is a simple way to determine where to mount jaws. It requires that you determine the diameter at which the jaw contact surface will be when the chuck is closed as just described. For both external and internal clamping, this diameter will be half the jaw stroke (in diameter) smaller than the workpiece diameter to be clamped - assuming you mount jaws while the chuck is in its closed position (master jaws toward chuck center).

Using a long boring bar

One simple and inexpensive way (though a little cumbersome) to easily mount each jaw in the same (and correct) serration is to use a boring bar as a *pointer*. The longer the boring bar, the more room you'll have to work. In essence, you'll be using the boring bar tip as a pointing device, synchronizing it with the X axis position display of the display screen.

One way to synchronize the boring bar with the position display is to bring the boring bar tip to a known diameter and then set the X axis display to this known diameter (most machines require that you do so on the *relative* display screen page). Since most three-jaw chucks have a through-hole, it makes a great *known diameter*. Simply measure the diameter of the through hole and keep it for future reference.

Say for your machine, there is a 1.500 inch diameter hole in the chuck. First bring the boring bar tip to this hole diameter. Then set the X axis display to **X1.500**. From this point, whenever you move the X axis, the X axis display will follow along, showing you the diameter at the boring bar's tip.

Actually mounting jaws

After determining the diameter at which you need the jaw contact surfaces to be when the chuck is closed (again, this is the workpiece diameter minus half the jaw stroke in diameter), and after you have calibrated the boring bar tip to the position display using the method shown above, move the boring bar in X until the X axis display is showing diameter you've calculated. Without moving X again, bring the boring bar in close to the chuck. This can be a little cumbersome, and again, the longer the boring bar, the more room you'll have to work.

Manually rotate the chuck (by hand) to bring one of the master jaws into close proximity to the boring bar tip. Place the jaw on the master jaw using the boring bar tip to target the jaw. You need the jaw's contact surface to be close to the boring bar tip. Now mount the jaw. See Figure 7.3 for an example that applies to our job. Repeat this for the other two jaws.

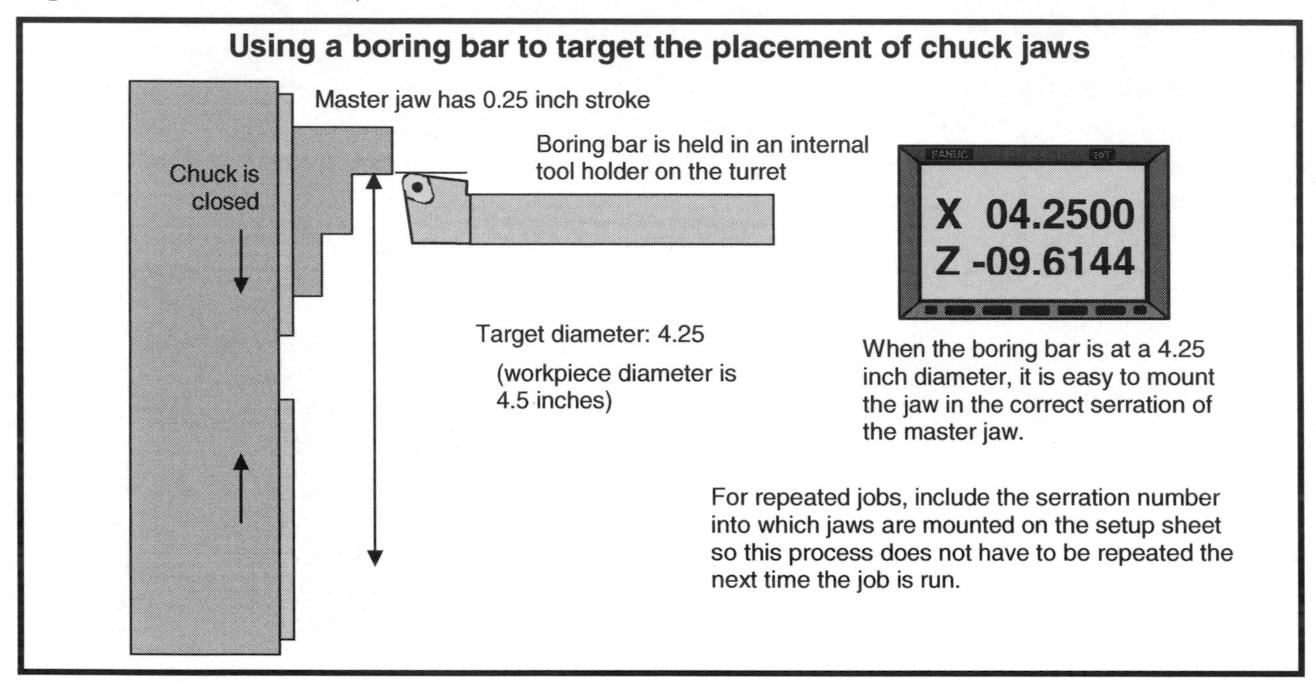

Figure 7.3 – Using a boring bar to target the placement of chuck jaws

A note about soft jaws

These techniques work great for hard jaws. But if you're going to be machining soft jaws once they are placed on the chuck, you'll need to do one more calculation. You must determine the diameter at which to place the *current* clamping surface of each jaw, while allowing for the amount of material you will be machining from each jaw.

First, do the calculation shown above to determine where the *finished* jaw surface will be. Next, determine how much material you will be removing from each jaw. You don't have to be perfect, but approximate how much material you'll be taking off each jaw. As you look at the jaw, say you think that about 0.100 inch must come off each jaw (0.200 diameter) in order to clean up the jaw for the new diameter it will be clamping on.

Finally, for external clamping, subtract this amount (0.200 in our case) from the diameter you previously calculated. If a 3.0 inch diameter workpiece must be held, your target diameter will be 2.55 (3.0 minus 0.25 minus 0.2). Once mounted, and when you have clamped on a diameter about half way through the stroke (as you must whenever machining soft jaws), this will allow about 0.100 of material to be removed from each jaw to bring the jaw clamping surfaces to the desired clamping diameter (3.000 in our case).

Note that for internal clamping, you must increase the jaw placement diameter by twice the amount of stock you expect to remove from each jaw.

Machining soft jaws during setup

If soft jaws are being used, most jobs require that they be *machined* after placement on the chuck to match the diameter they must grip on the workpiece. Some companies dedicate a set of soft jaws to each job (this can be justified when jobs are often repeated). If soft jaws are dedicated to a job, just a small amount of material must be removed from the jaw to make the gripping surfaces concentric and square with the chuck.

Other companies leave it to the setup person to determine which set of jaws (of many sets) will be used with each setup. In this case, there may be times when quite a bit of material must be removed from the soft jaws during machining.

If griping on an outside diameter, the jaws must be bored. If gripping on an inside diameter, they must be turned. If you have worked on an engine lathe (or any other style of manual lathe), you'll find that machining soft jaws for a CNC turning center is much the same as it is for any other style of lathe.

In order to properly machine jaws, the chuck must be in a clamped state and at the same clamping force that will be used to hold the workpiece. The master jaws should be about mid-way through their stroke during jaw machining, meaning the jaws must be gripping on something when they're machined. Figure 7.4 shows a jaws clamped on a ring for the jaw boring procedure.

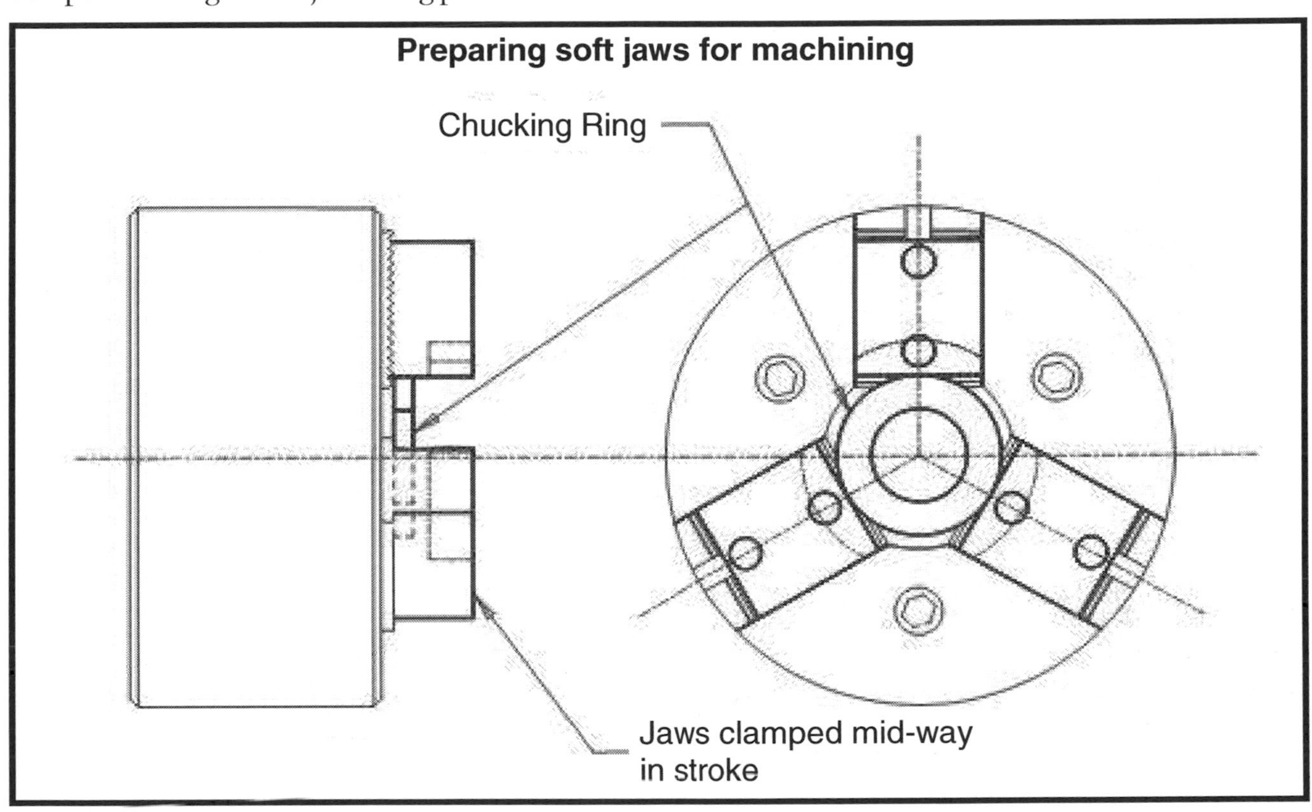

Figure 7.4 –Jaws clamped on a ring before machining

Because it is nearly impossible to determine the diameter a chucking ring or plug must be until the soft jaws are placed on the chuck, a great deal of setup time can be wasted while the setup person finds (or worse, makes) a plug of the appropriate size. One way to minimize this time is to make a set of chucking plugs to be used when ever jaws must be bored. A set of plugs could be made in about 0.125 inch increments, ranging from the smallest diameter needed to the largest diameter needed. Though making a set of chucking plugs for each turning center takes time, this time will be returned many times when you consider how much time many setup people waste during setups.

Anther alternative is to make one or two adjustable chuck clamping tools similar to the one shown in Figure 7.5.

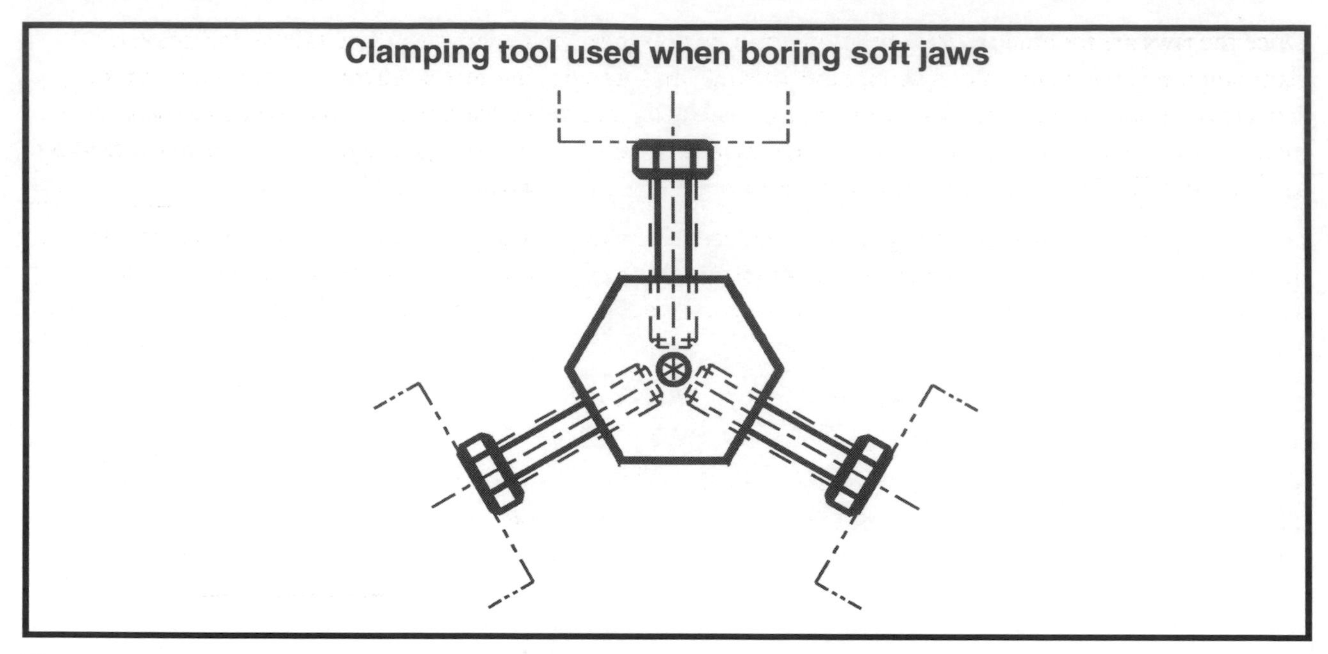

Figure 7.5 – Chuck clamping tool

The center piece for this tool can be made from hex stock. Three head screws are mounted 120 degrees apart to provide the clamping points for the jaws of the three-jaw chuck. How far the screws go into the center piece allow adjustment to determine the clamping diameter. And replacing screws with longer or shorter ones provides more adjustment. The hole in the center makes it easy to adjust the screws to their appropriate lengths. If used with a full scale drawing of concentric circles (though not to scale, this is shown in Figure 7.6), you can easily align the center hole with the cross hairs on the drawing to determine how to set the screws to get the appropriate clamping diameter.

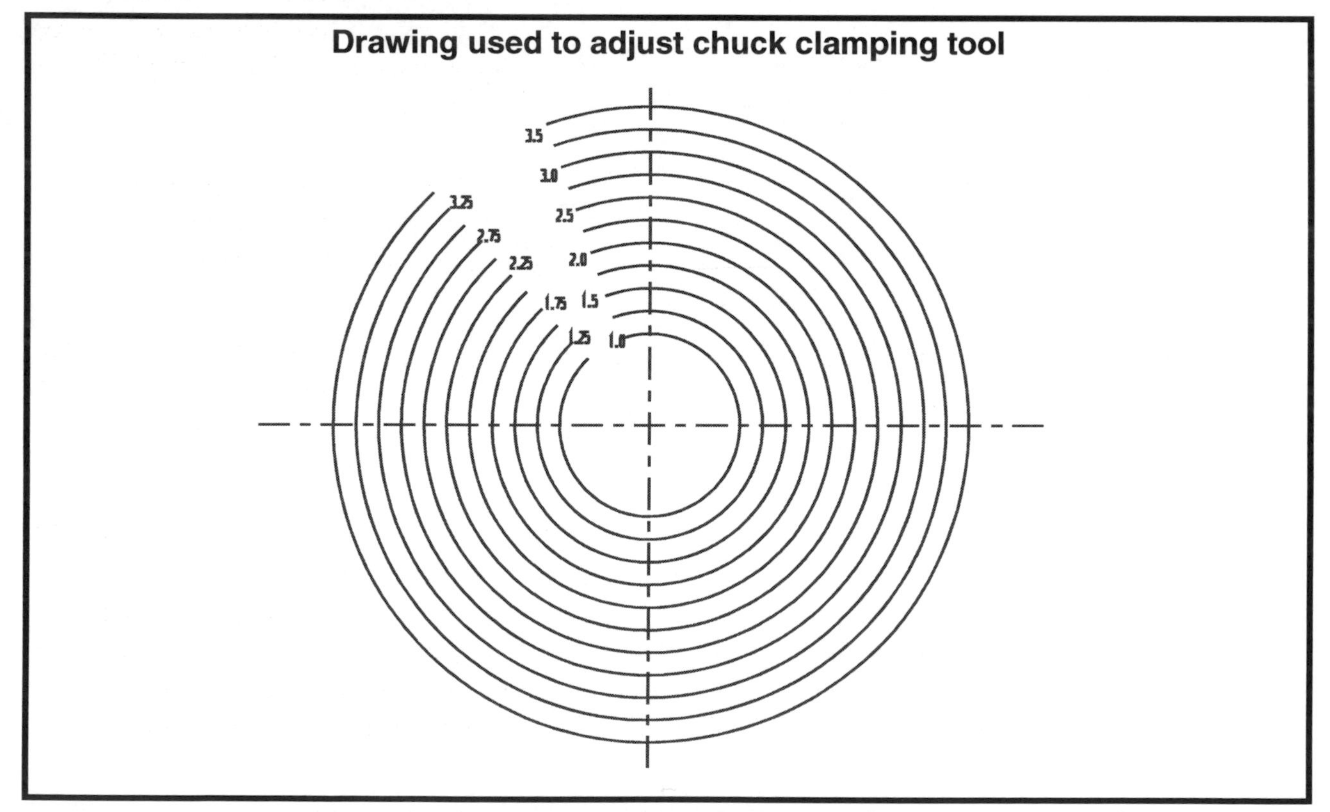

Figure 7.6 – Drawing to help set chuck clamping tool (drawing must be full scale)

Once the jaws are mounted to the chuck and clamped on the appropriate diameter with the appropriate clamping force, you're ready to machine the jaws. While some programmers provide a special program to actually bore or turn the soft jaws, most setup people are left completely on their own to machine the jaws manually. Using the manual functions of the machine (spindle activation, jog, and the handwheel), the turning tool or boring bar will be used to machine the jaws.

For external clamping (jaw boring), you should machine the jaws just *slightly smaller* than the diameter on which the jaws will be clamping. We recommend boring about 0.005 inch smaller. For internal clamping (jaw turning), you should machine the jaws about 0.005 *larger* than the diameter on which the jaws will be gripping. The final machined diameter is quite important to get adequate gripping area from each jaw. The depth to which you machine the jaws should be specified on the setup sheet and adequate to securely hold the workpiece.

Determining how the cutting tool is machining the jaws (diameter and length) can be difficult, especially in diameter. Since you're working with a set of three jaws that are 120 degrees apart, you cannot use a caliper to measure the diameter your tool has machined. For this reason, we recommend *calibrating* the axis position display screen with the tool's actual position. Once this is done, you can simply monitor the display screen to control the diameter being machined and how far into the jaws the tool is moving.

To calibrate, you must know the cutting tool's program zero assignment value in the X axis. We show how to determine program zero assignment values for a boring bar in Lessons Six and Seven. With the machine resting at its zero return position (reference position) in X, you can preset the relative position display screen to this value. From this point on, as you move the tool in the X axis, the position display screen page will show you the diameter the tool will machine. For the Z axis, bring the tip of the tool flush with the jaw face and preset the relative position display screen page's Z value to zero. From this point, as you move the Z axis, the Z axis position display will show you the position of the tool tip relative to the jaw face.

Other devices related to work holding setup

Though our example workpiece is adequately held and supported by a three jaw chuck alone, some jobs require additional work support. If using a tailstock, for example, the center must be placed in the quill and the tailstock body must be positioned and aligned. If using a steady rest, it must be positioned and aligned. If using a bar feeder, spacers must be placed and the new bar must be loaded.

Assemble cutting tools

Depending upon lot sizes and the availability of CNC personnel, your company may be performing some cutting-tool-related setup tasks *off line* (while the turning center is running workpieces). In essence, you can be getting ready for the next job while the machine is producing parts. Regardless of when cutting tools are assembled, this task must be completed before a setup can be finished.

Our example setup sheet assumes the setup person knows quite a bit about cutting tools. We simply name the tool – and the setup person is expected to know the components that make up the tool. We don't even specify the insert to be used with each tool (or its grade). Nor do we specify how far the boring bars should be protruding from the face of the tool holder. While this is typical of a single-page setup sheet, many companies do much better. They specify each component that is required in the tool. And they provide a drawing of each tool in the turret so the setup person can tell exactly how to mount the tool. This makes it easier for the setup person – and allows all components to be gathered by a person having lesser skills.

The actual task of assembling cutting tools is usually quite simple, requiring that you draw upon your basic machining practice skills. In most cases, you'll be using simple tools like Allen wrenches and screw-drivers, though some cutting tool manufacturers require the use of special hand tools to mount inserts. Ask an experienced person to demonstrate the assembly of the most common cutting tools used by your company.

A reminder about mounting inserts

From Lesson Thirteen, you know that many cutting tools used on CNC turning centers incorporate *inserts*. These inserts provide the cutting edge for cutting tools. If an eccentric pin is used to locate/clamp the insert into its tool holder, you know that there is an appropriate direction to turn the pin to ensure that the insert is properly clamped. You might want to review the topics *Consistently replacing inserts* and *Consistently indexing inserts* in Lesson Thirteen.

Load cutting tools into the turret and adjust coolant lines

Once assembled, cutting tools must be placed in the turret. The programmer must designate the tool station number for each cutting tool in the setup documentation as our example setup sheet shows.

Turret configurations vary from one turning center manufacturer to another. With some turrets, tools are mounted directly in the turret itself. External tools are placed in stations that have *wedge clamps* to hold them in place. Internal tools are mounted in holes in the turret face. Other turrets have external tool blocks, into which cutting tools are placed. The tool block is then mounted to the turret. Ask an experienced person to demonstrate the mounting of cutting tools into the turrets of your company's CNC turning centers.

Some turning centers make it very easy to adjust coolant flow to the tool tip. They use a coolant fitting that incorporates a ball joint that can be easily swiveled into place. Others have rather cumbersome coolant fittings and copper lines that must be mounted to the turret face and adjusted to get coolant to the tool tip (again, ask an experienced person to demonstrate this task).

Assign program zero for each new tool

You can learn more about program zero assignment in Lessons Five, Six, and Seven. You can learn more about geometry offsets in Lessons Seven and Twelve. You can learn more about trial machining in Lessons Eleven and Thirteen.

If you use our recommended methods (those shown in Lesson Seven), you should only have to perform the task of assigning program zero for tools that have just been placed in the turret. As long as you are using geometry offsets along with the work shift value to assign program zero, program zero assignments for tools still in the turret from a previous job will still be correct.

We provide a detailed description of how program zero is assigned in Lessons Six and Seven – and we won't repeat this description now. While you still need more specific procedures to do so (procedures that are shown in Lesson Twenty-Seven) and while you need hands on practice to master this task (experience we cannot provide in this text), you should be able to gain a good understanding of how program zero is assigned from Lessons Six and Seven. Ask an experienced person to demonstrate the task of program zero assignment for the most common types of cutting tools used by your company.

Enter tool nose radius compensation values (if the programmer uses this feature)

If the programmer uses tool nose radius compensation (described in Lesson Fourteen), two offset values must be entered for each tool that uses it. The programmer does have a way to *program* these offset settings (also shown in Lesson Fourteen), and if they do, you can skip this step.

If you will be expected to enter tool nose radius compensation values, it should be noted in the setup documentation. The setup sheet for our example job specifies the values to be entered for tools four and five.

The two values are entered into the R and T registers of the geometry offset display screen page (if your machine does not have geometry offsets, these values are entered on the wear offset page). In the R register, you enter the tool nose radius. In the T register, you enter the tool type number.

Some programmers do not explicitly specify the values you must enter (as our example setup sheet does). There is a chart provided in Lesson Fourteen that shows the tool type numbers (T register) for each kind of cutting tool. The only two you that you should try to remember are for turning tools (type 3) and boring bars (type 2).

Ask an experienced person in your company if tool nose radius compensation is used on a regular basis – and if you must enter tool nose radius compensation values during setup.

Load the CNC program/s

Almost all CNC using companies utilize some kind of *distributive numerical control (DNC) system* for the purpose of quickly loading programs into the CNC control.

Remember that some jobs require more than one program. A given job may, for instance, require a main program and several sub-programs. Additionally, some companies use setup programs (like jaw boring programs) to help with setup-related tasks. The setup documentation must specify the name of the program as well as where it can be found (as our example setup sheet does).

DNC systems include portable devices like notebook computers and portable floppy drive systems as well as more stationary devices like desktop computers. Though these devices vary in ease of use, quickness, and quality, any of them will be preferable to manually typing programs at the CNC control. The specific procedure to transfer programs to and from the CNC control is presented in Lesson Twenty-Seven. Ask an experienced person in your company to demonstrate how your company's DNC system is used.

For our example job, program number O0010 must be loaded. Our setup documentation specifies the location of program number O0010.

The physical tasks related to setup are now completed

At this point, you're (finally) ready to run the program. However, there could still be problems. How many problems you'll have and how severe they are depend upon several things. Generally speaking, new programs written manually by new programmers tend to present the most problems. But even proven programs (programs that have been successfully run before) can still have problems. Additionally, the setup person can make mistakes that will cause even a perfectly written program to fail. For these reasons, *all programs must be cautiously verified.* While proven programs tend to be easier to verify than new programs, you should never skip the program verification steps.

Verify the correctness of a new or modified program

This task is only necessary for new programs – or for programs that have been modified since the last time they were run. (CNC programmers must modify programs between production runs if the design engineer makes changes to the workpiece. *We recommend treating modified programs like new programs.*)

Verifying the correctness of a program is really the responsibility of the *CNC programmer*, and is usually done prior to making a setup. Indeed, it is usually done soon after the program is written. There are many *tool path verification systems* that allow a programmer to visually check the motions a CNC program will make even before the CNC program is loaded into the machine. Most are relatively inexpensive, so *there is really no excuse not to have one* – especially when you consider the expense of using a CNC machine to verify the correctness of a program.

Figure 7.7 shows a typical *solid-model-type* tool path verification screen for a sample program run on a Windows computer. These systems can be purchased for under about $200.00, which is extremely reasonable when you consider the amount of time and problems they save during program verification on a CNC turning center.

An off-line tool path verification system

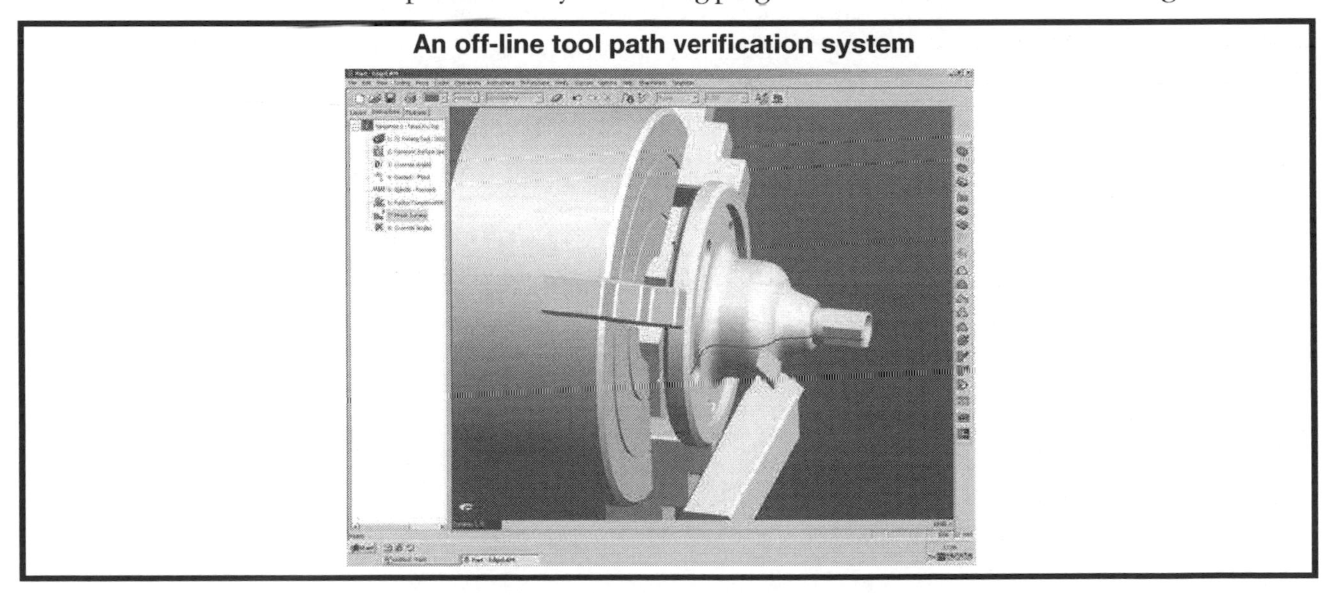

Figure 7.7 – An example of program verification done on a Windows computer

Once again, keep in mind that CNC-using-companies vary when it comes to what they expect of their CNC people. If you are a setup person in a product producing company, you probably won't have access to the program verification system, so you won't see the program verification being done off line. You'll have to accept the programmer's word that the program's tool path is correct.

Some CNC controls do include tool path verification capabilities that rival the best off line systems. If your machine has this capability, by all means, learn how to use it, especially if you can't see the program verification done off line. But if at all possible, this program verification step should be done *off line* (to save machine time). If mistakes are found while verifying the correctness of the program, the time required for *correcting* them can also be kept off line.

If you have neither of these tool path verification capabilities, you will not be able to see the movements that cutting tools in the program will make until you actually run the program on the machine. In this case, verifying the correctness of your programs will be extremely cumbersome, requiring that you cautiously check each movement the machine makes, using features like *dry run*, *single block*, and *feed hold*. When verifying programs in this manner, you may have to run the program several times just to begin to understand the program's movements. And you still won't be able to tell if all of the motions are truly correct.

Verify the correctness of the setup

Mistakes made during setup can cause even a perfect program to behave poorly. Consider a few of the possibilities: You could load a tool into the wrong turret station. You could measure or enter program zero assignment values incorrectly. You could forget to enter a tool nose radius compensation offset value. You could load the wrong program.

And by the way, there are certain programming mistakes that will not show up during tool path verification. If, for example, the programmer forgets to start the spindle (or if they have it running in the wrong direction) few tool path verification systems will catch the mistake. Additionally, most tool path verification systems will ignore offset use. If the program invokes the wrong geometry and/or wear offset, most tool path verification systems won't show it.

For these reasons, *you should run the program at least once without a workpiece.* This is commonly called a *dry run.* In fact, one of the program verification functions you have to help with program verification is called *dry run.* When the dry run switch is on, you will have control of how quickly the machine moves (even during rapid movements). A multi-position switch (usually the *feedrate override switch*) is used for this purpose. You'll also have a panic button that will stop all axis motion if you get nervous (called the *feed hold* button).

When you perform a dry run, again, there will be no workpiece in the setup and you'll have complete control of all motions. You'll be checking for very basic and serious mistakes – like collisions, spindle direction, turret stations, and general movements made by the program. If a mistake is found, you'll stop the cycle, diagnose and correct the problem, and do another dry run. You must not continue until you've seen the *entire program* run – and until you understand and agree with the movements made by the program. This may take several executions of the program.

A tip that will save a crash some day

Whenever you activate a cycle by pressing the cycle start button, *always have a finger ready to press the feed hold button.* If you don't have certain switches set properly, the machine will not behave as you expect, and you'll be ready to *immediately* stop the turret's motion when something goes wrong. For example, say you forget to turn on the dry run switch. You have feedrate override turned down to its minimum position, so you're *expecting* the machine to crawl during its first motion. But in reality, it will move at rapid (assuming rapid traverse override is set to one-hundred percent). If you have a finger ready to press the feed hold button, you'll be ready to stop the rapid motion as soon as you see that something is wrong. You won't have to look for the feed hold button – which takes time – time that may allow the machine to crash.

Dry running our example program

We remove the workpiece from the chuck, and turn on the dry run switch. We confirm that the machine is at a safe index position (the turret is far enough away from the chuck to allow it to index). We turn the feedrate override switch down to its lowest setting (usually ten percent), we place the mode switch in the auto or memory position (to run a program). We press the program check display screen key (this is the page that shows the distance-to-go function), and confirm that program number O0010 is the active program.

With a finger ready to press the feed hold button, we press the cycle start button. The first thing that happens is that the turret indexes to station number one – the rough face and turn tool. Then the spindle starts – sure enough, in the forward direction. From the program check page (which also shows absolute position of the cutting tool), we can see that the machine is moving, but barely. As we crank up the feedrate override switch, sure enough, we can see the X and Z axes moving. At first, this first XZ movement is pretty safe at first because the tool is a long way from the chuck. So we crank up the feedrate override switch until the tool gets closer to the chuck.

As the tool gets closer to the chuck, we slow the motion with feedrate override. That's better. Not so scary.

The tool begins its rough facing motion – and looks good. After the facing pass, the turret begins the motion to the safe index position – and we can crank up the motion rate with feedrate override.

The turret indexes to the 1.0 inch drill and the tool begins approaching in X and Z. As the tool gets closer we slow the motion. The tool's motions appear to be correct as it enters the hole and then retracts. It's now on its way back to the safe index position, so we crank up the motion rate.

Now the turret indexes to the rough boring bar. As it moves close, we slow the motion rate. It appears to be boring the hole correctly. Then it retracts and heads for the safe index position.

Next come the finish boring bar and finish turning tool. We use the same techniques with them and everything looks good.

Again, throughout this entire procedure we've kept a finger ready to press the feed hold button – but in our example, it was not required. However, consider this example. Say you're in the process of dry running a tool (like the drill in our example) and the tool continues moving toward the chuck during its approach motion. You're worried, since the tool is in a position that is only two inches or so from the chuck face – so you press the feed hold key. When you look at the distance-to-go display, it shows that the tool is going to move another five inches in the Z minus direction. Something is seriously wrong, and you just saved a crash!

Cautiously run the first workpiece

If the program is correct (you've seen it run on a tool path verification system or it is a program that has run before), if the process and cutting conditions are correct, and if you have made no mistakes during setup (the dry run looks good), you should be able to machine a good workpiece on your very first try.

Admittedly, there are a lot of *ifs* in the previous paragraph. And problems could still exist with *new* programs that you could not detect during the tool path verification and the dry run. If, for example, a small motion mistake is made in a tool path, you may not be able to spot it during the tool path verification. This, of course, will cause the workpiece to be machined incorrectly. This kind of mistake should not exist with proven programs – as long as the *current version* of the program is being used.

There could also still be mistakes in your *setup* that went undetected during the dry run. If, for example, you make a small measurement mistakes during program zero assignment measurement or entry (say, a program zero assignment is off by 0.25 inch), you probably won't catch it during a dry run.

The most dangerous time

You must still be very careful when running the first workpiece. During programming lessons, we recommend using a rapid approach distance of 0.1 inch (about 2.5 mm). While this is a safe and acceptable approach distance, it is difficult to tell during a dry run whether the tool tip truly stops 0.1 inch away from the work surface (since there is no workpiece in the chuck). If even a small mistake is made with program zero assignment, the result could be disastrous. For this reason, *you must be extremely careful with each tool's first approach movement*. This is the most dangerous movement of each tool. We recommend using a special procedure when running the first few commands of each tool. It assumes that the format shown in Lesson Sixteen is used to write the program. This procedure allows you to take full control of each tool's first approach movement.

Once you confirm that the machine is at its safe index position and that the correct program is active, you can follow this *safe approach procedure* to approach with each tool. The first workpiece, of course, must be loaded at this time.

1) Turn on single block and dry run.

2) Set the feedrate override switch to its lowest setting.

3) Press the program check display screen key (this is the page that shows distance-to-go)

4) Repeatedly press the cycle start button until the turret index occurs (because single block is turned on, each time you press cycle start only one command is executed).

5) Press cycle start (the spindle starts).

6) Press cycle start. The cutting tool begins its first approach movement. *This is the dangerous movement.* Be very careful as the tool gets close to the workpiece. When the tool comes to within about 0.5 inch of a work surface, stop the motion by pressing the feed hold button. Look at the distance-to-go display. Confirm that the amount of motion left in the approach movement is less than the distance between the tool tip and the workpiece. As long as it is, press the cycle start button to reactivate the cycle. Let the tool come to its final approach position and stop.

7) Turn off dry run. (*Never* let cutting tools machine a workpiece under the influence of dry run. Dry run tends to slow down rapid motions, but it *speeds up* cutting motions.)

8) The tool has safely approached. What you do next will depend upon whether you're running a new or a proven program. For a new program you must carefully step through the cutting tool's cutting motions. In this case, leave the single block switch on, and repeatedly press cycle start to step through the tool. For a proven program (are you *absolutely sure* that you're running the current version of the program?), turn off single block and press cycle start to let the tool machine the workpiece as it did the last time the program was run.

9) When the tool is finished, the machine will return to the safe index position. While the machine is moving to this position, turn on dry run and, if you turned it off, single block. Repeat the procedure, starting at step four.

For each tool in the program, you must take control of the tool's first approach movement with single block and dry run. Once you have seen the tool approach properly, you need not use this procedure (after running the tool for the first time). But whenever you're worried about a tool's first approach motion, this procedure will keep the tool from crashing into the workpiece.

Making sure the first workpiece is a good one

While there is a lot to think about when you run the first workpiece, it must always be your goal to make the very first workpiece being machined a good one. If there are any tiny motion-mistakes in a new program that you did not detect when you verified the tool path, it may be impossible to machine a good workpiece on your first try. The program must be correct, of course, in order for the first workpiece to be machined properly (as a *proven* program is).

We do consider the running of your first workpiece to be a *setup-related task*. If the first workpiece doesn't pass inspection, we also consider any corrections made to the program as well the running of the *second* workpiece to be setup-related tasks. Indeed, *the machine is still in setup until a workpiece passes inspection* and you can run acceptable workpieces.

Frankly speaking, if the program and process are correct, *there is no excuse for scrapping the first workpiece you machine.* If you use the techniques we recommend, each tool will be *forced* to machine properly – and when the program is finished, each workpiece surface being machined will be within it tolerance limits.

Look at step number one in the procedure shown above for cautiously approaching with each tool. This is the point at which you're going to *begin* verifying each tool's first approach. You must know which tool is going to be used next. If you must, refer to the setup documentation.

At this point in the procedure, you must consider what the cutting tool is going to do – and especially – consider the tolerances for dimensions that the cutting tool is going to machine. If tight tolerances must be held, you must use *trial machining* techniques.

Machining the first workpiece in our example job

Let's go through the entire example job and show how this is done. We'll assume that all five cutting tools have been placed in the turret during this setup. That is, none of the tools have been adjusted in previous setups. All need to be fine tuned in this setup.

Tool one: Tool number one is the rough facing tool. Before running it, we must stop to consider what this tool will be doing. As you know from the process, this tool will be leaving 0.005 inch of finishing stock on all faces and 0.08 inch of finishing stock on all diameters. While these values are not critical (they don't require trial machining), we must ensure that this tool leaves the appropriate amount of stock for finishing. We'll let this tool machine and then determine whether any offset values must be adjusted.

Before running the tool, we'll turn on the *optional stop* switch. Since there is an M01 word in the program at the end of every tool, this will make sure that the machine will stop (including spindle and coolant) when the tool is finished.

We use the safe approach procedure to ensure that this tool does not crash into the workpiece. It approaches properly. Since this is a new program, we leave the single block switch on – and while monitoring the program check page to see upcoming commands in the program and the distance-to-go display, we step through the tool's motions to rough face the workpiece. When the tool returns to the safe index position, the machine stops because the optional stop switch is on. This allows us to check the workpiece.

We measure the 2.5 inch diameter and find it to be 2.582. This diameter is only 0.002 inch larger than it should be at the current time. (Remember, this roughing tool is leaving 0.08 inch finishing stock on all diameters.) While some setup people will find this acceptable, we'll reduce the value of the X register of wear offset number one by 0.002 inches. This will force the 2.5 inch diameter on the *next* workpiece to be precisely 2.58 inch – allowing exactly 0.08 inch of finishing stock.

When we measure the length of the flange (1.75 overall workpiece length minus the 0.25 length of the 2.5 inch diameter), we find it to be 1.5007. It should currently be 1.5005 – again, leaving 0.005 finishing stock on this face. We reduce the Z register of offset number one by 0.002 so this length will come out perfectly on the next workpiece. (Note that you can also work with the overall workpiece length – 1.75 – to determine if this tool is machining properly, but it may be difficult to measure this dimension while the workpiece is in the machine.)

Tool two: This tool drills the 1.0 inch diameter hole. And again, we must stop to consider what this tool will be doing before the turret indexes to this tool (the turret index command also invokes the tool's offset). Since there is nothing critical about how this tool machines, we're going to let it run without modifying any offset values.

As with tool one, we use the safe approach procedure, and the tool approaches properly – stopping 0.1 inch from the face in Z and at the workpiece center in X. After turning off dry run and leaving single block on, we step through this tool's motion and let it machine the hole. After the tool returns to the safe index position, the machine stops.

We check the hole, confirming that it breaks through the workpiece and that it is close to a 1.0 inch diameter. Everything looks good.

Tool three: This tool rough bores the workpiece. And as with every tool, we must stop to consider what this tool will be doing before we allow it to run. Like tool one, this tool will be leaving finishing stock, so it is not critical. We won't trial machine. Instead, we'll let the tool run and then check to see if any adjustments are necessary.

After the safe approach, we step through the tool's motions, letting the tool rough machine the bore. When finished, the tool returns to the safe index position and the machine stops.

We measure the bore diameter and find it to be 1.462. Remember, this tool should be leaving 0.04 inch on all diameters, meaning the 1.5 inch bore diameter should currently be 1.460. Again, the small 0.002 inch deviation is not terrible, but we'll adjust the X value of wear offset number three, reducing it by 0.002 – the rough bore in the next workpiece will be perfect.

When it comes to the bore depth (1.25 according to the print), there are a couple of ways to determine if this tool is leaving the correct amount of finishing stock (again, 0.005 inch). If possible, we can measure the distance between this face and the other end of the workpiece (which is 1.75 overall length minus the 2.25 bore depth – a dimension of 0.5 inch). Leaving 0.005 finishing stock means this dimension should currently be 0.505 inch. But it may not be possible to measure this dimension while the workpiece is still in the machine.

We can also measure the current hole depth, but if we do, we must take into account the fact that the face of the workpiece is also in its rough state – and from tool one, we must remember that there is currently 0.007 inch of finishing stock on this face.

Say we decide to use a depth micrometer and measure the current bore depth. We find it to be 1.252. How much finishing stock is currently on the face of the bore? Since the face machined by tool one currently has 0.002 inch too much finishing stock, and since the current bore depth is 0.002 deeper than the specified hole depth, the current amount of finishing stock is precisely 0.005 – just what we want it to be. As you can see, adjusting depth (Z axis) offsets can be somewhat challenging.

Tool four: This tool finish bores the 1.5 inch diameter hole. And again, we must consider what it will be doing before the turret index. While the diameter machined by this tool is not critical, notice that the depth of this hole has a +/-0.001 tolerance. While this is not an extremely tight tolerance, it is possible to scrap the workpiece if this tool's geometry offset Z register is not correct. So we decide to trial machine for the hole depth.

To do so, we must adjust the Z axis value of offset number four (wear or geometry) by a small amount. Note that there is only 0.005 currently on this face, so the adjustment must be smaller than 0.005 – otherwise this tool will not machine at all. So we increase this offset value by 0.002 inch and let this tool run.

As usual, we use the safe approach procedure. The tool approaches properly and we allow it to step through its motions to machine the bore (again, with single block left on). When the tool returns to the safe index position, the machine will stop.

We measure the 1.5 inch diameter and find it to be 1.502. It's well within tolerance, but we reduce the X value of offset number four by 0.002 to make this diameter come out perfectly on the next workpiece.

For the Z axis (hole depth), we're faced with the same problem mentioned for tool three. If we measure the hole depth, we must consider the fact that there is currently an additional 0.002 on the rough face of this workpiece.

We measure the current hole depth and find it to be 1.255. If the outside face were to have the appropriate amount of finishing stock (0.005), the bore depth would be correct. However, the outside face has 0.002 additional finishing stock, meaning the bore is shallow by 0.002. The Z value of offset four must be reduced by 0.002.

In order to machine this bore to the appropriate depth on this workpiece, we must now re-run tool number four. This time there will be no need for the safe approach procedure – and we won't have to step through the tool's motion (we've seen this tool run once before). But we leave the optional stop switch on so the machine will stop when the tool is finished.

After re-running the tool, we find the current depth of the bore to be 1.257. The tool has machined the hole depth to precisely 1.25 deep (considering that there is still 0.007 left on the outside face).

Tool five: This tool finish faces the workpiece. And again, we must stop to consider what it will be doing before the turret indexes to station five. Notice the tight tolerance on the 2.5 inch diameter (+/-0.0005). We must trial machine to ensure that this diameter comes out correctly. So we increase the X value of offset five by 0.010.

While there is nothing critical about the faces machined by this tool, we must consider the fact that if this tool does not machine faces correctly, the hole depth will not come out to size. So we must also trial machine for the Z axis, increasing the Z value of offset five by 0.002 inch.

After turning single block and dry run back on, we use the safe approach procedure and step through this tool's motions to finish face the workpiece. When the tool returns to the safe index position, the machine will stop.

We measure the 2.5 inch diameter and find it to be 2.508 (good thing we trial machined). So we reduce the X value of offset five by 0.008.

For the Z axis, since the hole depth is critical, we'll use it to adjust the faces machined by tool five. We measure and find the hold depth to be 1.251. (If we measure the 1.75 overall workpiece length, we should find it to be 1.751). Tool five's Z axis offset must be reduced by 0.001 (again, it's a good thing we trial machined).

After re-running tool five, we find the 2.50 inch diameter and the 1.25 hole depth dimensions to be right on size.

Since all dimensions have been machined within tolerance limits, we have a workpiece that will pass inspection.

Move through the program one tool at a time

Notice how our recommended method works. As you machine the first workpiece, you stop to consider what each tool will be doing *before* allowing it to machine. You make the tool machine in an acceptable manner before moving on to the next tool. And when critical dimensions must be machined, this means you must trial machine. With this technique, your first workpiece will be an acceptable workpiece.

You may be tempted to run the entire program before checking anything. This is especially true if workpieces are plentiful and not very expensive – you may be willing to scrap a few workpieces to eventually get offsets properly adjusted. This can be a terrible mistake. Consider what would have happened in our example scenario.

We'll pick just one of the (several) problems. If we would have run the entire program before checking anything, the depth of the 1.5 diameter bore may have come out to 1.252. Which tool (the finish boring bar or the finish turning tool) is not correct? You may elect to make the change to the boring bar's offset, when in reality it is the finish turning tool that should be adjusted. In some cases, both of these tools might need an adjustment – but without checking what each tool has done right after it machines, you'll have no way of knowing.

Additionally, if you don't check what roughing tools have done, you won't know how much finishing stock they are leaving.

I'm amazed by how many experienced setup people don't follow these recommendations. While they do eventually get workpieces to pass inspection (actually they get pretty good at figuring out what must be done after a long period of trial-and-error), they don't have a logical approach.

And remember, the day will come when you're given five pieces of raw material and told to make five good workpieces (you cannot scrap even one workpiece).

Upcoming jobs

Remember that you must follow these methods only with new tools (tools that have just been placed in the turret). Once a tool is machining properly in one job, it will continue to machine properly in upcoming jobs.

In our example, it is likely that tools one and five (rough and finish facing tools) remain in the turret on a permanent basis. For every upcoming job, you can rest assured that they will machine properly in each axis. That is, the roughing tool will leave the appropriate amount of finishing stock on all diameters and faces – the finishing tool will machine diameters and faces within tolerance bands (dimensions machined by finishing tools will be perfect unless the cutting tool has experienced wear).

A note about our example job

The verifying of our example job went pretty well. While we did have to trial machine to get the first workpiece to come out correctly, we did not have to modify the program. In Lesson Twenty-Eight, we'll show another example job – but this time things won't go so smoothly.

First workpiece inspection

Once the first workpiece is machined, it must be inspected to confirm that it is within specifications. In some companies, the setup person is responsible for this inspection. In others, a separate person (an inspector in the quality control department) does the inspection. Regardless of who does the inspections, adjustments must be made if something is wrong, and the setup person is commonly responsible for making these adjustments. If the adjustments involve a program change, the programmer will be called to correct the program.

After the changes and the running of another workpiece, another inspection must be done. You cannot, of course, begin a production run until workpieces pass inspection.

Program optimizing

If your production run consists of running but two or three workpieces with relatively short cycle times, running good workpieces may be your only concern. But if you run very large lots, you must run good workpieces *efficiently*. For beginning setup people, this will require the help of an experienced programmer and/or an experienced setup person. However, many companies do expect their CNC setup people to perform basic program optimizing tasks related to cutting conditions (adjusting feeds, speeds, & depths of cut). More elaborate optimizing changes (minimizing rapid movements and changes in processing) are the responsibility of the CNC programmer.

Saving corrected version of the program

If the CNC program is being run for the first time, and especially if it has been prepared by a novice programmer, it is likely that many changes have been made during the program verification procedures (though our example job did not require any program changes). If you ever expect to see this job again (repeat business), or if you want to maintain records of what you have done, you must remember to save the program back to the DNC system. It is a good idea to update the documentation in the program to specify the fact that a revised program has been created (with messages in parentheses). While you're at it, include messages specifying the program's run time and the date the program is first used.

What will you be doing?

Again, you must confirm what (if any) setup-related responsibilities you will have as a CNC setup person or operator for your company. Note that some of the tasks just described (especially those related to cutting tools) will have to be repeated as tools get dull during the production run.

Production run documentation

With the setup completed and the first workpiece having passed inspection, you're ready to begin the production run. We're assuming at this point that you have more than one workpiece to machine. As with just about every facet of manufacturing, companies vary with regard to how many workpieces they commonly run per job. This will even vary within one company.

Generally speaking, CNC turning centers are most often applied to small to medium sized lots – from one to about one-thousand workpieces. While there are companies that dedicate their CNC turning centers to running one workpiece – day in and day out – the vast majority of CNC using companies, run a variety of jobs of varying lot sizes.

The number of workpieces commonly run has a big impact on how you approach production runs. Indeed, it has a big impact on how companies utilize CNC people. If a company consistently runs small lots of say, under ten workpieces, it won't take much time to complete each production run. Companies in this situation tend to have one person setup the job and run the job out (one person sets up the job and completes the production run). On the other hand, if a company consistently runs larger lots of say, five-hundred workpieces, it will take some time to complete the production run. These companies tend to have one person make the setup and another (lesser skilled person) run out the job.

If one person is doing everything, including programming, setting up, and running production, there won't be much of a need for production run documentation. This person will know what is intended since they planned the entire job. On the other hand, if tasks are divided (one person programs, another sets up, and yet another runs out the job), the need for good production run documentation is much greater.

As you have seen, most companies provide *setup* documentation in the form of a one-page setup sheet – and even this assumes one person is programming for a job and another is setting it up. However, many companies expect their setup people to relate what must be done to run out the job to the CNC operator. Admittedly, many of the tasks related to running production are quite basic – and very redundant if many workpieces must be produced. But adequate production-run documentation is essential if several people will be running out the job (like first, second, and third shift operators).

A note to programmers:

This text is providing you with information to master all three skills needed to use a CNC turning center – programming, setting up, and running production. However, depending upon your company, you may be responsible only for programming. In this case, you must provide adequate setup and production run documentation for other people to make setups and run production.

While you can probably rest assured that the setup person will have adequate basic machining practice skills to understand even minimal documentation, you can make no such assumption about CNC operators. Many companies hire people with limited (if any) basic machining practice skills to run CNC machines. Remember that you must direct your production run documentation to the *lowest skill level* of CNC operator in your company.

Many programmers do a terrible job in this regard. Even if your operators have pretty good basic machining practice skills, remember that they're not going to be nearly as familiar with the job as you are. Many programmers assume too much of their CNC operators. While experienced operators may eventually able to figure out what is expected of them, good production run documentation will minimize the time and effort required to do so.

Figure 7.8 shows the production run documentation for our example job as a one-page form. This form must answer any questions a CNC operator will have about the job. We'll be referring to this form as we describe the tasks related to completing a production run.

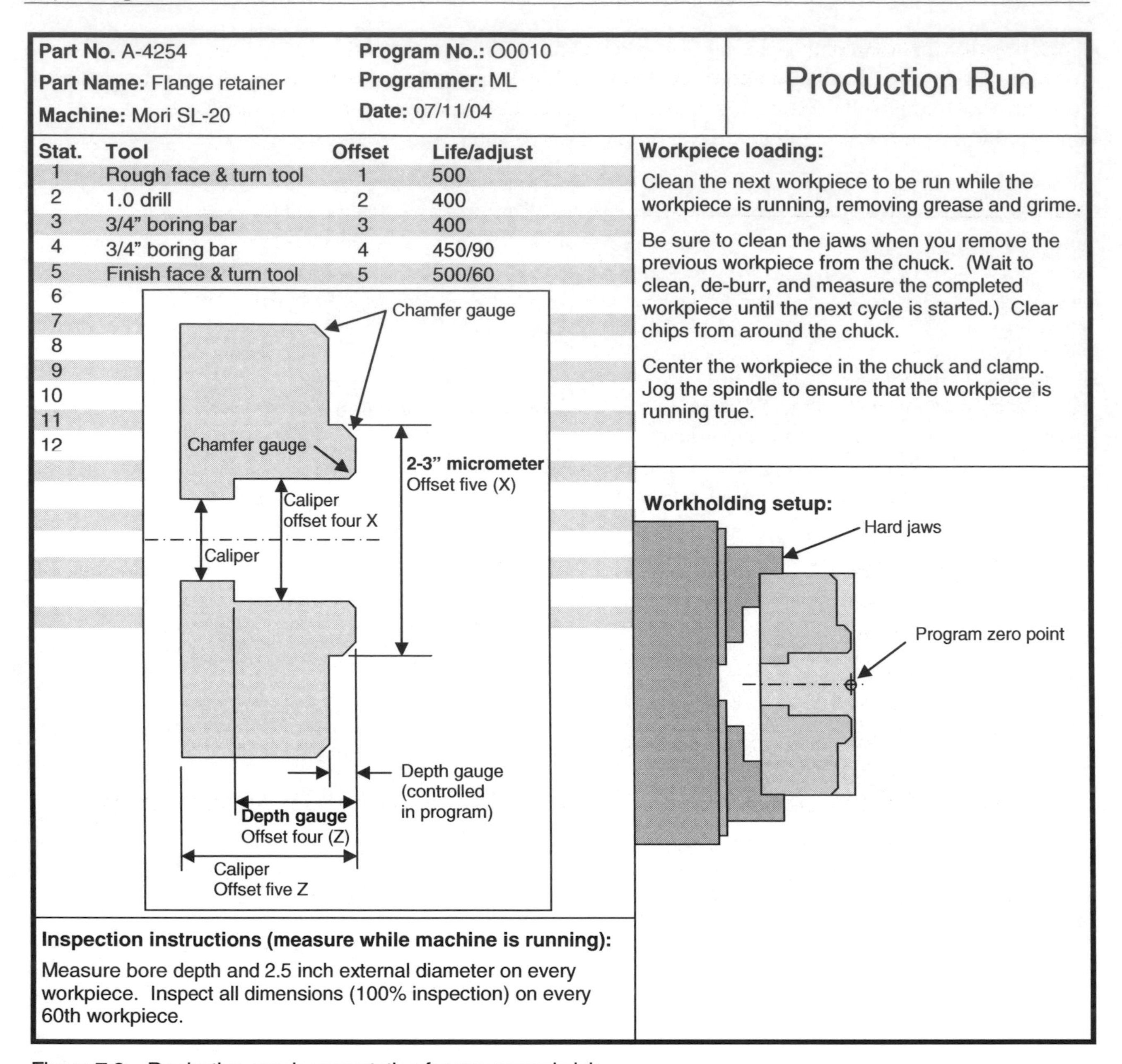

Figure 7.8 – Production-run documentation for our example job

Remove the previous workpiece

The task of workpiece loading begins with the removal of the most recently completed workpiece from the machine. This requires the operator to understand the workholding device being used to secure the workpiece – and should be described in the production run documentation.

For our example job, the operator must remove the workpiece from the chuck, which requires the simple pressing of a foot pedal. The operator must, of course be supporting the workpiece (by hand) when the jaws are opened. Workpiece removal usually requires that the location surfaces of the workholding device (jaws) be cleaned. Any chips that have been gathering around the workholding and work support devices must be cleared.

Note that some operators get into the (bad) habit of beginning to work with the workpiece just removed before loading the next workpiece. While the finished workpiece must be cleaned, de-burred and measured, the operator must set it aside until the machine is in cycle for the next workpiece. Admittedly, some turning center cycles are quite short, meaning they may not be able to complete all of these tasks while the machine is running. But operators must strive to keep as much of this work *off line* as possible.

Load the next workpiece

While loading workpieces is usually pretty simple, the setup documentation should include workpiece loading instructions. Notice that our example production run documentation form includes explicit loading instructions requiring the jogging of the spindle to confirm that the workpiece is concentric with the chuck. Like workpiece removal, workpiece loading requires an understanding of the workholding device being used.

Once the workpiece is loaded, the doors to the work area must be closed.

Activate the cycle

If several workpieces have just been run, this task simply involves pressing the *cycle start* button. The machine will begin running the next workpiece in the normal manner.

But if the operator has been away from the machine for any length of time (at lunch or on break), or if they are just beginning a shift, they must confirm that the machine is truly ready to run a workpiece. This involves checking (and setting) the current condition of several switches on the control panel. While this may not be a complete list, here are some of the switches that must be set.

- Dry run: off
- Single block: off
- Optional stop: off
- Machine lock (if available): off
- Block delete: As requested by programmer (probably off)
- Rapid traverse override: 100%
- Feedrate override: 100%
- Spindle override (if available): 100%
- Mode switch: Memory or Auto
- Display screen: Program mode (and correct active program is shown and cursor is at the beginning of the program)
- Machine position: At a safe index position

Again, with the machine ready to run a program, activating the cycle simply requires pressing the cycle start button. Whenever you press the cycle start button, *always* have a finger ready to press the feed hold button. If the machine behaves in an unexpected manner, you'll be ready to stop it.

Monitor the cycle

This may not be necessary for proven programs that you have seen run many times before. But if you're new to a program, it is a good idea to get familiar with the cycle – especially if you did not program the job or make the setup. This will help you understand the tools that are being used in the job and the general machining order.

For new jobs, you'll also want to confirm that cutting tools are machining properly for several workpieces. Certain cutting tool related problems, especially with speeds and feeds, may not present themselves when the first workpiece is being machined during setup. You'll need to stay alert while running the first few workpieces.

Clean and de-burr the workpiece

Once you have loaded the next workpiece and activated the cycle, you can begin working on the workpiece that has just been removed. It will probably be covered with coolant and debris. And it may have some razor sharp edges, so be careful handling it.

Most companies expect their CNC operators to clean and de-burr the workpieces they produce. Cleaning usually involves wiping the workpiece with a rag or shop towel. If you use any kind of air-blowing system to blow off the workpiece, *be extremely careful* – chips can fly anywhere. While you must wear eye-protection (all shops require it), nothing is protecting your ears, nose, mouth, etc.

While most companies strive to remove sharp edges in the machining cycle, there will almost always be some sharp edges on workpiece you remove from the machine. Again, be careful not to cut yourself as you handle newly removed workpieces.

A variety of hand tools is available to de-burr sharp edges, like files and hole de-burring tools. If any are unfamiliar to you, ask an experienced person to demonstrate them.

For our example workpiece, some of the surfaces may have sharp edges. These surfaces include where the drill breaks through and the face of the bored hole at the 1.0 inch diameter.

Perform specified measurements

Most companies expect their CNC operators to inspect the workpieces they machine. They also expect the results of these measurements to be recorded in some fashion. But companies vary when it comes to specific methods used to measure and record.

Many companies require one-hundred percent inspection on critical dimensions. And they use some kind of *statistical process control (SPC) system* to record the measurements. An SPC system usually incorporates a computer screen and keyboard at the measuring station for data entry.

Taking measurements, of course, requires basic machining practice skills. The operator must know how to use the various gauging tools needed to take the measurements. Production run documentation must specify the gauging tools to be used (as our example documentation does). If you are unfamiliar with the required gauges, you must ask to be shown how to use them. And you must practice with them to ensure that you can take accurate measurements.

For our example job, the gauges required for the two most critical dimensions are shown in bold. Also, the inspection instructions specify that you must measure these two dimensions for every workpiece. With all other dimensions, they must be checked (100%) for each 60th workpiece you produce.

Make offset adjustments to maintain size for critical dimensions (sizing)

The tasks shown so far must be performed in every cycle. The tasks shown from this point are only performed if and when they are required.

The tighter the tolerances a cutting tool must hold, the more likely it will be that it will not machine surfaces within the tolerance band for its entire life (at which time it must be replaced). With small lots, this probably won't present a problem. Every cutting tool will machine properly until the production run is completed.

But with larger lots, the wear a cutting tool experiences may cause the surfaces it machines to change by a small amount. And with tight tolerances, this may place the workpiece in jeopardy. In this case, the CNC operator must make an offset adjustment to keep the cutting tool machining properly.

With our example job, there are two very tight tolerances to hold: the 2.50 inch external diameter and the depth of the 1. 5 inch diameter bored hole. Say we have a large lot of two-thousand workpieces to produce. After the setup is made and the first workpiece passes inspection, each of these critical dimensions will be being machined at its target value (its mean value in our example).

Consider the 2.5 inch external diameter. It has a small +/-0.0005 tolerance and is machined by tool number five. Again, when you begin the production run, it will be coming out right on size – to 2.5 inches. But as the finish turning tool continues to machine workpieces, it will show signs of wear. After sixty workpieces have been machined, you find the 2.5 inch diameter is 2.5004. This dimension has changed, of course, due to tool wear. While this dimension is still within its tolerance band (barely), an adjustment must be made to bring this dimension back to the target dimension. If you are targeting the 2.5 inch mean value, an adjustment of -0.0004 must be made to the X register of wear offset number five.

Notice that the production run documentation tells the operator which measuring tool must be used to measure this diameter (a 2-3" micrometer), the offset that controls the diameter (wear offset five's X register), the approximate number of workpieces that can be produced before an adjustment is made (this can be determined only after gaining experience with the job), and the approximate number of workpieces that can be produced before the tool is dull (again, determined through experience). Armed with this documentation, an operator can easily maintain cutting tools – keeping them machining on size.

Replace worn tools

As with sizing adjustments, this task will only be required with large lots – when cutting tools will wear out before the job is completed.

When a cutting tool must be replaced, the same tasks required during the initial setup must be repeated. These tasks include assembly and possibly trial machining. These tasks are presented during our discussion of setup and will not be repeated here.

Keep in mind that if a cutting tool requires trial machining during the initial setup, it will require trial machining during replacement. Tools four and five in our example job (finish boring bar and finish facing tool) both require trial machining during setup. If these tools are replaced during the production run, they'll require trial machining again to ensure that the surfaces they machine come out to size.

The only exception to this statement has to do with tool touch off probes. If you use a properly calibrated tool touch off probe, during setup or during tool replacement, trial machining should not be necessary. See Lessons Six and Seven for more information about tool touch off probes.

All of this means, of course, that if a CNC operator is responsible for replacing tools during a production run, they must possess many of the same skills possessed by the setup person. For this reason, some CNC using companies do not expect their CNC operators to replace worn tools. Instead, setup people do so.

If possible, production run documentation should specify the expected life for each tool. Again, this requires some experience with the job (or at least, experience with the related cutting tools and the material being machined). In our example documentation, we specify an expected tool life for each tool – so the operator will know if the job can be completed before any tools wear out.

Clean the machine

Most companies expect their CNC operators to keep their machines clean. Every so often (commonly at the end of each shift), the operator will remove all chips from inside the machine. Chips machined during the shift will be dumped (from the chip disposal drum).

Preventive maintenance

Some CNC using companies expect their CNC operators to perform basic preventive maintenance tasks – like maintaining coolant levels, way lube levels, hydraulic oil levels, and filters. If this is required, instructions must be provided that describe the required procedures and their frequency.

Machine warm-ups

If machining extremely close tolerance workpieces, and if the machine is left idle for long periods (off shifts and weekends), you may notice changes in workpiece size during the machine's warm-up period. These changes are caused by thermal expansion of the machine's components (headstock, way systems, etc.) during warm-up. This is a very common problem with turning centers. It is not unusual, for example, for an external diameter to shrink over a series of workpieces by as much as *0.002 inch* during the first hour or so after power-up. With tight tolerances, this can wreak havoc with quality control.

If you notice substantial changes in workpiece size during the machine's warm-up period, you must allow the machine to warm up prior to starting production. Most companies have one CNC operator or setup person come in early to power up all the machines and run a special warm up program which activates the spindle and causes axis movements. Usually a 30-45 minute warm-up period is sufficient to ensure thermal stability.

Key points for Lesson Twenty-Four:

- There are only two general tasks that occur on CNC turning centers – machines are either in setup or they are running production.
- You must know how to perform the tasks needed to make setups – including trial machining.
- You must know how to perform each task that is required to complete a production run.

Practice running the first workpiece

Instructions: Study the following drawing paying particular attention to the dimensions and tolerances. Note that this drawing is not fully dimensioned. Only the dimensions and tolerances that are related to this assignment are provided. Next, study the process that will be used to machine this workpiece. Finally, answer the questions that follow.

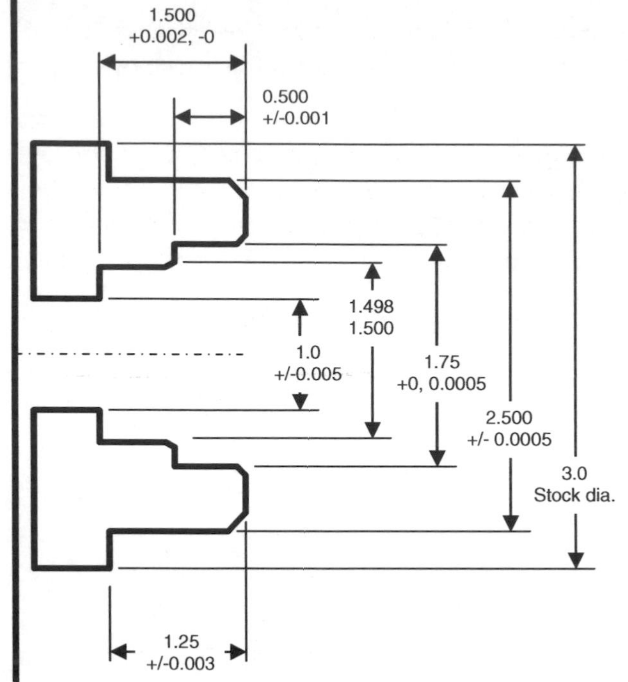

Process used to machine the workpiece:

Description	Tool	Station
Rough face and turn	Rough turning tool	1
Drill 1.0 hole	1.0 drill.	2
Rough bore	¾ boring bar	3
Finish face and turn	Finish turning tool	4
Finish bore	¾ boring bar	5

After assigning program zero, your ***geometry offsets*** look like this:

#	X	Z
1	-9.0223	-10.3371
2	-6.0227	-6.2377
3	-7.2377	-6.5444
4	-9.3665	-10.2336
5	-7.2345	-6.4455

Wear offsets are all currently zero, since you've just assigned program zero for each tool:

#	X	Z
1	00.0000	00.0000
2	00.0000	00.0000
3	00.0000	00.0000
4	00.0000	00.0000
5	00.0000	00.0000

As you've leaned in this lesson, you must consider each tool as you come to it in the program. You must make it cut to size (possibly using trial machining techniques) before going on to the next tool. For each tool in used in this process, answer the following questions.

A) For tool one only (skip this question for other tools): Assuming you're leaving 0.080 finishing stock on all diameters, what should the 2.500 diameter be after this tool machines?

B) For tool three only (skip this question for other tools): Assuming you're leaving 0.04 finishing stock on all diameters, what should the 1.75 diameter be after this tool machines?

C) Does this tool require trial machining? If so, describe how you would handle the related offset to trial machine. Be specific (which offset would you adjust prior to letting the tool cut and how much?).

D) If this tool does not require trial machining (you just let this tool cut), you'll still want to stop the machine when the tool is finished (with the optional stop switch) and check to see that the tool has machined properly. What will you be checking for with this tool?

Answers: **Tool 1:** A) 2.580, C) No, D) Check to be sure that the tool leaves 0.080 on all diameters and the right amount of stock on all faces. **Tool 2:** C) No, D) Check to be sure the drill breaks through. **Tool 3:** B) 1.710, C) No, D) Check to be sure that the tool leaves 0.040 on all diameters and the right amount of finishing stock on all faces. **Tool 4:** C) Yes, D) Increase the X register of offset four by 0.01 and the Z register by 0.002. Let the tool machine, then measure and adjust accordingly. **Tool 5:** C) Yes, D) Decrease the X register of offset five by 0.01 and increase the Z register by 0.002. Let the tool machine, then measure and adjust accordingly.

STOP! Do practice exercise number twenty-four in the workbook.

Lesson 25

Buttons And Switches On The Operation Panels

While there are many buttons and switches on a CNC turning center, you must try to learn the reason why each one exists. If you don't, you may be overlooking a helpful – if not necessary – machine function.

You now know the tasks required to setup and run a CNC turning center. During our discussions of these tasks, we have mentioned several specific buttons and switches. So you already know the function of some of the most important buttons and switches on the machine. But there are still some buttons and switches that you have not yet been exposed to. In Lesson Twenty-Five, we're going to introduce the buttons and switches that are found on a typical CNC turning center.

The two most important operation panels

Most turning centers have at least two distinct operation panels. We'll be calling them the *control panel* (designed by the control manufacturer) and the *machine panel* (designed by the machine tool builder).

For turning centers that have a Fanuc control, the control panel will be remarkably similar from one turning center to the next. There are not many variations, even among different Fanuc control models.

But since the machine panel is designed by the machine tool builder, turning centers that have been manufactured by different machine tool builders will have substantially different machine panels. To compound this problem, machine tool builders can't seem to agree on the machine panel functions needed by CNC setup people and operators.

While we can be pretty specific about the function of buttons and switches found on the control panel, we will be a little vague about machine panel buttons and switches. Also, there may be buttons and switches on your turning center's machine panel that we don't explain in this text. If you find one, be sure to reference the machine tool builder's operation manual to determine the function for the button or switch.

And by the way, *a proficient setup person or operator knows the function of all buttons and switches* on their machine/s. While some may be seldom or never used, you must not consider yourself fully capable of running a machine until you know the function of all buttons and switches. Again, if we don't cover a given button or switch in this text, you must read the machine tool builder's operation manual – or ask an experienced person – to gain an understanding of its use. Don't stop until you know why every button and switch is on your turning center.

The control panel buttons and switches

Figure 7.9 shows the control panel for one popular control model. Again, this operation panel is made by the control manufacturer (Fanuc in our case). For the most part, it is used in conjunction with the display screen (on the left side of this control panel). Notice the power buttons on the left side. The power-on button is usually the *second* function used to power up the machine (the first is usually a main breaker switch that is placed behind the machine). The power-off button is used, of course, as part of the procedure to turn off the machine.

As you know, the display screen is used to display all kinds of important information. We've already introduced the four most important display screen modes (position, program, offset, and program check). In this lesson, we'll discuss them in much greater detail.

On the right side of the control panel is the keyboard. Notice that this particular keyboard does not resemble the keyboard used with personal computers (some control panel keyboards do). With this keyboard, only the functions needed for CNC turning center usage are provided – along with some special functions that you will not find on a computer keyboard.

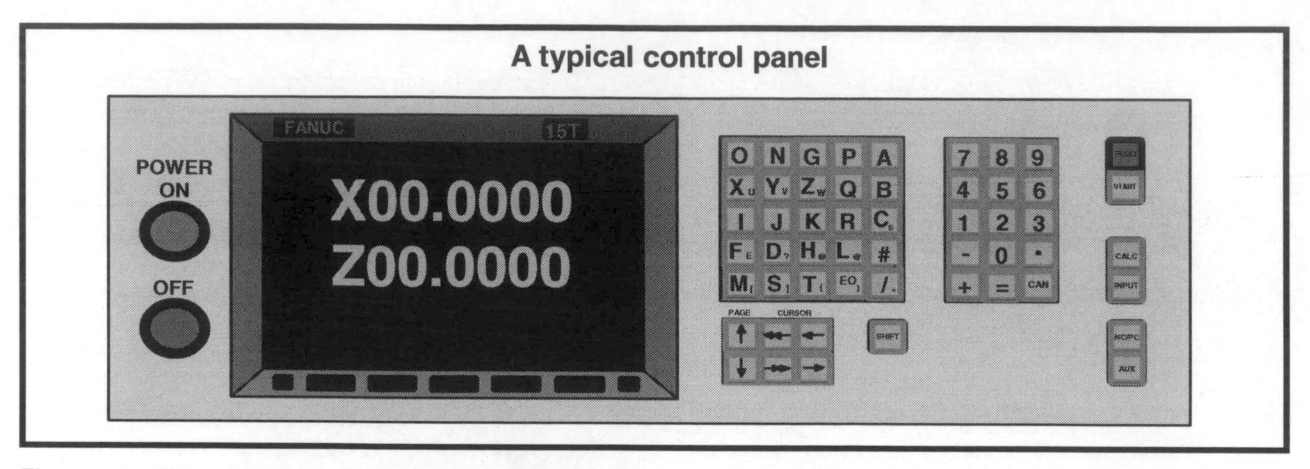

Figure 7.9 – The control panel for a popular control model

Display screen control keys (soft keys)

With most current CNC controls, these keys are located under the display screen (see the seven buttons under the display in Figure 7.9). They are called *soft keys*, and are used to manipulate what is shown on the display screen. They are a little difficult to describe because their functions change from one display page to another (which is why they're called *soft* keys). In most cases, when you press one, the functions of others will change. Their current functions will always be shown at the bottom of the display screen. Look at Figure 7.10 for an example.

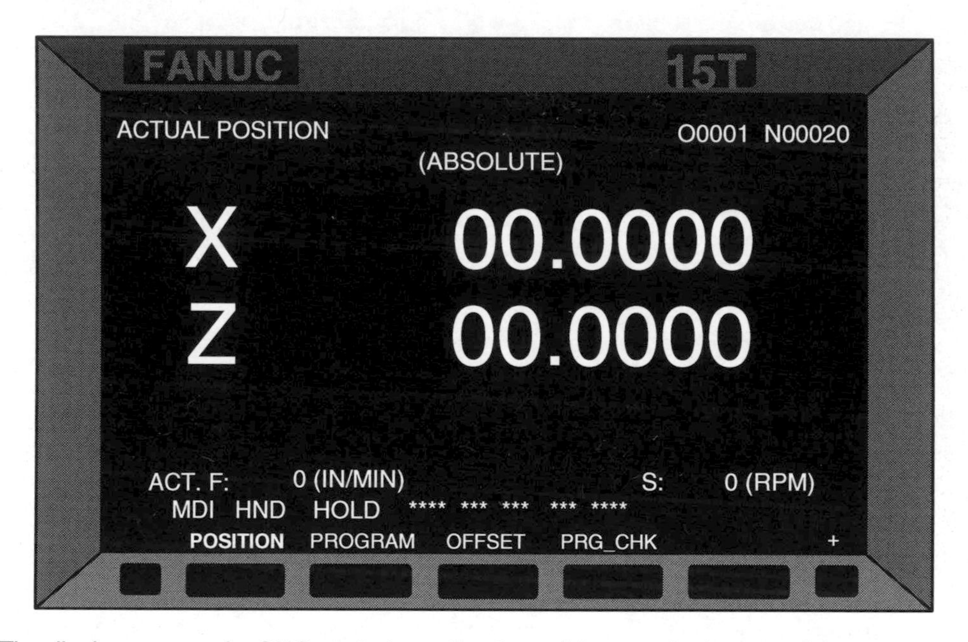

Figure 7.10 – The display screen of a CNC control – notice the soft keys at the bottom of the screen

At the present time, the position mode of the display screen is selected (notice that the word POSITION is bold above one of the soft keys). Right now, other display screen choices are PROGRAM, OFFSET, and PRG_CHK (program check).

Notice the two smaller keys on the right and left end of the soft keys. The key on the left will always bring back your most basic display screen choices (as are shown now). The key on the right (with the plus sign above it) will allow you to see more of what is in the current display screen mode.

While mastering the soft keys on the display screen takes practice, they are not at all difficult to master. Actually, the display screen modes are quite intuitive. Let's look at some of the most important display screen pages.

Position display pages

When you press the **POSITION** soft key, you will be shown the current position display page. With most controls, there are four position display pages, *absolute* (as is shown in Figure 7.10), *relative*, *machine*, and *all* (a combination of all pages along with the distance-to go display). To toggle from one position display page to another, simply continue to press the soft key under **POSITION**. Figure 7.11 shows the various position display pages (except the absolute position display page, which is shown in Figure 7.10).

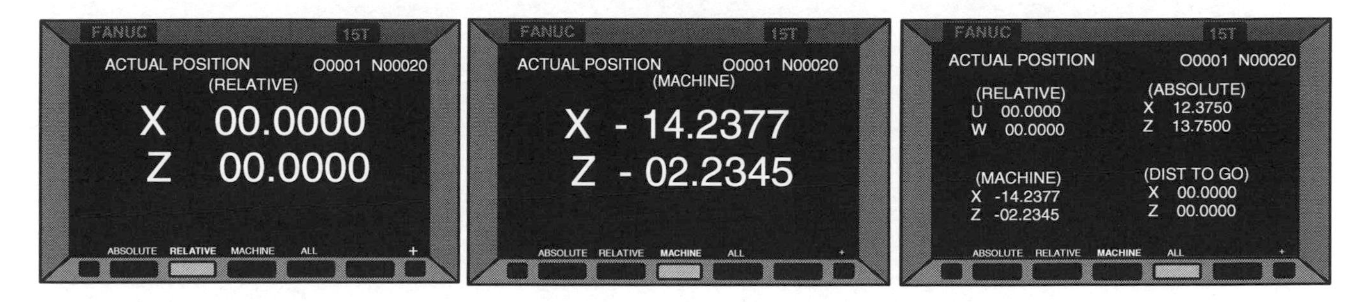

Figure 7.11 – Position display screen pages

If you've read the entire text to this point, you already know the meaning of each position display screen page. The *absolute* position display page shows the current position relative to program zero. The *relative* position display page is used to take measurements on the machine (like program zero assignment measurements), allowing you to set a point of reference for the measurement. The *machine* position page shows the current position relative to the zero return position. And the *all* position page additionally shows the distance-to-go display, which is extremely important when you are verifying programs.

Program display pages

When you press the soft key under **PROGRAM**, you'll be shown the current program display screen page. And again, to toggle through them, simply keep pressing the program soft key. What you'll see on each program display screen page depends upon which mode the machine is in. In the *edit* or *memory* mode, you'll see the currently active program. In the *manual data input* (MDI) mode, you'll see a special page that allows you to enter and execute MDI commands. Figure 7.12 shows the pages available in the program mode

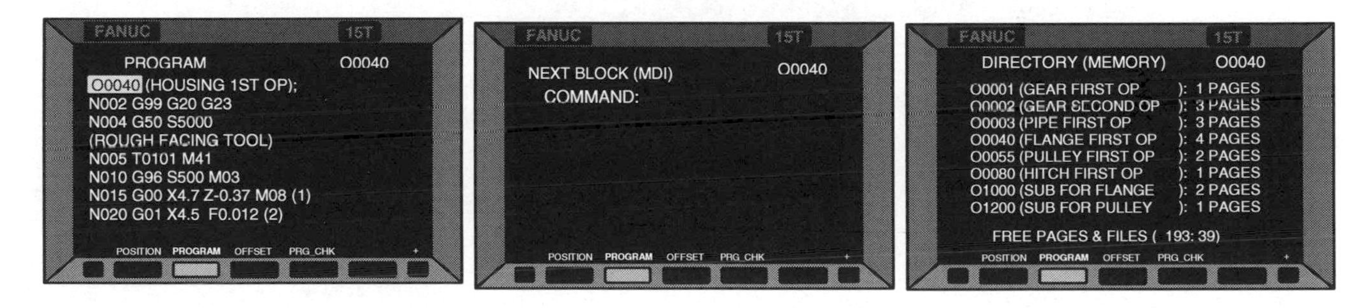

Figure 7.12 – Program display screen pages

The left-most illustration shows what you'll see in the edit or memory mode. The middle illustration shows what you'll see in the MDI mode. And the right-most illustration shows the directory page (also provided in the edit or memory mode), which allows you to see the programs that are currently in the machine.

Offset display pages

When you press the soft key under **OFFSET**, you'll see the current offset display screen page. And keep pressing this button to toggle through the various offset display screen pages. Figure 7.13 show them.

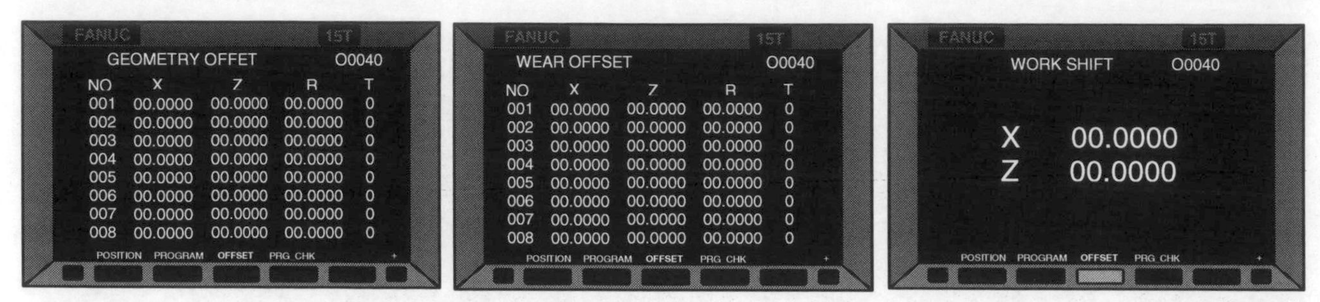

Figure 7.13 – Offset display screen pages

The left-most illustration shows the geometry offset display screen page, which is used to assign program zero. The middle illustration shows the wear offset display screen page, which is used to make sizing adjustments during a production run. (Once again, notice how similar these pages are – you must be careful to ensure that you are looking at the correct page.) And the right-most illustration shows the work shift display screen page, which is used to help assign program zero. Note that if your turning center does not have geometry offsets, only the illustration shown in the middle (wear offset page) will be available.

Once again, if you have read the entire text to this point, you should know the functions of these display screen pages.

Program check display pages

When you press the soft key under PRG_CHK, you will see the program check display screen page, as Figure 7.14 shows.

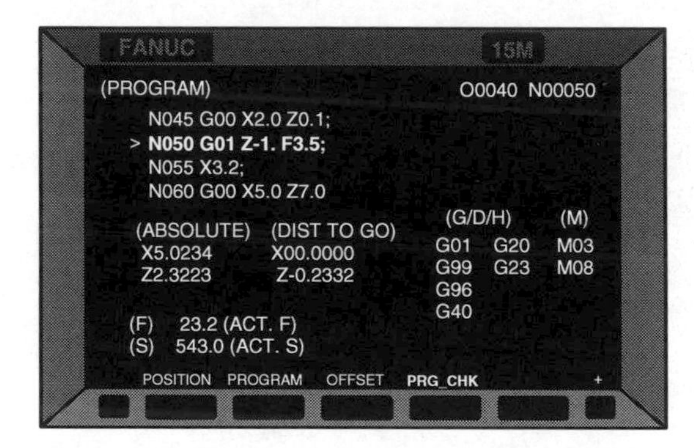

Figure 7.14 – Program check display screen page

As you know, this display screen page is used when you are verifying CNC programs. It shows a portion of the program, including the currently active command. It shows the most important currently instated G and M words, the machine's current position relative to program zero, and the very important distance-to-go display. Again, if you have read the entire text to this point, you know the importance of all of these functions.

Other display screen modes

There are some other display screen modes, but they are not of immediate importance to setup people and operators. From the main display screen page (that shows POSITION, PROGRAM, OFFSET, and PRG_CHK at the bottom, if you press the left-most soft key, you'll see display screen modes labeled SETTING, SERVICE, and ALARM.

The SETTING display screen pages are used to set the default modes for the machine. You can, for example, choose whether the machine will power up in the inch or metric mode from this page. The SERVICE display screen pages will allow you to view and change control parameters. And while the ALARM display screen page will be important to setup people and operators, if the machine does go into alarm state, this display screen page will be *automatically* shown (there is no need to press the soft key under ALARM when an alarm is sounded).

The keyboard

Now let's turn our attention to the keyboard (shown to the right of the display screen in Figure 7.9).

Letter Keys

This keypad allows character entry. Some control panel keyboards allow only those alpha keys needed for CNC programming (N, G, X, Z, etc.), while others provide a full character set (A through Z). With the keyboard shown in Figure 7.9, you must use a special *shift key* in order to gain access to lesser used characters (like parentheses and alphabet letters that are not used with CNC programming).

The slash key (/)

This key allows the entry of the slash code into programs. The slash code is the block delete character. Block delete is described in Lesson Twenty-Three.

Number keys

These keys allow numeric data entry, as is required when entering and modifying offsets and programs. Normally located close to the letter keypad, most CNC controls have these keys positioned in much the same way as they are on the keypad of an electronic calculator.

Decimal point key

This key allows numeric entry with a decimal point. Setting offsets and entering CNC programs are examples of when it will be needed.

The input key

This key is pressed to enter data. Examples of when this key is pressed include entering wear offsets, geometry offsets, and the work shift value. This key is *not* used for entering words and commands into a program. Instead the *insert* key (a soft key that is shown when the display screen is in the program mode) is used fro inserting words and commands.

Cursor control keys

The display screen often shows a *cursor* that designates the current entry position. Examples of its use include when working on the active program or when entering offset data.

How the cursor is moved will vary from one control model to another. With the control panel shown in Figure 7.9, the up and down arrows will move the cursor from page to page. The double arrows left and right will move the cursor from line to line. And the single arrows left and right will move the cursor left and right in the current line.

Program Editing Keys

Programs can be modified right at the machine using these keys. With most controls, they are soft keys that only appear when the display screen is in the program mode (and when the mode switch is set to edit). So you won't find them on the keyboard in Figure 7.9. See Lesson Twenty-Seven for step-by-step procedures for entering and modifying programs.

Insert key

Not to be confused with *input*, this program editing key allows new words and commands to be entered into a program. Most controls will insert the entered word or command *after* the cursor's position.

Alter key

This program editing key allows words in a program to be altered. After positioning the cursor to the *incorrect* word in the CNC program, enter the correct word, and press the alter key.

Delete key

This program editing key allows program words and commands to be deleted. Most CNC controls allow you to delete a word, a command, a series of commands, or an entire program.

Reset key

This very important key has three functions. First, in the *edit* mode, the reset key will send the cursor to the beginning of the program.

Second, in the *memory* or *auto* mode while executing programs, the reset key will clear the *look-ahead buffer* and stop execution of the program. This cancels the cycle's execution and is required when there is something wrong in the program and you wish to stop executing the program. This is common when verifying a new program. *Be careful with the reset key.* Again, the reset key will clear the look-ahead buffer. If the program is reactivated (by pressing cycle start) immediately after the reset key is pressed, the control will skip the commands that were in the look-ahead buffer. This, of course, can lead to serious problems.

Third, when the control is in alarm state, the reset key will *clear the alarm* as long as the problem causing the alarm has been corrected.

The machine panel

As stated, the machine panel is designed and built by the machine tool builder. And machine tool builders vary dramatically with regard to the functions they provide on the machine panel. They vary most with regard how much manual control they provide for machine functions (like spindle activation and turret indexing control). Figure 7.15 shows a typical machine panel for a turning center. As you look at this operator's panel, notice how many of its functions have been introduced in previous lessons.

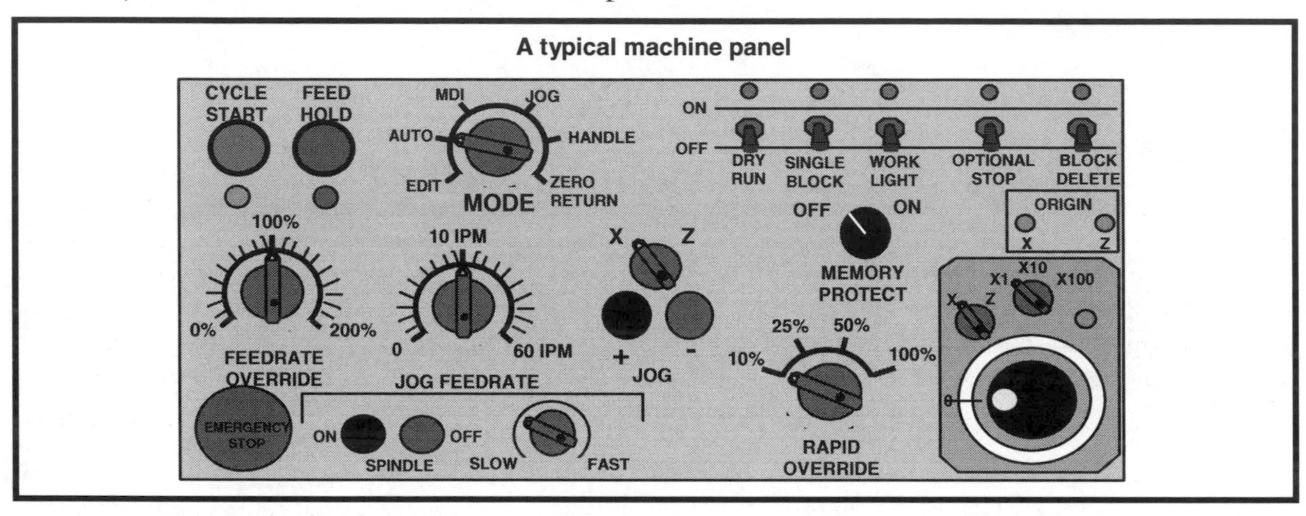

Figure 7.15 – A typical machine panel

Mode switch

The mode switch is the heart of a CNC turning center. It is so important that we devote an entire Key Concept to addressing its use. It must be *the very first switch you set when performing any function* on the machine. The mode switch must be positioned properly in order for the desired function to be performed. If it is not, the machine will not respond to your action. Again, we discuss the mode switch in great detail during Lesson Twenty-Six.

Cycle start button

This button has two functions. First, it is used to activate the active program when the machine is in the memory or auto mode. Second, the cycle start button can be used to activate manual data input (MDI) commands. We recommend that you to get in the habit of having a finger ready to press the feed hold button (which is usually in close proximity to the cycle start button) whenever you press the cycle start button.

Feed hold button

While a program (or MDI command) is being executed, this button allows you to halt axis motion. The cycle start button can be used to reactivate the cycle. Note that all other functions of the machine (coolant, spindle, etc.) will continue to operate even when the machine is in feed hold state.

You should think of feed hold as your *first panic button.* You should *always* have a finger on this button whenever you press the cycle start button. You should also keep a finger ready to press the feed hold button for the entire time that you are verifying a program.

Some people feel the *emergency stop button* should be considered the panic button. However, the emergency stop button will actually turn the power off to the machine tool, which sometimes causes more problems than it solves. For example, when the emergency stop button is pressed and the machine power is turned off, the axes of the machine will drift until a mechanical break can lock them in position. Axes bearing a great deal of weight (like the X axis of most slant bed turning centers) are most prone to drift. While the amount of drift is usually quite small (less than .010 inch in most cases), if a cutting tool is actually machining a workpiece when the emergency stop button is pressed, the X axis drift could cause damage to the tool and workpiece.

Feedrate override switch

This multi-position switch usually has two functions. Under normal circumstances, this switch allows you to change the programmed feedrate during cutting motions (like G01, G02, and G03). Notice we said *feedrate during cutting motions*. Under normal conditions, this switch allows no control of *rapid motion*. The feedrate override switch usually allows the programmed feedrate to be changed in 10% increments and will usually range from 0% through 200%. This means the operator can actually stop cutting motions (at the 0% setting) at one end of the feedrate override switch and double the programmed feedrate at the other.

If the program is written correctly, the entire cycle should run at 100% feedrate, meaning once a program is verified, you should be able to set this switch at 100% and leave it alone. But when you are verifying a program, running the very first workpiece, this switch can be very helpful to confirm that the programmed feedrate for each tool is correct.

For example, say a cutting tool has been cautiously brought into its first cutting position under the influence of single block and dry run (as is described in Lesson Twenty-Four). Now the cutting tool is ready to make its first machining motion. If you are in doubt as to whether the cutting conditions are correct (feeds and speeds), you can turn the feedrate override switch down to its lowest value. With single block still on but with dry run turned off, you can press the cycle start button to allow the tool to begin its cutting motion. As the tool begins to machine the workpiece, you can slowly crank up the feedrate override switch – shooting for the 100% setting. If the 100% position cannot be achieved – or if you end up with a position greater than 100%, the programmed feedrate must be changed. Again, a correct program will let you run the entire cycle at the 100% setting.

Most machine tool builders use the feedrate override switch for a *second* purpose. When the dry run switch is turned on, the feedrate override switch is used to take control of the motion rate for all movements (including rapid motions). This function is necessary during program verification, when you're running the program for the first time without a workpiece to check for setup-related mistakes. Some machine tool builders use the *jog feedrate switch* for this purpose.

Rapid traverse override switch

As the name implies, this function is used to slow the machine's rapid rate. As you know, the rapid rate on current model turning centers is very fast (some machines rapid at well over 1,500 inches per minute), and this switch brings some welcomed control over the very fast rapid rate.

With most machines, the rapid traverse override switch is a multi-position switch (as is the case with the one on our example machine panel). But we have seen some that are simple on/off switches. When this switch is on, rapid rate is slowed to about twenty-five percent of its normal rate. When it is off, rapid motions will at 100%.

With many machines, the rapid override switch has four settings, F0, 10%, 25%, and 100%. At the F0 setting, the machine will move at about 1% of its rapid rate when rapid movements are commanded.

Generally speaking, the rapid traverse override switch is used for program verification. It's nice to leave it at 10% or 25% while you're verifying programs. You can rest assured that the machine will never achieve its true rapid rate. This will give you more time to react should something go wrong.

Emergency stop button

This button will turn power off to the machine tool. Usually, power remains turned on to the control. See the description of the feed hold button for more information about how emergency stop is used.

Conditional switches

There are several on/off switches on the machine panel that control how the machine behaves during automatic and manual operation. They could be toggle switches (as our machine panel shows), locking push-buttons, or lighted buttons (a light within the button comes on when the function is on).

Frankly speaking, these functions are more related to the control panel than they are the machine panel, so they remain amazingly similar from one turning center to another, regardless of the machine tool builder. These switches are very important. If one improperly set, the machine will not perform as expected. You must get in the habit of checking each of these switches before a CNC program is executed.

Dry run on/off switch

As described in Lesson Twenty-Four, this conditional switch is used during program verification. When this switch is on, it gives you full control of the motion rate for all movements the machine makes (including cutting movements and rapid movements).

When the dry run switch is turned on, another switch (usually feedrate override) will act as a rheostat – and will allow you control of all motion rates. Dry run will slow rapid motions, but it will tend to speed up cutting motions. This means you should *never allow a cutting tool to machine a workpiece when dry run is turned on* (you'll have no idea as to what the actual feedrate is).

Single block on/off switch

As described in Lesson Twenty-Four, this conditional switch will force the machine to stop after it executes each command. To reactivate the cycle (execute the next command), you must push the cycle start button.

This switch is helpful during program verification. With a new program, you must cautiously check each motion the machine makes, especially movements that have a cutting tool approach the workpiece. With single block turned on, you can rest assured that the machine will stop at the end of each command, giving you a chance to check the motion just made and to check the program to see what is going to happen next. It is commonly used with the distance-to-go display, which shows how much further the machine is going to move in the current command.

Block delete on/off switch (also called optional block skip)

The applications for block delete are described in Lesson Twenty-Three. The block delete switch works in conjunction with slash codes (/) in the program. If the control sees a slash code in the program, it will look to the position of the block delete switch. If the switch is on, the control will ignore the words to the right of the slash code. If the block delete switch is off, the control will execute the words to the right of the slash code. This function is used to give the operator a choice between one of two possibilities. (See Lesson Twenty-Three to learn more about block delete applications.)

Optional stop on/off switch

This conditional switch works in conjunction with an M01 word in the program. When the control sees an M01, it looks to the position of the optional stop switch. If the switch is on, the control will stop the execution of the program and turn off certain machine functions (like spindle and coolant). You must press the cycle start button to reactivate the program. If the optional stop switch is off, the control will ignore the M01 and continue executing the program.

In the programming format shown in Lesson Sixteen, we recommend placing an M01 in the program at the end of each tool. This gives you the ability to stop the machine at the end of each tool by simply turning on the optional stop switch. Stopping at the end of each tool is important during program verification (as described in Lesson Twenty-Four) to let you check what one tool has done before going on to the next.

Buttons and switches for manual functions

The machine panel for all CNC turning centers will include some buttons and switches related to manual control of the machine's functions. These buttons and switches vary dramatically from one builder to the next. It seems no two builders can agree on what manual functions a setup person or operator needs.

Axis jogging controls
On our example machine panel (Figure 7.15), notice the multi-position switch labeled jog feedrate, the two push-buttons (plus and minus) and the axis selector (X and Z). These functions are active when the machine is in a manual mode (jog or zero return on the mode switch). To jog an axis, after selection a manual mode, you will first select the axis to be jogged (with the X/Z switch). Then you will select the motion rate (with the jog feedrate switch). Finally, you will press the desired direction button (plus or minus). Some turning centers utilize a joystick instead of pushbuttons to activate motion. As you press the button or hold the joystick, the selected axis will move in the selected direction and at the selected feedrate. This, of course, requires that you know which axis you want to move and which way is plus and which way is minus (as is described in Lesson One).

Jogging an axis is often required. Applications include measuring program zero assignment values, machining soft jaws, sending the machine to its zero return position, and moving an axis into a convenient position so a cutting tool can be inspected or replaced.

Positioning is not very precise with jog functions. For this reason, most machine tool builders also provide a *handwheel* on the control panel.

Handwheel controls
On the example machine panel, notice the handwheel in the lower-right corner of the panel. The handwheel is active when the mode switch is set to handwheel (with many machines, a light next to the handwheel will come on when the handwheel is active). Also notice the axis selector switch (X/Z) and the rate selector switch (X1, X10, and X100).

To use the handwheel, after you select the handwheel mode, you first place the axis selector to the desired position (X or Z). Then you select the motion rate. In the X1 (times one) position, each increment of the handwheel is 0.0001 inch (or .001 mm), and motion is very slow – barely detectable except with the position displays. In the X10 (times ten) position, each increment is .001 inch (or 0.010 mm), and motion is faster – easily detectable. In the X100 (times one hundred) position, each increment is 0.010 inch (or 0.100 mm), and quite fast.

Using the handwheel is often required after using the jog function to manually move an axis. Again, the jog function does not allow precise positioning control. So you will commonly use jog to bring an axis close to its intended position, and then use the handwheel for the rest of the motion. Applications include measuring program zero assignment values and machining soft jaws.

Spindle control
Manual spindle control is required, for example, when measuring program zero assignment values if your machine does not have a tool touch off probe. It is also required when manually machining soft jaws.

Some turning centers do allow manual control of the spindle. With our example machine panel, for example, you have a speed selector rheostat and buttons to start and stop the spindle. But notice that you cannot select the spindle direction (with this machine panel). Some machine panels don't provide *any* manual spindle control.

For any machine function that you must activate manually – but for which there are no manual controls – you must use the manual data input mode (described in Lesson Twenty-Six).

Turret index control
Notice that there are no manual controls for turret indexing on our example machine panel. Some machine tool builders do provide a multi-position switch to select the turret station and a push-button to activate the turret index. You will have to manually index the turret on a regular basis (when changing or indexing inserts, for example).

For any machine function that you must activate manually – but for which there are no manual controls – you must use the manual data input mode (described in Lesson Twenty-Six).

Indicator lights and meters

Most CNC turning centers have a series of lights and meters that allow the operator to monitor the condition of important machine functions.

Spindle rpm and horsepower meters

Most turning centers have two meters that show you key information about the spindle. The first meter is the rpm meter, which shows you how fast the spindle is currently rotating. Some machines show spindle speed on the display screen (on the program check page). The second spindle-related meter is a *load meter*. If the machine has this meter, you will be able to monitor the amount of stress on the spindle drive system. Usually this meter shows a percentage-of-load, ranging from 0% through 150%. This means you can easily tell to what extent a machining operation is taxing the spindle of the machine.

Axis drive-motor horsepower meter

This meter, if equipped, allows the operator to see how much horsepower is being drawn by any of the machine axes drive motors. Usually there is only one meter and you must select which axis (X or Z) you wish to monitor by a two-position switch. Like the spindle horsepower meter, this meter allows you to see how much stress is on a drive motor during a machining operation.

Cycle indicator lights

Most CNC machines have two indicator lights to show whether the machine is in cycle. One is above (or close to) the cycle start button and will stay on as long as the machine is in cycle. The other is above (or close to) the feed hold button and comes on when the machine is in feed hold state (when you press the feed hold button).

Zero return position indicator lights

All turning centers have indicator lights that come on when an axis is at its zero return position. These lights are commonly labeled *axis origin lights*. If a program is planned to start from the zero return position, these lights can be helpful to an operator. They will tell you whether it is safe to activate the cycle.

Optional stop indicator light

Some turning centers have an indicator light that is close to the optional stop switch. If the machine has been halted by an optional stop in the program (M01), this indicator light comes on to tell you why the machine has halted operation.

Other buttons and switches on the machine panel

Remember that the machine panel on your particular machine may have more buttons and switches than we have described – especially if the machine is equipped with special accessories (like a bar feeder, tool touch off probe, tailstock, or steady rest). When you come across a button or switch that you don't recognize, be sure to reference the machine tool builder's operation manual to find out what it does.

Other operation panels on your turning center

Many machine tool builders include additional operation panels. Because a bar feeder is mounted quite a distance from the main operation panels, for example, many machine tool builders place a special control panel right next to the bar feeder.

STOP! Do practice exercise number twenty-five in the workbook.

Key Concept

8

Know The Three Basic Modes Of Operation

When it comes right down to it, every button and switch on the machine can be divided into one of three categories. It is related to manual mode, manual data input mode (MDI), or program operation mode.

Key Concept Number Eight is a short, one-lesson Key Concept:

26: The three modes of operation

We mentioned the mode switch in Lesson Twenty-Five. You know it is the most important switch on a CNC turning center – and it is the first switch you should set when you perform any operation on the machine.

Lesson 26

The Three Modes Of Operation

The most common operation mistake is having the mode switch in the wrong position. Fortunately, this mistake will not cause serious problems. The machine will not respond to your action.

You know the mode switch of a turning center is a multi-position switch. The actual positions of a typical mode switch include edit, memory (or auto), jog, zero return, and handwheel. Though most mode switches have at least five actual positions, there are really only three basic modes of operation. Helping you master these three modes will be the focus of Lesson Twenty-Six.

The manual mode

In the manual mode, a CNC turning center behaves much like a conventional engine lathe. The manual mode positions of the mode switch include *jog* (sometimes called *manual*) mode, *handwheel* mode, and *zero return* mode.

You know from previous lessons that you will often need to perform manual functions. Examples include measuring the program zero assignment values, when machining soft jaws, and when replacing dull tools.

With the manual mode, you will press a button, turn a handwheel, or activate a switch that will cause an *immediate response* from the machine. An axis will move, the spindle will start, the coolant will come on, or some other machine function will respond to your action.

As you know from Lesson Twenty-Five, the *jog* mode switch position will allow you to manually move a selected axis. After selecting the jog mode, you select the axis to move (X or Z), select the motion rate (with the jog feedrate switch) and press a button or hold a joystick that corresponds to the direction you want the axis to move (plus or minus). The machine will immediately respond by moving the selected axis at a selected motion rate in the selected direction.

In similar fashion, the *handwheel* mode switch position will allow you to move an axis with the handwheel. Again, after placing the mode switch to the handwheel position, you select the axis to move (X or Z), select the rate of motion (with **X1**, **X10**, or **X100**), and turn the handwheel. The machine will respond by moving the selected axis.

In the *zero return* mode switch position, you can manually send the machine to its zero return position. Turning centers vary when it comes to the actual procedure needed to send an axis to its zero return position. With most machines, after placing the mode switch to the zero return position, you select the axis to be zero returned (X or Z) and then press the plus button. Hold it until the axis reaches its zero return position (and the axis origin light for the axis comes on).

In any of the manual mode switch positions, certain other buttons and switches may be active. Some turning centers allow manual control of the spindle and coolant. Some allow manual control of the turret index function. The related buttons and switches will be aptly named and placed on the machine panel.

But as you know from Lesson Twenty-Five, there are usually some machine functions that you do need to control manually that the machine tool builder has not provided buttons or switches to control. The turret index function is a classic example. Many turning centers do not allow you to manually index the turret.

For functions that you need to activate manually but for which you have no manual buttons and switches, you must use the manual data input (MDI) mode switch position.

The manual data input mode

This mode includes two positions on the mode switch, the *edit* position, and the *manual data input* position (MDI). With both positions, the operator will be using the display screen and keyboard to enter data.

Though these two mode switch positions have substantial differences, we consider them together for two reasons. First, both mode switch positions provide manual capabilities that can be done in a more automatic way. With the edit mode switch position, an operator can enter CNC programs into the controls memory. This can also be accomplished by loading the program a distributive numerical control (DNC) system. With the MDI mode switch position, CNC commands are entered and executed just like CNC program commands.

Second, both mode switch positions involve entering information through the control panel keyboard. With the edit position of the mode switch, a program is entered or modified with the keyboard. With the MDI position, CNC commands are entered and executed with the keyboard.

The manual data input (MDI) mode switch position

Manual data input (MDI) mode switch position is used for two reasons. First, it is used to perform manual functions that cannot be done by any other means. Again, machine tool builders vary with regard to what can be done in a completely manual manner. For those machine functions of which you have no manual control, you must use the MDI mode.

Second, MDI mode can be used to perform certain manual function of which you do have manual control, but can do faster or easier in the MDI mode. A manual zero return on most machines, for example, takes much more time and effort to complete than a zero return done in the MDI mode.

Commanding an MDI zero return

The MDI mode requires that you know the CNC command to activate the manual function you want to perform. This is the same command used in a CNC program to activate the manual function. If, for example, you want to use the MDI mode to command a zero return in both axes, you must know the program command needed to do so, which happens to be

```
G28 U0 W0;
```

When this command is entered and executed in the MDI mode, the machine will rapid to the zero return position in both axes (simultaneously).

Notice the semi-colon (;) at the end of the G28 command example. This is the character that most (Fanuc) controls use to represent an *end-of-block*. There is a key on the keyboard labeled EOB, which stands for end-of block. When you enter an MDI command for most controls, it must end with the end-of-block character. With these control models, when you are finished entering an MDI command, you must remember to press the EOB key prior to inserting the command into the MDI buffer.

If you want only the X axis to be sent to the zero return position (which is the tool change position on most vertical turning centers), the command will be

G28 U0;

The complete procedure to give an MDI command

The step-by-step procedure to use the MDI mode with current model Fanuc controls is as follows:

1) Place the mode switch to MDI

2) Press the soft key under PROGRAM (the MDI display screen page is shown)

3) Using the keyboard, enter the MDI command you wish to execute (for the zero return, this is G28 U0 W0). There is no need for spaces between each word.

4) Press the EOB key (this places a semi-colon at the end of the command).

5) Press the soft key under *Insert* (as soon as you started typing the MDI command, the soft keys changed to include the insert key). When you press the insert soft key, the command moves up into the active MDI buffer.

6) Press the *start button* on the control panel keyboard or the *cycle start button* on the machine panel. Because the cycle start button is closer to the feed hold button, and since we recommend that you always have a finger ready to press feed hold when you execute an MDI command, you should get in the habit of using the *cycle start button* to activate MDI commands. As soon as you press this button, the machine will perform the commanded action. In our example, the machine will rapid to the zero return position in both axes.

Commanding an MDI turret index

We've mentioned several times to this point that some turning centers provide no manual control of the turret index function. If you want to manually index the turret with these machines, you must use the MDI mode to do so. Just as in a CNC program, however, the machine must be at a safe index position. And again, you must know the CNC word related to your machine's turret. For most turning centers, a four-digit *T word* causes the turret index. The first two digits specify the turret station number and geometry offset number. The second two digits specify the wear offset number. Since you won't need a wear offset when commanding a manual turret index, you should specify two zeros as the second two digits. The command

T0400;

cause the turret to index to station four. Again, notice the last two zeros. This ensures that a wear offset will not be instated.

Commanding spindle activation with MDI

While some turning centers provide adequate manual control of the spindle, others do not. If yours does not, you can use the MDI mode to activate the spindle. As you know from the programming lessons, G96 specifies constant surface speed mode and G97 specifies rpm mode. Most applications for starting the spindle in the manual data input mode require that you specify an rpm, meaning you'll be using G97. An *S word* is used to specify the desired speed (in rpm if G97 is selected). An M03 will start the spindle in the forward direction (needed for right-hand tools) and an M04 will start the spindle in the reverse direction (needed for left-hand tools). When you want to stop the spindle, an M05 must be commanded. The command

G97 S700 M03;

will start the spindle at 700 rpm in the forward direction. The command

M05;

will stop the spindle. Again, enter these commands in step three of the procedure shown above.

Other times when MDI is used

Again, any time you need to perform a manual function, you can use MDI to do so. Any machine function can be activated in MDI mode – as can any G code. Truly, if a command works in a program, it will work in the MDI mode. Some examples include

- Coolant (M08 and M09)
- Low and high spindle range (M41 and M42 with most machines)
- Door open and door close (if the machine has automatic doors)
- Chip conveyor on and off (if the chip conveyor is programmable)
- Switching between inch and metric modes (G20 and G21)

Can you make motion commands with MDI?

The MDI mode can even be used to machine a workpiece. Since just about all CNC commands that run in a program will work in the MDI mode (including G00, G01, G02, and G03), you can make machining commands in the same way they are commanded in a CNC program. You can even enter several commands at a time – just remember to enter the EOB key at the end of each command.

However, you must be extremely careful when using MDI to command axis motion. While some operators get very good at making motion commands with MDI, your command/s will be executed just as you enter them. If you make a mistake while entering a CNC command in the MDI mode, the results could be disastrous. You will have no chance to verify your MDI commands as you can with a CNC program. And MDI commands cannot be saved. As soon as you press the cycle start button to execute MDI commands, they will be lost.

The edit mode switch position

With the *edit* mode switch position, you can enter new programs (which can be time-consuming) or modify programs that are currently in the machine's memory. Aptly named, editing functions include *insert*, *alter*, and *delete*. *Insert* allows you to enter new words and commands into a program. *Alter* allows you to modify words in the program. And *delete* allows you to delete a word, a command, a series of commands, or an entire program.

Some machines have a feature called *memory protect* that can be used to keep the operator from making changes in a program. A special *key* (like the key to the door of your home) is used to turn this function on and off. If the memory protect function is turned on and the key is removed from the machine, you will not be able to modify programs.

In order to modify programs, you must understand programming words and commands. If you have read the programming lessons in this text, you should have a good understanding of programming. Without this understanding, an operator cannot make safe and correct changes to a CNC program (which is the reason many companies use the memory protect function – they have CNC operators that are unfamiliar with programming).

The step-by-step procedures to use program editing functions is shown are Lesson Twenty-Three. But we want to give a few examples.

All program editing procedures begin with:

1) Place the mode switch to edit.

2) Press the soft key under program (the active program is shown). Be sure the active program is the one you want to edit.

3) Turn off the memory protect function (if the machine has this function). This requires a key.

You will notice a cursor somewhere on the page (a highlighted or back-lit word in the program). Before performing any editing function you must to move the cursor to the desired location. One way to do so is with the cursor control keys (described during the control panel discussions in Lesson Twenty-Five). Cursor control keys consist of a series of arrow keys that allow you to move the cursor – one word, one command, or one page at a time.

Another way to move the cursor is to use the *forward-search* and *backward search* functions. You can type the CNC word to which you want the cursor positioned and then press the soft key under forward-search or backward-search (depending upon the cursor's current position). The machine will bring the cursor to the next occurrence of the word you typed. Again, procedures for editing programs, including cursor movement, are shown in Lesson Twenty-Seven.

Once the cursor is at the desired position, you will be able to insert a new word or command after the current cursor position, modify the word on which the cursor is placed, or delete words or commands starting from the cursor's current position.

Say, for example, the cursor is currently at the beginning of the program (on the O word). You want to change the first feedrate word in the program to F0.014. If you can see the F word on the current page of the display screen, the easiest way to position the cursor may be to use the arrow keys. Or you could type F and press the forward search soft key. The control will scan to the next occurrence of an F word – which is the F word you want to change. With the cursor on the F word to be changed, type F0.014 and press the *alter* soft key. The feedrate word will be changed to F0.014.

To make a program in memory the active program (to call up a program)

We have been referring to the *active program*. CNC controls can hold several programs, but only one of them will be the active program. A program must be the active program in order for you to modify it or run it. To make a program the active program, first follow the three steps above. Then:

4) Type the letter address O (letter O, *not* number zero) and the program number for the program in memory that you want to make active.

5) Press the soft key under forward search. As long as the program you've typed is in the machine's memory, it will be shown on the display screen. It is now the active program. If the program number you've typed is not in the machine's memory, an alarm will be sounded.

To enter a new program

Entering programs using the display screen and keyboard is time-consuming. Hopefully your company has a distributive numerical control (DNC) system to eliminate this time consuming task.

First, follow the three steps given above. Then:

4) Press the letter address O key (letter O, *not* the number zero) and type program number to be entered.

5) Press the soft key under insert.

6) Press the EOB key and press the soft key under insert.

7) Type the first command of your program followed by the EOB key and then press the soft key under insert.

8) Enter the rest of the commands in the program, ending each by pressing the EOB key and then the soft key under insert.

What if I make a mistake when typing?

There will always be some kind of back-space key. With most Fanuc controls, it is the *cancel* key. When you press it while entering a command, the cursor on the command entry line will back up one space, deleting the last character you typed.

Remember that a CNC control makes a very expensive typewriter. Many machines cannot be running a workpiece at the same time a program is entered (unless the machine has a feature called *background edit*). Even with those controls that do have background edit, it is somewhat cumbersome to enter programs while the machine is running production. Also, the control panel for most machines is not mounted in a comfortable position (most are mounted in a vertical attitude). Your arm will soon get tired when typing in this position.

The program operation mode

The third mode of operation involves actually *running programs*. With current CNC controls, there is only one program operation mode. It will be called *memory*, *auto*, or *automatic* mode on the mode switch. The machine must be in this mode in order to run the active program in memory. Note that older machines have another program operation mode switch position called *tape mode*. This mode switch position only applies to machines that have a *tape reader*. No CNC machines made today do.

When in the memory or auto mode, the *cycle start button* is used to activate the program and *feed hold button* can be used to stop axis motion at any time during the cycle. (Again, keep a finger ready to press the feed hold button whenever you press the cycle start button.) When the cycle start button is pressed, the control will begin executing the program from the cursor's current position.

Several conditional switches (discussed in Lessons Twenty-Four and Twenty-Five) determine how the machine will behave in the program operation mode. The *dry run* switch gives the operator control of the motion rate. *Single block* forces the control to execute only one command at a time. *Optional stop* (when on) will cause the control to stop the program's execution when an M01 word is executed in the program. *Block delete* (when on) will cause the control to skip words to the right of the slash code (/). (Again, see Lesson Twenty-Five for more information about these conditional switches.)

To run the active program from the beginning

This procedure will only make sense if you understand the presentations made in Lessons Twenty-Four and Twenty-Eight. You must thoroughly understand these presentations in order to safely run a CNC program. If you don't, it is likely that you will incorrectly set one of the switches listed in this procedure – and the results will be disastrous.

1) Place the mode switch to the program operation mode (memory or auto)
2) Press the soft key under program. The active program will be displayed. Be sure the active program is the one you want to run and that the cursor is on the program number.
3) Check the position of all conditional switches (dry run, single block, optional stop, etc.).
4) Place the feedrate override switch to 100% (assuming the program is proven).
5) Place the rapid override switch to the desired position (100% if running production).
6) Be sure the machine is in an appropriate position for the program to begin (the safe index position).
7) With a finger ready to press the feed hold button, press the cycle start button. The program will run, behaving in accordance with how the conditional switches are set.

Key points for Lesson Twenty-Six:

- Though there are more than three positions on the mode switch, there are really only three basic modes of operation – manual, manual data input, and program operation. All buttons and switches can be placed into one of these three categories.
- Manual mode is used to perform an immediate action.
- Manual data input mode is used to use the display screen and keyboard to enter and modify commands.
- Program operation mode is used to run programs.

STOP! Do practice exercise number twenty-six in the workbook.

Understand The Importance Of Procedures

Running a CNC turning center requires little more than following a series of step-by-step procedures. The trick lies in knowing when a given procedure is required. From the material presented to this point, you should now know when each procedure is needed.

Key Concept Number Nine is another short, one-lesson Key Concept:

27: The key operation procedures

From what has been presented to this point, you should be pretty comfortable with what you must do to make setups and run production – at least in theory. But when you step up to a CNC turning center for the first time, you'll probably be quite intimidated. Things you thought you knew won't seem so clear.

This will happen because you are still lacking hands-on experience – experience that we cannot provide in this text. Though we can't provide hands-on training, we can provide you with a description of the *procedures* that you are going to need in order to run your turning center.

Procedures will help you get familiar with your turning center. For example, if you need to power-up the machine at the beginning of your shift and make it ready to run production, what will you do?

Without a machine start-up procedure – or at least someone to demonstrate how this procedure is performed – you'll be lost. And trying to operate a CNC turning center without being sure of what you are doing can be very dangerous.

Even with a person available to help, there can be problems. This person may not be available every time you need them. And they will soon tire of repeating procedures that they feel you should have memorized by now. While you will *eventually* memorize often-used procedures, it may take you longer than it took the person helping you. And this can be frustrating – for both of you.

When you are shown a procedure for the first time, *write it down*. This will keep you from having to keep asking someone for help every time you need to perform the procedure. In Lesson Twenty-Seven, we do provide quick reference sheets for nine of Fanuc's most popular control models. Though you may find them quite helpful, they're just intended to get you started. You will surely come across machine functions for which we have not provided a procedure. You'll have to develop one – for yourself – and for others that come after you.

Lesson 27

The Key Operation Procedures

Step-by-step procedures keep you from having to memorize every function that you must perform on your CNC turning center. You will soon memorize procedures for tasks that you perform on a regular basis – but written procedures will help you perform less common tasks.

We divide the procedures needed for CNC turning center usage into five categories:

- Manual procedures
- Setup procedures
- Manual data input (MDI) procedures
- Program editing procedures
- Program operation procedures

On the pages that follow, we're providing four *quick reference sheets* to help you with the most commonly used procedures. These quick reference sheets will help with some of Fanuc's most popular control models. Here is a list of procedures shown in each category:

Manual procedures:

To power up the machine
To do a manual zero return
To manually start the spindle
To manually jog the axes
To use the handwheel
To manually index the turret
To load tools into the turret

Setup procedures:

To set or reset the relative position display
To enter and modify geometry and wear offsets
To enter and modify the work shift offset

MDI procedures:

To use MDI to index the turret
To use MDI to activate the spindle
To use MDI to do a zero return

Program editing procedures:

To enter a program through the keyboard
To load a program from a DNC system
To save a program to a DNC system
To see a directory of programs
To delete a program
To call up a program (make it the active program)
To search within a program
To alter a word in a program
To delete a word in a program
To insert a word in a program

Program operation procedures:

To run a verified program

In addition we have shown some other important procedures in this text, including:

To measure program zero assignment values – Lesson Six
To mount and machine jaws – Lesson Twenty-Four
To verify programs – Lessons Twenty-Four and Twenty-Eight

In Lesson Twenty-Eight, we show two more important procedures:

To cancel the cycle
To re-run a tool

The goal with any step-by-step procedure is to keep you from having to memorize. But you must, of course, understand *when* and *why* to perform each procedure. If you have read the entire text – or if you have read the sections recommended in the introduction to Key Concept Seven, you should be pretty comfortable with the when and why. For example, say you're beginning a new setup and you've just mounted hard jaws on the chuck and cutting tools in the turret. Now you must measure program zero assignment values. According to the general procedure in Lesson Six, if you don't have a tool touch off probe, you must machine a small amount of material from the workpiece. This means you must start the spindle and manually machine the workpiece.

As you're standing in front of the machine – scratching your head – you remember that we've provided procedures to manually start the spindle and jog the axes. There is also a procedure to use the handwheel. With these step-by-step procedures, even a newcomer can perform the needed tasks. But again, you must know when and why to perform them.

If you come across a task for which we have not supplied a step-by-step procedure, get someone to help you. As they demonstrate what it takes to perform the task, write down – in step-by-step fashion – what they do. You'll have a procedure for the *next time* you must perform the task – and you won't have to ask for help. Figure 9.1 shows a form you can use to create your own procedures.

Procedure name: ____________________ **Machine:** ____________________

Description: ____

____

Procedure:

1) ____________	9) ____________
2) ____________	10) ____________
3) ____________	11) ____________
4) ____________	12) ____________
5) ____________	13) ____________
6) ____________	14) ____________
7) ____________	15) ____________
8) ____________	16) ____________

Notes: ____

____

Figure 9.1 – Procedure form

The next eight pages provide you with quick reference sheets for some popular Fanuc control models. But if you have other Fanuc controls, or if you have turning centers with controls that are not made by Fanuc, *make your own quick reference sheet*. It is not difficult to do – make step-by-step procedures for the same list of tasks we have shown in our quick reference sheets.

Fanuc 0T and 3T Quick Reference For Key Procedures

To power-up the machine:

Required before doing anything else on the machine.

1. Turn on main breaker (Usually located at rear of machine).
2. Press control power on button.
3. Press machine ready or hydraulic on button. (not necessary on some machines.) Be sure that emergency stop button is *not locked in.*
4. Follow sequence to do a manual zero return. (All Fanuc controls require that you do a manual zero return as part of the power-up procedure.)

To manually jog the axes:

Required when taking measurements on the machine, when performing manual machining operations, and in general, whenever you must move an axis.

1. Place the mode switch to a manual mode (jog, manual, zero return, etc.).
2. Place axis selector switch to desired axis (X or Z).
3. Place the jog feedrate switch to the desired motion rate.
4. Using the plus or minus buttons or joystick, move the axis in the desired direction and amount.

Jogging the machine does not provide precise positioning. Use the handwheel when more precise positioning is needed.

To load tools into the turret

Required during setup.

1. Be sure the turret is in a safe index position and index the turret to the desired turret station.
2. Using the clamping system supplied by the machine tool builder, remove the previous tool and load the new one.

Machines vary when it comes to turret design. You must reference your machine tool builder's operation manual to learn more about loading tools in the turret.

To enter and modify the work shift value

Required if using geometry offsets to assign program zero – when the face of the chuck is the point of reference for geometry offset entries in Z.

1. The mode switch can be in any position.
2. Press the offset key until the work shift page appears.
3. Using the cursor control up and down arrow keys, position the cursor to the register to enter (Z).
4. Type the value of the offset and press input.

To do a manual zero return:

Required right after power up. The zero return position is commonly the safe index position for some turning centers. The machine may have to be at this position prior to running a program.

1. Place mode switch to zero return.
2. Using the axis select switch select the X axis.
3. Using the minus direction push button or joystick, move the machine about two to three inches in the minus direction (watch for obstructions).
4. Using the plus direction push button or joystick, hold plus until zero return origin indicator light comes on.
5. Repeat steps 2-4 for the Z axis.

To use the handwheel

Required when precise axis motion is needed - as when measuring program zero assignment values and manually machining soft jaws.

1. Place mode switch to handwheel.
2. Place axis select switch to desired position (X, Y, Z, or B/C).
3. Place handwheel rate switch to desired position (X1, X10, or X100).
4. Using handwheel rotate plus or minus to cause desired motion.

Use the relative position display when taking measurements.

To set or reset the relative position display

Required when taking measurements on the machine.

To reset (set to zero)

1. Mode switch can be in any position.
2. Press the POS key and page down until the relative position display page appears (without the dots).
3. Type the letter address of the axis you wish to reset (X or Z). That letter will start flashing on the axis display.
4. Press the cancel key (CAN).

To set to a specific value

1. Place the mode switch to MDI.
2. Press the PRGRM key.
3. Type G50 and press input.
4. Type the letter address of the axis you wish to set and the value – then press input.
5. Press start or cycle start.
6. Press the POS key. The value you set will be displayed on the absolute page (with the "dots".).

To manually start the spindle:

Required whenever you must manually measure program zero assignment values, when machining soft jaws, or when you must manually machine a workpiece.

1. Place the mode switch to a manual mode (jog, manual, zero return, etc.).
2. Select the desired direction (forward or reverse) with the appropriate switch.
3. If available, select the desired speed in rpm with the appropriate switch.
4. Press the spindle on button.
5. To stop spindle, press the spindle stop button.

Not all machines have manual controls for the spindle. This procedure will only work for those that do. You can also use the MDI procedure to start the spindle. Some machines allow you to manually start and stop the spindle, but not select speed or direction.

To manually index the turret

Required when loading and inspecting cutting tools.

1. Be sure the turret is in a safe index position.
2. Place the station selector switch to the desired position (usually between one and twelve).
3. Place the mode switch to a manual mode (jog, manual, zero return, etc.).
4. Press the turret index button.

Not all turning centers provide a way to manually index the turret. Many require that you use the MDI procedure to do so.

To enter and modify geometry or wear offsets

Modifying geometry offsets is required during setup to enter initial offset values and when trial machining. Modifying wear offsets is required during a production run when sizing is needed.

1. The mode switch can be in any position.
2. Press the offset key until the desired offset page appears (geometry or wear).
3. Using the cursor control up and down arrow keys, position the cursor to the offset number to enter.
4. Type the letter address and the value of the offset and press input.

Using X and Z will overwrite the current offset value . Using U (for X) and W (for Z) will modify the value of the current offset value.

Fanuc 0T and 3T Quick Reference For Key Procedures

To use MDI to index turret

Required whenever a manual tool change is needed – as when loading tools or changing inserts.

1. Place the mode switch to MDI.
2. Press the prgrm key.
3. Type T and the number of the tool followed by two trailing zeros (like T0200).
4. Press the input key.
5. Press the start key or cycle start button. (Turret indexes to specified station.)

This procedure assumes the machine is in a safe index position.

To enter a program using the keyboard

A program must reside in the machine's memory before it can be run. This is one way to load a program – but it is time consuming.

1. Place the mode switch to edit.
2. Press the prgrm key.
3. Type the letter O and the program number of the program to be entered.
4. Press insert key.
5. Press the EOB key and press the insert key.
6. Typing one word at a time and pressing insert after each word, enter the balance of the program. You must remember to end each command with EOB.

To see a directory of programs

Required when you want to know which programs are in the machine.

1. Place the mode switch to edit.
2. Press the prgrm key.
3. Type P1 and press input.

To search within a program

Required when you must edit a program.

1. Place mode switch to edit.
2. Press prgrm key.
3. Press the reset key to return the program to the beginning (not always required).
4. Type the word you wish to search to.
5. Press the down arrow key (cursor forward). The control will find the first occurrence of the word you typed.

You can also use the cursor control keys to move the cursor (up and down arrows).

To use MDI to activate the spindle

Required when manually measuring program zero assignment values or when machining soft jaws.

1. Place the mode switch to MDI.
2. Press the prgrm key.
3. Type G97 and press the input key.
4. Type S and the desired rpm (example: S500 = 500 rpm) and press the input key.
5. Type M03 for forward or M04 for reverse and press the input key.
6. Press start key or cycle start button. (spindle starts)
7. To stop spindle, type M05 and press input, then press start key or cycle start button.

This procedure assumes a spindle range has been selected.

To load a program from a DNC system

Required during setup to load the program/s needed for the job.

1. Place mode switch to edit.
2. Press the prgrm key.
3. Be sure the DNC system is connected the machine.
4. Type O (letter, not zero) and the program number to be loaded.
5. Press the input key.
6. Go to the DNC system and send the program.

To delete a program

Required when you're finished using a program.

1. Place mode switch to edit.
2. Press prgrm key.
3. Type letter O and the program number to be deleted.
4. Press the delet key.

You won't be asked for confirmation, so be careful with this procedure.

To alter a word in a program

Required when you must edit a program.

1. Place mode switch to edit.
2. Press prgrm key.
3. Search to the word to be altered.
4. Type the new word.
5. Press the alter key.

To insert a word in a program

Required when you must edit a program.

1. Place mode switch to edit.
2. Press prgrm key.
3. Search to the word just prior to the word you want to insert.
4. Type the word you wish to insert.
5. Press the insrt key.

To use MDI to do a zero return

This procedure is faster than the manual procedure.

1. Place mode switch to MDI.
2. Press prgrm key.
3. Check that the machine can go straight home with no interference.
4. Type G28 and press input.
5. Type U0 and press input.
6. Type W0 and press input.
7. Press start key or cycle start button.

You can omit axes that you don't want to zero return.

To save a program to a DNC system

Required after program verification if changes have been made to the program.

1. Get the DNC system ready to receive a program.
2. Place mode switch to edit.
3. Press prgrm key.
4. Type the letter O and the program number of the program to be sent.
5. Press the start key.

To call up a program (make it the active program)

Required when you want to work on another program in the control.

1. Place mode switch to edit.
2. Press prgrm key.
3. Type the letter O and the program to be searched to.
4. Press the down arrow key (cursor forward).

To delete a word in a program

Required when you must edit a program.

1. Place mode switch to edit.
2. Press prgrm key.
3. Search to the word to be deleted.
4. Type the new word.
5. Press the delet key.

To run a verified program

Required during a production run.

1. Load part into setup.
2. Send the machine to its starting point (usually the tool change position).
3. Place the mode switch to edit.
4. Press the prgrm key.
5. Press the reset key (check the active program number).
6. Check position of all conditional switches (dry run, etc.).
7. Place the mode switch to memory or auto.
8. Place the feedrate override switch to 100 per cent.
9. Press cycle start button.

Other important procedures you can find in this text:

To measure program zero assignment values: Lesson Six
To mount and machine jaws: Lesson Twenty-Four
To verify programs: Lessons Twenty-Four and Twenty-Eight
To re-run tools: Lesson Twenty-Eight

Fanuc 6T and 6TB Quick Reference For Key Procedures

To power-up the machine:

Required before doing anything else on the machine.

1. Turn on main breaker (Usually located at rear of machine).
2. Press control power on button.
3. Press machine ready or hydraulic on button. (not necessary on some machines.) Be sure that emergency stop button is *not locked in.*
4. Follow sequence to do a manual zero return. (All Fanuc controls require that you do a manual zero return as part of the power-up procedure.)

To manually jog the axes:

Required when taking measurements on the machine, when performing manual machining operations, and in general, whenever you must move an axis.

1. Place the mode switch to a manual mode (jog, manual, zero return, etc.).
2. Place axis selector switch to desired axis (X or Z).
3. Place the jog feedrate switch to the desired motion rate.
4. Using the plus or minus buttons or joystick, move the axis in the desired direction and amount.

Jogging the machine does not provide precise positioning. Use the handwheel when more precise positioning is needed.

To load tools into the turret

Required during setup.

1. Be sure the turret is in a safe index position and index the turret to the desired turret station.
2. Using the clamping system supplied by the machine tool builder, remove the previous tool and load the new one.

Machines vary when it comes to turret design. You must reference your machine tool builder's operation manual to learn more about loading tools in the turret.

To do a manual zero return:

Required right after power up. The zero return position is commonly the safe index position for some turning centers. The machine may have to be at this position prior to running a program.

1. Place mode switch to zero return.
2. Using the axis select switch select the X axis.
3. Using the minus direction push button or joystick, move the machine about two to three inches in the minus direction (watch for obstructions).
4. Using the plus direction push button or joystick, hold plus until zero return origin indicator light comes on.
5. Repeat steps 2-4 for the Z axis.

To use the handwheel

Required when precise axis motion is needed - as when measuring program zero assignment values and manually machining soft jaws.

1. Place mode switch to handwheel.
2. Place axis select switch to desired position (X, Y, Z, or B/C).
3. Place handwheel rate switch to desired position (X1, X10, or X100).
4. Using handwheel rotate plus or minus to cause desired motion.

Use the relative position display when taking measurements.

To set or reset the relative position display

Required when taking measurements on the machine.

To reset (set to zero)

1. Mode switch can be in any position.
2. Press the POS key and page down until the relative position display page appears.
3. Type the letter address of the axis you wish to reset (X or Z). That letter will start flashing on the axis display.
4. Press the origin key.

To set to a specific value

1. Place the mode switch to MDI.
2. Press the cmnd key and press the down arrow key until Next Block/MDI appears at the top of the CRT screen.
3. Type G50 and press input.
4. Type the letter address of the axis you wish to set and the value – then press input.
5. Press start or cycle start.
6. Press the POS key. The value you set will be displayed on the absolute page.

To manually start the spindle:

Required whenever you must manually measure program zero assignment values, when machining soft jaws, or when you must manually machine a workpiece.

1. Place the mode switch to a manual mode (jog, manual, zero return, etc.).
2. Select the desired direction (forward or reverse) with the appropriate switch.
3. If available, select the desired speed in rpm with the appropriate switch.
4. Press the spindle on button.
5. To stop spindle, press the spindle stop button.

Not all machines have manual controls for the spindle. This procedure will only work for those that do. You can also use the MDI procedure to start the spindle. Some machines allow you to manually start and stop the spindle, but not select speed or direction.

To manually index the turret

Required when loading and inspecting cutting tools.

1. Be sure the turret is in a safe index position.
2) Place the station selector switch to the desired position (usually between one and twelve).
3. Place the mode switch to a manual mode (jog, manual, zero return, etc.).
4. Press the turret index button.

Not all turning centers provide a way to manually index the turret. Many require that you use the MDI procedure to do so.

To enter and modify geometry or wear offsets

Modifying geometry offsets is required during setup to enter initial offset values and when trial machining. Modifying wear offsets is required during a production run when sizing is needed.

1. The mode switch can be in any position.
2. Press the offset key until the desired offset page appears (geometry or wear).
3. Using the cursor control up and down arrow keys, position the cursor to the offset number to enter.
4. Type the letter address and the value of the offset and press input.

Using X and Z will overwrite the current offset value . Using U (for X) and W (for Z) will modify the value of the current offset value.

Fanuc 6T and 6TB Quick Reference For Key Procedures

To use MDI to index turret

Required whenever a manual tool change is needed – as when loading tools or changing inserts.

1. Place the mode switch to MDI.
2. Press the comnd key and press the down arrow key until Next Block/MDI appears at the top of the CRT screen.
3. Type T and the number of the tool followed by two trailing zeros (like T0200).
4. Press the input key.
5. Press the start key or cycle start button. (Turret indexes to specified station.)

This procedure assumes the machine is in a safe index position.

To enter a program using the keyboard

A program must reside in the machine's memory before it can be run. This is one way to load a program – but it is time consuming.

1. Place the mode switch to edit.
2. Press the prgrm key.
3. Type the letter O and the program number of the program to be entered.
4. Press insert key.
5. Press the EOB key and press the insert key.
6. Typing one word at a time and pressing insert after each word, enter the balance of the program. You must remember to end each command with EOB.

To see a directory of programs

Required when you want to know which programs are in the machine.

1. Place the mode switch to edit.
2. Press the prgrm key.
3. Press the cancel key and then press the origin key.

To search within a program

Required when you must edit a program.

1. Place mode switch to edit.
2. Press prgrm key.
3. Press the reset key to return the program to the beginning (not always required).
4. Type the word you wish to search to.
5. Press the down arrow key (cursor forward). The control will find the first occurrence of the word you typed.

You can also use the cursor control keys to move the cursor (up and down arrows).

To use MDI to activate the spindle

Required when manually measuring program zero assignment values or when machining soft jaws.

1. Place the mode switch to MDI.
2. Press the comnd key and press the down arrow key until Next Block/MDI appears at the top of the CRT screen.
3. Type G97 and press the input key.
4. Type S and the desired rpm (example: S500 = 500 rpm) and press the input key.
5. Type M03 for forward or M04 for reverse and press the input key.
6. Press start key or cycle start button. (spindle starts)
7. To stop spindle, type M05 and press input, then press start key or cycle start button.

This procedure assumes a spindle range has been selected.

To load a program from a DNC system

Required during setup to load the program/s needed for the job.

1. Place mode switch to edit.
2. Press the prgrm key.
3. Be sure the DNC system is connected the machine.
4. Type O (letter, not zero) and the program number to be loaded.
5. Press the read key.
6. Go to the DNC system and send the program.

To delete a program

Required when you're finished using a program.

1. Place mode switch to edit.
2. Press prgrm key.
3. Type letter O and the program number to be deleted.
4. Press the DEL key.

You won't be asked for confirmation, so be careful with this procedure.

To alter a word in a program

Required when you must edit a program.

1. Place mode switch to edit.
2. Press prgrm key.
3. Search to the word to be altered.
4. Type the new word.
5. Press the alter key.

To insert a word in a program

Required when you must edit a program.

1. Place mode switch to edit.
2. Press prgrm key.
3. Search to the word just prior to the word you want to insert.
4. Type the word you wish to insert.
5. Press the insrt key.

To use MDI to do a zero return

This procedure is faster than the manual procedure.

1. Place mode switch to MDI.
2. Press the comnd key and press the down arrow key until Next Block/MDI appears at the top of the CRT screen.
3. Check that the machine can go straight home with no interference.
4. Type G28 and press input.
5. Type U0 and press input.
6. Type W0 and press input.
7. Press start key or cycle start button.

You can omit axes that you don't want to zero return.

To save a program to a DNC system

Required after program verification if changes have been made to the program.

1. Get the DNC system ready to receive a program.
2. Place mode switch to edit.
3. Press prgrm key.
4. Type the letter O and the program number of the program to be sent.
5. Press the punch key.

To call up a program (make it the active program)

Required when you want to work on another program in the control.

1. Place mode switch to edit.
2. Press prgrm key.
3. Type the letter O and the program to be searched to.
4. Press the down arrow key (cursor forward).

To delete a word in a program

Required when you must edit a program.

1. Place mode switch to edit.
2. Press prgrm key.
3. Search to the word to be deleted.
4. Press the DEL key.

To run a verified program

Required during a production run.

1. Load part into setup.
2. Send the machine to its starting point (usually the tool change position).
3. Place the mode switch to edit.
4. Press the prgrm key.
5. Press the reset key (check the active program number).
6. Check position of all conditional switches (dry run, etc.).
7. Place the mode switch to memory or auto.
8. Place the feedrate override switch to 100 per cent.
9. Press cycle start button.

Other important procedures you can find in this text:

To measure program zero assignment values: Lesson Six
To mount and machine jaws: Lesson Twenty-Four
To verify programs: Lessons Twenty-Four and Twenty-Eight
To re-run tools: Lesson Twenty-Eight

Fanuc 10T, 11T and 15T Quick Reference For Key Procedures

To power-up the machine:

Required before doing anything else on the machine.

1. Turn on main breaker (Usually located at rear of machine).
2. Press control power on button.
3. Press machine ready or hydraulic on button. (not necessary on some machines.) Be sure that emergency stop button is *not locked in.*
4. Follow sequence to do a manual zero return. (All Fanuc controls require that you do a manual zero return as part of the power-up procedure.)

To manually jog the axes:

Required when taking measurements on the machine, when performing manual machining operations, and in general, whenever you must move an axis.

1. Place the mode switch to a manual mode (jog, manual, zero return, etc.).
2. Place axis selector switch to desired axis (X or Z).
3. Place the jog feedrate switch to the desired motion rate.
4. Using the plus or minus buttons or joystick, move the axis in the desired direction and amount.

Jogging the machine does not provide precise positioning. Use the handwheel when more precise positioning is needed.

To load tools into the turret

Required during setup.

1. Be sure the turret is in a safe index position and index the turret to the desired turret station.
2. Using the clamping system supplied by the machine tool builder, remove the previous tool and load the new one.

Machines vary when it comes to turret design. You must reference your machine tool builder's operation manual to learn more about loading tools in the turret.

To enter and modify the work shift value

Required if using geometry offsets to assign program zero – when the face of the chuck is the point of reference for geometry offset entries in Z.

1. The mode switch can be in any position.
2. Press the soft key under offset key until the work shift page appears.
3. Using the cursor control up and down arrow keys, position the cursor to the register to enter (Z).
4. Type the value of the offset and press input.

To do a manual zero return:

Required right after power up. The zero return position is commonly the safe index position for some turning centers. The machine may have to be at this position prior to running a program.

1. Place mode switch to zero return.
2. Using the axis select switch select the X axis.
3. Using the minus direction push button or joystick, move the machine about two to three inches in the minus direction (watch for obstructions).
4. Using the plus direction push button or joystick, hold plus until zero return origin indicator light comes on.
5. Repeat steps 2-4 for the Z axis.

To use the handwheel

Required when precise axis motion is needed - as when measuring program zero assignment values and manually machining soft jaws.

1. Place mode switch to handwheel.
2. Place axis select switch to desired position (X, Y, Z, or B/C).
3. Place handwheel rate switch to desired position (X1, X10, or X100).
4. Using handwheel rotate plus or minus to cause desired motion.

Use the relative position display when taking measurements.

To set or reset the relative position display

Required when taking measurements on the machine.

To reset (set to zero)

1. Mode switch can be in any position.
2. Press the extreme left soft key until position appears at the bottom of the CRT screen.
3. Press the soft key under position until the relative display page is shown.
4. Type the letter address of the axis you wish to reset (X or Z). That letter will start flashing on the axis display.
5. Press the soft key under origin.

To set to a specific value

1. Mode switch can be in any position.
2. Press the extreme left soft key until position appears at the bottom of the CRT screen.
3. Press the soft key under position until the relative display page is shown.
4. Type the letter address for the axis you wish to set and type its value.
5. Press the soft key under preset.

To manually start the spindle:

Required whenever you must manually measure program zero assignment values, when machining soft jaws, or when you must manually machine a workpiece.

1. Place the mode switch to a manual mode (jog, manual, zero return, etc.).
2. Select the desired direction (forward or reverse) with the appropriate switch.
3. If available, select the desired speed in rpm with the appropriate switch.
4. Press the spindle on button.
5. To stop spindle, press the spindle stop button.

Not all machines have manual controls for the spindle. This procedure will only work for those that do. You can also use the MDI procedure to start the spindle. Some machines allow you to manually start and stop the spindle, but not select speed or direction.

To manually index the turret

Required when loading and inspecting cutting tools.

1. Be sure the turret is in a safe index position.
2) Place the station selector switch to the desired position (usually between one and twelve).
3. Place the mode switch to a manual mode (jog, manual, zero return, etc.).
4. Press the turret index button.

Not all turning centers provide a way to manually index the turret. Many require that you use the MDI procedure to do so.

To enter and modify geometry or wear offsets

Modifying geometry offsets is required during setup to enter initial offset values and when trial machining. Modifying wear offsets is required during a production run when sizing is needed.

1. The mode switch can be in any position.
2. Press the extreme left soft key until Offset appears at the bottom of the CRT screen.
3. Press the soft key under offset until the desired offset page appears (geometry or wear).
4. Using the cursor control up and down arrow keys, position the cursor to the offset number to enter.
5. Type the value of the offset and press input.

You can overwrite the current offset setting by pressing the soft key under input – or you can modify its current value by your entry by pressing the soft key under input +.

Fanuc 10T, 11T, and 15T Quick Reference For Key Procedures

To use MDI to index turret

Required whenever a manual tool change is needed – as when loading tools or changing inserts.

1. Place the mode switch to MDI.
2. Press the extreme left soft key until program appears at the bottom of the screen.
3. Press the soft key under program until MDI appears at top of screen.
4. Type T and the number of the tool followed by two trailing zeros (like T0200).
5. Press the EOB key and press the soft key under insert.
6. Press the start key or cycle start button. (Turret indexes to specified station.)

This procedure assumes the machine is in a safe index position.

To enter a program using the keyboard

A program must reside in the machine's memory before it can be run. This is one way to load a program – but it is time consuming.

1. Place the mode switch to edit.
2. Press the extreme left soft key until program appears at the bottom of the screen – then press the soft key under program.
3. Type the letter O and the program number of the program to be entered.
4. Press the soft key under insert.
5. Press the EOB key and press the soft key under insert.
6. Typing one command at a time and ending each command with the EOB key, enter the balance of the program.

To see a directory of programs

Required when you want to know which programs are in the machine.

1. Place the mode switch to edit.
2. Press the extreme left soft key until program appears at the bottom of the screen – then press the soft key under program.
3. Press the soft key under program again.

To search within a program

1. Place mode switch to edit.
2. Press the extreme left soft key until program appears at the bottom of the screen – then press the soft key under program.
3. Press the reset key to return the program to the beginning (not always required).
4. Type the word you wish to search to.
5. Press soft key under fwd search. The control will find the first occurrence of the word you typed.

To use MDI to activate the spindle

Required when manually measuring program zero assignment values or when machining soft jaws.

1. Place the mode switch to MDI.
2. Press the extreme left soft key until program appears at the bottom of the screen.
3. Press the soft key under program until MDI appears at top of screen.
4. Type G97 and press the soft key under insert.
5. Type S and the desired rpm (example: S500 = 500 rpm) and press the soft key under insert.
6. Type M03 for forward or M04 for reverse, press the EOB key, and press the soft key under insert.
7. Press start key or cycle start button. (spindle starts)
8. To stop spindle, type M05, the EOB key, and press the soft key under insert. Then press the cycle start button.

To load a program from a DNC system

1. Place the mode switch to edit.
2. Press the extreme left soft key until program appears at the bottom of the screen – then press the soft key under program.
3. Be sure the DNC system is connected the machine.
4. Type O (letter, not zero) and the program number to be loaded.
5. Press the soft key under read.
6. Go to the DNC system and send the program.

To delete a program

1. Place mode switch to edit.
2. Press the extreme left soft key until program appears at the bottom of the screen – then press the soft key under program.
3. Type letter O and the program number to be deleted.
4. Press the soft key under delete.

To alter a word in a program

1. Place mode switch to edit.
2. Press the extreme left soft key until program appears at the bottom of the screen – then press the soft key under program.
3. Search to the word to be altered.
4. Type the new word.
5. Press soft key under alter.

To insert a word in a program

1. Place mode switch to edit.
2. Press the extreme left soft key until program appears at the bottom of the screen – then press the soft key under program
3. Search to the word just prior to the word you want to insert.
4. Type the word you wish to insert.
5. Press soft key under insert.

To use MDI to do a zero return

This procedure is faster than the manual procedure.

1. Place the mode switch to MDI.
2. Press the extreme left soft key until program appears at the bottom of the screen.
3. Press the soft key under program until MDI appears at top of screen.
4. Check that the machine can go straight home with no interference.
5. Type G28 U0 W0, press the EOB key and then press the soft key under insert.
6. Press start key or cycle start button.

You can omit axes that you don't want to zero return.

To save a program to a DNC system

1. Get the DNC system ready to receive a program.
2. Place mode switch to edit.
3. Press the extreme left soft key until program appears at the bottom of the screen – then press the soft key under program.
4. Type the letter O and the program number of the program to be sent.
5. Press the soft key under punch.

To call up a program (make it the active program)

Required when you want to work on another program in the control.

1. Place mode switch to edit.
2. Press the extreme left soft key until program appears at the bottom of the screen – then press the soft key under program.
3. Type the letter O and the program to be searched to.
4. Press the soft key under fwd search.

To delete a word in a program

1. Place mode switch to edit.
2. Press the extreme left soft key until program appears at the bottom of the screen – then press the soft key under program.
3. Search to the word to be deleted.
4. Press the soft key under delete.

To run a verified program

1. Load part into setup.
2. Send the machine to its starting point (usually the tool change position).
3. Place the mode switch to edit.
4. Press the extreme left soft key until program appears at the bottom of the screen – then press the soft key under program.
5. Press the reset key (check the active program number).
6. Check position of all conditional switches (dry run, etc.).
7. Place the mode switch to memory or auto.
8. Place the feedrate override switch to 100 per cent.
9. Press cycle start button.

Other important procedures you can find in this text:

To measure program zero assignment values: Lesson Six
To mount and machine jaws: Lesson Twenty-Four
To verify programs: Lessons Twenty-Four and Twenty-Eight
To re-run tools: Lesson Twenty-Eight

Fanuc 16T and 18T Quick Reference For Key Procedures

To power-up the machine:

Required before doing anything else on the machine.

1. Turn on main breaker (Usually located at rear of machine).
2. Press control power on button.
3. Press machine ready or hydraulic on button. (not necessary on some machines.) Be sure that emergency stop button is *not locked in.*
4. Follow sequence to do a manual zero return. (All Fanuc controls require that you do a manual zero return as part of the power-up procedure.)

To manually jog the axes:

Required when taking measurements on the machine, when performing manual machining operations, and in general, whenever you must move an axis.

1. Place the mode switch to a manual mode (jog, manual, zero return, etc.).
2. Place axis selector switch to desired axis (X or Z).
3. Place the jog feedrate switch to the desired motion rate.
4. Using the plus or minus buttons or joystick, move the axis in the desired direction and amount.

Jogging the machine does not provide precise positioning. Use the handwheel when more precise positioning is needed.

To load tools into the turret

Required during setup.

1. Be sure the turret is in a safe index position and index the turret to the desired turret station.
2. Using the clamping system supplied by the machine tool builder, remove the previous tool and load the new one.

Machines vary when it comes to turret design. You must reference your machine tool builder's operation manual to learn more about loading tools in the turret.

To enter and modify the work shift value

Required if using geometry offsets to assign program zero – when the face of the chuck is the point of reference for geometry offset entries in Z.

1. The mode switch can be in any position.
2. Press the soft key under offset key until the work shift page appears.
3. Using the cursor control up and down arrow keys, position the cursor to the register to enter (Z).
4. Type the value of the offset and press input.

To do a manual zero return:

Required right after power up. The zero return position is commonly the safe index position for some turning centers. The machine may have to be at this position prior to running a program.

1. Place mode switch to zero return.
2. Using the axis select switch select the X axis.
3. Using the minus direction push button or joystick, move the machine about two to three inches in the minus direction (watch for obstructions).
4. Using the plus direction push button or joystick, hold plus until zero return origin indicator light comes on.
5. Repeat steps 2-4 for the Z axis.

To use the handwheel

Required when precise axis motion is needed - as when measuring program zero assignment values and manually machining soft jaws.

1. Place mode switch to handwheel.
2. Place axis select switch to desired position (X, Y, Z, or B/C).
3. Place handwheel rate switch to desired position (X1, X10, or X100).
4. Using handwheel rotate plus or minus to cause desired motion.

Use the relative position display when taking measurements.

To set or reset the relative position display

Required when taking measurements on the machine.

To reset (set to zero)

1. Mode switch can be in any position.
2. Press the key labeled position until the relative display screen is shown.
3. Type the letter address of the axis you wish to reset (X or Z). That letter will start flashing on the axis display.
4. Press the soft key under origin.

To set to a specific value

1. Mode switch can be in any position.
2. Press the key labeled position until the relative display screen is shown.
3. Type the letter address for the axis you wish to set and type its value.
4. Press the soft key under preset.

To manually start the spindle:

Required whenever you must manually measure program zero assignment values, when machining soft jaws, or when you must manually machine a workpiece.

1. Place the mode switch to a manual mode (jog, manual, zero return, etc.).
2. Select the desired direction (forward or reverse) with the appropriate switch.
3. If available, select the desired speed in rpm with the appropriate switch.
4. Press the spindle on button.
5. To stop spindle, press the spindle stop button.

Not all machines have manual controls for the spindle. This procedure will only work for those that do. You can also use the MDI procedure to start the spindle. Some machines allow you to manually start and stop the spindle, but not select speed or direction.

To manually index the turret

Required when loading and inspecting cutting tools.

1. Be sure the turret is in a safe index position.
2) Place the station selector switch to the desired position (usually between one and twelve).
3. Place the mode switch to a manual mode (jog, manual, zero return, etc.).
4. Press the turret index button.

Not all turning centers provide a way to manually index the turret. Many require that you use the MDI procedure to do so.

To enter and modify geometry or wear offsets

Modifying geometry offsets is required during setup to enter initial offset values and when trial machining. Modifying wear offsets is required during a production run when sizing is needed.

1. The mode switch can be in any position.
2. Press the key labeled offset/setting.
3. Press the soft key under offset until desired offset page appears (geometry or wear).
4. Using the cursor control up and down arrow keys, position the cursor to the offset number to enter.
5. Type the value of the offset and press input.

You can overwrite the current offset setting by pressing the soft key under input – or you can modify its current value by your entry by pressing the soft key under input +.

Fanuc 16T and 18T Quick Reference For Key Procedures

To use MDI to index turret

Required whenever a manual tool change is needed – as when loading tools or changing inserts.

1. Place the mode switch to MDI.
2. Press the key labeled program until MDI appears at the top of the display screen.
3. Type T and the number of the tool followed by two trailing zeros (like T0200).
4. Press the EOB key and press the soft key under insert.
5. Press the start key or cycle start button. (Turret indexes to specified station.)

This procedure assumes the machine is in a safe index position.

To enter a program using the keyboard

A program must reside in the machine's memory before it can be run. This is one way to load a program – but it is time consuming.

1. Place the mode switch to edit.
2. Press the key labeled program until the word program appears at the top of the display screen.
3. Type the letter O and the program number of the program to be entered.
4. Press the soft key under insert.
5. Press the EOB key and press the soft key under insert.
6. Typing one command at a time and ending each command with the EOB key, enter the balance of the program.

To see a directory of programs

Required when you want to know which programs are in the machine.

1. Place the mode switch to edit.
2. Press the key labeled program until the word program appears at the top of the display screen.
3. Press the soft key under program again.

To search within a program

Required when you must edit a program.

1. Place mode switch to edit.
2. Press the extreme left soft key until program appears at the bottom of the screen – then press the soft key under program.
3. Press the reset key to return the program to the beginning (not always required).
4. Type the word you wish to search to.
5. Press soft key under fwd search. The control will find the first occurrence of the word you typed.

To use MDI to activate the spindle

Required when manually measuring program zero assignment values or when machining soft jaws.

1. Place the mode switch to MDI.
2. Press the key labeled program until MDI appears at the top of the display screen.
3. Type G97 and the speed you want (example: S500 = 500 rpm) then press the soft key under insert.
4. Type M03 for forward or M04 for reverse, press the EOB key, and press the soft key under insert.
5. Press start key or cycle start button. (spindle starts)
6. To stop spindle, type M05, the EOB key, and press the soft key under insert. Then press start key or cycle start button.

To load a program from a DNC system

Required during setup to load the program/s needed for the job.

1. Place the mode switch to edit.
2. Press the key labeled program until the word program appears at the top of the display screen.
3. Be sure the DNC system is connected the machine.
4. Type O (letter, not zero) and the program number to be loaded.
5. Press the soft key under read.
6. Go to the DNC system and send the program.

To delete a program

Required when you're finished using a program.

1. Place mode switch to edit.
2. Press the key labeled program until the word program appears at the top of the display screen.
3. Type letter O and the program number to be deleted.
4. Press the soft key under delete.

You won't be asked for confirmation, so be careful with this procedure.

To alter a word in a program

1. Place mode switch to edit.
2. Press the key labeled program until the word program appears at the top of the display screen.
3. Search to the word to be altered.
4. Type the new word.
5. Press soft key under alter.

To insert a word in a program

1. Place mode switch to edit.
2. Press the key labeled program until the word program appears at the top of the display screen.
3. Search to the word just prior to the word you want to insert.
4. Type the word you wish to insert.
5. Press soft key under insert.

To use MDI to do a zero return

This procedure is faster than the manual procedure.

1. Place the mode switch to MDI.
2. Press the key labeled program until MDI appears at the top of the display screen.
3. Check that the machine can go straight home with no interference.
4. Type G28 U0 W0, press the EOB key and then press the soft key under insert.
5. Press start key or cycle start button.

You can omit axes that you don't want to zero return.

To save a program to a DNC system

Required after program verification if changes have been made to the program.

1. Get the DNC system ready to receive a program.
2. Place mode switch to edit.
3. Press the key labeled program until the word program appears at the top of the display screen.
4. Type the letter O and the program number of the program to be sent.
5. Press the soft key under punch.

To call up a program (make it the active program)

Required when you want to work on another program in the control.

1. Place mode switch to edit.
2. Press the key labeled program until the word program appears at the top of the display screen.
3. Type the letter O and the program to be searched to.
4. Press the soft key under fwd search.

To delete a word in a program

Required when you must edit a program.

1. Place mode switch to edit.
2. Press the key labeled program until the word program appears at the top of the display screen.
3. Search to the word to be deleted.
4. Press the soft key under delete.

To run a verified program

Required during a production run.

1. Load part into setup.
2. Send the machine to its starting point (usually the tool change position).
3. Place the mode switch to edit.
4. Press the key labeled program until the word program appears at the top of the display screen.
5. Press the reset key (check the active program number).
6. Check position of all conditional switches (dry run, etc.).
7. Place the mode switch to memory or auto.
8. Place the feedrate override switch to 100 per cent.
9. Press cycle start button.

Other important procedures you can find in this text:

To measure program zero assignment values: Lesson Six
To mount and machine jaws: Lesson Twenty-Four
To verify programs: Lessons Twenty-Four and Twenty-Eight
To re-run tools: Lesson Twenty-Eight

Key points for Lesson Twenty-Seven:

- Running a CNC turning center is little more than following a series of procedures. The trick is knowing when and why to perform a given procedure.
- Writing down procedures will keep you from having to ask someone for help every time you need to perform the procedure.
- Write down a procedure for every task you must perform on your turning center.

STOP! Do practice exercise number twenty-seven in the workbook.

You Must Know How To Safely Verify Programs

You cannot begin a production run until the CNC program is verified and a workpiece passes inspection. Safely verifying programs will be the focus of Key Concept Number Ten.

Key Concept Number Ten is a short, one-lesson Key Concept:

28: Program verification

You know that program verification is part of setup. Indeed, we provide some pretty good explanations in Lesson Twenty-Four about how programs are verified. But since program verification is the most dangerous part of running a CNC machine, we want to spend more time discussing it. Lesson Twenty-Eight will show you how to safely verify programs, even if serious mistakes have been made while making the setup and/or writing the program.

Safety priorities

When verifying new programs, remember CNC machines will follow programmed instructions *precisely as they are given*, even if there are mistakes in the program. In Lesson Nine, we present some of the most common programming mistakes. With the exception of basic syntax (program formatting) mistakes, the machine will rarely alert you when a mistake has been made. While verifying any new program – especially a manually written program – you must be ready for just about anything. If you make a mistake in the program which tells the machine to rapid a tool into the workpiece, the machine will follow your commands and do so, causing what is commonly referred to as a *crash*.

And even with a proven program, remember that you can make serious mistakes during setup. Mistakes with tool loading, program zero assignment value measurement, and offset entry can be every bit as serious as programming mistakes. So you must also be very careful when running programs that you have run before.

We cannot overstress the need for using safe procedures and staying alert when working with CNC equipment. While we are not trying to scare you, we do want to instill in you a very high level of respect for your very powerful and potentially dangerous machine tool.

There are three levels of priority that you must adhere to when you work with any machine tool, including CNC turning centers.

Operator safety

The first priority must be your safety. You must take every opportunity to ensure your safety and the safety of the people around you. The verification procedures we provide stress operator safety as the number one priority. As time goes on and you start gaining experience, your tendency will be to short cut these procedures in order to save some time. We urge you to avoid this tendency. When you begin relaxing your guard, you open the door to very dangerous situations.

Compare this to a person that enjoys snow skiing. The first few times a person goes skiing, they tend to be very careful, and rarely does a new skier out for the first time get seriously injured. It is only after a skier gains confidence that they become bold and careless. More experienced skiers sustain serious injuries than do beginners.

In similar fashion, most entry level setup people and operators tend to be very careful when running a CNC machine. It is only after they gain some experience that some people become bolder and less careful. This tendency is inspired by the need to work faster. Again, don't compromise safety in order to go faster. As you gain experience, you'll naturally gain the ability to perform more efficiently. You don't have to shortcut safety-related procedures to do so.

Machine tool safety

The second safety priority is the CNC turning center itself. Every operator must do their best to ensure that no damage to the machine can occur. Obviously, CNC machine time is very expensive. When a CNC machine goes down for any reason, the actual cost of repairing the machine is usually very small compared to the lost production time.

There is no excuse for machine downtime caused by operation mistakes. If the verification procedures we give are followed, you can ensure that the machine will not be placed in dangerous situations. While no method is completely failsafe, our recommendations will truly minimize the potential for machine damage.

Workpiece safety

The third safety priority is making acceptable workpieces. The effort that companies put forth to ensure zero scrap varies from one company to another. One company may be machining extremely expensive material, like titanium or stainless steel. The raw material for large workpieces can be expensive –especially for workpieces that require machining operations *prior to* the CNC turning center operation. For this company, *anything* they can do to minimize the potential for even one scrap workpiece will be done.

In another company, the raw material cost may be very low. Consider machining a small 5/8 inch long, 1/2 inch diameter piece of steel bar. The total cost of raw material may be less than 0.50 cents. In this company, the setup person's time may be much more valuable than the time it takes to run one (the first) workpiece. This company may be less concerned with attaining zero scrap. Some companies in this situation even supply the setup person with extra pieces (commonly called *practice parts*). They don't expect every workpiece machined to be a good one.

While things can happen during a production run that will cause scrap workpieces, the most critical time is during the machining of the very *first* workpiece (during setup). Knowing this, and by knowing which workpiece attributes are the most critical, a CNC setup person can minimize the potential for scrap workpieces by using *trial machining techniques* (as described in Lessons Eleven, Thirteen, and Twenty-Four). If you consider what each tool in the program will be doing when you come to it – and if you use trial machining techniques when appropriate – each tool will machine the workpiece properly. When you're finished with the last tool in the program, you'll have a good workpiece that will pass inspection.

There are people in our industry that feel that trial machining is wasteful. The people that feel this way tend to come from companies that machine very *inexpensive* raw material. They do not care about scrapping the first few workpieces as long as they *eventually* learn enough about offset settings to get a workpiece to come out to size.

But regardless of whether your company believes in trial machining or not, *all* setup people should have the ability to machine the very first workpiece correctly for the times when it is necessary to do so. Even if workpiece material cost is very low, the day will come when you have five pieces of raw material and you must machine five good workpieces.

Lesson 28

Program Verification

The most dangerous time for a CNC setup person is when verifying programs. You must stay alert and be ready for anything. You must master program verification procedures – they must truly become second nature.

In Lesson Twenty-Four, we show all of the tasks related to making setups and running production. We do so by using an example job to stress how each task is done – including program verification. In Twenty-Eight, we're going to do so again – but we'll only discuss tasks related to program verification. The setup and program we show in Lesson Twenty-Four is perfect – there are no mistakes. By comparison, the setup and program we use *in this lesson* contain many mistakes. Most of the mistakes are typical of mistakes a beginner is likely to make.

While you may be able to spot some of the mistakes in our example job as soon as you see them, remember that when *you* make mistakes, you won't know it (if you did, you wouldn't make the mistake). Our objective is to show you how to catch even very serious mistakes as you verify CNC programs – and of course – to catch them *before a crash can occur.*

Two more procedures

In Lessons Twenty-Four and Twenty-Seven, we show several procedures that are related to verifying programs, including how to perform a dry run, how to cautiously run the first workpiece using a special approach procedure for each tool, how to trial machine, and how to run a verified program. These procedures are extremely important, and you must master them. But they don't show you how to *handle problems* when mistakes are found. Here are two procedures that are needed when you do find mistakes.

Canceling the CNC cycle

You know that feed hold is your panic button. As you're running a program, you can press it any time you are worried. It causes all axis motion to stop. If you find that nothing is wrong, you can press the cycle start button to continue. *But what if something is wrong?* – Something that is so serious that you cannot allow the program to continue.

Say for example, during the running of the first workpiece, you are allowing a tool to approach the workpiece for the first time. You are using the safe approach procedure shown in Lesson Twenty-Four, so you have dry run and single block turned on, and you're controlling the tool's motion rate with feedrate override. As the tool gets within an inch or so from the work surface, you stop the cycle. You're worried that the tool will not stop at a position 0.1 inch above the work surface. So you look at the distance to go page. It says the Z axis is still going to move another 1.3 inches. As you look at the tool's position relative to the work surface, you can easily tell that the tool cannot move another 1.3 inches in Z without contacting the workpiece (you've just saved a crash).

To cancel the cycle

You cannot, of course, allow the program to continue. You must find and correct the problem. But before you go any further, you must *cancel the cycle.* Here is the procedure to do so:

1) Press the reset key on the control panel. For most machines, this will stop the spindle and coolant (if they're on).

2) Place the mode switch to edit (if the spindle and coolant don't stop when the reset button is pressed, they will now).

3) Select the program display screen page. The cursor is somewhere in the middle of the program.

4) Press reset again (this sends the cursor back to the program number at the beginning of the program).

5) Send the machine to a safe index position (like the zero return position). This can be done manually or by using an MDI command.

6) Find and correct the problem that caused you to have to cancel the cycle.

7) When the problem is corrected, follow the procedure to re-run the current tool.

To re-run a tool

This procedure assumes that you can identify the appropriate *restart command* for the tool you want to re-run. It also assumes the programmer has used the format shown in Lesson Sixteen that uses geometry offsets to assign program zero. As we show in Lesson Sixteen, the restart command with our given format is the command that contains the turret index word (like T0404). This series of commands should help to clarify which command is the restart command:

```
.
.
.
N125 Z-1.245 (13)
N130 X0.95 (14)
N135 G00 Z0.1 (15)
N140 X5.0 Z6.0
N145 M01

(FINISH BORING BAR)
N150 T0404 M42 ←←←← This command is the restart command for tool number four.
N155 G96 S500 M03
N160 G41 G00 X1.625 Z0.1 M08 (16)
N165 G01 Z0 F0.005 (17)
N170 X1.5 Z-0.0625 (18)
N175 Z-1.25 (19)
.
.
.
```

Here is the procedure to re-run a tool. The machine must be at a safe index position when this procedure is used (if you canceled the cycle with the procedure just shown, it will be):

1) Consider the condition of single block (if you've seen the tool run before, turn it off, if not, turn it on).

2) Consider the condition of dry run (if you've seen the tool run (completely) before, turn it off, if not, turn it on).

3) Consider the condition of feedrate override (if you seen the tool run before, set it to 100%. If not, set it at its lowest setting.

4) Consider the condition of optional stop (if you want the machine to stop when the tool is finished, as is normally the case when verifying programs, turn it on. If not, turn it off).

5) Place the mode switch to edit.

6) Press the program display screen key. The display will show the active program.

7) Press the reset key. This places the cursor on the first word in the program (the program number).

8) Scan to the restart command. To do so, type the sequence number for the restart command and press the forward search key.

9) Place the mode switch to memory or auto.

10) Press the cycle start button to activate the cycle. If you've seen the tool run before (single block is off), the tool will perform its machining operation. As long as you have the optional stop switch turned on, the machine will stop when the tool is finished so you can check what the tool has done.

11) If you have not seen this tool run before (single block is on), you must press the cycle start button repeatedly until the machine begins to move. You'll use the feedrate override switch to control motion rate.

12) Once the tool has successfully approached the workpiece, turn off dry run. (*Never* let cutting tools machine a workpiece under the influence of dry run. Dry run tends to slow down rapid motions, but it *speeds up* cutting motions.)

13) The tool has safely approached. What you do next will depend upon whether you're running a new or a proven program. For a new program you must carefully step through the cutting tool's cutting motions. In this case, leave the single block switch on, and repeatedly press cycle start to step through the tool. For a proven program (are you *absolutely sure* that you're running the current version of the program?), turn off single block and press cycle start to let the tool machine the workpiece as it did the last time the program was run.

14) When the tool is finished, the machine will return to its tool change position.

Verifying a job that contains mistakes

Remember, there are mistakes contained in the setup and program we are about to show.

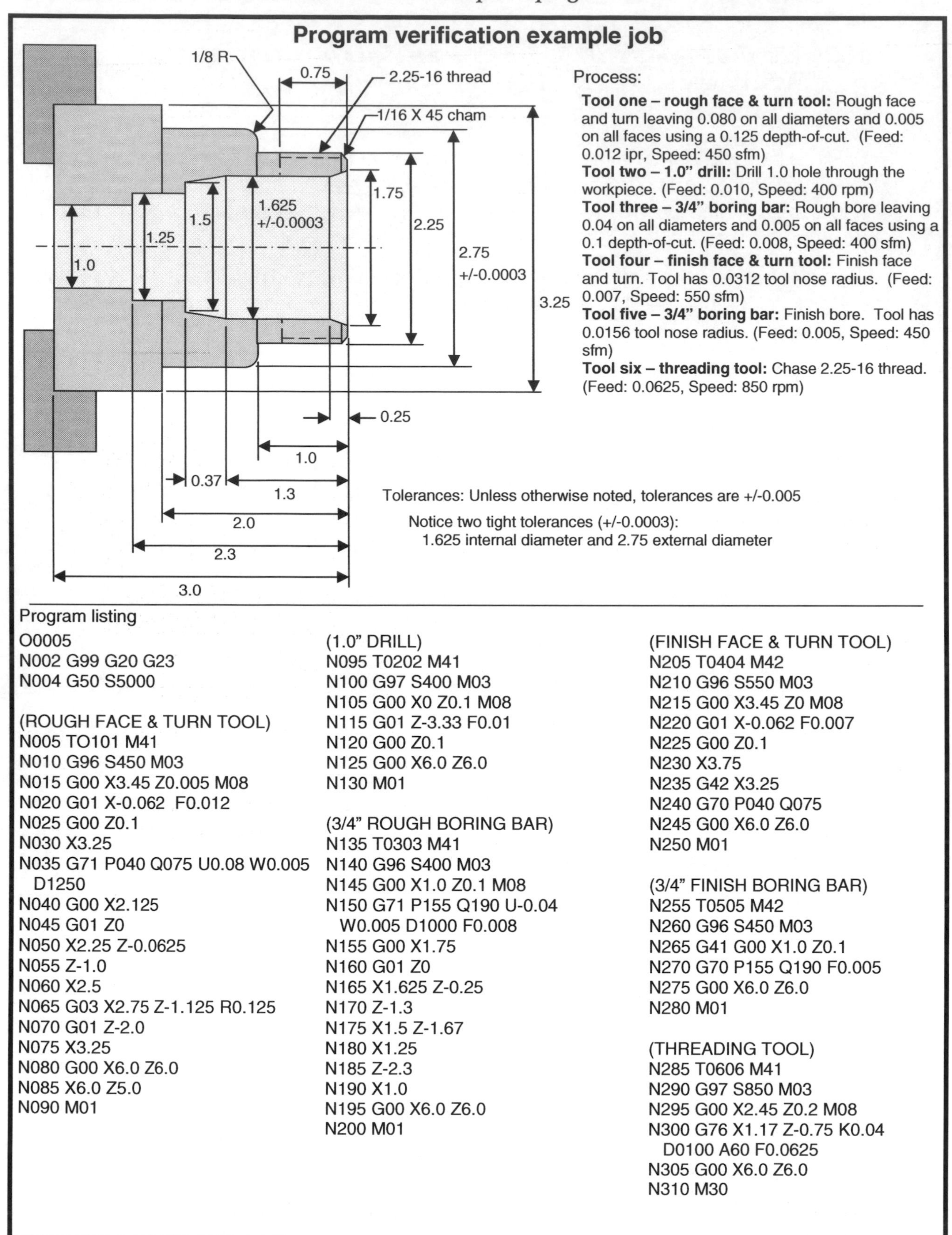

Process:

Tool one – rough face & turn tool: Rough face and turn leaving 0.080 on all diameters and 0.005 on all faces using a 0.125 depth-of-cut. (Feed: 0.012 ipr, Speed: 450 sfm)
Tool two – 1.0" drill: Drill 1.0 hole through the workpiece. (Feed: 0.010, Speed: 400 rpm)
Tool three – 3/4" boring bar: Rough bore leaving 0.04 on all diameters and 0.005 on all faces using a 0.1 depth-of-cut. (Feed: 0.008, Speed: 400 sfm)
Tool four – finish face & turn tool: Finish face and turn. Tool has 0.0312 tool nose radius. (Feed: 0.007, Speed: 550 sfm)
Tool five – 3/4" boring bar: Finish bore. Tool has 0.0156 tool nose radius. (Feed: 0.005, Speed: 450 sfm)
Tool six – threading tool: Chase 2.25-16 thread. (Feed: 0.0625, Speed: 850 rpm)

Tolerances: Unless otherwise noted, tolerances are +/-0.005

Notice two tight tolerances (+/-0.0003):
1.625 internal diameter and 2.75 external diameter

Program listing

```
O0005
N002 G99 G20 G23
N004 G50 S5000

(ROUGH FACE & TURN TOOL)
N005 TO101 M41
N010 G96 S450 M03
N015 G00 X3.45 Z0.005 M08
N020 G01 X-0.062  F0.012
N025 G00 Z0.1
N030 X3.25
N035 G71 P040 Q075 U0.08 W0.005
  D1250
N040 G00 X2.125
N045 G01 Z0
N050 X2.25 Z-0.0625
N055 Z-1.0
N060 X2.5
N065 G03 X2.75 Z-1.125 R0.125
N070 G01 Z-2.0
N075 X3.25
N080 G00 X6.0 Z6.0
N085 X6.0 Z5.0
N090 M01

(1.0" DRILL)
N095 T0202 M41
N100 G97 S400 M03
N105 G00 X0 Z0.1 M08
N115 G01 Z-3.33 F0.01
N120 G00 Z0.1
N125 G00 X6.0 Z6.0
N130 M01

(3/4" ROUGH BORING BAR)
N135 T0303 M41
N140 G96 S400 M03
N145 G00 X1.0 Z0.1 M08
N150 G71 P155 Q190 U-0.04
  W0.005 D1000 F0.008
N155 G00 X1.75
N160 G01 Z0
N165 X1.625 Z-0.25
N170 Z-1.3
N175 X1.5 Z-1.67
N180 X1.25
N185 Z-2.3
N190 X1.0
N195 G00 X6.0 Z6.0
N200 M01

(FINISH FACE & TURN TOOL)
N205 T0404 M42
N210 G96 S550 M03
N215 G00 X3.45 Z0 M08
N220 G01 X-0.062 F0.007
N225 G00 Z0.1
N230 X3.75
N235 G42 X3.25
N240 G70 P040 Q075
N245 G00 X6.0 Z6.0
N250 M01

(3/4" FINISH BORING BAR)
N255 T0505 M42
N260 G96 S450 M03
N265 G41 G00 X1.0 Z0.1
N270 G70 P155 Q190 F0.005
N275 G00 X6.0 Z6.0
N280 M01

(THREADING TOOL)
N285 T0606 M41
N290 G97 S850 M03
N295 G00 X2.45 Z0.2 M08
N300 G76 X1.17 Z-0.75 K0.04
  D0100 A60 F0.0625
N305 G00 X6.0 Z6.0
N310 M30
```

Figure 10.1 – Print and process for example job

Figure 10.1 shows the print, process, and program for our example job. Notice the two tight tolerances (the 1.625 internal diameter and the 2.75 external diameter – both tolerances are +/-0.0003). Figure 10.2 shows the setup sheet.

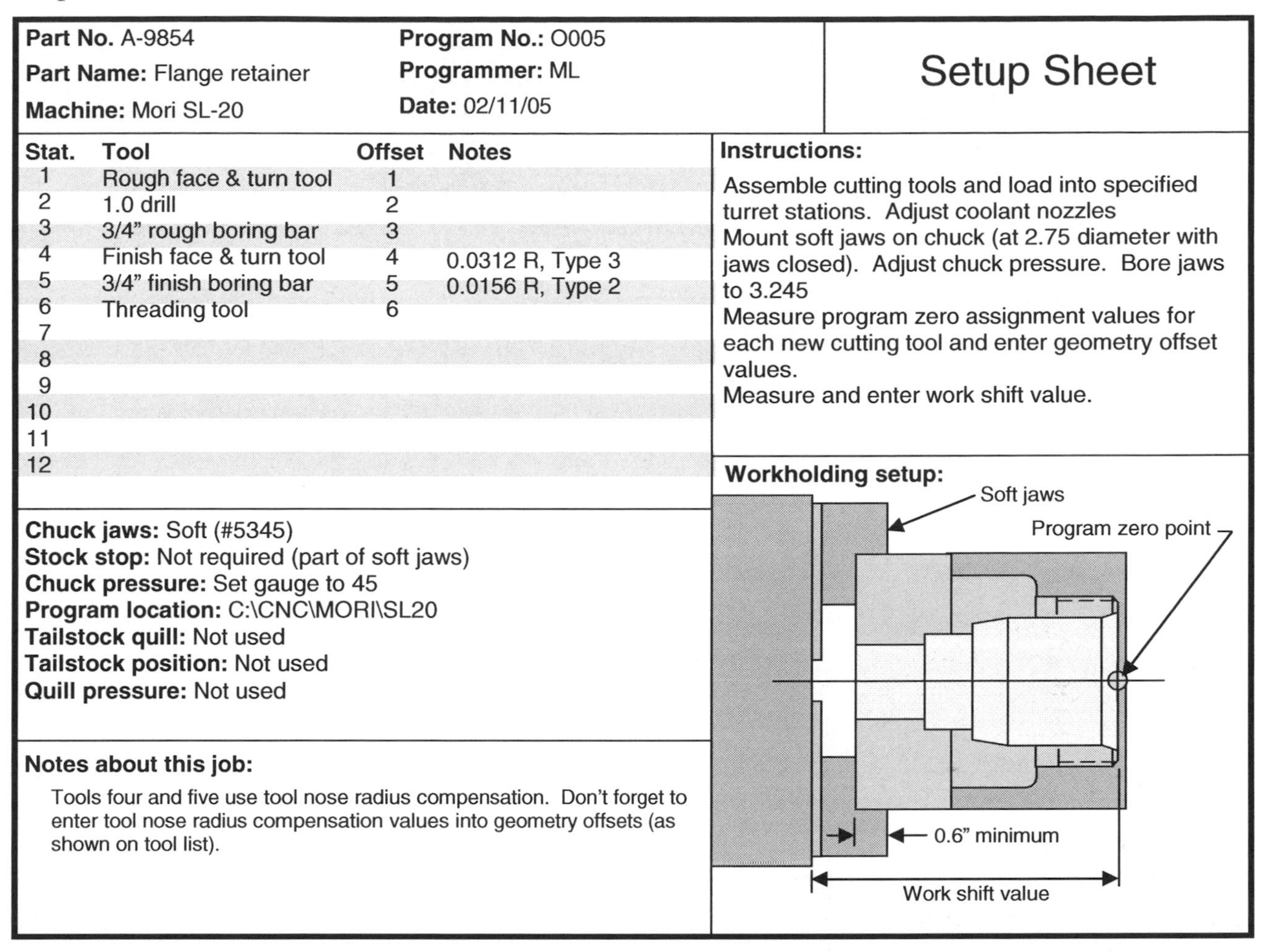

Part No. A-9854
Part Name: Flange retainer
Machine: Mori SL-20

Program No.: O005
Programmer: ML
Date: 02/11/05

Setup Sheet

Stat.	Tool	Offset	Notes
1	Rough face & turn tool	1	
2	1.0 drill	2	
3	3/4" rough boring bar	3	
4	Finish face & turn tool	4	0.0312 R, Type 3
5	3/4" finish boring bar	5	0.0156 R, Type 2
6	Threading tool	6	
7			
8			
9			
10			
11			
12			

Instructions:

Assemble cutting tools and load into specified turret stations. Adjust coolant nozzles
Mount soft jaws on chuck (at 2.75 diameter with jaws closed). Adjust chuck pressure. Bore jaws to 3.245
Measure program zero assignment values for each new cutting tool and enter geometry offset values.
Measure and enter work shift value.

Chuck jaws: Soft (#5345)
Stock stop: Not required (part of soft jaws)
Chuck pressure: Set gauge to 45
Program location: C:\CNC\MORI\SL20
Tailstock quill: Not used
Tailstock position: Not used
Quill pressure: Not used

Workholding setup:

Notes about this job:

Tools four and five use tool nose radius compensation. Don't forget to enter tool nose radius compensation values into geometry offsets (as shown on tool list).

Figure 10.2 – Setup sheet for example job

Before you can begin verifying a program, of course, the *physical tasks* related to making the setup must be completed. Here is a checklist of setup-related tasks in the approximate order that setups are made. This is the same set of tasks shown in Lesson Twenty-Four.

- ❑ Tear down previous setup an put everything away
- ❑ Gather the components needed to make the setup
- ❑ Make the workholding setup
- ❑ Assemble cutting tools
- ❑ Load cutting tools into the machine's turret
- ❑ Assign program zero for each new tool
- ❑ Enter tool nose radius compensation values
- ❑ Load the CNC program/s
- ❑ Verify the correctness of a new or modified program
- ❑ Verify the correctness of the setup
- ❑ Cautiously run the first workpiece – ensure that it passes inspection
- ❑ If necessary, optimize the program for better efficiency (new programs only)
- ❑ If changes to the program have been made, save corrected version of the program

We'll say all of the cutting tools have just been placed in the turret (none were in the turret from the previous job) – and that no other tools are currently in the turret. After you measure program zero assignment values, the geometry offset, and wear offset, and work shift pages look like this:

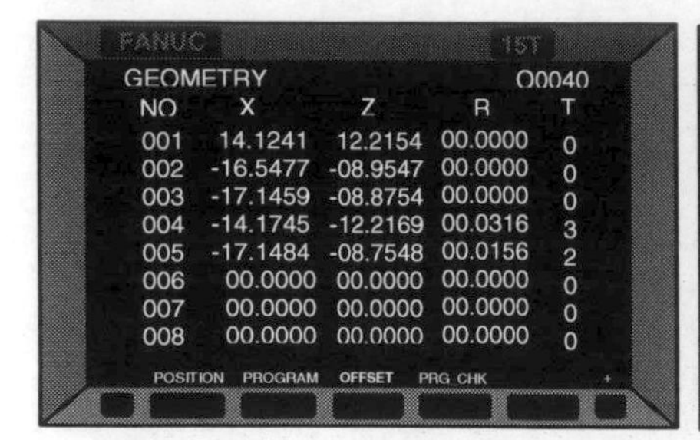

FANUC 15T

GEOMETRY O0040

NO	X	Z	R	T
001	14.1241	12.2154	00.0000	0
002	-16.5477	-08.9547	00.0000	0
003	-17.1459	-08.8754	00.0000	0
004	-14.1745	-12.2169	00.0316	3
005	-17.1484	-08.7548	00.0156	2
006	00.0000	00.0000	00.0000	0
007	00.0000	00.0000	00.0000	0
008	00.0000	00.0000	00.0000	0

POSITION PROGRAM OFFSET PRG CHK +

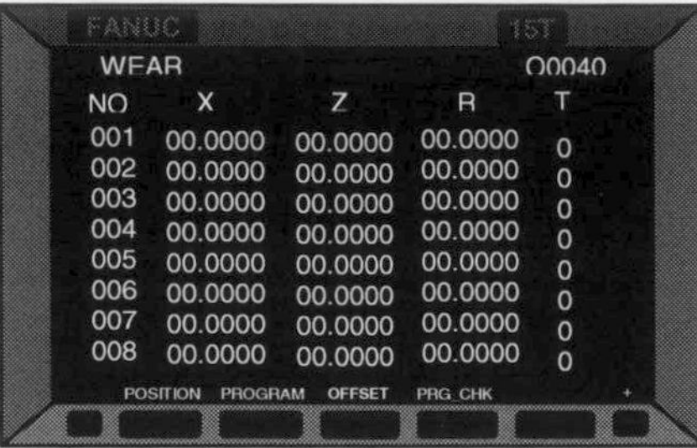

FANUC 15T

WEAR O0040

NO	X	Z	R	T
001	00.0000	00.0000	00.0000	0
002	00.0000	00.0000	00.0000	0
003	00.0000	00.0000	00.0000	0
004	00.0000	00.0000	00.0000	0
005	00.0000	00.0000	00.0000	0
006	00.0000	00.0000	00.0000	0
007	00.0000	00.0000	00.0000	0
008	00.0000	00.0000	00.0000	0

POSITION PROGRAM OFFSET PRG CHK +

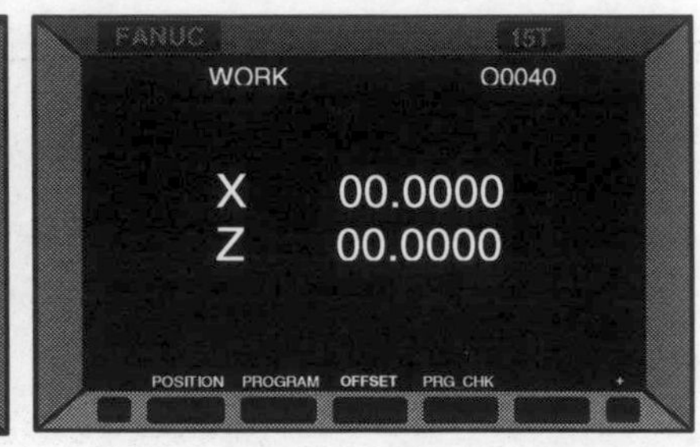

Loading the program

At the eighth step in the checklist shown above, you're ready to load the program. You follow the procedure shown in Lesson Twenty-Seven, and load the program for this job (O0005). But when you finish loading the program, you notice that something very strange has happened. As you look at the program shown on the display screen, it begins like this:

```
O101
M41
N010 G96 S450 M03
N015 G00 X3.45 Z0.005 M08
N020 G01 X-0.062 F0.012
N025 G00 Z0.1
N030 X3.25
N035 G71 P040 Q075 U0.08 W0.005
  D1250
N040 G00 X2.125
.
.
.
```

It continues through to the end of the program. When you look at the directory display screen page, you notice that *there is* a program named O0005. When you call it up, here's what you see:

```
O0005
N002 G99 G20 G23
N004 G50 S5000

(ROUGH FACE & TURN TOOL)
N005 T
```

Obviously, something is wrong. Before reading on, can you determine what it is (by looking at the program listing for the job)?

Here's a hint: The problem is in line N005.

Actually, this is a very common beginners' mistake. The person typing this program used the upper case letter O in the T word of line N005, typing TO101 instead of T0101 (the T word must be T-zero-one-zero-one, not T-oh-one-zero-one). As you know, the only letter O in a program (other than in comments) is the program number.

As this program is being loaded, the machine sees the letter O in line N005 and thinks you want to begin loading *another* program. Since the number following the letter O in line N005 is 101, the balance of the program being loaded will be placed in a program named O0101. When the program loading process is finished, *two* programs have been loaded, O0005 and O0101. And since O0101 is the last program loaded, it is the active program when the program loading process is completed.

To correct this mistake, of course, you must change line N005 from TO101 to T0101 (in the program being loaded – at the DNC system). You must then delete both programs O0101 and O0005 from memory and repeat the program loading process. When you do, the program will be loaded properly (there will be no other program-loading problems).

Note that a good off-line program verification system will find and display this problem well before you're ready to load the program into the machine – but since this is such a common mistake, we wanted to show you what will happen when you type the letter O in a program instead of a number zero.

The dry run to check for setup mistakes

We'll say this program has been verified on an offline program verification system, and the general motions made by the program look pretty good to the person performing the off-line verification. So during the dry run, we're looking, first and foremost, for setup mistakes.

With the setup complete and the program loaded, you're ready to do the dry run. At this point, of course, you have removed the workpiece from the setup. You close the jaws (on nothing) so the program can run.

You follow the procedure given in Lesson Twenty-Four for doing a dry run. You press the cycle start button and the turret indexes to **tool number one** – the rough face and turn tool. But in the first XZ motion command (line N015) the machine over-travels in the plus X and Z directions. The machine is going in the wrong direction in both the X and Z axes. Before reading on, can you determine what the problem is? Hint: The problem is *not* with the program.

On the geometry offset display screen page (shown above), the values stored in geometry offset number one have been entered as *positive* values. The machine thinks program zero is on the positive side of the zero return position. As you know from Lesson Seven, geometry offsets must be specified *from* the zero return position *to* program zero. Since the zero return position is at the extreme plus end of each axis for this machine (and this is the case with most machines), geometry offset values will be *negative* in each axis.

To correct this problem, you must change the X and Z values of geometry offset number one to negative values (retype them). You must also move the machine off its limits in X and Z, clear the alarm (by pressing reset), and send the machine to a safe index position. You do so and now you are ready to continue. (By the way, since this mistake is a *setup mistake*, no program verification system can find it off line.)

Since you are still working with the first tool, you simply rerun the program (there is no need to restart the tool). You are still doing a dry run, so you still have full control of all motions with the feedrate override switch. And you're keeping a finger ready to press feed hold at all times.

You crank up the feedrate override switch, and the tool begins its approach movement. It approaches properly in the X axis and continues moving in Z. It appears the tool is moving too far in Z. The tool tip looks to be well past where the workpiece will be when it is loaded into the chuck. You're worried – so you press the feed hold button and look at the distance-to-go display. It says the tool is still going to move another 3.9874 inches in the Z minus direction. You've just saved a crash. You follow the procedure to cancel the cycle. Now you must determine what is wrong. Before reading on, can you spot what is wrong? Hint: The problem is not in the program.

When you check the work shift page, you notice that the Z value is zero. During setup, you were supposed to measure and enter the work shift value – but you forgot. At this point, the machine thinks the chuck face is the program zero point for the program! To correct the problem, of course, you must measure and enter the work shift value.

You have already cancelled the cycle, and you're still working with the first tool, so you simply re-run the program. This time, it approaches properly and the rest of the motions for this tool look fine. The tool eventually returns to the safe index position.

The turret indexes to **tool number two** (the 1.0" drill). It moves in X and Z to approach the hole to be drilled in the center of the workpiece. It appears to machine the hole properly and returns to the safe index position.

The turret now indexes to **tool three** (the rough boring bar) and then begins the approach. This tool looks good as it approaches and makes the motions to rough bore the workpiece.

The turret indexes to **tool four** (the finish face and turn tool). It approaches properly and motions look good. The same is true **tool five** (the finish boring bar).

Now the turret indexes to **tool six** (the threading tool) and begins its approach. But it's going the wrong way in each axis and over-travels. You cancel the cycle and move the machine off its over-travel limits. Then you clear the alarm. You must now find the problem. Before reading on, can you tell what's wrong?

When you look at the geometry offset page (shown above), you notice that the X and Z registers are currently zero. You must have forgotten to measure the program zero assignment values for the threading tool. You do so now. You follow the procedure to re-run tool number six (its restart command is line N285) and this time it runs fine.

During this dry run, you have found some mistakes that, if not detected, would have caused crashes. And you did so in a very safe manner. Again, some setup people will eventually skip the dry run step in order to try to save some time – especially after they have run several consecutive jobs without mishap. But as you have seen in this example, doing so can be a terrible mistake.

Cautiously running the first workpiece

With the dry run complete, you load the first workpiece. You will follow the safe approach procedure for each tool. You will keep the optional stop switch on so the machine will stop after every tool – letting you check and see what the tool has done. And before the turret indexes to the next tool, you'll consider what the tool will be doing. If appropriate, you'll make the tool trial machine.

(Remember, when the turret indexes to a tool station, the tool's offsets are instated. Any trial machining offset adjustments you make must be done *before* the turret is allowed to index.)

Tool one – the rough face and turn tool: Before activating the cycle, you think about what this tool will be doing. It's a roughing tool, so the surfaces it machines are not critical. You won't trial machine. But it is important that this tool leaves the appropriate amount of finishing stock (0.080 on all diameters and 0.005 on all faces). So after this tool machines, you'll measure how much stock is on diameters and faces. If necessary, you'll adjust the tool's offset to ensure that the appropriate amount of finishing stock is left. We recommend making this initial adjustment in the geometry offset so the wear offset registers can begin the production run at zero.

Tool two – the 1.0" drill: The surfaces machined by this tool have no critical dimensions. So you'll let the drill machine the workpiece without trial machining. When it's finished, you check that the drill has broken through and that it has machined a 1.0 inch diameter.

Tool three – the 3/4" rough boring bar: Again, this is a roughing tool, so the surfaces it machines are not critical. You won't trial machine. But it is important that this tool leaves the appropriate amount of finishing stock (0.040 on all diameters and 0.005 on all faces). So after this tool machines, you'll measure how much stock is on diameters and faces. If necessary, you'll adjust the tool's offset to ensure that the appropriate amount of finishing stock is left. We recommend making this initial adjustment in the geometry offset so the wear offset registers can begin the production run at zero.

Tool four – the finish face and turn tool: This tool *is* machining a critical dimension (the 2.75 diameter). So you decide to trial machine. You increase the X value of offset number four by 0.01. This will force the tool to leave a little excess stock on all diameters. When the tool is finished, the machine stops at the M01 in the program. You measure the 2.75 inch diameter and adjust the offset accordingly. Then you follow the procedure to re-run this tool. (We recommend using geometry offsets for trial machining. Again, this will allow the wear offsets to begin the production run at zero.)

Tool five – the finish boring bar: This tool is also machining a critical dimension (the 1.625 inch diameter). So you decide to trial machine. You decrease the X value of offset number five by 0.01. This will force the tool to leave a little excess stock on all diameters. When the tool is finished, the machine stops at the M01 in the program. You measure the 1.625 inch diameter and adjust the offset accordingly. Then you follow the procedure to re-run this tool. (Again, we recommend using geometry offsets for trial machining.)

Tool six – the threading tool: This tool will be machining the thread on the workpiece. It is quite critical that the thread is machined to the appropriate diameter – so you decide to trial machine. You increase the X value of offset number six by 0.010. When the tool is finished and the machine stops at the M30 (this is the last tool), you measure the thread diameter (possibly with a thread gauge or over pins) to determine how much adjustment is necessary. You then reduce the X value of offset six accordingly and re-run the tool.

This example should give you a very good idea of what you are in for as a CNC setup person. While (hopefully) there shouldn't be as many mistakes to find and correct in your jobs as there are in our example job, *you must be extremely careful when verifying programs.*

STOP! Do practice exercise number twenty-eight in the workbook.

INDEX

Turning Center Programming Quick Reference Card

G Codes

Code	Description	Status	Initial-ized	Modal
G00	Rapid motion	Std	Yes	Yes
G01	Straight line cutting motion	Std	No	Yes
G02	CW circular cutting motion	Std	No	Yes
G03	CCW circular cutting motion	Std	No	Yes
G04	Dwell	Std	No	No
G09	Exact stop check, one shot	Std	No	No
G10	Offset input by program	Opt	No	No
G17	XY plane selection- (Y axis	Std	Yes	Yes
G18	XZ plane selection- machines	Std	No	Yes
G19	YZ plane selection- only)	Std	No	Yes
G20	Inch mode	Std	Yes	Yes
G21	Metric mode	Std	No	Yes
G22	Stored stroke limit instating	Opt	No	Yes
G23	Stored stroke limit cancel	Opt	Yes	Yes
G27	Zero return check	Std	No	No
G28	Zero return command	Std	No	No
G29	Return from zero return	Std	No	No
G30	Second reference point return	Opt	No	No
G31	Skip cutting for probe	Opt	No	No
G40	Cancel tool nose radius comp.	Std	Yes	Yes
G41	Tool nose radius comp. left	Std	No	Yes
G42	Tool nose radius comp. right	Std	No	Yes
G50	1) Spindle speed limiter	Std	No	Yes
	2) Program zero designator	Std	No	Yes
G54	Instate fixture offset #1	Std	Yes	Yes
G55	Instate fixture offset #2	Std	No	Yes
G56	Instate fixture offset #3	Std	No	Yes
G57	Instate fixture offset #4	Std	No	Yes
G58	Instate fixture offset #5	Std	No	Yes
G59	Instate fixture offset #6	Std	No	Yes
G61	Exact stop check mode	Std	No	Yes
G64	Normal cutting (cancel G61)	Opt	No	Yes
G65	Custom macro call	Opt	No	No
G66	Custom macro modal call	Opt	No	Yes
G67	Cancel custom macro call	Opt	Yes	No

Code	Description	Status	Initial-ized	Modal
G70	Finishing cycle	Std	No	Yes
G71	Rough turn/bore cycle	Std	No	Yes
G72	Rough facing cycle	Std	No	Yes
G73	Pattern repeating cycle	Std	No	Yes
G74	Peck drilling cycle	Std	No	Yes
G75	Grooving cycle	Std	No	Yes
G76	Threading cycle	Std	No	Yes
	G80-G89 live tooling only			
G80	Cancel canned cycle	Std	Yes	Yes
G81	Drilling cycle	Std	No	Yes
G82	Counterboring cycle	Std	No	Yes
G83	Deep hole peck drilling cycle	Std	No	Yes
G84	Right hand tapping cycle	Std	No	Yes
G85	Reaming cycle	Std	No	Yes
G86	Boring cycle	Std	No	Yes
G87	Back boring cycle	Std	No	Yes
G88	Boring cycle	Std	No	Yes
G89	Boring cycle with dwell	Std	No	Yes
G90	One pass turn/bore cycle	Std	No	Yes
G92	One pass treading cycle	Std	No	Yes
G94	One pass facing cycle	Std	No	Yes
G96	Constant surface speed mode	Std	No	Yes
G97	RPM mode	Std	Yes	Yes
G98	Feed per minute mode	Std	No	Yes
G99	Feed per revolution mode	Std	Yes	Yes

Notes about G codes: 1) Machine tool builders vary dramatically with regard to which G codes they make standard. 2) Parameters control the initialized state of certain G code groups (like G20-G21). 3) Not all control models include all G codes shown in this list.

Common M codes

Code	Description	Status	Initial-ized	Modal
M00	Program stop	Std	No	No
M01	Optional stop	Std	No	No
M02	End of program (no rewind)	Std	No	No
M03	Spindle on forward (CW)	Std	No	Yes
M04	Spindle on reverse (CCW)	Std	No	Yes
M05	Spindle off	Std	No	Yes
M06	Bar feeder command	Std	No	No
M07	Mist coolant	Opt	No	Yes
M08	Flood coolant	Std	No	Yes
M09	Coolant off	Std	Yes	Yes
M30	End of program (rewinds)	Std	No	No
M98	Subprogram call	Std	No	No
M99	Subprogram return	Std	No	No

Other M codes you may have

Code	Description	Status	Initial-ized	Modal
	Jaw open command			
	Jaw close command			
	Tailstock body forward			
	Tailstock body back			
	Tailstoc quill forward			
	Tailstock quill back			
	Low jaw clamping pressure			
	High jaw clamping pressure			

As with G codes, M code numbers vary dramatically from one machine tool builder to another. Be sure to check the M codes list that comes with your machine to see what other M codes you may have.

Turning Center Programming Quick Reference Card	
Letter Addresses Used With Programming	
O	This word is used to designate programs in the control's memory. O0001 through O9999 can be used. The program number is always the very first word of the CNC program. No decimal point is allowed with this word.
N	This word designates a sequence number (also called *block* number). It is used for line identification only, and does not have to be in the CNC program at all. To keep programs organized, beginners should include sequence numbers and place them in a logical order. N0001 (or N1) through N9999 can be used. No decimal point is allowed with this word.
G	This word specifies a preparatory function. Preparatory functions prepare the control for what is coming in the current command or future commands (many G codes are modal). Though there are a few exceptions, G codes commonly range from G00 through G99 and normally do not include a decimal point. For a full list of G codes, see the reverse side of this quick reference card.
X **U**	These words designate movement along the X axis. X designates absolute movement, U designates incremental movement. In inch mode, a the smallest increment of programming is 0.0001 inch. A fixed format will be used if a decimal point is not specified in this word (X10000 will be taken as 1 inch in the inch mode or 10 mm in metric mode). A decimal point should be included with this word. Example: X10.375 Secondary use: X is used with a dwell command to specify the time of dwell. G04 X0.5 is a 5 second dwell.
Z **W**	These words designate movement along the Z axis. Z designates absolute movement, W designates incremental movement. In inch mode, a the smallest increment of programming is 0.0001 inch. A fixed format will be used if a decimal point is not specified in this word (Z10000 will be taken as 1 inch in the inch mode or 10 mm in the metric mode). Get in the habit of specifying this word with a decimal point. Example: Z-0.437 Secondary use: Z is used with hole machining canned cycles to specify the bottom position of the hole.
C	For turning centers that have a rotary axis built into the spindle (live tooling machines), this word is used to specify positioning along the rotary axis. The smallest increment of programming is 0.001 degree. As with X, Z, U and W, beginners should program these words with a decimal point. A position of 45 degrees is specified as C45.0.
R	Used to designate the radius of a circular movement. All decimal point related points apply to the R word.
I **K**	Can be used to designate the center of radius being formed with circular commands, but beginners should concentrate on using the R word. These words specify the distance and direction from the start point to the center of the arc in X and Z respectively.
F	Used to designate feedrate. If G98 is instated, F designate feed per minute. If G99, F designates feed per revolution. G20 and G21 also affect the programmed feedrate. If G20/G99 are instated, F0.012 designates a feedrate of 0.012 inches per revolution. If G21/G99 are instated, F0.5 designates 0.5 mm per revolution. G98 G20 F30.0 designates thirty inches per minute.
S	Used to designate spindle speed. If G96 is instated, S designates a constant surface speed. If G97 is instated, S designates an RPM. G20 and G21 also affect the programmed spindle speed. If G20/G96 are instated, S300 designates 300 surface feet per minute (SFM). If G21/G96 is instated, S300 designates 300 meters per minute. G97 always designates RPM. G97 S300 designates 300 RPM.
T	For most turning centers, T is a four digit, two function word. The first two digits designate the tool station and geometry offset to be used. The second two digits specify the wear offset. For example, T0101 designates that the turret index to station one, use geometry offset number one (both from the first two digits), and instate wear offset number one. To avoid confusion, you should make the tool station number the same as the wear offset number for most applications.
M	This word specifies one of a series of two digit miscellaneous functions. No decimal point is allowed with this word. For a full list of M codes, see your machine tool builders programming manual.
E	This word can be used with threading (instead of F) to specify the pitch of the thread to six digits (E0.083333).
L	Used to specify the number of times a subprogram must be executed.
P	P specifies the program number to call with subprogramming techniques. No decimal point is allowed with P. Secondary use: P can also be used to specify the length of time in a dwell command.
D	Specifies the depth of cut in multiple repetitive cycles.
/	Optional block skip word (also called block delete). When programmed at the beginning of a command, this word causes the control to look at an on/off switch on the control panel. If the switch is on, the command is skipped.

Made in the USA
Lexington, KY
26 November 2019